# POWER DISTRIBUTION SYSTEM RELIABILITY

## BOOKS IN THE IEEE PRESS SERIES ON POWER ENGINEERING

*Principles of Electric Machines with Power Electronic Applications*, Second Edition
M. E. El-Hawary

*Pulse Width Modulation for Power Converters: Principles and Practice*
D. Grahame Holmes and Thomas Lipo

*Analysis of Electric Machinery and Drive Systems*, Second Edition
Paul C. Krause, Oleg Wasynczuk, and Scott D. Sudhoff

*Risk Assessment for Power Systems: Models, Methods, and Applications*
Wenyuan Li

*Optimization Principles: Practical Applications to the Operations of Markets of the Electric Power Industry*
Narayan S. Rau

*Electric Economics: Regulation and Deregulation*
Geoffrey Rothwell and Tomas Gomez

*Electric Power Systems: Analysis and Control*
Fabio Saccomanno

*Electrical Insulation for Rotating Machines: Design, Evaluation, Aging, Testing, and Repair*
Greg Stone, Edward A. Boulter, Ian Culbert, and Hussein Dhirani

*Signal Processing of Power Quality Disturbances*
Math H. J. Bollen and Irene Y. H. Gu

*Instantaneous Power Theory and Applications to Power Conditioning*
Hirofumi Akagi, Edson H. Watanabe and Mauricio Aredes

*Maintaining Mission Critical Systems in a 24/7 Environment*
Peter M. Curtis

*Elements of Tidal-Electric Engineering*
Robert H. Clark

*Handbook of Large Turbo-Generator Operation and Maintenance*, Second Edition
Geoff Klempner and Isidor Kerszenbaum

*Introduction to Electrical Power Systems*
Mohamed E. El-Hawary

*Modeling and Control of Fuel Cells: Distributed Generation Applications*
M. Hashem Nehrir and Caisheng Wang

*Power Distribution System Reliability: Practical Methods and Applications*
Ali A. Chowdhury and Don O. Koval

# POWER DISTRIBUTION SYSTEM RELIABILITY

## Practical Methods and Applications

Ali A. Chowdhury

Don O. Koval

IEEE PRESS SERIES ON POWER ENGINEERING

IEEE

IEEE Press

WILEY

A JOHN WILEY & SONS, INC., PUBLICATION

IEEE Press
445 Hoes Lane
Piscataway, NJ 08854





Published by John Wiley & Sons, Inc., Hoboken, New Jersey. 
Published simultaneously in Canada



***Library of Congress Cataloging-in-Publication Data is available.***

ISBN 978-0470-29228-0

Printed in the United States of America

10 9 8 7 6 5 4 3 2 1

*To my wife Razia, daughter Fariha, late parents Hesamuddin Ahmed and Mahfuza Khatun, late elder brother Ali Hyder, and late older sister Chemon Ara Chowdhury*

*—Ali A. Chowdhury*

*To my wife Vivian, my mother Katherine, and late father Peter Koval*

*—Don. O. Koval*

# CONTENTS

# PREFACE

Historically, the attention to distribution reliability planning was proportional to the operating voltage of utilities and the primary focus was on generation and transmission reliability studies. It has, however, been reported in the technical literature that approximately 80% of the customer interruptions occur due to the problems in the distribution system. Under the new era of deregulation of power utilities, the focus has shifted to distribution systems to economically provide a reliable service. There are not many textbooks in the world dealing with topics in power distribution reliability planning and operation. We found that many of the theoretical examples presented in the literature were not representative of actual distribution systems. These anomalies raise the question of their credibility in modeling these systems. There are reliability programs for calculating customer reliability indices. The details and the assumptions, however, made in some of these computer programs are not revealed. We found in many cases the results of these programs were incorrect. The basic intention of this book is to provide the theory and detailed longhand calculations and their assumptions with many examples that are required in planning and operating distribution system reliably (i.e., reliability cost versus reliability worth) and to validate the results generated by commercial computer programs.

This book evolved from many practical reliability problems and reports written by us while working for various utilities (e.g., Alberta Power Ltd, BC Hydro, SaskPower, and MidAmerican Energy Company) in North America over the past 40 years. Some of the book materials evolved from the content of the reliability courses taught by Dr. Don Koval at the University of Alberta. The book has been written for senior-level undergraduate and graduate-level power engineering students, as well as practicing engineers in the electric power utility industry. It can serve as a complete textbook for either a one-semester or two-semester course.

It is impossible to cover all aspects of distribution system reliability in a single book. The book attempts to include the most important topics of fundamentals of probability and statistics, reliability principles, applications of simple reliability models, engineering economics, reliability analysis of complex network configurations, designing reliability into industrial and commercial power systems, application of zone branch reliability methodology, equipment outage statistics, historical assessment, deterministic planning criteria, important factors related to distribution standards, standards for re-regulated distribution utility, customer interruption cost models for load point reliability assessment, value-based predictive reliability assessment, isolation and restoration procedures, meshed distribution system layout, radial feeder reconfiguration analysis, distributed generation, models for spare equipment, and voltage sags and surges at industrial and commercial sites that are routinely dealt by distribution engineers in planning, operating and designing distribution systems. The special feature of this book is that many of the numerical examples are based on actual utility data and are presented throughout all chapters in an easy-to-understand manner. Selected problem sets with answers are provided at the end of the book to enable the reader to Review and self-test the material in many of the chapters of the book. The problems range from straightforward applications, similar to the examples in the text, to quite challenging problems requiring insight and refined problem-solving skills. We strongly believe that the book will prove very useful to power distribution engineers in their daily engineering functions of planning, operating, designing, and maintaining distribution systems.

## ACKNOWLEDGMENTS

We are grateful to Dr. Fred VanEldik, professor emeritus, University of Alberta for his editing skills and valuable suggestions in the writing of this book. We are most grateful to numerous colleagues and friends: Yakout Mansour, president and CEO of California Independent System Operator; John Propst, Brent Hughes, and Peter Hill of BC Hydro; Charles Heising, an independent consultant; Doug Hollands of SaskPower; Dr. Roy Billinton of the University of Saskatchewan; Dr. James McCalley of the Iowa State University; Dr. Ward Jewell of the Wichita State University; Dr. S.S. Venkata of the University of Washington; Dr. Anil Pahwa of the Kansas State University; Dr. Chanan Singh of the Texas A&M University; Dr. Armando M. Leite Da Silva of Universidade Federal de Itajubá; Dr. Gomaa Hamoud of Hydro One; Dr. Damir Novosel, Dr. Richard Brown, James Burke, and H. Lee Willis of Quanta Technology; James Averweg, Richard Polesky, Tom Mielnik, Brian Shell, Dan Custer, James Hettrick, and James Mack of MidAmerican Energy; R.M. Godfrey of SNC LAVALIN; Cheryl Warren of National Grid; C.V. Chung of Seattle City Light; John Vitagliano of Canadian Electricity Association; Pat O'Donnell, independent consultant; J.P. Ratusz, Andy Swenky, Angie Kirkwood, and Lance Barker of EPCOR; Tony Palladino and Murray Golden of Atco Electric; Ibrahim Ali Khan of IK Power Systems Solutions; Roger Bergeron of IREQ; Dr. Turan Gonen of Sacramento State; Bill Braun of Owens Corning; Robert Arno of EYP Mission Critical Facilities, Inc.; Ariel Malanot of ABB, Switzerland; David Mildenberger of AltaLink, Laverne Stetson of University of Nebraska, Lincoln, Glenn Staines of Stantec; Lou Heimer and Michele Ransum of Public Works and Government Services Canada; Darcy Braun of ETAP; Dr. Costas Vournas of the National Technical University, Athens; Professor George E. Lasker, president of IIAS; Dr. Mohamed Hamza, president of IASTED; and the members of the Gold Book Working Group (IEEE Standard 493-2007) for their keen interest and invaluable suggestions over the many years. We express our kindest appreciations and gratitude to Dr. Mohammed E. El-Hawary, series editor; Jeanne Audino, project editor; and Steve Welch, acquisitions editor of IEEE Press for their constant encouragement and deep interest in our manuscript.

We are particularly grateful to all the undergraduate and graduate students in the Department of Electrical and Computer Engineering at the University of Alberta for their valuable suggestions, research works, and validation of many of the reliability concepts over the years. Particular thanks go to Cameron Chung, Cindy Zhang, Joseph Dong, Tahir Siddique, Imran Khan, Catalin Statineanu, Jack Zheng, Jianguo Qiu, Mihaela Ciulei, Haizhen Wang, Kai Yao, Bin Shen, Meina Xiao, Xiaodong Liu, Ming Wu, Vikas Gautam, Zhengzhao Lu, Shrinivasa Binnamangale, Aman Gill, Sukhjeet Toor, Fatima Ghousia, Delia Cinca, Faraz Akhtar, Tushar Chaitanya, and many other undergraduate and graduate students. Our sincere thanks go to Pamela McCready of California Independent System Operator for meticulously proofreading the entire manuscript.

Finally, our deepest appreciations go to our wives and family for their limitless patience and understanding while we were working on this book.

ALI A. CHOWDHURY
DON O. KOVAL

*Folsom, California*
*Edmonton, Alberta, Canada,*
*March 2009*

# 1

# OUTLINE OF THE BOOK

## 1.1 INTRODUCTION

Reliability is an abstract term meaning endurance, dependability, and good performance. For engineering systems, however, it is more than an abstract term; it is something that can be computed, measured, evaluated, planned, and designed into a piece of equipment or a system. Reliability means the ability of a system to perform the function it is designed for under the operating conditions encountered during its projected lifetime.

Historically, a power system has been divided into three almost independent areas of operation as follows:

1. *Generation System*: facilities for the generation of electricity from economical energy sources.
2. *Transmission System*: transportation system to move large energy blocks from generation facilities to specific geographical areas.
3. *Distribution System*: within a specific geographical area distribute the energy to individual consumers (e.g., residential, commercial, industrial, etc.).

---

*Power Distribution System Reliability.* By Ali A. Chowdhury and Don O. Koval

Ideally, a power system's reliability from the viewpoint of consumers means uninterrupted supply of power from the generation, transmission, or distribution systems. In reality, the key indicators of a power system's reliability for consumers are the frequency and duration of interruptions at their point of utilization (i.e., their load point). From an engineering viewpoint, the question is how do you determine mathematically the frequency and duration of load point interruptions? The "how to, assessment for distribution systems with practical examples is the subject of this book.

## 1.2 RELIABILITY ASSESSMENT OF POWER SYSTEMS

The basic function of a power system is to supply its customers with electrical energy as economically and as reliably as possible. There were some simple applications of probability methods to calculations of generation reserve capacity since 1940s; however, the real interest in power system reliability evaluation started to take off only after 1965, most notably influenced by the New York City blackout that year. Reliability mathematics is constantly evolving to accommodate technical changes in operations and configurations of power systems. At present, renewable energy sources such as wind and photovoltaic systems have a significant impact on the operation of generation, transmission, and distribution systems.

At present, deregulation is forcing electric utilities into uncharted waters. For the first time, the customer is looking for value-added services from their utilities or they will start shopping around. Failure to recognize customer needs has caused a great number of business failures in numerous industries. The electric industries' movement toward a competitive market forces all related businesses to assess their focus, strengths, weaknesses, and strategies. One of the major challenges to electric utilities is to increase the market value of the services they provide with the right amount of reliability and to lower their costs of operation, maintenance, and construction to provide customers electricity at lower rates. For any power system supplying a specific mix of customers, there is an optimum value of reliability that would result in lowest combined costs. Quantitative value-based reliability planning concepts presented in this book are an attempt to achieve this optimum reliability in power systems.

### 1.2.1 Generation System Reliability Assessment

In evaluating generation capacity adequacy, the commonly accepted definition of failure is "loss of load„ which is an outage due to capacity inadequacy. The reliability is defined in terms of the loss of load probability in a given time interval, usually a year, or the loss of load expectation (LOLE) in days per year. For a loss of load to occur, the system capacity has to fall to a level due to scheduled maintenance and/or forced outages of other generating units by a margin exceeding the spinning reserve to meet the system peak load. Even then, there may not be an outage because the system load is not always at its peak. To calculate the amount of time when the capacity cannot meet the actual load of the time, the load duration curve has to be brought into the picture. The most commonly used generation reliability index of LOLE can be calculated if all parameters, namely, forced outage rates of different generating units, the load forecast, the load duration curve, the spinning reserve, and the other refinements deemed necessary (e.g., reliability of the transmission system), are known. Significant research has gone into developing reliability assessment tools and models applied to generating capacity adequacy over the past four decades. Electric utilities are routinely performing probabilistic assessments of generation reserve margin requirements using the sophisticated tools based on *Monte Carlo simulation* and *contingency enumeration* approaches. Recent developments in generating capacity adequacy assessment include, but are not confined to, novel models for energy limited units such as wind, solar, geothermal, and other exotic energy technologies and merchant plant modeling as well as capacity market design models for deregulated markets. The system planning engineer can then decide if the level of reliability is adequate and also determine the effect of alternative actions such as increasing the spinning reserve, adding a generating unit, and changing the maintenance schedules and interconnections with other areas.

### 1.2.2 Transmission System Reliability Assessment

In earlier reliability works on generation capacity adequacy assessments, only the energy production systems were considered. The transmission and distribution systems were ignored. In a mathematical sense, the transmission and distribution systems were implicitly assumed to be perfectly reliable, which in reality was not true. Determining the probability of system capacity outage levels based on the forced outage rates of the generators alone will lead to overly optimistic results.

The transmission system consists of high-voltage transmission lines and terminal stations including different equipment and control. The average forced failure rate and outage duration of each component such as line sections, transformers, and circuit breakers of the transmission system can be computed and the reliability of a load point can be calculated using an appropriate reliability model.

The load point reliability depends on the reliability of the individual component; however, it also depends on other factors. The two most important factors are system configuration and environment. The transmission system is a network of lines and equipment. Failure of one component does not necessarily render the system failure. There is a lot of inherent redundancy in other parts of the transmission system. Another factor in transmission system reliability is the weather and environment under which it is subjected to operate. The failures of many outdoor components are caused by lightning, snow, high winds, and so on. In addition, failures are not always independent as generally assumed in statistical calculations. The failure of one component may increase the chance of failure of another. One type of such dependent failures is the common-mode failure, that is, failure of more than one component due to the same cause, which generally happens more often in inclement weather than in fair weather. In the analysis of transmission system reliability, therefore, different failure rates are assigned to different weather conditions, and the dependency of failures, at least in adverse weather, has to be taken into account. It can be seen that a detailed analysis can be very complex and it gets more complex when the composite generation and transmission system is taken together. The use of powerful computers is almost mandatory for any system reliability analysis. Significant works have been done in probabilistic assessments of transmission systems to augment the current deterministic criteria in planning and designing of transmission systems.

### 1.2.3 Distribution System Reliability Assessment

The application of reliability concepts to distribution systems differs from generation and transmission applications in that it is more customer load point oriented instead of being system oriented, and the local distribution system is considered rather than the whole integrated system involving the generation and transmission facilities. Generation and transmission reliability also emphasizes capacity and loss of load probability, with some attention paid to components, whereas distribution reliability looks at all facets of engineering: design, planning, and operations. Because the distribution system is less complex than the integrated generation and transmission system, the probability mathematics involved is much simpler than that required for generation and transmission reliability assessments.

It is important to note that the distribution system is a vital link between the bulk power system and its customers. In many cases, these links are radial in nature that makes them vulnerable to customer interruptions due to a single outage event. A radial distribution circuit generally uses main feeders and lateral distributors to supply customer energy requirements. In the past, the distribution segment of a power system received considerably less attention in terms of reliability planning compared to generation and transmission segments. The basic reason behind this is the fact that generation and transmission segments are very capital intensive, and outages in these segments can cause widespread catastrophic economic consequences for society.

It has been reported in the literature that more than 80% of all customer interruptions occur due to failures in the distribution system. The distribution segment has been the weakest link between the source of supply and the customer load points. Though a single distribution system reinforcement scheme is relatively inexpensive compared to a generation or a transmission improvement scheme, an electric utility normally spends a large sum of capital and maintenance budget collectively on a huge number of distribution improvement projects.

At present, in many electric utilities, acceptable levels of service continuity are determined by comparing the actual interruption frequency and duration indices with arbitrary targets. For example, monthly reports on service continuity statistics produced by many utilities contain the arbitrary targets of system reliability indices for performance comparison purposes. It has long been recognized especially in the deregulated market environment that rules of thumb and implicit criteria cannot be applied in a consistent manner to the very large number of capital and maintenance investment and operating decisions that are routinely made. Though some reliability programs with limited capabilities are available, virtually no utilities perform distribution system expansion studies using probabilistic models. Unlike bulk transmission system that is subject to North American Electric Reliability Council's deterministic criteria in planning and designing the transmission systems, the distribution system is not subject to any established planning standards. Distribution utilities are required only to furnish historical distribution system performance indices to regulatory agencies.

There are ample opportunities for distribution utilities to judiciously invest in distribution system expansion activities to meet the future load growth by using the probabilistic reliability methods that would eliminate the risk of over/underinvestment in the system while providing the optimum service reliability at the right cost. The reluctance of electric utilities to use the reliability methods in planning and designing distribution systems is due to the prevailing perception that it requires sophisticated probabilistic computer tools and trained engineers in power system reliability engineering. This book intends to eliminate this misperception and presents practical probabilistic reliability models for planning and designing distribution systems.

The applications of the developed reliability models presented in this book are illustrated using hand calculations that require no sophisticated computer tools and virtually little or no knowledge of probability mathematics. Problem sets and answers are provided at the end of the book to test the reader's ability to solve reliability problems in distribution systems.

## 1.3 ORGANIZATION OF THE CHAPTERS

Two approaches to reliability evaluation of distribution systems are normally used, namely, historical assessment and predictive assessment. Historical assessment involves the collection and analysis of distribution system outage and customer interruption data. It is essential for electric utilities to measure actual distribution system reliability performance levels and define performance indicators to assess the basic function of providing cost-effective and reliable power supply to all customer types. Historical assessment generally is described as measuring the past performance of a system by consistently logging the frequency, duration, and causes of system component failures and customer interruptions. Predictive reliability assessment, however, combines historical component outage data and mathematical models to estimate the performance of designated configurations. Predictive techniques therefore rely on two basic types of data to compute service reliability: component reliability parameters and network physical configurations.

This book deals with both historical and predictive distribution system reliability assessments. Simple and easy-to-use practical reliability models have been developed and their applications illustrated using practical distribution system networks. Virtually all reliability calculations have been performed by hand and no sophisticated computer programs are necessary. A simple but realistic live distribution system has been frequently used to illustrate the application of different reliability models developed and presented in this book. For the convenience of the readers, the mathematical reliability models and formulas relevant to particular applications have been repeated in chapters where necessary to maintain the flow of understanding the models and concepts. Each chapter is independent of other chapters, and cross-referencing different chapters is not required to understand the new concepts presented in a particular chapter. The applications of the novel concept of reliability cost–reliability worth or commonly known as the value-based reliability model are extensively discussed and illustrated with many numerical examples in this book.

The book is organized as follows:

Chapter 1 presents the basic definition of the term "reliability, and its application to power systems. The current state of the reliability methodology applications in generation, transmission, and distribution segments of the power system is briefly described.

Chapters 2 and 3 very briefly describe fundamentals of probability theories and reliability principles. Although the basic probability and reliability models presented with numerical examples in Chapters 2 and 3 are available in many textbooks, these models are repeated in these chapters to help the readers understand the models that will be used extensively in the later chapters of this book. The majority of systems in the real world do not have a simple structure or are operated by complex operational logic. For solving complex networks or systems, additional modeling and evaluation techniques are required to evaluate the reliability of such networks or systems. Chapters 2 and 3 also include models to assess the reliability complex network configurations. The basic models for complex network solutions have been illustrated using numerical examples.

Chapter 4 illustrates the applications of the probability and statistical models presented in Chapters 2 and 3 using simple numerical examples in distribution system planning and designing. Distribution system planners will be able to utilize the probability and statistical models by using hand calculations in real-life situations.

Chapter 5 presents the basic engineering economics models. The economics concepts and models related to distribution system planning and design are illustrated with numerous simple examples. The novel value-based reliability model presented in later chapters is based on economic theories discussed in Chapter 5.

In Chapter 6, the basic models for complex network solutions are illustrated using numerous numerical examples. Chapter 6 introduces models to assess the reliability complex network configurations. Some of the common methodologies in practice are (1) state enumeration methods (event-space methods), (2) network reduction methods, and (3) path enumeration methods.

In Chapter 7, a description is given of how to make quantitative reliability and availability predictions for proposed new configurations of industrial and commercial power distribution systems. Several examples are worked out, including a simple radial system, a primary selective system, and a secondary selective system. The simple radial system that was analyzed had an average number of forced hours of downtime per year that was 19 times larger than a secondary selective system; the failure rate was 6 times larger. The importance of two separate power supply sources from the electric utility provider has been identified and analyzed. This approach could be used to assist in cost–reliability trade-off decisions in the design of power distribution systems.

Chapter 8 presents a zone branch methodology that overcomes many of these limitations and applies the methodology to a large industrial plant power system configuration. There are many methods available for evaluating the frequency and duration of load point interruptions within a given industrial power system configuration. However, as systems become larger and more interconnected, these existing methods can become computationally bound and limited in their ability to assess the impact of unreliable protective equipment and unreliable protection coordination schemes on individual load point reliability indices within a given plant configuration. These methods also may not often account for complex isolation and restoration procedures within an industrial plant configuration that are included in the zone branch reliability methodology.

Chapter 9 deals with the types of data needed for distribution system's predictive reliability assessments and presents typical distribution component outage statistics in urban and rural environments for use in predictive reliability analysis. This database is the result of comprehensive synthesis of a large number of industry data available in different technical publications. The distribution system is an important part of the total electric supply system as it provides the final link between a utility's bulk transmission system and its ultimate customers. All quantitative reliability assessments require numerical data. Historical assessment generally analyzes discrete interruption events occurring at specific locations over specific time periods. Predictive assessment determines the long-term behavior of systems by combining component failure rates and the duration of repair, restoration, switching, and isolation activities for the electric utility's distribution system for given system configurations to calculate average reliability performance. Accurate component outage data are therefore the key to distribution system predictive performance analysis. In addition to the physical configuration of the distribution network, the reliability characteristics of system components, the operation of protection equipment, and the availability of alternative supplies with adequate capacity also have a significant impact on service reliability.

In Chapter 10, the methodology used to assess the historical reliability performance of a practical utility's electric distribution system is outlined. Included in the discussion is an overview of the process used to collect and organize the required interruption data as well as a description of the performance indices calculated for use in the causal assessment. Various components of reliability performance assessment are described, including reliability indices, comparison between years of operation, comparisons of the averages at different levels of the system, and outage cause and component failures. The application of the calculated performance statistics in planning, operating, and maintaining distribution systems is also described.

Chapter 11 provides a brief overview of current deterministic planning practices in utility distribution system planning and design. The chapter also introduces a probabilistic customer value-based approach to alternative feed requirements planning for overhead distribution networks to illustrate the advantages of probabilistic planning.

Chapter 12 identifies a number of pertinent factors and issues taken into account in establishing distribution reliability standards and illustrates the issues and factors considered in using historical reliability performance data. Actual utility data are used in the illustrations. The development of standard distribution reliability metric values, for example, System Average Interruption Frequency Index (SAIFI), System Average Interruption Duration Index (SAIDI), and Customer Average Interruption Duration Index (CAIDI), against which all utilities can compare performance, can be problematic without strict adherence to a national or international standard (e.g., IEEE Standard 1366). This issue has been discussed in Chapter 11. At present, there are many differences between data collection processes and characteristics of utility systems to make comparisons against such standard metric values impossible for many utilities. Rather, the development of uniform standard metric values, which utilities compare to their own historical reliability performance indices, is more practical. If cross-comparisons between utilities are desirable, a number of issues and factors associated with individual utilities must be taken into consideration when establishing distribution reliability standards.

Chapter 13 identifies a number of factors and issues that should be considered in generating a PBR (performance-based rate making) plan for a distribution utility. A brief analysis of cause contributions to reliability indices is also performed and presented in this chapter. The historic reliability-based PBR framework developed in this chapter will find practical applications in the emerging deregulated electricity market. In an attempt to reregulate the distribution segment of an electric power system, public utility commissions (PUCs) in a number of states in the United States are increasingly adopting a reward/penalty framework to guarantee acceptable electric supply reliability. This reward/penalty framework is commonly known as PBR. A PBR framework is introduced to provide distribution utilities with incentives for economic efficiency gains in the competitive generation and transmission markets. A distribution utility's historical reliability performance records could be used to create practical PBR mechanisms. The chapter presents actual reliability performance history from two different utilities to develop PBR frameworks for use in a reregulated environment. An analysis of financial risk related to historic reliability data is presented by including reliability index probability distributions in a PBR plan.

Chapter 14 presents the basic concepts and applications for computing load point customer reliability indices and interruption costs. Case studies showing the applications of load point reliability index calculations including customer interruption costs in distribution system planning are described in detail. The practical distribution system used in this chapter to illustrate the computation of the load point customer interruptions costs has been extensively applied in Chapters 15, 16 and 19 for demonstrating value-based predictive system planning methods, probabilistic distribution network isolation, and restoration procedures and for determining distributed generation (DG) equivalence to replace a distribution feeder requirement.

Chapter 15 presents a series of case studies of an actual industrial load area supplied by two feeder circuits originating from two alternate substations. A basic conclusion of this chapter is that expansion plans of an industrial distribution system can be optimized in terms of reliability by using an economic criterion in which the sum of both the industrial facility interruptions and the utility system costs is minimized. Society is becoming increasingly dependent on a cost-effective reliable electric power supply. Unreliable electric power supplies can be extremely costly to electric utilities and their customers. Predictive reliability assessment combines historical outage data and mathematical models to estimate the performance of specific network and system configurations. Chapter 15 has expanded the customer interruption cost methodology presented in Chapter 14 and applied to a practical distribution in illustrating the value-based assessment of proposed modifications to an existing industrial distribution system configuration to minimize the costs of interruptions to both the utility and the utility's industrial customers.

Chapter 16 presents a new restoration methodology for distribution system configurations that maximizes the amount of load that can be restored after a grid blackout, substation outage, and distribution feeder line section outages and evaluates the cost of load point interruptions considering feeder islanding and substation capacity constraints. Several case studies with restoration tables have been presented and discussed to clearly reveal the impact of distribution system capacity constraints on load point reliability indices and the cost of load point interruptions. A recent report on the U.S.–Canada blackout on August 14, 2003 revealed that the duration of restoring the Eastern Interconnect to a normal operating configuration was lengthy and complicated. One of the difficulties in modeling a power system is to represent the significant changes in loading patterns that present themselves during the restoration process after a major outage. The capacity of the equipment may be adequate during normal operating conditions; however, it may be severely compromised during restoration procedures, particularly the restoration of thousands of distribution system feeder circuits.

Chapter 17 presents a customer cost–benefit probabilistic approach to designing meshed urban distribution systems. The customer value-based reliability methodology is illustrated using a practical urban distribution system of a Canadian utility. Achieving high distribution reliability levels and concurrently minimizing capital costs can be viewed as a problem of optimization. Using mathematical models and simulations, a comparison of design concepts can be performed to compute the optimal feeder section length, feeder loading level, and distribution substation transformer loading level. The number of feeder ties and feeder tie placement in a meshed network are also optimized through the models. The overall outcome of this analysis is that capital costs can then be directed toward system improvements that will be most cost-effective in improving distribution system reliability.

Chapter 18 discusses a reliability methodology to improve the radial distribution feeder reliability performance normally prevailing in a rural environment using a simple illustrative feeder configuration. As indicated earlier, historical distribution feeder reliability assessment generally summarizes discrete interruption events occurring at specific locations over specific time periods, whereas predictive assessment estimates the long-term behavior of systems by combining component failure rates and repair (restoration) times that describe the central tendency of an entire distribution of possible values with feeder configurations. The outage time due to component failures can substantially be reduced by protection and sectionalizing schemes. The time required to isolate a faulted component by isolation and switching action is known as switching or restoration time. The provision of alternative supply in radial networks normally enhances the load point reliability. Fuses usually protect the lateral distributors connected to the customers.

Chapter 19 delves into a reliability model for determining the DG equivalence to a distribution facility for use in distribution system planning studies in the new competitive environment. The primary objective of any electric utility company in the new competitive environment is to increase the market value of the services by providing the right amount of reliability and, at the same time, lower its costs of operation, maintenance, and construction of new facilities to provide customers its services at lower rates. The electric utility company will strive to achieve this objective by many means, one of which is to defer the capital distribution facility requirements in favor of a DG solution by an independent power producer (IPP) to meet the growing customer load demand. In this case, the distribution capital investment deferral credit received by the IPP will depend on the incremental system reliability improvement rendered by the DG solution. In other words, the size, location, and reliability of the DG will be based on the comparable incremental reliability provided by the distribution solution under considerations.

Chapter 20 discusses probabilistic models developed based on Poisson probability distribution for determining the optimal number of transformer spares for distribution transformer systems. To maintain adequate service reliability, a distribution utility needs to maintain a certain number of distribution equipment in its inventory as spare equipment. The outage of a transformer is a random event, and the probability mathematics can best describe this type of failure process. The developed models have been described by using illustrative 72 kV distribution transformer systems. Industry average catastrophic transformer failure rate and a 1-year transformer repair or procurement time have been used in examples considered in the chapter. Among the models developed for determining the optimum number of transformer spares, the statistical economics model provides the best result as it attempts to minimize the total system cost including the cost of spares carried in the system.

Chapter 21 deals with service quality issues in terms of voltage sags and surges. A voltage sag may be caused by a switching operation involving heavy currents or by the operation of protective devices (including autoreclosers) resulting from faults. These events may emanate from the consumer's systems or from the public supply network. Voltage sags and short supply interruptions may disturb the equipment connected to the supply network and cause a consumer interruption. The conclusions of this chapter are that some of the inconveniences created by power quality problems are made worse by the fact that restarting an industrial process may take from a few minutes to a few hours. This chapter attempts to answer many questions asked by a utility's industrial customers. The answers presented in Chapter 21 are based on the statistical characteristics of the Canadian National Power Quality Survey.

## 1.4 CONCLUSIONS

This chapter has introduced the basic definition of the term "reliability, in a more generic form. The application of reliability techniques to power systems performance assessment was discussed briefly. Power generation system reliability evaluation by using the reliability techniques using the 1 day in 10 years loss of load expectation criterion is an accepted practice in the electric power industry. Reliability assessments in transmission systems have made great strides in recent years, and sophisticated computer models are available for large-scale transmission system assessments. With the recent movement toward competition in the electric energy market, increasing attention is being paid to the utilization of probabilistic reliability techniques in distribution system assessments and performance-based rate makings. This book is an attempt to achieve the objective of providing distribution planning engineers simple and easy-to-use reliability models that can be applied in routine distribution system cost–benefit enhancement planning without resorting to sophisticated computer tools. The reliability concepts and models developed and illustrated with practical system examples do not require knowledge of probability mathematics, and virtually all reliability assessment tasks can be performed by hand calculations. It is important to note that the book does not purport to cover every known and available method in distribution system reliability planning, as it would require a text of infinite length.

## REFERENCES


1. R. Billinton and R. N. Allan, *Reliability Evaluation of Power Systems*, 2nd edition, Plenum Press, New York, 1996.
2. J. Endrenyi, *Reliability in Electric Power Systems*, John Wiley & Sons, Ltd., Chichester, 1978.
3. T. Gönen, *Electric Power Distribution System Engineering*, McGraw Hill, New York, 1986.
4. J. J. Burke, *Power Distribution Engineering: Fundamentals and Applications*, Marcel Dekker, Inc., New York, 1994.
5. R. E. Brown, *Electric Power Distribution Reliability*, Marcel Dekker, Inc., New York, 2002.
6. G. J. Anders, *Probability Concepts in Electric Power Systems*, John Wiley & Sons, Inc., New York, 1990.

# 2

# FUNDAMENTALS OF PROBABILITY AND STATISTICS

## 2.1 CONCEPT OF FREQUENCY

### 2.1.1 Introduction

In recent years, there has been a significant increase in public awareness on the subject of probability and statistics. At present, nearly all high school mathematics courses introduce some elementary level of probability and statistics topics to many students, while at the university level, many liberal arts disciplines such as geography and sociology require some knowledge of probability and statistical mathematics from college-bound students. Moreover, probability and statistical mathematics are being increasingly used by almost all academic disciplines. There are relatively few science or social science disciplines that do not require knowledge of probability and statistics. This chapter will introduce some basic theories associated with probability and statistics.

It is a well-known fact that things in nature exhibit variations. People have different heights, earn different incomes, and machines turn out parts that are not perfectly identical—the list can go on infinitely. To analyze the data, we divide them into groups and count the number of occurrences in each group. Consider two examples. In a bag of

*Power Distribution System Reliability.* By Ali A. Chowdhury and Don O. Koval

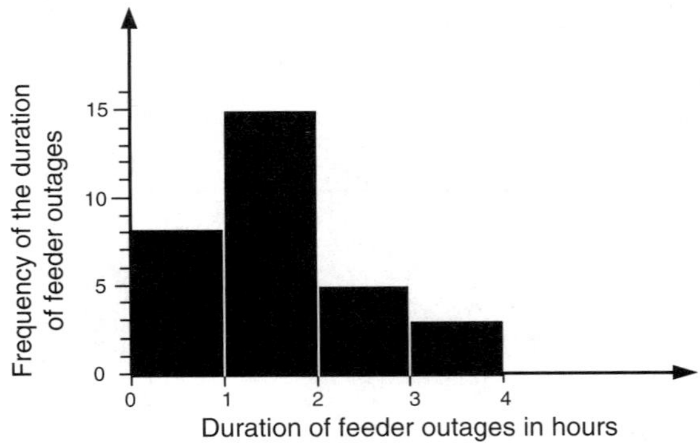


Figure 2.1. Frequency histogram of duration of feeder outages.

marbles, there are five blue ones, seven red ones, and three white ones; second, the duration of distribution feeder outages for a particular substation lasted 8 times between 0 and 1 h, 15 times between 1 and 2 h, 5 times between 2 and 3 h, and 3 times between 3 and 4 h, as illustrated in Fig. 2.1. In the first example, the groups are classified by a qualitative characteristic, the color of the marble. We have the information on how many marbles of each color there are, but that is the end. There is no relationship between the groups. The second example is different. The groups are classified by a quantitative characteristic, the duration of the outages, and there is a quantitative relationship between the groups. It is this kind of classification that lends itself to analysis. The groups are called classes and the number in each group is called the frequency. This frequency can also be converted to relative frequency in percentage of the total population. The classification of a group of items by some quantitative characteristic is called a frequency distribution.

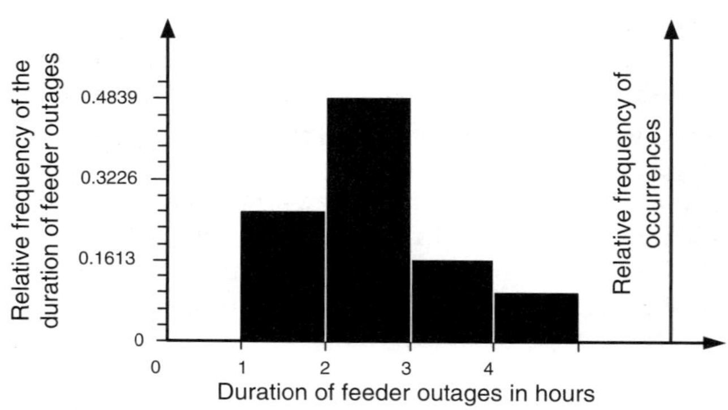


Figure 2.2. Relative frequency histogram of duration of feeder outages.

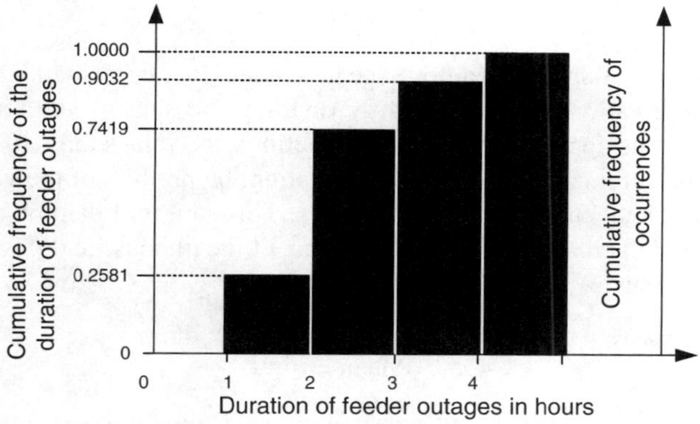


Figure 2.3. Cumulative frequency histogram of duration of feeder outages.

### 2.1.2 Concept of Class

The classes in the duration of feeder outages example have a class width of 1 h. The width can be made narrower to have a more detailed description of the duration of feeder outages resulting in more classes. In general, the classes have equal widths and are consecutive.

There are also classes that are discrete numbers instead of intervals, for example, age of students (age rounded off to integers). As long as the discrete numbers are arranged in some consecutive order, they form a frequency distribution that can be analyzed systematically.

### 2.1.3 Frequency Graphs

The relationship between frequency or relative frequency and class can be shown graphically as illustrated in Fig. 2.2. If shown as a bar graph, it is a *histogram*. Histograms can also be used for qualitatively defined classes. If the midpoints of consecutive classes are joined together with a line, it becomes a line graph. Sometimes the line graph is smoothened, and an approximate, continuous frequency distribution is obtained. Line graphs have no meaning if the classes are not quantitatively related.

### 2.1.4 Cumulative Frequency Distribution Model

Instead of the frequency of a class, the sum of frequencies of all proceedings or subsequent classes can be shown as illustrated in Fig. 2.3. Cumulative frequency distributions have a lot of applications, one of which is the load duration curve used in generation capacity adequacy studies.

## 2.2 IMPORTANT PARAMETERS OF FREQUENCY DISTRIBUTION

The basic objective of constructing a frequency distribution is to analyze the pattern of variation of a phenomenon. This pattern can be defined by several parameters.

### 2.2.1 Mean

*Mean* refers to the arithmetic mean or expected value. It is computed by summing the values of all observations or items and by dividing the sum by the total number of observations or items. In most frequency distributions, the values fall into different class intervals, and the summation is done by calculating the product of the value of a class and its frequency and summing over all classes. This sum will then be divided by the total frequency. Mathematically, if $X_i$ is the value of the $i$th class and $f_i$ is the frequency of the $i$th class, then

$$\text{Mean} = \frac{\sum f_i x_i}{\sum f_i} \tag{2.1}$$

The mean represents the average value of each item in that frequency distribution, such as the average height of a group of students in a class, the average income of employees in a company, and so on.

### 2.2.2 Median

*Median* is the value of the middle item when all the items are arranged in either ascending or descending order. It is the 50% point of the spectrum; so there are an equal number of items on both sides of the median.

### 2.2.3 Mode

*Mode* is the value in a frequency distribution that occurs most often, that is, the value of the class with the highest frequency. When represented in a graph form, it is the class value corresponding to the highest point of the curve.

### 2.2.4 Standard Deviation

*Standard deviation* is a measure of the extent of variation in a frequency distribution. It is defined as the square root of the average of squared deviations of the frequency distribution. A deviation is the difference between the value of an item and the mean value, and it could be negative. The squared deviation is the square of that and is always positive. The average of squared deviations is obtained by summing all the squared deviations in the frequency distribution and dividing by the number of items. Mathematically,

$$\text{Standard deviation} = \sigma = \sqrt{\frac{\sum f_i (x_i - \tilde{x})^2}{\sum f_i}}, \quad \text{where } \tilde{x} \text{ is the mean} \tag{2.2}$$

The standard deviation is therefore a measure of the spread of the items about the mean. For example, the three numbers 20, 25, and 30 have a mean of 25. The mean gives a fairly good approximation of the three individual numbers. The numbers 5, 10, and 60 also have a mean of 25, but the mean does not come close to giving an indication of what the individual numbers are. The standard deviation tells the story; in the former case, it is 4.08, in the latter case, it is 24.83.

## Problem 2.1

The following chart shows the seniority of 40 workers at a plant:

| Seniority | 1 | 2 | 3 | 4 | 5 | 6 | 8 | 9 | 11 | 15 | 18 | 20 |
|---|---|---|---|---|---|---|---|---|---|---|---|---|
| Number | 2 | 2 | 4 | 6 | 6 | 10 | 3 | 2 | 1 | 2 | 1 | 1 |

What is the average seniority?
What is the standard deviation?

*Solution:*

$$\text{Average seniority} = \bar{x} = \frac{\sum f_i x_i}{\sum f_i}$$

$$= \frac{(2\times 1)+(2\times 2)+(4\times 3)+(6\times 4)+\cdots+(1\times 20)}{2+2+6+6+\cdots+1}$$

$$= 6.33$$

$$\text{Standard deviation} = \sigma = \sqrt{\frac{\sum f_i(x_i-\bar{x})^2}{\sum f_i}}$$

$$= \sqrt{\frac{2(1-6.33)^2+2(2-6.33)^2+4(3-6.33)^2+\cdots+1(18-6.33)^2+1(20-6.33)^2}{2+2+4+\cdots+1+1}}$$

$$= 4.19$$

### 2.2.5 Variance

*Variance* is the square of the standard deviation and has more direct applications in some statistical analyses than the standard deviation.

## 2.3 THEORY OF PROBABILITY

### 2.3.1 Concept

*Probability* in simple terms is a measure of how likely it is for an event to happen or take place. The approach used is normally a relative frequency approach, that is, the number of outcomes in which the event of interest will take place expressed as a percentage (or decimal) of the total number of possible outcomes assuming implicitly that all outcomes are equally likely.

There are two kinds of situations usually encountered in this method. The first one is when the number of outcomes is finite, and the probability is known exactly, for example, the probability of rolling a "six" in a backgammon game in which two dice are used is 5/36. The second situation is when the number of outcomes is infinite, such as the probability of having a sunny day on Chinese New Year's Day. The total number of outcomes is all the Chinese New Year's Day from the beginning of the world to eternity, which is infinite and impossible to count, so the probability can only be estimated from a limited account of past data. If in the past 10 years, 8 years have a sunny Chinese New Year's Day, the probability of having a sunny day on the next Chinese New Year's Day will be estimated to be 0.8. If, however, records of the past 25 years were used, it might be found that 22 years had sunny Chinese New Year's Day, giving a probability of 0.88. In this example, there is no exact probability as there was for the rolling dice.

### 2.3.2 Probability Laws and Theorems

There are many laws and theorems pertaining to probability. The examples listed below are some of the most fundamental and most frequently used. No rigorous mathematical derivations are given.

1. The probability of an event occurring and probability of that event not occurring always add up to 1.

$$P(A) + P(\bar{A}) = 1.0 \tag{2.3}$$

2. The probability of event $A$ or event $B$ or both occurring is equal to the probability of event $A$ occurring plus the probability of event $B$ occurring minus the probability of both events occurring simultaneously.

$$P(A \cup B) = P(A) + P(B) - P(A, B) \tag{2.4}$$

3. The probability of two independent events both occurring is equal to the product of the individual probabilities.

$$P(A \cap B) = P(A) \cdot P(B) \tag{2.5}$$

4. The probability of event $A$ given that event $B$ has occurred is equal to the probability of $A$ and $B$ both occurring divided by the probability of event $B$ occurring.

$$P(A|B) = \frac{P(A \cap B)}{P(B)} \tag{2.6}$$

This is known as the conditional probability. No independence between $A$ and $B$ is assumed. In fact, rearranging the terms gives the probability of both events $A$ and $B$ occurring when $A$ and $B$ are not independent.

Some examples will clarify the last two theorems. Suppose there are 400 boys and 400 girls in a school and suppose one-quarter of the students wear glasses. The probability that a student picked at random will be a girl wearing glasses is $1/2 \times 1/4 = 1/8$ according to the third theorem. The theorem applies because the two attributes are independent.

Now suppose 300 of the boys and 100 of the girls are interested in computer games. The school has 400 students out of 800 who like computer games. However, if a student is picked at random, the probability of finding a boy who is interested in computer games is not $400/800 \times 400/800 = 0.25$. It should be $300/800 = 0.375$ from first principles. The Product Rule does not apply here because the two events, being a boy and being interested in computer games, are not independent—boys seem to be more interested in computer games than girls. Instead, the conditional probability of the fourth theorem should be used.

$$\begin{aligned} P(\text{boy} \cap \text{likes computer games}) &= P(\text{boy}|\text{computer games}) \\ &\quad \times P(\text{likes computer games}) \\ &= 300/400 \times 400/800 \\ &= 0.375 \end{aligned}$$

It does not matter which event is the dependent one and which event is the independent one. The results will be identical:

$$\begin{aligned} P(\text{likes computer games} \cap \text{boy}) &= P(\text{likes computer games}|\text{boy}) \times P(\text{boy}) \\ &= 300/400 \times 400/800 \\ &= 0.375 \end{aligned}$$

## 2.4 PROBABILITY DISTRIBUTION MODEL

### 2.4.1 Random Variable

Most probability and statistical problems involve a number that can vary between a range of values. This number is the value of the item under consideration and is determined by a random process and hence it is called a random variable. The random variable can take on any value within the range, but the probability that it will assume a certain value varies depending on what value it assumes. Using the backgammon example again, the possible rolls of the two dice are from "two" to "twelve" but probability of each roll is not the same as illustrated in Fig. 2.4. If the probability is plotted against the roll, the graph given in Fig. 2.5 will be obtained.

The graph is called a probability distribution. This probability distribution is a discrete one because the values of the rolls can only be integers. If the random variable is continuous and can take on any value within the range, then it will be a continuous probability distribution.

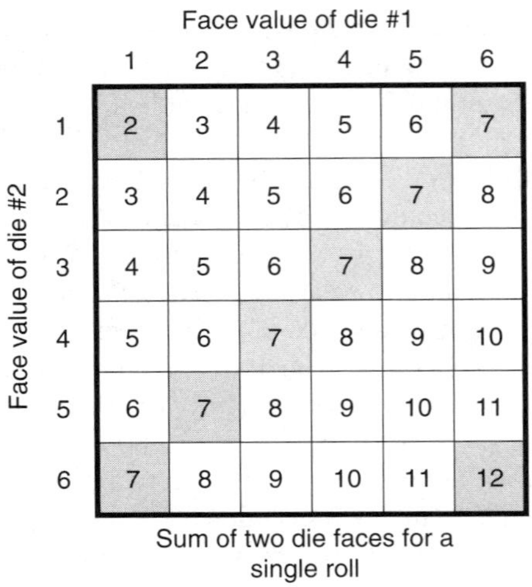


Figure 2.4. Sum of two die faces for a single roll as a function of the face value of the die #2.

## 2.4.2 Probability Density Function

In a discrete probability distribution, the ordinate of the random variable represents the probability that the random variable will take on that particular value, for example, the bar height of the roll "seven" in Fig. 2.5 is 0.1667 (one-sixth), which is the probability of rolling a "seven." The sum of the heights of all the bars is 1. This representation runs into difficulty with continuous probability distributions. Since the random variable can assume an infinite number of possible values, the sum of all these probabilities will

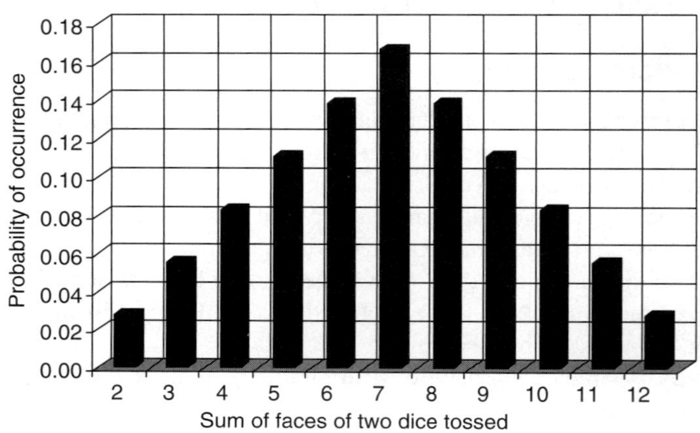


Figure 2.5. Probability distribution of the sum of faces of two dice tossed.

add up to infinity. The way to overcome this problem is by introducing the probability density function.

In a probability density function, the ordinate of the random variable $x$ represents the probability density and not the probability itself. The probability is represented by the area under the curve, so the probability of $x$ falling between $A$ and $B$ is the area under the curve between $x = A$ and $x = B$, and the probability of the random variable being equal to a certain value exactly is zero because the area of a line is zero. The area under the entire curve is, of course, equal to 1. The probability density function can usually be represented by a mathematical expression, for example,

$$f(x) = \frac{1}{a}\,\mathrm{e}^{-x/a} \tag{2.7}$$

The area under the curve from point $A$ to point $B$ can be found by integration.

$$\text{Probability } (A < x < B) = \int_A^B f(x)\,\mathrm{d}x \tag{2.8}$$

For application purposes, there are tables for the calculation of area under the curve as long as the end limits are known.

For discrete probability distributions, there are no continuous curves, just bars at the discrete values that the random variable may take on. The heights of the bars are the respective probabilities and no integration is necessary. For ascertaining cumulative probabilities, however, it is required to calculate the heights of the bars separately and sum them up whereas with a continuous probability density function, all that has to be done is integration between the proper limits. For common discrete probability distributions, there are standard tables for cumulative probabilities.

### 2.4.3 Parameters of Probability Distributions

The mean of a probability distribution is the average value of the random variable. It is analogous to the mean of a frequency distribution. For a discrete distribution, the mean value of the random variable is given by

$$\bar{x} = \sum P(x_i)x_i \tag{2.9}$$

For a continuous probability distribution,

$$\bar{x} = \int x{\cdot}f(x)\,\mathrm{d}x \tag{2.10}$$

The mean is not necessarily in the middle of the range of possible $x$'s. However, there is an equal chance for the random variable $x$ to fall on the lower side of $x$ as on the higher side. In a continuous probability distribution, this means the area under the curve is divided into two equal halves at $x = \bar{x}$.

The standard deviation is defined in the same way as that for frequency distributions.

$$\sigma = \sqrt{\sum (x_i - \bar{x})^2 P(x_i)} \tag{2.11}$$

For a continuous distribution, this becomes

$$\sigma = \sqrt{\int (x - \bar{x}^2) f(x)\, dx} \tag{2.12}$$

## 2.4.4 The Binomial Distribution

An example will illustrate this probability distribution very clearly. Consider the probability of rolling a fair die and getting a "six" two times out of three tosses. Work from first principle:

Probability of rolling a "six" = 1/6
Probability of not rolling a "six," that is, $X = 5/6$

To get two "sixes" out of three tosses, there are three ways:

6 6 $X$, Probability = 1/6 × 1/6 × 5/6 = 5/216
6 $X$ 6, Probability = 1/6 × 5/6 × 1/6 = 5/216
$X$ 6 6, Probability = 5/6 × 1/6 × 1/6 = 5/216

The probabilities of the three sequences are the same. Each consists of the probability of rolling a "six" raised to the power 2 (the number of "sixes" required) times the probability of not rolling a "six" raised to the power of 1 (the number of "non-sixes" required). The number of sequences is the number of possible combinations of two objects out of three. Multiply the three terms together and we get the required probability. Total probability = $3 \times (1/6)^2 \times (5/6) = 5/72$.

The process can be generalized by the binomial theorem as follows:

$$P(x) = nC_x p^x (1-p)^{n-x} \tag{2.13}$$

where $n$ is the number of trials, $x$ is the number of successful trials required, $nC_x = n!/(n - x)!x!)$ is the number of combinations of $x$ objects out of $n$, and $p$ is the probability of success.

This term is readily recognized as the $p^x$ term in the expansion of the binomial term $[p + (1 - p)]^n$. For the above example,

$$(1/6 + 5/6)^3 = (1/6)^3 + 3(1/6)^2(5/6) + 3(1/6)(5/6)^2 + (5/6)^3$$

This is no coincidence. In fact, the first term represents the probability of rolling three "sixes," the second term two "sixes" and one "non-six," the third term one "six"

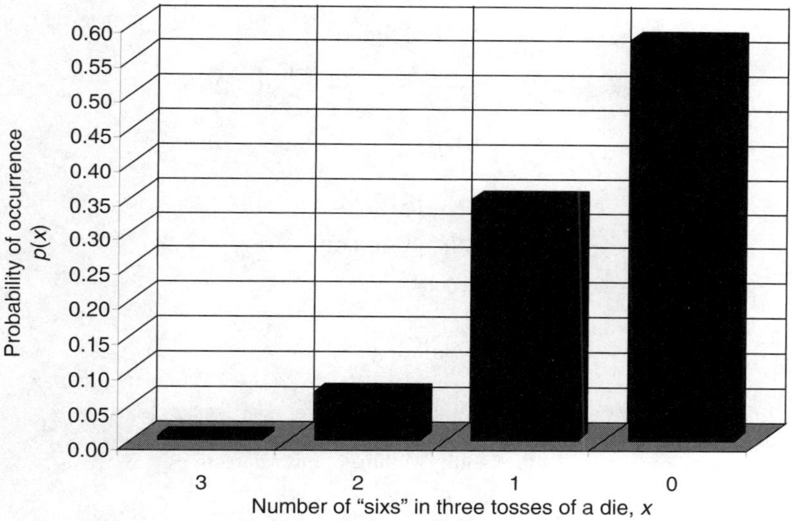


Figure 2.6. Binomial probability distribution $n = 3$, $p = 1/6$.

and two "non-sixes," and the fourth term three "non-sixes." Since that covers all possibilities, the sum of all these probabilities must be 1. That is automatically true because $[p + (1 - p)]^n$ is always equal to 1 regardless of what $p$ and $n$ are. The plot of $P(x)$ versus $x$ for all the $n$ terms is a binomial probability distribution and is shown in Fig. 2.6.

Here, $n$ and $p$ are the parameters of the distribution that determine the shape of the binomial distribution. The mean of the distribution is $\sum P(x_i)x_i = p$ and the standard deviation $\sigma = \sqrt{p(1-p)}$. Calculation of individual terms of the binomial distribution is not too difficult with a calculator. Calculations of the cumulative probability can be tedious, but there are tables available. Also for large $n$, the binomial distributions can be approximated by other distributions.

## Problem 2.2

It is known that 5% of the insulators are defective. What is the probability of finding three or more defective insulators in a string of five?

*Solution:*
Defective rate $p = 0.2$.
Probability of finding three or more defectives in five is given by

$$P(x) = nC_x p^x (1-p)^{n-x} \qquad \text{(Section 2.4.4)}$$

$$\begin{aligned}P(3) &= 5C_3(0.2)^3(0.8)^{5-3}\\ &= \frac{5\times4\times3\times2\times1}{(3\times2\times1)\times(2\times1)}(0.2)^3(0.8)^2\\ &= 0.0011\end{aligned}$$

$$\begin{aligned}P(4) &= 5C_4(0.2)^4(0.9)^{5-4}\\ &= 5(0.2)4\times(0.8)\\ &= 0.0064\end{aligned}$$

$$\begin{aligned}P(5) &= 5C_5(0.2)^5(0.8)^0\\ &= 0.00032\end{aligned}$$

Probability of three or more defectives

$$\begin{aligned}&= P(3)+P(4)+P(5)\\ &= 0.00782\\ &= 0.782\%\end{aligned}$$

## Problem 2.3

If there are three good insulators in a string of four, for 72 kV line, the probability of flashover is quite small (0.02%). On a 72 kV line, salvaged insulators with 10% defectives are used. What is the probability that a string will have fewer than three good insulators? The engineering manager decides to add one unit to each string. How does that help?

*Solution:*
If percentage of defectives is 10, probabilities of getting zero, one, or two good insulators in a string of five are

$$P(0) = {}_4C_0(0.8)^0(0.2)^4=(0.2)^4=0.0016$$

$$P(1) = {}_4C_1(0.8)^1(0.2)^3=4\times0.8\times0.008=0.0256$$

$$P(2) = {}_4C_2(0.8)^2(0.2)^2=0.1536$$

$$P(0)+P(1)+P(2) = 0.0016+0.0256+0.1536 = 0.1808 = 18.08\%$$

If length of string is increased to 5,

$$\begin{aligned}P(0)&={}_5C_0(0.8)^0(0.2)^5=0.00032\\ P(1)&={}_5C_1(0.8)^1(0.2)^4=0.0064\\ P(2)&={}_5C_2(0.8)^2(0.2)^3=0.0512\end{aligned}$$

The probability of having less than three good insulators is

$$P(0)+P(1)+P(2)=0.0579=5.79\%$$

The probability of inadequate insulation drops from over 18.08% to 5.79%.

### 2.4.5 The Poisson Distribution

*Poisson distribution* is a discrete probability distribution with an infinite number of possible points for the random variable. The probability that the random variable will take on a value $x$ is given by

$$P(x)=\frac{\mu^x\,\mathrm{e}^{-\mu}}{x!} \tag{2.14}$$

where $\mu$ is a parameter of the distribution. Indeed, it is the mean and the standard deviation that are $\sqrt{\mu}$. The Poisson distribution describes the probability of occurrence of a random event for a specified number of times within a given interval of time or scope. Although the average number of occurrences is $\mu$ in the long run, there is always a chance that for a particular interval, the number of occurrences is something other than $\mu$. For example, during a lightning storm, there are, say, two strokes per minute on an average, that is, $\mu=2$; but for any given minute, there is always a chance that there are 0, 1, 2, 3, . . . strokes. In fact,

$$\begin{aligned}P(0)&=13.5335\%\\P(1)&=27.0671\%\\P(2)&=27.0671\%\\P(3)&=18.0447\%\\P(4)&=9.0224\%\\P(5)&=3.6089\%\end{aligned}$$

Like the binomial distribution, the Poisson distribution can be approximated by continuous probability distributions. It is used as an approximation for the binomial distribution in many cases. There are tables giving the cumulative probabilities.

## Problem 2.4

The failure of power transformers is assumed to follow a Poisson probability distribution. Suppose on average, a transformer fails once every 5 years. What is the probability that it will not fail in the next 12 months? That it will fail once in the next 24 months?

*Solution:*

$$\begin{aligned}\text{Failure rate}&=\text{once in 5 years}\\&=0.2/\text{year}\end{aligned}$$

Number of expected failures in 12 months = 0.2.

Probability of having zero failures is given by

$$P(0) = \frac{(0.2)^0 \, e^{-0.2}}{0!}$$
$$= 0.8187$$

Number of expected failures in 24 months $= 0.2 \times 2 = 0.4$.
Probability of having exactly one failure in that period is

$$P(1) = \frac{(0.4)^1 \, e^{-0.4}}{1!}$$
$$= 0.2681$$

### 2.4.6 The Exponential Distribution

The exponential distribution is a continuous probability density function (i.e., the area indicates the probability) given by the formula

$$f(x) = \lambda \, e^{-\lambda x} \tag{2.15}$$

where $\lambda$ is a parameter of this probability function. It extends from 0 to $\infty$ and is illustrated in Fig. 2.7.

The exponential distribution describes a probability that decreases exponentially with increasing $x$. That probability is indicated by the area under the curve to the right of $x$, which extends to $\infty$ as illustrated in Fig. 2.8.

$$R(x) = \int_x^\infty \lambda \, e^{-\lambda x} \, dx = e^{-\lambda x} \tag{2.16}$$

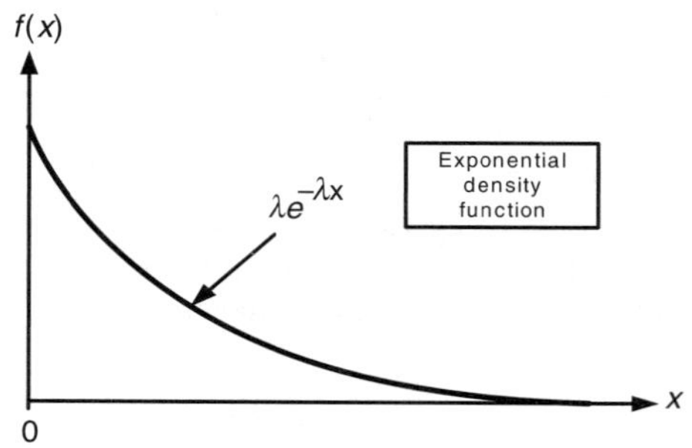


Figure 2.7. Exponential distribution.

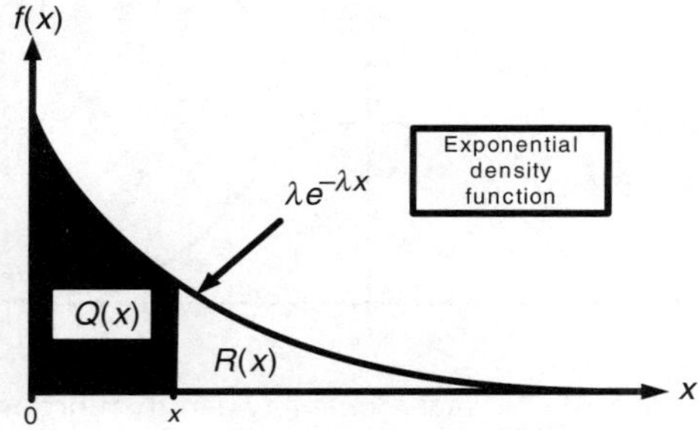


Figure 2.8. Areas under the exponential density function. *Note:* $Q(x) = 1 - R(x)$ because the total area under the density function equals 1.

where $R(x)$ is the probability that the random variable is greater than $x$ and $Q(x)$ is the probability that the random variable is less than or equal to $x$.

The mean of the exponential distribution can be found from the formula

$$\mu = \int_0^\infty x\lambda\, e^{-\lambda x}\, dx = \frac{1}{\lambda} \tag{2.17}$$

The standard deviation is given by

$$\sigma = \int_0^\infty \left(x - \frac{1}{\lambda}\right)^2 \lambda\, e^{-\lambda x}\, dx = \frac{1}{\lambda} \tag{2.18}$$

The parameter λ and the mean 1/λ all have significant physical meanings when the exponential distribution is applied to reliability assessments.

### 2.4.7 The Normal Distribution

*Normal distribution* is the most widely used probability distribution due to the fact that most things that are phenomena in nature tend to follow this distribution. It is a good approximation for many other distributions such as the binomial when the population is large. It is a continuous distribution; hence, the curve is the probability density function that takes on a symmetrical bell shape as illustrated in Fig. 2.9. The mathematical formula for the probability density function is

$$f(x) = \frac{1}{\sqrt{2\pi\sigma}}\, e^{-(x-\mu)^2/2\sigma^2} \tag{2.19}$$

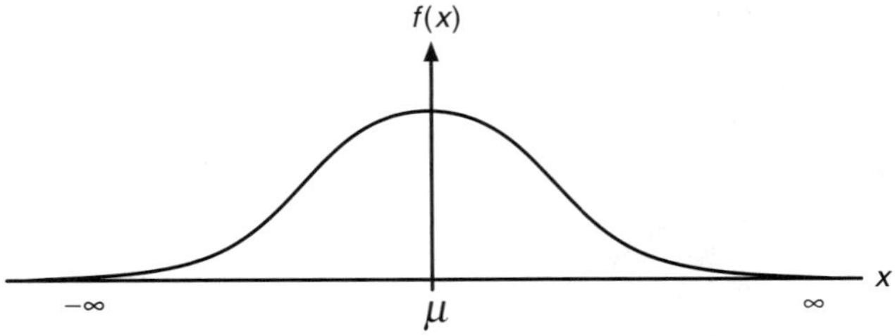


Figure 2.9. Normal probability density function.

There are two parameters with this distribution, μ and σ. It can be proved that the mean is μ and the standard deviation is σ. Being symmetrical, the mean μ naturally coincides with the midpoint of the bell-shaped curve. The area under the curve represents probability. The curve extends from $-\infty$ to $\infty$; however, the areas at the tail ends are negligible. Over 99% of the area falls within $\pm 3\sigma$, that is, three standard deviations from the mean. Very often, instead of using the actual value of $x$, measurement is done in terms of standard deviations from the mean and is called $z$. The mean becomes zero on this normalized scale as shown in Fig. 2.10. If the mean value of the normal curve is set at zero and all deviations are measured from the mean in terms of standard deviations, the equation for the normal curve in standard form for $Y$ becomes

$$Y = \left(\frac{1}{\sqrt{2\pi}}\right) e^{-z^2/2}, \quad \text{where} \quad z = (x - \mu)/\sigma \tag{2.20}$$

For example, suppose $\mu = 520$ and $\sigma = 11$. A value of $x = 492.5$ is 2.5 standard deviations below the mean. On the normalized scale, $x = 492.5$ simply becomes $z = -2.5$. There are tables for computing the area under a normal distribution curve, and these tables are all based on the normalized scale.

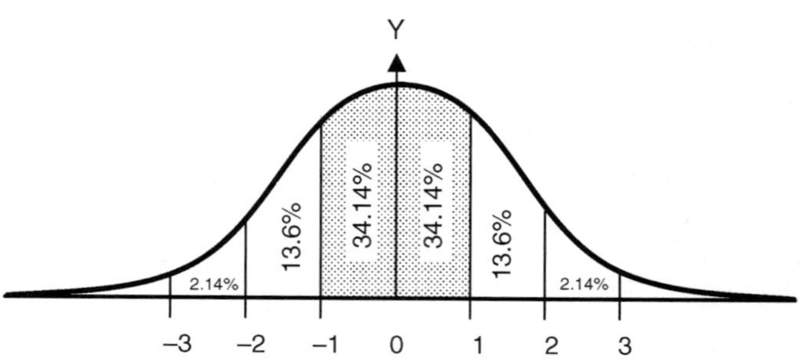


Figure 2.10. Normalized normal probability density function. The area from $z = -1$ to $z = 1$ is 68.28%; from $z = -2$ to $z = 2$ is 95.48%; and from $z = -3$ to $z = 3$ is 99.76%.

## Problem 2.5

In a normal distribution, what percentage is within 1.6 standard deviations from the mean?

*Solution:*
Refer to the normal distribution shown in Table 2.1.

$$z = 1.6 \text{ corresponds } to\ 0.4452$$

This is the shaded area.

The question says "within 1.6 standard deviations from the mean," which includes the other side of the mean (i.e., area corresponding to $z = -1.6$).

So the area within 1.6 standard deviation from the mean

$$= 0.4452 \times 2$$
$$= 0.8904$$
$$= 89.04\%$$

## 2.5 SAMPLING THEORY

### 2.5.1 Concepts of Population and Sample

In statistics, the totality of things, persons, events, or other items under study is called the *population*. There is certain information about the population that needs to be ascertained. This information can be collected from the entire population; however, this is often impractical and sometimes impossible, so sampling is used. A sample is a part of the population selected so that inferences can be made from it about the entire population.

### 2.5.2 Random Sampling Model

In order for the information provided by the sample to be an accurate representation of the population, there are two fundamental requirements, namely, the sample must be a part of the population and there must be no bias in selecting the sample, that is, it should be a random sample.

To achieve the accuracy desired, there are other principles and rules to follow, such as the sample size to use and the techniques to select the sample. It must be recognized that there is always the probability of error in sampling because part of the population has been missed out. The error, however, can be predicted and controlled. If decisions are made based on sampling information, the probability of error will be known and the risk can hence be gauged. Sampling methods can be designed to suit the need. On the contrary, a complete census is not always free of error in practice, and these errors are hard to predict and control.

### 2.5.3 Sampling Distributions

When a sample is selected and the characteristics of interest of each unit in the sample are observed or tested, a set of statistics such as mean, standard deviation, and so on

TABLE 2.1. Areas for Standard Normal Probability Distribution

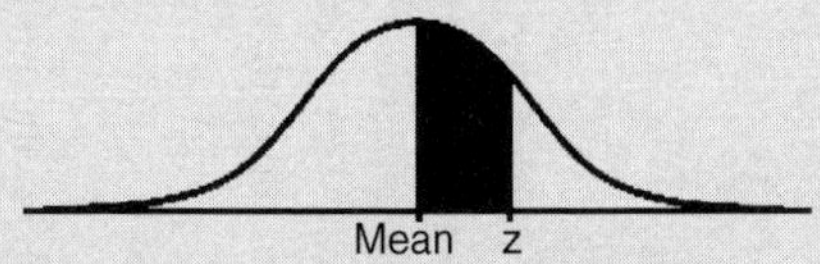


| z | 0.00 | 0.01 | 0.02 | 0.03 | 0.04 | 0.05 | 0.06 | 0.07 | 0.08 | 0.09 |
|---|---|---|---|---|---|---|---|---|---|---|
| 0.0 | 0.0000 | 0.0040 | 0.0080 | 0.0120 | 0.0160 | 0.0199 | 0.0239 | 0.0279 | 0.0319 | 0.0359 |
| 0.1 | 0.0398 | 0.0438 | 0.0478 | 0.0517 | 0.0557 | 0.0596 | 0.0636 | 0.0675 | 0.0714 | 0.0753 |
| 0.2 | 0.0793 | 0.0832 | 0.0871 | 0.0910 | 0.0948 | 0.0987 | 0.1026 | 0.1064 | 0.1103 | 0.1141 |
| 0.3 | 0.1179 | 0.1217 | 0.1255 | 0.1293 | 0.1331 | 0.1368 | 0.1406 | 0.1443 | 0.1480 | 0.1517 |
| 0.4 | 0.1554 | 0.1591 | 0.1628 | 0.1664 | 0.1700 | 0.1736 | 0.1772 | 0.1808 | 0.1844 | 0.1879 |
| 0.5 | 0.1915 | 0.1950 | 0.1985 | 0.2019 | 0.2054 | 0.2088 | 0.2123 | 0.2157 | 0.2190 | 0.2224 |
| 0.6 | 0.2257 | 0.2291 | 0.2324 | 0.2357 | 0.2389 | 0.2422 | 0.2454 | 0.2486 | 0.2518 | 0.2549 |
| 0.7 | 0.2580 | 0.2612 | 0.2642 | 0.2673 | 0.2704 | 0.2734 | 0.2764 | 0.2794 | 0.2823 | 0.2852 |
| 0.8 | 0.2881 | 0.2910 | 0.2939 | 0.2967 | 0.2995 | 0.3023 | 0.3051 | 0.3078 | 0.3106 | 0.3133 |
| 0.9 | 0.3159 | 0.3186 | 0.3212 | 0.3238 | 0.3264 | 0.3289 | 0.3315 | 0.3340 | 0.3365 | 0.3389 |
| 1.0 | 0.3413 | 0.3438 | 0.3461 | 0.3485 | 0.3508 | 0.3531 | 0.3554 | 0.3577 | 0.3599 | 0.3621 |
| 1.1 | 0.3643 | 0.3665 | 0.3686 | 0.3708 | 0.3729 | 0.3749 | 0.3770 | 0.3790 | 0.3810 | 0.3830 |
| 1.2 | 0.3849 | 0.3869 | 0.3888 | 0.907 | 0.3925 | 0.3944 | 0.3962 | 0.3980 | 0.3997 | 0.4015 |
| 1.3 | 0.4032 | 0.4049 | 0.4066 | 0.4082 | 0.4099 | 0.4115 | 0.4131 | 0.4147 | 0.4162 | 0.4177 |
| 1.4 | 0.4192 | 0.4207 | 0.4222 | 0.4236 | 0.4251 | 0.4265 | 0.4279 | 0.4292 | 0.4306 | 0.4319 |
| 1.5 | 0.4332 | 0.4345 | 0.4357 | 0.4370 | 0.4382 | 0.4394 | 0.4406 | 0.4418 | 0.4429 | 0.4441 |
| 1.6 | 0.4452 | 0.4463 | 0.4474 | 0.4484 | 0.4495 | 0.4505 | 0.4515 | 0.4525 | 0.4535 | 0.4545 |

(Continued)

TABLE 2.1. (*Continued*)

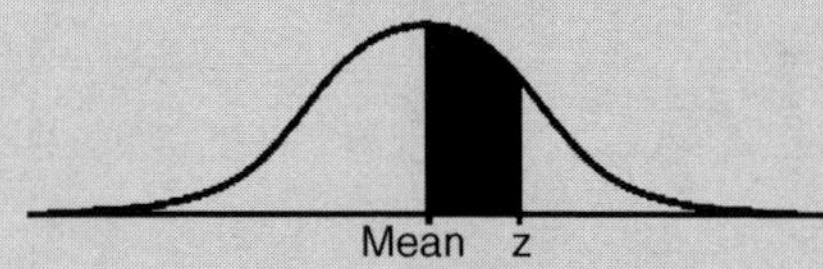


| $z$ | 0.00 | 0.01 | 0.02 | 0.03 | 0.04 | 0.05 | 0.06 | 0.07 | 0.08 | 0.09 |
|---|---|---|---|---|---|---|---|---|---|---|
| 1.7 | 0.4554 | 0.4564 | 0.4573 | 0.4582 | 0.4591 | 0.4599 | 0.4608 | 0.4616 | 0.4625 | 0.4633 |
| 1.8 | 0.4641 | 0.4649 | 0.4656 | 0.4664 | 0.4671 | 0.4678 | 0.4686 | 0.4693 | 0.4699 | 0.4706 |
| 1.9 | 0.4713 | 0.4719 | 0.4726 | 0.4732 | 0.4738 | 0.4744 | 0.4750 | 0.4756 | 0.4761 | 0.4767 |
| 2.0 | 0.4772 | 0.4778 | 0.4783 | 0.4788 | 0.4793 | 0.4798 | 0.4803 | 0.4808 | 0.4812 | 0.4817 |
| 2.1 | 0.4821 | 0.4826 | 0.4830 | 0.4834 | 0.4838 | 0.4842 | 0.4846 | 0.4850 | 0.4854 | 0.4857 |
| 2.2 | 0.4861 | 0.4864 | 0.4868 | 0.4871 | 0.4875 | 0.4878 | 0.4881 | 0.4884 | 0.4887 | 0.4890 |
| 2.3 | 0.4893 | 0.4896 | 0.4898 | 0.4901 | 0.4904 | 0.4906 | 0.4909 | 0.4911 | 0.4913 | 0.4916 |
| 2.4 | 0.4918 | 0.4920 | 0.4922 | 0.4925 | 0.4927 | 0.4929 | 0.4931 | 0.4932 | 0.4934 | 0.4936 |
| 2.5 | 0.4938 | 0.4940 | 0.4941 | 0.4943 | 0.4945 | 0.4946 | 0.4948 | 0.4949 | 0.4951 | 0.4952 |
| 2.6 | 0.4953 | 0.4955 | 0.4956 | 0.4957 | 0.4959 | 0.4960 | 0.4961 | 0.4962 | 0.4963 | 0.4964 |
| 2.7 | 0.4965 | 0.4966 | 0.4967 | 0.4968 | 0.4969 | 0.4970 | 0.4971 | 0.4972 | 0.4973 | 0.4974 |
| 2.8 | 0.4974 | 0.4975 | 0.4976 | 0.4977 | 0.4977 | 0.4978 | 0.4979 | 0.4979 | 0.4980 | 0.4981 |
| 2.9 | 0.4981 | 0.4982 | 0.4982 | 0.4983 | 0.4984 | 0.4984 | 0.4985 | 0.4985 | 0.4986 | 0.4986 |
| 3.0 | 0.49865 | 0.4987 | 0.4987 | 0.4988 | 0.4988 | 0.4989 | 0.4989 | 0.4989 | 0.4990 | 0.4990 |
| 4.0 | 0.4999683 | | | | | | | | | |

Illustration: For $z = 1.93$, shaded area is 0.4732 out of a total area of 1.

is obtained. The basic purpose of sampling is to deduce the corresponding statistics of the population from these sample statistics. To do so, it is important to know how representative a sample statistic is of the corresponding population statistic, that is, how it deviates from the population statistic. For any one sample, the sample statistic could be above, below, or even right on the population statistic; however, if many samples are drawn and the sample statistic for each sample is worked out, it will be found that their values are distributed in a certain pattern that can be described by a probability distribution. This is called the sampling distribution of that statistic. Sampling distributions have different forms, depending on the statistic in question and the population from which the samples are drawn; however, when the sample size is large enough, they all become approximate normal distributions. This very useful result is known as the central limit theorem.

### 2.5.4 Concept of Confidence Limit

The mean of the sampling distribution is obtained from all possible samples. The number of all possible samples is enormous; thereby, it is a much more reliable estimator for the population statistic than the value that is based on just one sample. This sampling distribution mean is an unknown quantity, since the sampler is not going to draw all these samples—as the sampler will be drawing only one sample. A relation exists, however, between the single sample statistic and the sampling distribution mean when the sampling distribution is a normal distribution. The sample statistic is a point on the normal curve, and the probability that it is within a given distance from the sampling distribution mean is known. For example, there is a 99% chance that it is within 2.58 standard deviations, a 95% chance that it is within 1.96 standard deviations, and so on. This percentage, a measure of the accuracy of the prediction, is known as the confidence level. Note that a higher confidence level is offset by a wider range, and a narrower range is accompanied by a lower confidence level. In other words, there is a trade-off between accuracy and precision.

The parameter that is required in linking the sample statistic to the sampling distribution mean is the standard deviation of the sampling distribution. This parameter $\sigma_x$ depends on the population size $N$, sample size $n$, and the standard deviation of the population $\sigma$ and is given by

$$\sigma_x = \sqrt{\frac{N-n}{N-1}}\frac{\sigma}{\sqrt{n}} \tag{2.21}$$

Frequently, the standard deviation of the population $N$ is not known, and the standard deviation of the sample, $s$, is used as a substitute. The value of $s$ can always be worked out once the sample is selected.

### 2.5.5 Estimation of Population Statistic

The basic purpose of sampling is to find out something about the population from a sample. Thus far, we have established the relationship between the sample statistic and the mean of the sampling distribution of that statistic. The next step is to establish the link between the sampling distribution mean and the value of that statistic for the entire population. For the two most frequently used statistics, namely, the mean and the proportion, this relationship is very simple. The sampling distribution mean coincides with the population statistic. Hence, the process of going from a single sample to the estimation of the population statistic is complete.

## Example 2.1

Out of 30,000 insulators, a sample of 3000 was drawn and 30 were found to be defective. What is the percentage defective of the population at 99% confidence level?

Proportion of defectives in the sample $p = 0.03$

Standard deviation of sample $s = \sqrt{0.03 \times 0.97} = 0.171$

Standard deviation of sampling distribution $\sigma_x = \sqrt{\dfrac{30,000 - 3000}{30,000 - 1}} \times \dfrac{0.171}{\sqrt{3000}}$

$$= 0.003$$

99% confidence level $\Rightarrow$ 2.58 standard deviations from the mean.

Percentage defective in population $= 0.03 \pm 2.58 \times 0.003 = 3 \pm 0.77\%$.

## Problem 2.6

Twenty ground rods were tested in a station area with 500 ground rods. The sample results are as follows:

| | | | | | | | | | |
|---|---|---|---|---|---|---|---|---|---|
| 17 | 25 | 32 | 10 | 6 | 5 | 8 | 9 | 12 | 17 |
| 46 | 64 | 83 | 70 | 10 | 15 | 2 | 8 | 29 | 11 |

What is the 90% confidence interval?

*Solution:*
The mean and standard deviation of the sample are computed:

$$\bar{x} = \frac{\sum f_i x_i}{\sum f_i} = \frac{17 + 25 + 32 + \cdots + 29 + 11}{20}$$

$$= 23.95$$

$$s = \sqrt{\frac{\sum f_i (x_i - 23.95)^2}{\sum f_i}} = \sqrt{\frac{(17 - 23.95)^2 + (25 - 23.95)^2 + (11 - 23.95)^2}{20}}$$

$$= 22.98$$

To estimate the population mean from the sample statistics, the following formulas are used:

$$\text{Population mean} = \text{sample mean} \pm z\sqrt{\frac{N-n}{N-1}}\frac{s}{\sqrt{n}} \qquad \text{(Section 2.5.5)}$$

For 90% confidence level, $z = 1.645$

$$\begin{aligned}\text{So the 90\% confidence interval} &= \pm 1.645\sqrt{\frac{500-20}{500-1}} \times \frac{22.98}{\sqrt{20}} \\ &= \pm 8.29\end{aligned}$$

## Problem 2.7

One hundred poles out of 2000 were tested and five were found to be rotten. We want to state that we are 90% confident that the percentage of rotten poles does not exceed a certain value. What is that percentage?

*Solution:*
The sample proportion (of defectives) and standard deviations are computed:

$$p = \frac{5}{100} = 0.05$$

$$\begin{aligned}\text{Standard deviation} &= \sqrt{0.05(1-0.05)} \\ &= 0.2179\end{aligned}$$

The population defective proportion is given by

$$\bar{p} \leq p + z\sqrt{\frac{N-n}{N-1}}\frac{s}{\sqrt{n}}$$

Since we are interested in the upper limit only, the confidence limit is a one-sided one; at 95% confidence, $z = 1.96$.

$$\begin{aligned}\bar{p} &\leq 0.05 + 1.96\sqrt{\frac{2000-100}{2000-1}} \cdot \frac{0.2179}{\sqrt{100}} \\ &= 0.0894 = 8.94\%\end{aligned}$$

### 2.5.6 Computation of Sample Size

It is intuitively clear that a larger sample gives a better estimate of the population, but it also costs more, so the choice of sample size is a compromise between accuracy and cost. Sample size is related to accuracy through the following formulas:

$$\bar{x} = x \pm z\sigma_x \tag{2.22}$$

$$\sigma_x = \sqrt{\frac{N-n}{N-1}}\frac{\sigma}{\sqrt{n}} \tag{2.23}$$

Rearranging,

$$|\bar{x} - x| = z\sigma_x \tag{2.24}$$

$$h = z\sqrt{\frac{N-n}{N-1}}\frac{\sigma}{\sqrt{n}} \tag{2.25}$$

The term $|\bar{x} - x|$ is the magnitude of the difference between the sample statistic and the true population statistic, which is called the *half width* of the confidence interval and denoted by $h$. It represents the maximum possible error (within the confidence level); for instance, in Example 2.1 in Section 2.5.5, percentage defective in the population (at 99% confidence level) is $4 \pm 1.5\%$. That means the maximum error is 1.5%; however, it could be less. $z$ depends on the confidence level desired; for example, for 95% confidence level, $z = 1.96$. Standard deviation $\sigma$ of the population is unknown and has to be estimated.

So after determining the maximum error $h$ that can be tolerated and the confidence level desired, the sample size $n$ required can be calculated provided that the population size $N$ is known and the population standard deviation $\sigma$ is known or estimated.

$$n = \frac{\sigma^2}{(h^2/z^2) + (\sigma^2/N)} \tag{2.26}$$

If $N$ is very large, this is simplified to

$$n = \frac{(\sigma z)^2}{h^2} \tag{2.27}$$

## Example 2.2

In ground testing, we want the error to be within 5 Ω at 95% confidence level. The standard deviation of ground rod resistances is 8 Ω, from previous years' experience. If we go into a station area with 450 rods, how many should be tested?

$$\begin{aligned} N &= 450 \\ \sigma &= 8 \\ h &= 5 \\ z &= 1.96 \\ n &= \frac{8^2}{(5^2/1.96^2) + (8^2/450)} = 9.6 \end{aligned}$$

## Problem 2.8

An opinion poll was taken to find out the percentage of the population, which supports the Green Party. If the result is to be within five percentage points at a 95% confidence level, what sample size should be used? (Assume that the support is around 40%.)

*Solution:*
The sample size for an infinite population is given by

$$n = \left(\frac{\sigma z}{h}\right)^2 \qquad \text{(Section 2.5.6)}$$

$\sigma$ is usually not known and has to be estimated as follows:
In the case of proportions,

$$\sigma = \sqrt{p \times (1-p)}$$

If $p$ is estimated to be 0.4, $\sigma = 0.6197$.
$z$ for 95% confidence is 1.96.

$$n = \left(\frac{0.6197 \times 1.96}{0.05}\right)^2$$

$$= 590$$

## Problem 2.9

The average weight of a batch of 3000 screws is determined by sampling. If the standard deviation of the weights is 0.2 g, what sample size has to be used so that we are 99% sure that the sampling result is within 0.04 g of the true average?

*Solution:*
Sample size for a finite population is given by

$$n = \frac{\sigma^2}{(h^2/z^2) + (\sigma^2/N)}$$

$$= \frac{(0.2)^2}{[(0.04)^2/(2.575)^2] + [(0.2)^2/3000]} \qquad (z \text{ for } 99\% \text{ confidence is } 2.575)$$

$$= 157$$

## 2.6 STATISTICAL DECISION MAKING

A statistical decision is simply a decision based on the result of random sampling. There is a possibility that the decision made is wrong, but the probability of making the wrong decision, that is, the risk, is known. Also, the decision-making process can be structured to control the errors.

### 2.6.1 Procedure of Decision Making

For ease and clarity of analysis and discussion, the statistical decision-making process usually takes the following form.

A hypothesis is stated. A random sample is drawn to test the hypothesis. Two alternative actions are possible:

Action 1 = accept the hypothesis (and act accordingly)
Action 2 = reject the hypothesis (and act accordingly)

A decision rule has to be made to lead from the sampling result to the actions.

---

## Example 2.3

It is suspected that the poles on a power line are rotting. If more than 25% are rotten, a changeout program should be initiated.

| | |
|---|---|
| Hypothesis: | More than 25% of the poles are rotten |
| Sampling: | 50 poles out of 2000 are checked |
| Actions: | 1. Hypothesis is true—changeout required<br>2. Hypothesis not true—no changeout required |
| Decision rule: | If sample shows more than 10 rotten poles conclude Action 1, otherwise conclude Action 2. |

### 2.6.2 Types of Error

There are two kinds of errors possible:

- Type I error: rejecting the hypothesis when it is true.
- Type II error: accepting the hypothesis when it is false.

Using the above example of changing out rotten poles, we want a changeout if the *real* percentage of rotten poles is 25% or more, but our decision rule is that if the proportion of rotten poles in the sample is more than $10/50 = 0.2$, we have to do a changeout. When the population proportion of rotten poles is 0.25, the population standard deviation is $\sqrt{0.25 \times (1-0.25)} = 0.433$. If we draw samples from this population, the mean of the sampling distribution will also be 0.25, and the standard

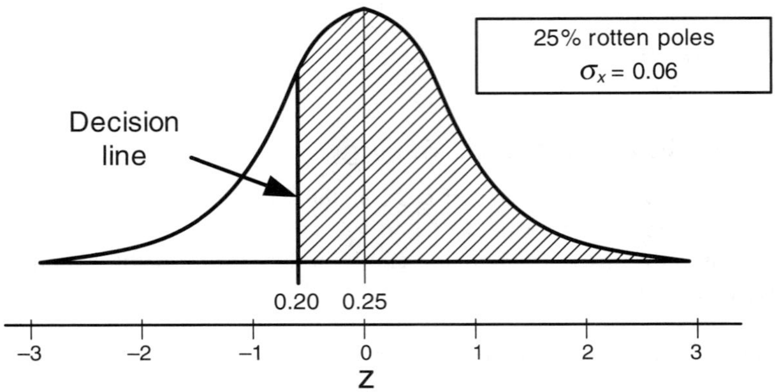


Figure 2.11. Sampling distribution $\sigma_x = 0.06$.

deviation of the sampling distribution will be

$$\sigma_x = \sqrt{\frac{2000 - 50}{2000 - 1}}\frac{0.433}{\sqrt{50}} = 0.06$$

There is an 80% chance that our sample proportion will be higher than 0.2 (shaded area in Fig. 2.11), leading us to conclude that 25% of the population is rotten and a changeout is needed (correct decision). Twenty percent of the time, our sample proportion will fall below 0.2 and we will conclude erroneously that the population proportion of rotten poles is less than 25% (Type I error). Note that if the decision rule is changed, say, to 13 rotten poles in 50 instead of 10 in 50, all these percentages will be different. Also note that the above calculation is for a population with 25% rotten poles. For different percentages of rotten poles in the population, the above calculation can be repeated to obtain different sets of probabilities of making correct decisions and making wrong decisions. The probabilities for populations with 20% and 30% rotten poles are listed in Table 2.2 and shown in Fig. 2.12 as an illustration.

TABLE 2.2. Probabilities for Populations with 20% and 30% Rotten Poles

| | Probability of Actions | | |
|---|---|---|---|
| Percentage of Rotten Poles in Population | Action 1 | Action 2 | Error Probability |
| 20 | 50% | 50% | Type II 50% |
| 25 | 80% | 20% | Type I 20% |
| 30 | 94% | 6% | Type I 6% |

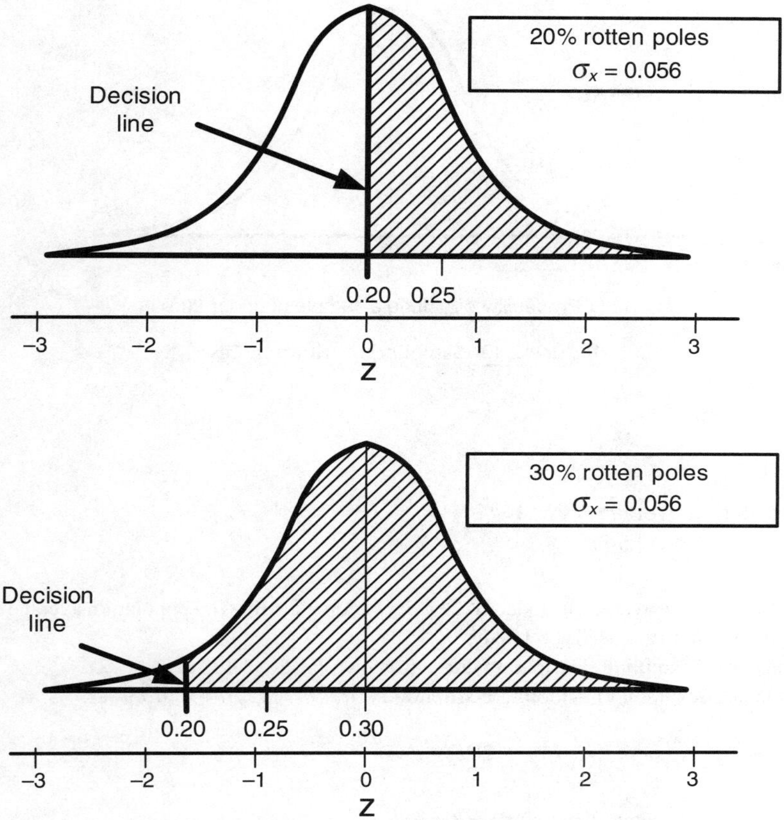


Figure 2.12. Sampling distribution $\sigma_x = 0.056$ for 20% and 30% rotten poles.

## Problem 2.10

A utility's policy is that if the average ground rod resistance in a station area is over 25 Ω, it will have to be retested the following year. In a station area with 1000 rods, 40 were tested. To be on the safe side, the area manager decided that if the average of these 40 rods was over 20 Ω, the station area would be marked for retest the next year. Records from past years indicated that the standard deviation of rod resistances was 12 Ω. Using the area manager's rule, what was the Type I error when the actual average resistance of the area was 20 Ω? What was the Type II error when the actual average resistance was 23 Ω? What would be the respective errors if the retest criterion was set at 25 Ω (instead of 20 Ω)?

*Solution:*

$$\text{Population} = 1000$$
$$\text{Population standard deviation} = 12\,\Omega$$
$$\text{Sample size} = 40$$

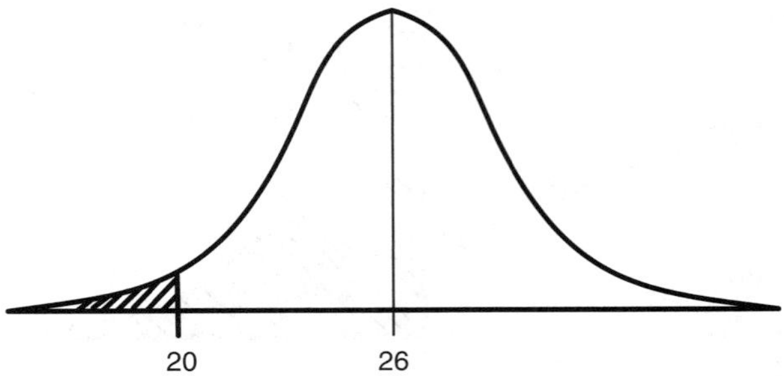


Probability of getting a sample of under 20 W

Figure 2.13. Sampling distribution Case I.

*Case I*

Population average = 26 Ω
Decision rule = 20 Ω

that is, if the sample average is under 20 Ω, we will conclude that the population average is under 25 Ω, which is not true—a Type I error.

The Case I sampling distribution curve is shown in Fig. 2.13.

Standard deviation of sampling distribution $= \sigma/\sqrt{n} = 12/\sqrt{40} = 1.897\ \Omega$.

$$20\ \Omega \rightarrow z = \frac{20-26}{1.897} = -3.16$$

(from normal distribution tables for $z = 3.16$)

Probability of getting a sample of under 20 Ω = shaded area = 0.08%.

*Case II*

Population average = 23 Ω
Decision rule = 20 Ω

that is, if the sample average is over 20 Ω, we will conclude that the population average is over 25 Ω, which is not true—a Type II error.

The Case II sampling distribution curve is shown in Fig. 2.14.

Standard deviation of sampling distribution $= \sigma/\sqrt{n} = 12/\sqrt{40} = 1.897\ \Omega$.

$$20\ \Omega \rightarrow z = \frac{20-23}{1.897} = -1.58$$

(from normal distribution tables for $z = 1.58$)

Probability of getting a sample of under 20 Ω = shaded area = 0.443 + 0.5 = 94.3%.

The implication of shifting the decision rule from 25 to 20 Ω is that we will practically never let anything over 25 Ω slip through, while wrongly rejecting many populations in the range of 20–25 Ω as shown in Figs. 2.15 and 2.16 for Case I and Case II.

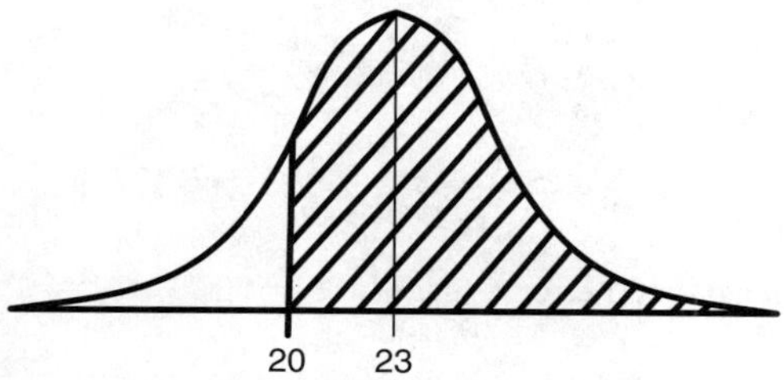


Figure 2.14. Sampling distribution Case II.

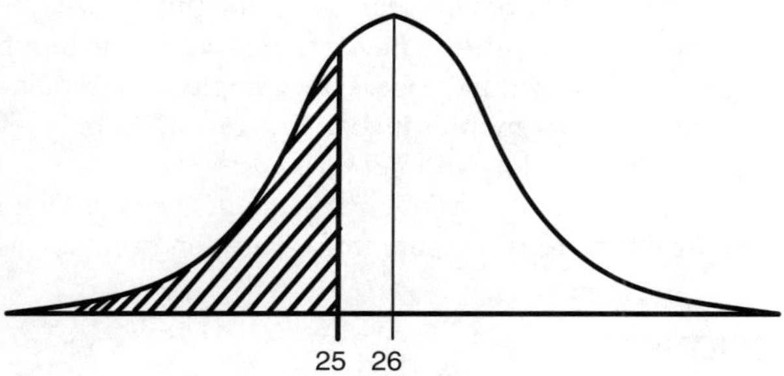


Figure 2.15. Sampling distribution Case I when the decision rule is set at 25 Ω.

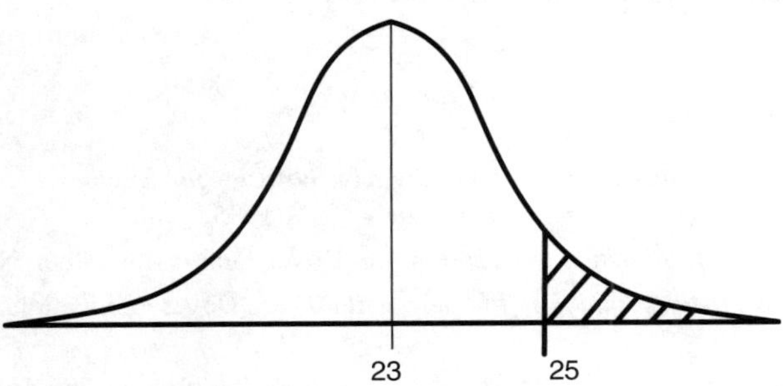


Figure 2.16. Sampling distribution Case II if the decision rule is set at 25 Ω.

If the decision rule is set at 25 Ω,
In Case I:

$$25\,\Omega \Rightarrow z = \frac{25-26}{1.897} = 0.527$$

Probability of making Type I error = shaded area = 30%.
In Case II:

$$25\,\Omega \Rightarrow z = \frac{25-23}{1.897} = 1.054$$

Probability of making Type I error = shaded area = 14.6%.

### 2.6.3 Control of Errors

Looking at the figures in Section 2.6.1, one will realize that the decision rule is actually the decision line on the sampling distribution curve. If that line is moved to the right, Type I error will increase, but Type II error will decrease. Sometimes one type of error is more serious than another. The decision rule can then be adjusted so as to minimize that type of error but the other type of error will increase. For example, in acceptance sampling, we can make the accept/reject criteria more stringent. As a result, not only fewer bad lots will slip through but also more good lots will be rejected.

There is a way to reduce both types of errors simultaneously. That is by increasing the sample size, but again, there is a compromise between accuracy and cost.

## 2.7 CONCLUSIONS

This chapter has introduced the basic concepts and principles of some commonly used probability theories in reliability assessments using simple numerical examples. Applications of different distributions in solving some typical every day problems were illustrated in an easy-to-understand manner. Statistical analysis of real problems plays a vital role in every day decision making in many businesses and industries. The probability theories will be enhanced via frequent examples in later chapters, and their importance in real-world situations will be made obvious.

## REFERENCES

1. R. E. Walpole and R. H. Myers, *Probability and Statistics for Engineers and Scientists*, 3rd edition, Macmillan Publishing Company, New York, 1985.
2. R. Langley, *Practical Statistics*, revised edition, Dover Publications, Inc., New York, 1970.
3. J. Mandel, *The Statistical Analysis of Experimental Data*, Dover Publications, Inc., New York, 1984.
4. R. E. Barlow and F. Proschan, *Statistical Theory of Reliability and Life Testing: Probability Models*, Holt, Rinehart and Winston, Inc., New York, 1975.
5. J. Endrenyi, *Reliability in Electric Power Systems*, John Wiley & Sons, Ltd., Chichester, 1978.

6. R. Ramakumar, *Engineering Reliability: Fundamentals and Applications*, Prentice-Hall, Inc., Upper Saddle, 1993.
7. T. Gönen, *Electric Power Distribution System Engineering*, McGraw Hill, New York, River, NJ, 1986.
8. R. Billinton and R. N. Allan, *Reliability Evaluation of Engineering Systems: Concepts and Techniques*, 2nd edition, Plenum Press, New York, 1992.
9. N. R. Mann, R. E. Schafer, and N. D. Singpurwalla, *Methods for Statistical Analysis of Reliability and Life Data*, John Wiley & Sons, Inc., New York, 1977.

# 3

# RELIABILITY PRINCIPLES

## 3.1 FAILURE RATE MODEL

### 3.1.1 Concept and Model

Since every piece of equipment in a system will eventually fail if it is in service for a long period, there is a failure rate associated with each one. For some items, the failure rate is quite significant while for others it could be extremely low.

Failure rate is defined as the number of expected failures per unit in a given time interval. It is just an expected value—the actual number of failures in any given time interval may differ from this; for example, a computer with a failure rate of 12 failures per year does not necessarily have one failure every month. For a group of equipment or components, the number of expected failures is equal to the number of units in the group times the failure rate. In calculating the failure rate of a group of units, the total operating time of the units should be used instead of the chronological time. The formula is

$$\text{Failure rate, } \lambda = \frac{\text{number of failures}}{\text{total operating time of units}} \tag{3.1}$$

*Power Distribution System Reliability.* By Ali A. Chowdhury and Don O. Koval

## Example 3.1

Ten transformers were tested for 500 h each and four transformers failed after the following test time periods:

one failed after 50 h
one failed after 150 h
two failed after 400 h

What is the failure rate for these types of transformers?

Total operating time of units

$$= (1 \times 50 + 1 \times 150 + 2 \times 400 + 6 \times 500) \text{ unit h}$$
$$= 4000 \text{ unit h}$$

$$\lambda = \frac{4}{4000} = 0.001 \text{ failures/unit h}$$

This is the failure rate of each of these transformers. If there are 1000 transformers in the system, we can expect $1000 \times 0.001 = 1$ failure/h somewhere in the system.

## Problem 3.1

Thirty motors were tested for 200 h. Five motors failed during the test. The failures occurred after the following test times:

| | |
|---|---|
| Motor 1 | 60 h |
| Motor 2 | 71 h |
| Motor 3 | 157 h |
| Motor 4 | 160 h |
| Motor 5 | 170 h |

What is the estimated failure rate?

*Solution*:

Total number of unit operating hours

$$= 60 + 71 + 157 + 160 + 170 + 25 \times 200$$
$$= 5618 \text{ unit h}$$

$$\text{Failure rate, } \lambda = \frac{\text{number of failures}}{\text{total unit operating time}}$$
$$= 5/5168 \text{ unit h}$$
$$= 0.00092 \text{ failures/h}$$

### 3.1.2 Concept of Bathtub Curve

The life of equipment usually has the following three major distinguishable periods:

1. Infant mortality period
2. Useful life period
3. Wear-out period

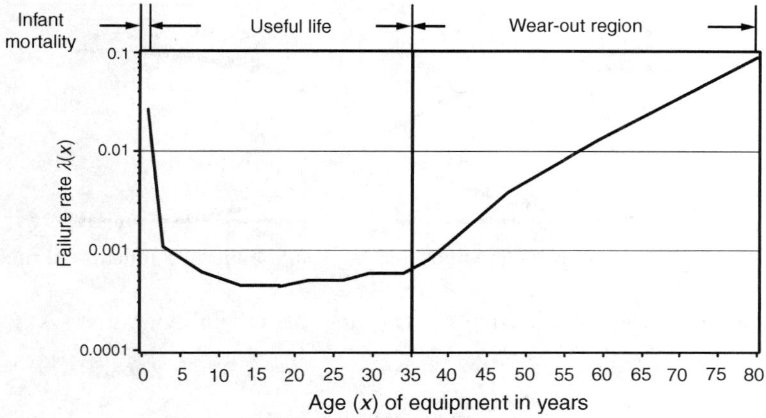


Figure 3.1. Bathtub curve failure rate versus time.

In the infant mortality period, the failure rate is high due to the presence of weak spots from the manufacturing process such as poor workmanship, substandard components, and so on. As these weaknesses are manifested one by one by the stress of operation, the failure rate keeps decreasing until a low constant level is reached. The equipment then enters the useful life period where failures are due to chance and occur at random times. Although failures occur at unpredictable moments and irregularly, in sufficiently long periods of equal length (say, a year), the same number of failures occur. Failures in this period are also independent of the age of the equipment. This period eventually ends when the components of the equipment start to wear out. From this time on, the failure rate rises rather rapidly due to deterioration. Actually, chance failures still occur; but the overall failure rate is dominated by the wear-out process.

If the failure rate is plotted against time, a bathtub-shaped curve will result as illustrated in Fig. 3.1.

Most reliability work deals with the useful life period when the failure rate is constant and the exponential distribution applies. The wear-out period is of some interest too and is usually modeled by either the normal distribution or the Weibull distribution, one of the several possible distributions.

## 3.2 CONCEPT OF RELIABILITY OF POPULATION

### 3.2.1 Theory of First Principles

Reliability is the probability of not failing in a specified time interval. Applied to a population of equipment, it is equal to the proportion of the original population that survives after that given time as illustrated in Fig. 3.2. If the original population is $N_o$ and

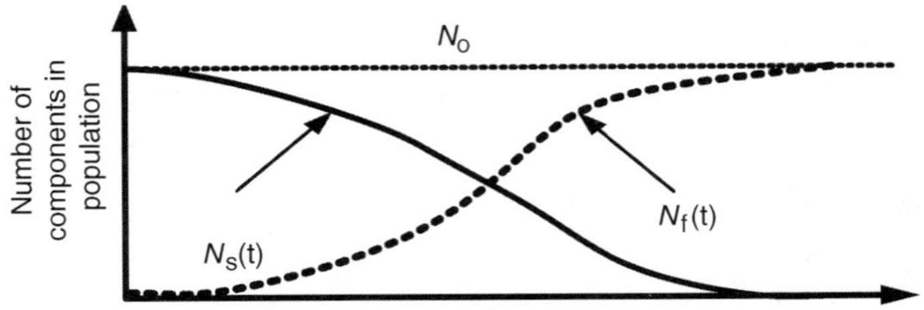


Figure 3.2. Number of failing and surviving components at a function of time.

$N_f$ of them fail after time $t$, leaving $N_s$ surviving, the reliability at time $t$ is

$$R(t) = \frac{N_s(t)}{N_o} = \frac{N_o - N_f(t)}{N_o} \tag{3.2}$$

where $N_o$ is the original number of components in the population, $N_s(t)$ is the number of components surviving at time $t$, and $N_f(t)$ is the number of components failed at time $t$.

If the failure rate is constant, then the number of failures in successive equal time intervals (say, successive years) will be decreasing because the population is also decreasing. In other words, the population will decrease at an ever slower rate. Since reliability is the surviving population expressed as a proportion of the original population, the rate of decrease of reliability also gets ever slower with time.

Because the actual number of failures in a time period depends on where that period is with respect to the starting date, that is, the age of the population, such problems are usually performed in a stepwise fashion, as illustrated in Table 3.1.

The annual reliability, obtained by expressing the population at the end of the year as a proportion of the population, is roughly constant, under a constant failure rate. The actual number of failures per year is decreasing because the population is decreasing. The last column "Reliability" is the probability that a unit will survive the given number of years. It is equal to the number of surviving units divided by the original population as stated in the formula $R(t) = N_s/N_o$, and it can also be derived by multiplying the annual reliability of the year in question by the cumulative reliability of the year before. This "step method" is especially useful when there are discontinuities in the population. In the above-mentioned example, if 9 units were shut down for economic reasons at the end of Year 4 so that only 56 (instead of 65) were in service at the beginning of Year 5 and again 6 failed leaving

TABLE 3.1. Stepwise Illustration of Failures with the Age of Population

| Year | Population | Failures | Annual Reliability | Cumulative Reliability |
|---|---|---|---|---|
| 0 | 100 | 0 | 1.000 | 1.0 |
| 1 | 90 | 10 | 0.900 | 0.9 |
| 2 | 81 | 9 | 0.900 | 0.81 |
| 3 | 73 | 8 | 0.901 | 0.73 |
| 4 | 65 | 8 | 0.890 | 0.65 |
| 5 | 59 | 6 | 0.908 | 0.59 |

50 units at the end of Year 5, the calculation of reliability after 5 years will be difficult using the $N_s/N_o$ formula. If $N_s$ is taken to be 59, the reliability will be 0.59, while if $N_s$ is taken to be 50, the reliability will be 0.50. Both are wrong because the former assumes that those nine units would all have survived had they been in service and the latter assumes those nine units shut down for economic reasons would all have survived had they been in service and the latter assumes those nine units would all have failed. The correct way is to calculate the annual reliability of Year 5 and multiply it by the (cumulative) reliability of Year 4. This will give

$$\text{Annual reliability of Year 5} = 50/56 = 0.893$$

$$\text{Reliability after 5 years} = 0.65 \times 0.893 = 0.58$$

## Problem 3.2

Two hundred capacitors were installed and at the end of each year, the number of surviving units was tallied.

| End of | Number of Units Remaining |
|---|---|
| Year 1 | 196 |
| Year 2 | 188 |
| Year 3 | 179 |
| Year 4 | 175 |
| Year 5 | 169 |

Based on these figures, what is the reliability of the capacitors for 5 years? The annual reliability of Year 4? Assuming the reliability function is exponential, that is, $R(t) = e^{-t}$, what is the failure rate for this formula?

*Solution*:

$$\begin{aligned}\text{Reliability for 5 years} &= 169/200\\ &= 0.845\end{aligned}$$

$$\begin{aligned}\text{Annual reliability of Year 4} &= \frac{\text{number of units surviving at the end of Year 4}}{\text{number of units at the beginning of Year 4}}\\ &= 175/179\\ &= 0.9777\end{aligned}$$

$$R(t) = e^{-\lambda t}$$

$$\begin{aligned}\text{For 5 years, } R(t) &= e^{-\lambda 5} \text{ (from formula)}\\ &= 0.845 \text{ (from actual data)}\end{aligned}$$

$$e^{-5\lambda} = 0.845$$

$$\begin{aligned}-5\lambda &= ln(0.845)\\ &= -0.1684\end{aligned}$$

$$\begin{aligned}\lambda &= \frac{-0.1684}{-5}\\ &= 0.0337\end{aligned}$$

### 3.2.2 Reliability Model

The relationship between failure rate and reliability does not have to be computed from first principles every time. Some very simple formulas can be derived for components with a constant failure rate.

Let the population at the beginning of observation be $N_o$ and the failure rate be $\lambda$. After time $t$, some units will have failed and some will still be operating. Let them be denoted as $N_f$ and $N_s$, respectively. Naturally, $N_f$ increases with time and $N_s$ decreases with time. The time rate of increase of $N_f$ is the number of expected failures per unit time for the existing population at that moment and is equal to the failure rate times the number of units in the existing population, that is,

$$\frac{dN_f}{dt} = \lambda N_s$$

$N_f$ and $N_s$ vary with time, but they always add up to the original population $N_o$, that is,

$$N_f + N_s = N_o$$

As stated in the previous section, reliability is equal to the number of surviving units divided by the original population, that is,

$$R(t) = \frac{N_s}{N_o}$$

Combining these three equations together, we get

$$\begin{aligned}
R(t) &= \frac{N_s}{N_o} = 1 - \frac{N_f}{N_o} \\
\frac{dR(t)}{dt} &= \frac{-1}{N_o}\frac{dN_f}{dt} \\
&= -\lambda \frac{N_s}{N_o} \\
&= -\lambda R(t) \\
\int \frac{1}{R} dR &= -\int \lambda \, dt \\
\ln R(t) &= -\lambda t \\
R(t) &= e^{-\lambda t}
\end{aligned} \tag{3.3}$$

$R(t)$ is the probability of surviving for time $t$. As with continuous probabilities, a probability density function is derived such that the area under the curve represents the probability. Because $R(t)$ decreases with time, the probability density function $f(t)$ is derived for its complement $Q(t)$, which is the probability of failure in time $t$ as shown in Fig. 3.3.

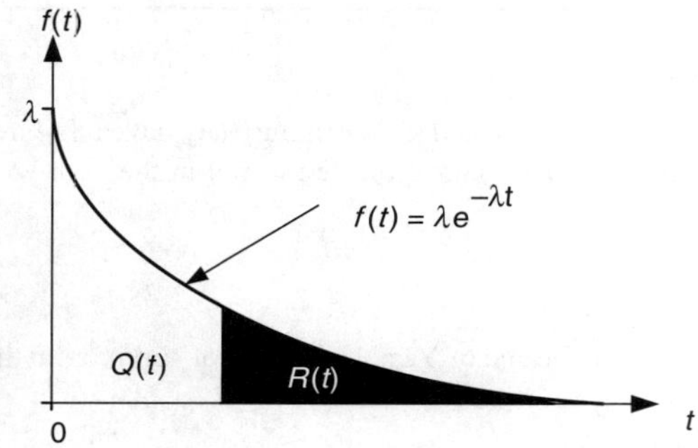


Figure 3.3. Exponential density function and $R(t)$ and $Q(t)$.

$$Q(t) = 1 - R(t) = 1 - e^{-\lambda t}$$
$$f(t) = dQ(t) = \lambda e^{-\lambda t}\, dt \tag{3.4}$$

Things that have a constant failure rate all follow this exponential failure probability.

## Problem 3.3

Ten thousand new oil circuit reclosers (OCRs) are put in service. They have a constant failure rate of 0.1 per year. How many units of the original 10,000 will still be in service after 10 years? How many of the original will fail in Year 10?

*Solution*:
Probability of survival is given by

$$R(t) = e^{-\lambda t}$$

In 10 years, probability of survival

$$R(10) = e^{-0.1\times 10} = e^{-1.0} = 0.3679$$

Out of 10,000 original units
10,000 × 0.3679 = 3679 should survive

Number of failures in Year 10

$$= (\text{number of survivors after Year 9}) - (\text{number of survivors after Year 10})$$
$$= 10,000 \times e^{-0.1\times 9} - 3679$$
$$= 1000 \times e^{-0.9} - 3679$$
$$= 4066 - 3679$$
$$= 387$$

## Problem 3.4

One thousand lightning arresters are installed. Assuming they have a failure rate of 0.05 per year, how many units (of the original batch) are expected to fail in the 10th year of service?

*Solution*:

$$
\begin{aligned}
&\text{Number of failures in Year 10}\\
&= \text{number of survivors at the end of Year 9} - \text{number of survivors at the end of Year 10}\\
&= (e^{-\lambda 9} - e^{-\lambda 10}) \times 1000\\
&= (e^{-0.45} - e^{-0.50}) \times 1000\\
&= (0.6376 - 0.6065) \times 1000\\
&= 31.1\\
&\approx 31
\end{aligned}
$$

### 3.2.3 The Poisson Probability Distribution

The exponential reliability function $R(t) = e^{-\lambda t}$ can be derived as a special case of the Poisson probability distribution. The Poisson distribution states that if the expected value of a variable is $\mu$, then the probability that the variable will assume a value of $x$ (an integer) is given by

$$P(x) = \frac{e^{-\mu}\mu^x}{x!} \tag{3.5}$$

In reliability assessments, if $\lambda$ is the failure rate, the number of expected failures in time $t$ will be $\lambda t$. The probability of having $x$ failures in time $t$:

$$P(x) = \frac{e^{-\lambda t}(\lambda t)^x}{x!} \tag{3.6}$$

The Poisson probability distribution comes from the identity

$$
\begin{aligned}
1 &= e^{-\lambda t} e^{\lambda t}\\
&= e^{-\lambda t}\left(1 + \lambda t + \frac{(\lambda t)^2}{2!} + \frac{(\lambda t)^3}{3!} + \cdots\right)
\end{aligned} \tag{3.7}
$$

Applied to reliability, each successive term represents the probability of having successive number of failures in time $t$.

Reliability is the probability of having no failures in time $t$; that is, it is the first term of that series

$$R(t) = P(0) = \frac{e^{-\lambda t}(\lambda t)^0}{0!} = e^{-\lambda t} \tag{3.8}$$

which is the exponential reliability function derived from first principles earlier.

### 3.2.4 Reliability of Equal Time Steps

When a component is in its useful life period where failures occur at random and the failure rate is constant, its reliability is independent of age; that is, the probability of failure during any interval $t$ is independent of the prior operating time as long as it is operating soundly when the interval of interest begins. It does not mean that a piece of equipment has equal probability of failure in every year of service; actually, the probability of failure in a specific year keeps decreasing for each successive year. However, given that a unit is in working condition at a given instant of time, the probability of failure for equal lengths of time from that instant would be the same regardless of prior operating time. For example, if the failure rate of a piece of equipment is 0.1 per year, the probability of failure in the next 5 years is as follows:

Year 1 = 9.5%
Year 2 = 8.6%
Year 3 = 7.8%
Year 4 = 7.1%
Year 5 = 6.4%

However, if the equipment is still in working condition at the end of Year 4, the probability of failing in the next year (Year 5) will be 9.5%. The fact that it is Year 5 of the original observation does not enter the picture.

This is because the probability of failure in Year 5, looking from the beginning, is an *a priori* probability. The unit may fail before Year 5. The probability of failing in exactly Year 5, no sooner, no later, is only 6.4%. However, if the unit is still working at the end of Year 4, the probability of failing in the immediate next year (Year 5) becomes an *a posteriori* probability. Something has already been achieved, namely, the survival for the first 4 years. From that vantage point, Year 5 becomes Year 1.

Stated in more general terms: If $P$ is the probability that a unit will fail in the interval $T$ to $T+t$, given that the unit is in working conditions at time $T$, then $P$ is independent of $T$ and depends only on the length of interval $t$.

## Proof

Probability of failing in interval $T$ to $T+t$ (looking from the beginning)
= (probability of surviving up to $T$) × (probability of failing in $T$ to $T+t$)
$= \mathrm{e}^{-\lambda t} \times P$

But probability of failing in the interval $T$ to $T+t$ is also given by $Q(T+t) - Q(T)$.

The two probabilities are the same, that is,

$$\begin{aligned} \mathrm{e}^{-\lambda T} \times P &= Q(T+t) - Q(T) \\ &= 1 - \mathrm{e}^{-\lambda(T+t)} - (1 - \mathrm{e}^{-\lambda T}) \\ &= \mathrm{e}^{-\lambda T} - \mathrm{e}^{-\lambda(T+t)} \\ &= \mathrm{e}^{-\lambda T}(1 - \mathrm{e}^{-\lambda t}) \end{aligned}$$

$$P = (1 - \mathrm{e}^{-\lambda t})$$

which is the probability of failing in time 0 to $t$ and is independent of $T$.

This property of equal reliability for equal time intervals independent of prior operation is true only for components and equipment following the exponential distribution of failure, which is applicable in the useful life period characterized by a constant failure rate. It is a very important property because it means all components that are still working are equally reliable and can be treated equally regardless of their in-service dates. This property ceases when the wear-out stage sets in.

## 3.3 MEAN TIME TO FAILURES

The exponential reliability function is a continuous probability density function with respect to time, so there exists an expected value for the function that may be considered as the average time value for the entire function. As the reliability function is actually a failure density function, the average time for the function is the average time for a failure to occur and is known as the mean time to failures or MTTF.

The expected value of a probability density function is given by

$$E(x) = \int x f(x)\,\mathrm{d}x \tag{3.9}$$

In our case, this becomes

$$\mathrm{MTTF} = \int_0^{\infty} t\lambda \mathrm{e}^{-\lambda t} \mathrm{d}t = 1/\lambda \tag{3.10}$$

It can be proven that the MTTF can also be obtained by integrating the reliability function over the entire range, that is,

$$\mathrm{MTTF} = \int_0^{\infty} R(t) \mathrm{d}t \tag{3.11}$$

This simplifies the calculation in most cases. For the simple exponential distribution, it becomes

$$\mathrm{MTTF} = \int_0^{\infty} \mathrm{e}^{-\lambda t} \mathrm{d}t = 1/\lambda \tag{3.12}$$

This mean time between failures turns out to be the reciprocal of the failure rate $\lambda$. This result is true for the exponential reliability functions only. The probability of failure from $t = 0$ to MTTF is

$$Q(\mathrm{MTTF}) = 1 - \mathrm{e}^{-\lambda/\lambda} = 1 - \mathrm{e}^{-1} = 0.632$$

So although the MTTF is the average time for a failure to occur, the probability of failure in the first MTTF interval is 63.2% and not 50%. For low failure rates, the MTTF could be very long, far exceeding the normal life span of the system or equipment itself.

This is possible because the MTTF is the average time elapsed before a failure occurs provided that the equipment is in the useful life period, so if the useful life period is long enough, one may get to see a failure, otherwise the equipment may enter the wear-out stage before a failure occurs, and the MTTF will no longer apply. For example, a power transformer may have an MTTF of 500 years while it is in its useful life period, but the useful life period may last only 60 years and the transformer starts having failures due to old age rather than having random failures.

## Problem 3.5

There are 10 generators in a generating station. The units are assumed to have a failure rate of 0.02 per year. What is the mean time to failures in that station?

*Solution*:

$$\text{Mean time to failures} = \frac{1}{\lambda}$$

$$\lambda_{\text{station}} = 0.02 \times 10 = 0.2 \text{ per } year$$

$$\text{MTTF}(station) = \frac{1}{0.2} = 5 \text{ years}$$

## 3.4 RELIABILITY OF COMPLEX SYSTEMS

### 3.4.1 Series Systems

Most systems are made up of subsystems or components. This is true for things as simple as a torchlight to something as an aeroplane or the power system. Functionally, these subsystems or components are arranged in series or parallel connections or in combinations of the two.

The term "series" refers to the functional relationship of the components and not the physical connection. A series connection means every component in the series is required for the system to function. Schematically, one has to pass through every element to go from the input side to the output side as shown in Fig. 3.4.

Each component in a system has its own failure rate and reliability, and the failure rate and reliability of the system depends on that of the individual components. Reliability is the probability of functioning in the given time interval, so the reliability of a series system is the probability that every component will function simultaneously in that given time interval. If the failures of the components are not dependent, then the

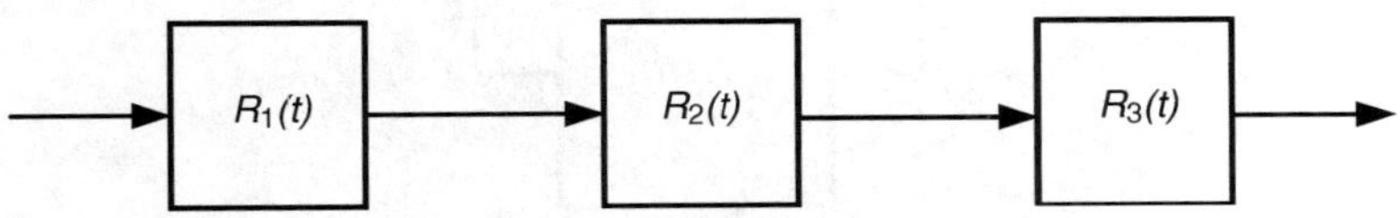


Figure 3.4. A series connection.

probability of all the components functioning is simply the product of the probabilities of individual components functioning, which is the product law of probability.

$$R_{\text{system}}(t) = R_1(t) \times R_2(t) \times R_3(t) \times \cdots \times R_n(t) \quad (3.13)$$

If the components have exponential failure probabilities with corresponding failure rates, $\lambda_1$, $\lambda_2$, $\lambda_3$, and so on, then the system reliability

$$\begin{aligned} R_{\text{system}} &= e^{-\lambda_1 t} \times e^{-\lambda_2 t} \times e^{-\lambda_3 t} \times \cdots \times e^{-\lambda_n t} \\ &= e^{-(\lambda_1 + \lambda_2 + \lambda_3 + \cdots + \lambda_n)t} \end{aligned} \quad (3.14)$$

The sum $(\lambda_1 + \lambda_2 + \lambda_3 + \cdots + \lambda_n)$ is a constant. It is the composite failure rate of the series system. From the system failure rate $\lambda$, the mean time between failures of the system can be calculated:

$$\text{MTTF}_{\text{system}} = 1/\lambda \quad (3.15)$$

### 3.4.2 Parallel Systems

A parallel system in a reliability sense means only one of the components in the parallel connections has to work in order that the system will function. Schematically, there are several alternative paths to go from the input side of the system to the output side as shown in Fig. 3.5.

Since only one element in the parallel connection is required for the system to work, the other elements are redundant. It is the redundancy that makes the overall system

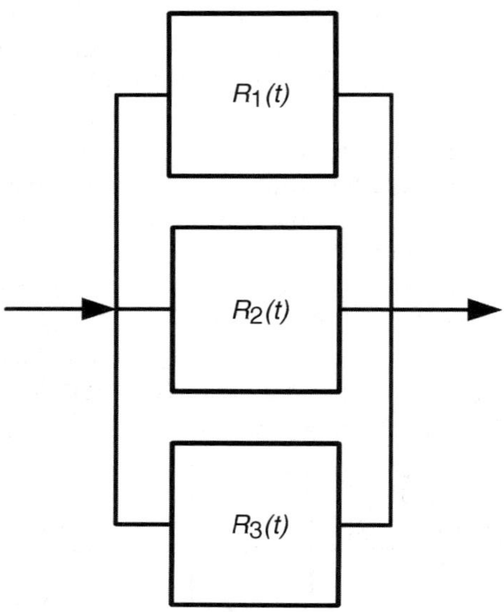


Figure 3.5. A parallel connection.

reliability high, because for the system to fail, all elements in the parallel connection have to fail. If the failures are independent, the probability that all of them will fail will be equal to the product of the failure probabilities of all individual elements. If $Q(t)$ is the probability of failure in a given period, that is, the unreliability, then

$$Q(t)_{\text{system}} = Q_1(t) \times Q_2(t) \times Q_3(t) \times \cdots \times Q_n(t) \tag{3.16}$$

If all the elements have the same $Q(t)$'s

$$\begin{aligned} Q(t)_{\text{system}} &= Q^n(t) \\ R(t)_{\text{system}} &= 1 - Q(t)_{\text{system}} \end{aligned} \tag{3.17}$$

If all the elements of the parallel system have constant failure rates and hence exponential reliability functions, as is usually assumed, the reliability of the parallel system will be

$$\begin{aligned} Q_{\text{system}}(t) &= Q(t)^n \\ R(t)_{\text{system}} &= 1 - Q(t)_{\text{system}} \\ &= 1 - (1 - e^{-\lambda_1 t}) \times (1 - e^{-\lambda_2 t}) \times (1 - e^{-\lambda_3 t}) \cdots (1 - e^{-\lambda_n t}) \end{aligned} \tag{3.18}$$

If all $\lambda$ are the same,

$$R(t)_{\text{system}} = 1 - (1 - e^{-\lambda t})^n \tag{3.19}$$

This system reliability is not an exponential function as is the case in series systems. Consequently, the failure rate of a parallel system is not constant but varies with time. The mean time to failure, however, is still a constant. Obviously, the relationship MTTF $= 1/\lambda$ does not hold any more. To illustrate this, consider two identical units in parallel as shown in Fig. 3.6.

$$\begin{aligned} \text{Reliability of system} &= 1 - (1 - e^{-\lambda t})^2 \\ &= 2e^{-\lambda t} - e^{-2\lambda t} \end{aligned}$$

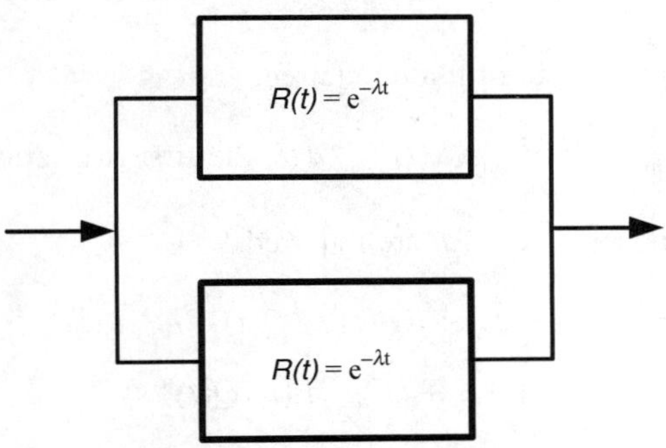


Figure 3.6. A parallel connection of identical components.

$$\text{Failure rate of system} = \frac{1}{R(t)}\frac{\mathrm{d}R(t)}{\mathrm{d}t}$$
$$= \frac{2\lambda(\mathrm{e}^{-\lambda t} - 1)}{\mathrm{e}^{-\lambda t} - 2}$$

$$\text{MTTF of system} = \int_0^\infty R(t)\mathrm{d}t$$
$$= \frac{3}{2\lambda}, \text{ a constant}$$

In general, when $n$ identical units are in parallel,

$$\text{MTTF} = \frac{1}{\lambda} + \frac{1}{2\lambda} + \frac{1}{3\lambda} + \cdots + \frac{1}{n\lambda}$$

### 3.4.3 Partially Redundant Systems

If a system requires $x$ components out of $n$ to work for it to function, and all those components have identical failure rates and no distinction is drawn about the individual components, the reliability of the overall system will be determined by applying the binomial probability distribution.

Let the reliability of each component be $R$.

$$\text{Unreliability} = 1 - R$$
$$(R(t) + Q(t))^n = R(t)^n + n\,R(t)^{n-1}Q(t) + nC_2R(t)^{n-2} + \cdots + n\,R(t)Q(t)^{n-2} + Q(t)^n \tag{3.20}$$

If all $n$ components are required to work in order that the system will function (i.e., a series system),

$$R(t)_{\text{system}} = R(t)^n, \text{ the first term of the series}$$

If $(n - 1)$ components out of $n$ are required,

$$R(t)_{\text{system}} = R(t)^n + n\,R(t)^{n-1}Q(t), \text{ the first two terms of the series}$$

In general, if $x$ components out of $n$ are required,

$$R(t)_{\text{system}} = R(t)^n + n\,R(t)^{n-1}Q(t) + nC_2R(t)^{n-2}Q(t)^2 + \cdots + nC_{n-x}R(t)^xQ(t)^{n-x} \tag{3.21}$$

that is, up to the term containing $R(t)^x$.

If only one out of $n$ components is required in order that the system will function (i.e., a parallel system),

$$\begin{aligned} R(t)_{\text{system}} &= R(t)^n + nR(t)^{n-1}Q(t) + \cdots + nR(t)Q(t)^{n-1} \\ &= (R(t)+Q(t))^n - Q(t)^n \\ &= 1 - Q(t)^n \end{aligned} \tag{3.22}$$

So the binomial expansion of $(R(t)+Q(t))^n$ is the most general expression that covers both series and parallel systems. More complex systems should be reduced to simpler combinations whenever possible and analyzed sequentially.

## Problem 3.6

What is the reliability of the following system as shown in Fig. 3.7?

$$R(t)_{\text{A}} = R(t)_{\text{B}} = R(t)_{\text{C}} = 0.7$$

$$R(t)_{\text{D}} = 0.90$$

$$R(t)_{\text{E}} = 0.80$$

*Solution*:
Reliability of system

$$\begin{aligned} &= R(t)_{\text{A//B//C}} \times R(t)_{\text{D}} \times R(t)_{\text{E}} \\ &= (1 - Q(t)_{\text{A}} \times Q(t)_{\text{B}} \times Q(t)_{\text{C}} \times Q(t)_{\text{D}} \times Q(t)_{\text{E}}) \quad \text{(Section 3.4.2)} \end{aligned}$$

$$\begin{aligned} Q(t)_{\text{A}} = Q(t)_{\text{B}} = Q(t)_{\text{C}} &= 1 - R(t)_{\text{A}} \\ &= 1 - 0.7 \\ &= 0.3 \end{aligned}$$

$$R(\text{system}) = (1 - 0.3 \times 0.3 \times 0.3) \times (0.90) \times (0.80) = 0.701$$

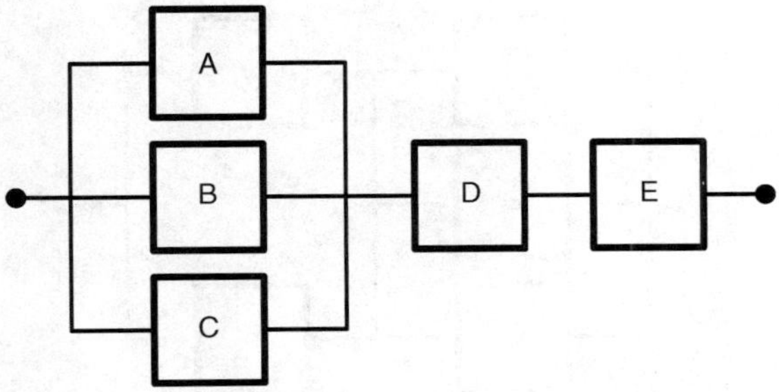


Figure 3.7. A parallel–series connected system.

### 3.4.4 Bayes' Theorem

Some systems are made up of different subsystems connected in ways that are neither series nor parallel nor partially redundant. However, if a schematic diagram of that system can be established, and the reliabilities of the individual components known, that system can be analyzed by means of the Bayes' theorem. In a reliability context, it can be stated as

$$\begin{aligned} R(\text{system}) &= R(\text{under condition } 1) \times P(\text{condition } 1) \times R(\text{under condition } 2) \\ &\quad \times P(\text{condition } 2) \times R(\text{under condition } 3) \times P(\text{condition } 3) + \cdots \end{aligned} \tag{3.23}$$

Given

$$P(\text{condition } 1) + P(\text{condition } 2) + \cdots = 1$$

(i.e., all possibilities are exhausted)

*Note*: The time functionality is dropped; that is, $R(t)$ is expressed as $R$ for a particular mission time.

In most applications, there are only two conditions: condition 1 is that a certain component $X$ is good and condition 2 is that component $X$ is bad. In this way, one can zero in on a critical component in a complex system.

$$\begin{aligned} R_{\text{system}} &= (R_{\text{system}} \text{ given component } X \text{ is good}) \cdot R(X) \\ &\quad + R_{\text{system}} \text{ given component } X \text{ is bad}) \cdot Q(X) \end{aligned} \tag{3.24}$$

The theorem can be applied repeatedly so that a complicated system can be broken down into parts solvable by standard formulas.

An example will illustrate the use of this theorem using Fig. 3.8 in the following.

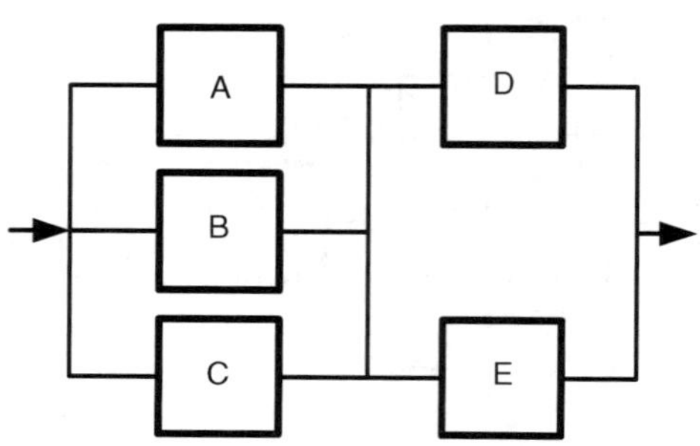


Figure 3.8. A five-component system configuration.

## Example 3.2

The system is composed of two subsystems connected in series, that is, subsystem 1 consists of components A, B, and C connected in parallel represented by (A//B//C) and subsystem 2 consists of components D and E connected in parallel represented by (D//E). The reliability of system is

$$\begin{aligned} R_{\text{system}} &= R(\text{A}//\text{B}//\text{C}) \times R(\text{D}//\text{E}) \\ &= (1 - Q_A Q_B Q_C)(1 - Q_D Q_E) \\ &= 1 - Q_A Q_B Q_C - Q_D Q_E + Q_A Q_B Q_C Q_D Q_E \end{aligned}$$

Using the Bayes' theorem, a component, say $B$, is chosen

$$\begin{aligned} R_{\text{system}} &= (R_{\text{system}},\ \text{given B is good}) \times R_B + (R_{\text{system}},\ \text{given B is bad}) \times Q_B \\ &= R(\text{D}//\text{E}) \times R_B + R(\text{A}//\text{C}) \times R(\text{D}//\text{E}) Q_B \\ &= (1 - Q_D Q_E)(1 - Q_B) + (1 - Q_A Q_C)(1 - Q_D Q_E) Q_B \\ &= 1 - Q_D Q_E - Q_B + Q_D Q_E Q_B + Q_B - Q_D Q_E Q_B - Q_A Q_C Q_B \\ &\quad + Q_A Q_B Q_C Q_D Q_E \\ &= 1 - Q_D Q_E - Q_A Q_B Q_C + Q_A Q_B Q_C Q_D Q_E \end{aligned}$$

## Problem 3.7

Two diodes are connected in parallel as shown in Fig. 3.9.

A diode may fail in one of the two ways of short circuiting (s.c.) or open circuiting (o.c.). Given

$$p(\text{o.c.}) = 0.1$$
$$p(\text{s.c.}) = 0.2$$

What is the probability that two diodes connected in parallel will work?

*Solution*:
This problem is an illustration of the use of Bayes' theorem.

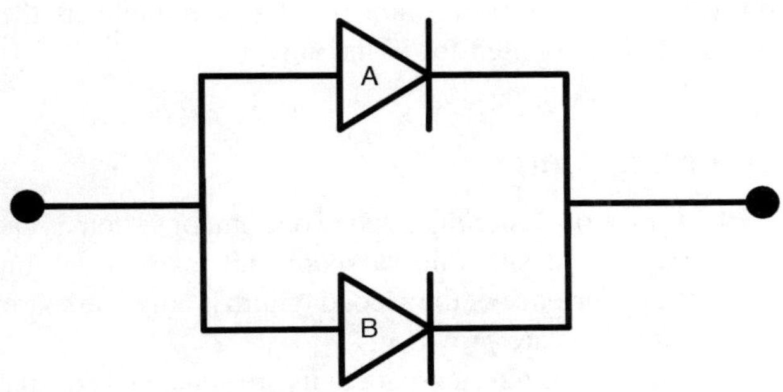


Figure 3.9. A parallel diode system configuration.

Bayes' theorem can be written as

$$R(\text{system}) = R\,(\text{under condition 1}) \times P(\text{condition 1}) + R\,(\text{under condition 2}) \\ \times P(\text{condition 2}) + R\,(\text{under condition 3}) \times P\,(\text{condition 3}) + \cdots$$

In this problem, there are three conditions, namely, open circuit, short circuit, and good.

$$R(\text{system}) = R(\text{diode A o.c.}) \times P(\text{diode A o.c.}) + (\text{diode A s.c.}) \times P(\text{diode A s.c.}) \\ \times R(\text{diode A good}) \times P(\text{diode A good})$$

When diode A is an open circuit, system reliability depends on diode B:

$$R = P(\text{diode B good}) = 0.7$$

When diode A is a short circuit, system is always down:

$$R = 0$$

When diode A is good, system is good unless diode B is short:

$$R = P(\text{diode B good}) + P(\text{diode B o.c.}) = 0.8$$

$$R(\text{system}) = (0.7 \times 0.1) + (0 \times 0.2) + (0.8 \times 0.7) = 0.63$$

## 3.5 STANDBY SYSTEM MODELING

### 3.5.1 Background

A standby system is an arrangement where one or more units are standing by ready to take over the operation when an operating unit fails. The spare can be provided either for a part of the system or for the entire system.

A standby unit is idle before taking over the operation and is therefore not under the stress of operation and is assumed to have 100% reliability while in the standby mode. In reality, a unit is under some stress in standby. Also, the sensing and switching devices for putting the standby units into operation are not 100% reliable. If these factors are significant, they should be accounted for in the analysis.

### 3.5.2 Spares for One Unit

The most basic form of spare provisioning is when one unit or system is backed up by one or more spare units. When the unit fails, a spare unit takes over immediately. The operation will stop at the occurrence of the second failure if only one spare is provided, at the $(n+1)$th failure if $n$ spares are provided.

If the failure rates of the original and spare units are equal and constant, as is usually assumed, the spare will act as a continuation of the original unit as if the original unit had failed and recovered immediately. The takeover by a "brand new" unit does not make the system any more reliable than when the original unit was operating because of the "memory-less" property of exponential reliability functions. As far as the system is concerned, the reliability is the same whether that particular component is the original unit or a replacement as long as their failure rates are the same, and the system fails when the $(n+1)$th failure occurs to that component, when there are $n$ spares. The probability of system failure is viewed as the probability of the $(n+1)$th failure of the system and not the probability of the first failure of the $n$th spare unit.

### 3.5.3 Spares for Multiple Interchangeable Units

Very often there is a group of interchangeable units backed by a number of interchangeable spares. A spare can be used wherever a failure occurs in the system until all spares are used up. Reliability is defined as the probability of having a spare when needed during the given period of time. For simplicity, all units and all spares are assumed to have equal and constant failure rates.

The problem is similar to that of the last section in that we are trouble-free until there is one more failure than the number of spares. It is not defined as system failure because the group of units does not necessarily form a closely related "system," and losing one unit is not necessarily a failure of the "system"; but the two situations are analogous. Unreliability is thus the probability of having one more failure than the number of spares. The difference from the single-unit problem is, of course, that failure could happen to any one of the units in the population, including replacements.

Again, the memory-less characteristics of the exponential reliability function dictate that replacements are indistinguishable from original units, and the probability that a certain number of failures in the population will occur in that period follows the Poisson distribution.

From the Poisson probability distribution, probabilities for having a specific number of failures in time $t$ is given by the series

$$
\begin{aligned}
1 &= P(0) + P(1) + P(2) + \cdots + P(n) \\
1 &= \mathrm{e}^{-\lambda t}\left(1 + \lambda t + \frac{(\lambda t)^2}{2!} + \frac{(\lambda t)^3}{3!} + \cdots\right)
\end{aligned}
\tag{3.25}
$$

With $n$ spares, the system can endure $n$ failures; therefore, system reliability is the probability of having 0 to $n$ failures:

$$R(t) = \mathrm{e}^{-\lambda t}\left(1 + \lambda t + \frac{(\lambda t)^2}{2!} + \frac{(\lambda t)^3}{3!} + \cdots + \frac{(\lambda t)^n}{n!}\right)$$

Unreliability, which is the probability of running out of spares,

$$Q(t) = 1 - R(t)$$

Failure rate of the standby system is a time-varying function, but the mean time between failures is a constant.

$$\begin{aligned}\text{MTTF} &= \int_0^\infty R(t)\mathrm{d}t \\ &= \int_0^\infty \mathrm{e}^{-\lambda t}\left(1+\lambda t+\frac{(\lambda t)^2}{2!}+\cdots+\frac{(\lambda t)^n}{n!}\right)\mathrm{d}t \\ &= \frac{n+1}{\lambda}\end{aligned} \tag{3.26}$$

Compared to a parallel system of $(n+1)$ components, the MTTF of which is

$$\text{MTTF} = \frac{1}{\lambda}+\frac{1}{2\lambda}+\frac{1}{3\lambda}+\cdots+\frac{1}{n\lambda}+\frac{1}{(n+1)\lambda} \tag{3.27}$$

the standby system has longer MTTFs.

If the failure rate of an individual unit is $\lambda$ and there are $N$ units, the expected number of failures in time $t$ will be $N\lambda t$

$$1 = \mathrm{e}^{-N\lambda t}\left(1+N\lambda t+\frac{(N\lambda t)^2}{2!}+\frac{(N\lambda t)^3}{3!}+\cdots\right)$$

As before, if there are $n$ spares, probability of having a spare when needed over time $t$ is

$$R = \mathrm{e}^{-N\lambda t}\left(1+N\lambda t+\frac{(N\lambda t)^2}{2!}+\frac{(N\lambda t)^3}{3!}+\cdots+\frac{(N\lambda t)^n}{n!}\right) \tag{3.28}$$

Probability of not having a spare when needed is

$$Q = 1-R$$

## Problem 3.8

We have three spare transformers supporting 100 single-bank stations. If the failure rate of a transformer in service is 0.1 per year, what is the probability of having no spares available in a year?

*Solution*:

When there are three spares, the system can tolerate three failures. Use the Poisson probability series

$$R = e^{-N\lambda t}\left(1 + N\lambda t + \frac{(N\lambda t)^2}{2!} + \frac{(N\lambda t)^3}{3!}\right)$$

$$N = 100$$

$$\lambda = 0.10$$

$$t = 1$$

$$N\lambda t = 10$$

$$R = e^{-10}\left(1 + 10 + \frac{10^2}{2} + \frac{10^3}{6}\right)$$

$$= 0.01034$$

Probability of having no spares

$$= 1 - R$$

$$= 1 - 0.01034$$

$$= 0.98966$$

## 3.6 CONCEPTS OF AVAILABILITY AND DEPENDABILITY

### 3.6.1 Mean Time to Repair

When a piece of equipment or a system fails, it is out of service. Until it is repaired or substituted by another unit, its service will be unavailable to the user. Even if spare units are available, it will take some time to do the replacement. The time required to restore service, whether by repair or replacement, can be called "repair time."

Although failures are generally viewed as instantaneous events while repairs are continuous processes during the repair time, for analysis purposes, the two are analogous. A failure takes a unit from an "up" state to a "down" state and it takes on average the MTTF for it to take place. A repair takes a unit from a "down" state to an "up" state, and it takes on the average, the mean time to repair (MTTR) to do it. The failure rate is the reciprocal of mean time between failures; so in a similar fashion, a repair rate $r$ equal to the reciprocal of repair time can be defined. Also, just as the constant failure rate leads to an exponential failure probability, the constant repair rate leads to an exponential repair function. What it means is that the probability of equipment being repaired within a time $t$ is given by an exponential function even though the long-term repair time is a constant. This is quite reasonable as different kinds of failure require different repair times.

The introduction of the repair activity makes the system truly dynamic because the two opposing forces of failure and repair drive the system back and forth between the "up" and "down" states. In reality, systems have more than just the "up" and "down" states; they have multiple states due to failure of $x$ out of $n$ units, spare provisions, derating of some equipment, and many other factors. The analysis can be very complicated, but the principle of failure rates and repair rates leading from one state to another still applies.

### 3.6.2 Availability Model

When a system fails, it will be out of service for some time until it is repaired or replaced. Even for systems with spare units, the system can be "down" if a failure occurs when no more spares are available. The percentage of time that the system is functioning is called the availability of the system. It is usually expressed as

$$\text{Availability} = A = \frac{\text{total hours of operation in 1 year}}{8760} \tag{3.28}$$

$$\text{Unavailability} = \bar{A} = \frac{\text{total hours of down time in 1 year}}{8760} \tag{3.29}$$

Since on average, it takes a time interval equal to the MTTF for the system to fail and a time interval equal to the MTTR for the system to be operational again, the availability, defined as uptime/(uptime + downtime), can be expressed as

$$A = \frac{\text{MTTF}}{\text{MTTF} + \text{MTTR}} \tag{3.30}$$

and unavailability can be expressed as

$$\bar{A} = \frac{\text{MTTR}}{\text{MTTF} + \text{MTTR}} \tag{3.31}$$

For systems that can be treated as a single component with a constant failure rate $\lambda$ and repair rate $\mu$, the availability

$$A = \frac{\text{MTTF}}{\text{MTTF} + \text{MTTR}} = \frac{1/\lambda}{1/\lambda + 1/\mu} = \frac{\mu}{\lambda + \mu} \tag{3.32}$$

and unavailability

$$\bar{A} = \frac{\lambda}{\lambda + \mu} \tag{3.33}$$

### 3.6.3 Markov Model

When more than two possible conditions are possible for a system, the analysis becomes complicated. The best way to analyze such a system is by state–space analysis using the Markov process. Each possible system condition is called a *state*. If the rate of entry into and departure from each state is known, the condition of the system at any stage can be calculated.

## Example 3.3

A system consists of an operating unit and an identical spare with failure rate λ and repair rate μ. Three states are possible as shown in Fig. 3.10.

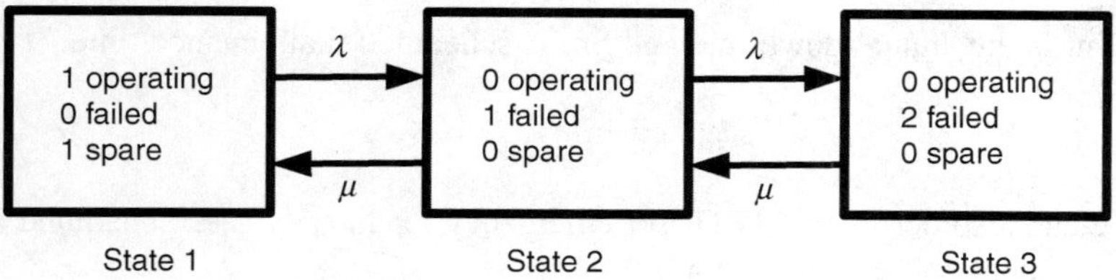


Figure 3.10. A model for a unit with a spare.

The probability of being in any one state can be calculated by using the Markov process model. It is time dependent but settles to a steady-state value. Multiplying this probability by the rate of departure from that state gives the frequency of encountering that state, and the reciprocal of the departure rate from that state gives the duration of each encounter. In this example, State 3 is the failed state. The steady-state probability of being in the failed state is by definition the system unavailability.

System unavailability $= P(3)$ as calculated from the Markov process model
System availability $= 1 - P(3)$
Frequency of system failure $= P(3) \times 2\mu$
Average duration of each system failure $= 1/2\mu$

The calculation of the conditions of a dynamic system with many states requires matrix differential equations. To obtain a time-dependent solution is extremely difficult for all but the simplest systems. However, the states of the system at any time can be obtained using an iterative process called the Markov chain that is well suited to computer applications.

The steady-state conditions, in particular, are quite simple. In fact, if the failure rates and repair rates of various components are constants, the steady-state probability of any state will also be a constant. The steady-state probability $P(3)$ in the above example is

$$P(3) = \frac{\lambda^2}{\lambda^2 + 2\lambda\mu + 2\mu^2} \tag{3.34}$$

### 3.6.4 Concept of Dependability

When we talk about unavailability, we include all the time when the system is not operating. This includes failures as well as scheduled maintenance. In a period of time when no maintenance is scheduled, the system operation may still be interrupted due to failures, for example, the operation of hydraulic generators in winter months.

The availability of the system in a period intended for continuous operation is called dependability. It can be calculated by subtracting the scheduled maintenance time from the annual system downtime and dividing it by 8760.

$$\bar{D} = \frac{\mathrm{Tm} - \mathrm{Sm}}{8760} \tag{3.35}$$

where Tm is the annual downtime and Sm is scheduled maintenance time.

$$\text{Dependability, } D = 1 - \bar{D} \tag{3.36}$$

One can also derive an MTTR for emergency repair only (i.e., scheduled maintenance excluded) and use the formulas

$$\begin{aligned} D &= \frac{\mathrm{MTTF}}{\mathrm{MTTF} + \mathrm{MTTR(emergency-repair)}} \\ \bar{D} &= 1 - D \end{aligned} \tag{3.37}$$

### 3.6.5 Design Considerations

For systems such as a mainframe computer or a generating station, the availability or dependability is of prime importance. For a designer of a complicated system or machine, very often, there is a specified availability. He will have to pick some values for MTTF and MTTR to meet the required availability. Improvement of MTTF and MTTR often works against each other, because MTTF is limited by the state of art or constituent components, and further extension of MTTF can only be accomplished through redundancy techniques that complicate the system, while MTTR is proportional to the complexity of the system. The designer has to choose the right trade-off between the two. Of the two, MTTF usually has less room for maneuvering so the designer has to work on the MTTR by applying maintainability engineering techniques to achieve the required availability. The following monograph is a design tool based on the equations in the last section.

## 3.7 RELIABILITY MEASUREMENT

### 3.7.1 Concept

Reliability engineering is a discipline in which reliability is treated in a quantitative way, that is, it is observable and measurable. Reliability itself is defined as the probability of functioning or surviving for a given period of time, so the measurement of reliability involves these two aspects: functioning and time. The measurement of time is straightforward while measurement of functioning or survival is usually in the negative sense, that is, the recording of failures or retirements.

Reliability data come from two sources: past records of failure and repair and reliability tests. Actually, past operations can be viewed as tests on the actual system, and the records as test results. From these past records and test results, the critical parameters can be established and reliability evaluated.

### 3.7.2 Accuracy of Observed Data

Reliability data are collected from past records or current tests. If the purpose is just to analyze past performance then a complete set of past records will suffice. However, many times, the date is used for reliability predictions of present and future systems. In those cases, past records or reliability life test results, no matter how complete, are merely samples of all possible reliability data of similar equipment used in the past, present, and future over the entire time spectrum.

From statistical theory, values derived from a sample are not the same as the true values for the entire population and the deviations depend on the sample size. For example, if there were three failures among the 10 transformers purchased 5 years ago, the estimated failure rate would be 3/50 per year. If another utility had 120 failures out of 100 similar transformers in the last 25 years, we would place more confidence in their failure rate.

As with all sampling, although the sample mean is not the true mean, a statement can be made that the true mean is within a certain range from the sample mean for a given percentage of time. This is known as the confidence level. In reliability work, one is usually concerned with how worse the true mean could be, that is, a one-sided confidence limit. However, if a two-sided confidence level is desired, it can easily be calculated.

### 3.7.3 Confidence Limit of Failure Rate

In most analyses, it is assumed that the equipment or system is in its useful life period. In this period, the single most important parameter is the failure rate, because other parameters are derived from it, for example,

$$R = e^{-\lambda t}$$

$$\text{MTTF} = \frac{1}{\lambda}$$

There is always an uncertainty associated with the failure rate obtained from a limited amount of data. As mentioned above, the concern is usually on the upper confidence limit, that is, how much higher it could be over the estimated value.

If $r$ failures are observed in a period of time $t$, the upper confidence limit is computed by equating the probability of having 0 to $r$ failures in time $t$ to the complement of the confidence level, that is

$$P(0 \text{ to } r \text{ failures}) = 1 - \text{confidence level}$$

According to the Poisson distribution,

$$P(0 \text{ to } r \text{ failures}) = \mathrm{e}^{-\lambda t}\left(1+\lambda t+\frac{(\lambda t)^2}{2!}+\frac{(\lambda t)^3}{3!}+\cdots+\frac{(\lambda t)^r}{r!}\right) \quad (3.38)$$

If the confidence level is $\alpha$ (in decimals, not percentage),

$$\mathrm{e}^{-\lambda t}\left(1+\lambda t+\frac{(\lambda t)^2}{2!}+\frac{(\lambda t)^3}{3!}+\cdots+\frac{(\lambda t)^r}{r!}\right) = 1-\alpha \quad (3.39)$$

## Example 3.4

If one failure was observed over a year, the 95% upper confidence limit, will be given by

$$\begin{aligned} P(0 \text{ to } r \text{ failures}) &= \mathrm{e}^{-\lambda t}(1+\lambda t) = 1-0.95 \\ \mathrm{e}^{-\lambda}(1+\lambda) &= 0.05 \\ &= 4.50 \text{ per year} \end{aligned}$$

What it means is that although the estimated failure rate from the observed data is 1 per year, the true failure rate could be anything. However, 95% of the time, it will be less than 4.50 per year.

The above formulas are based on a predetermined observation time $t$, for example, a year. If the observation is truncated after the $r$th failure, the Poisson distribution will not hold true for $r$ failures in time $t$ because the observation time is arbitrarily truncated, and no time is allowed for a possible $(r+1)$th failure to occur. However, the Poisson distribution will apply to $(r-1)$ failures in time $t$, because just before the $r$th failure occurs a time interval of $t$ has elapsed with only $(r-1)$ failures recorded, that is, the upper limit of the failure rate at a confidence level is given by

$$\mathrm{e}^{-\lambda t}\left(1+\lambda t+\frac{(\lambda t)^2}{2!}+\frac{(\lambda t)^3}{3!}+\cdots+\frac{(\lambda t)^{r-1}}{(r-1)!}\right) = 1-\alpha \quad (3.40)$$

This is the same formula with one less term.

### 3.7.4 Chi-Square Distribution

The Poisson cumulative probability discussed in this section is not easy to solve when the number of terms exceeds two or three. Nowadays, it can be solved with a computer. However, the traditional and fairly accurate way is by relating it to the chi-square distribution for which tabulated values are available.

The chi-square distributions are a family of continuous probability distributions with a random variable known as $\chi^2$ (chi-square), which can take on any value between 0 and $\infty$ as shown in Fig. 3.11. The chi-square distribution has a parameter called the degree of freedom. For each degree of freedom, there is a separate curve. Since the chi-square distribution is a continuous probability distribution, the area under the curve indicates probability. For each curve (i.e., degree of freedom), the area to the right of a $\chi^2$ value represents the probability of exceeding this value (Table 3.2). Conversely, given the

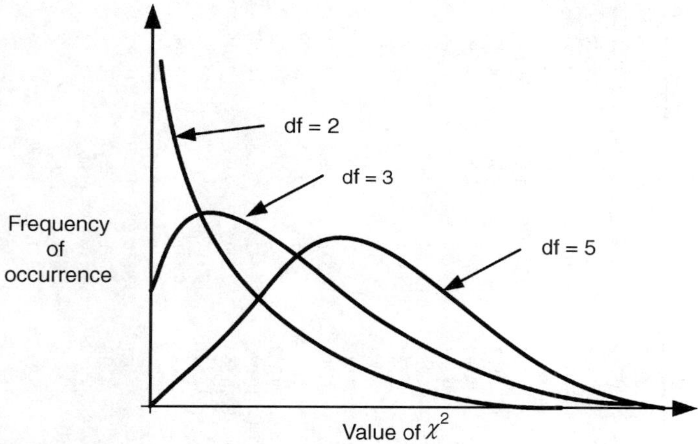


Figure 3.11. Chi-squared distributions for difference degrees of freedom.

degree of freedom and the area to the right of a $\chi^2$ value, that $\chi^2$ value can be uniquely determined. It is denoted by $\chi^2_{w,n}$, where $w$ is the area to the right of $\chi^2$ and $n$ the degree of freedom.

The Poisson probability of $n$ terms is equal to the probability of exceeding a $\chi^2$ value in the curve with $2n$ degrees of freedom, so if the cumulative probability is equal to $(1-\alpha)$, the corresponding $\chi^2$ value will be $\chi^2_{1-\alpha,2n}$. It should be noted that

$$e^{-\lambda t}\left(1+\lambda t+\frac{(\lambda t)^2}{2!}+\frac{(\lambda t)^3}{3!}+\cdots+\frac{(\lambda t)^r}{r!}\right) \tag{3.41}$$

has $(r+1)$ terms, so the corresponding $\chi^2$ value is $\chi^2_{1-\alpha,2r+2}$.

It can be proven mathematically that when a Poisson cumulative probability is equal to the area to the right of a $\chi^2$ value, then $\lambda t=\chi^2/2$.

$$\lambda=\frac{\chi^2}{2t}=\frac{\chi^2}{2r}\frac{r}{t}=\frac{\chi^2}{2r}\lambda_{\text{est}} \tag{3.42}$$

So for an observation that ends with $r$ failures in a predetermined time $t$,

$$\lambda=\frac{1}{2r}\chi^2_{1-\alpha,2r+2} \tag{3.43}$$

For an observation that ends with the $r$th failure,

$$\lambda=\frac{1}{2r}\chi^2_{1-\alpha,2r} \tag{3.44}$$

Figure 3.12 illustrates chi-squared distribution for a degree of $2r+2$. The confidence limit of failure rate is determined by multiplying the estimated failure rate $(r/t)$ by the factor $(1/2r)\chi^2$.

## TABLE 3.2. Chi-SquareValues

| Degrees of Freedom (df) | Probability of a Deviation Greater than $\chi^2$ | | | | | | | | | | | | |
|---|---|---|---|---|---|---|---|---|---|---|---|---|---|
| | 0.01 | 0.02 | 0.05 | 0.10 | 0.20 | 0.30 | 0.50 | 0.70 | 0.80 | 0.90 | 0.95 | 0.98 | 0.99 |
| 1 | 6.635 | 5.412 | 3.841 | 2.706 | 1.642 | 1.074 | 0.455 | 0.148 | 0.0642 | 0.0158 | 0.00393 | 0.000628 | 0.000157 |
| 2 | 9.210 | 7.824 | 5.991 | 4.605 | 3.219 | 2.408 | 1.386 | 0.713 | 0.446 | 0.211 | 0.103 | 0.0404 | 0.0201 |
| 3 | 11.341 | 9.837 | 7.815 | 4.642 | 3.665 | 3.665 | 2.366 | 1.424 | 1.005 | 0.584 | 0.352 | 0.185 | 0.115 |
| 4 | 13.277 | 11.668 | 9.488 | 7.779 | 5.989 | 4.878 | 3.357 | 2.195 | 1.649 | 1.064 | 0.711 | 0.429 | 0.297 |
| 5 | 15.086 | 13.388 | 11.070 | 9.236 | 7.289 | 6.064 | 4.351 | 3.000 | 2.343 | 1.610 | 1.145 | 0.752 | 0.554 |
| 6 | 16.812 | 15.033 | 12.592 | 10.645 | 8.558 | 7.231 | 5.348 | 3.828 | 3.070 | 2.204 | 1.635 | 1.134 | 0.872 |
| 7 | 18.475 | 16.622 | 14.067 | 12.017 | 9.803 | 8.383 | 6.346 | 4.671 | 3.822 | 2.833 | 2.167 | 1.564 | 1.239 |
| 8 | 20.090 | 18.168 | 15.507 | 13.362 | 11.030 | 9.524 | 7.344 | 5.527 | 4.594 | 3.490 | 2.733 | 2.032 | 1.646 |
| 9 | 21.666 | 19.679 | 16.919 | 14.684 | 12.242 | 10.656 | 8.343 | 6.393 | 5.380 | 4.168 | 3.325 | 2.532 | 2.088 |
| 10 | 23.209 | 21.161 | 18.307 | 15.987 | 13.442 | 11.781 | 9.342 | 7.267 | 6.179 | 4.865 | 3.940 | 3.059 | 2.558 |
| 11 | 24.725 | 22.618 | 19.675 | 17.275 | 14.631 | 12.899 | 10.341 | 8.148 | 6.989 | 5.578 | 4.575 | 3.609 | 2.053 |
| 12 | 26.217 | 24.054 | 21.026 | 18.549 | 15.812 | 14.011 | 11.340 | 9.034 | 7.807 | 6.304 | 5.226 | 4.178 | 3.571 |
| 13 | 27.688 | 25.472 | 22.362 | 19.812 | 16.985 | 15.119 | 12.340 | 9.926 | 8.634 | 7.042 | 5.892 | 4.765 | 4.107 |
| 14 | 29.141 | 26.873 | 23.685 | 21.064 | 18.151 | 16.222 | 13.339 | 10.821 | 9.467 | 7.790 | 6.571 | 5.368 | 4.660 |
| 15 | 30.578 | 28.259 | 24.996 | 22.307 | 19.311 | 17.322 | 14.339 | 11.721 | 10.307 | 8.547 | 7.261 | 5.985 | 5.229 |
| 16 | 32.000 | 29.633 | 26.296 | 23.542 | 20.465 | 18.418 | 15.338 | 12.624 | 11.152 | 9.312 | 7.962 | 6.614 | 5.812 |
| 17 | 33.409 | 30.995 | 27.587 | 24.769 | 21.615 | 19.511 | 16.338 | 13.531 | 12.002 | 10.085 | 8.672 | 7.255 | 6.408 |
| 18 | 34.805 | 32.346 | 28.869 | 25.989 | 22.760 | 20.601 | 17.338 | 14.440 | 12.857 | 10.865 | 9.390 | 7.906 | 7.015 |
| 19 | 36.191 | 33.687 | 30.144 | 27.204 | 23.900 | 21.689 | 18.338 | 15.352 | 13.716 | 11.651 | 10.117 | 8.567 | 7.633 |
| 20 | 37.566 | 35.020 | 31.410 | 28.412 | 25.038 | 22.775 | 19.337 | 16.266 | 14.578 | 12.443 | 10.851 | 9.237 | 8.260 |
| 21 | 38.932 | 36.343 | 32.671 | 29.615 | 26.171 | 23.858 | 20.337 | 17.182 | 15.445 | 13.240 | 11.591 | 9.915 | 8.897 |
| 22 | 40.289 | 37.659 | 33.924 | 30.813 | 27.301 | 24.939 | 21.337 | 18.101 | 16.314 | 14.041 | 12.338 | 10.600 | 9.542 |
| 23 | 41.638 | 38.968 | 35.172 | 32.007 | 28.429 | 26.018 | 22.337 | 19.021 | 17.187 | 14.848 | 13.091 | 11.293 | 10.196 |
| 24 | 42.980 | 40.270 | 36.415 | 33.196 | 29.553 | 27.096 | 23.337 | 19.943 | 18.062 | 15.659 | 13.848 | 11.992 | 10.856 |
| 25 | 44.314 | 41.566 | 37.652 | 34.382 | 30.675 | 28.172 | 24.337 | 20.867 | 18.940 | 16.473 | 14.611 | 12.697 | 11.524 |
| 26 | 45.642 | 42.856 | 38.885 | 35.563 | 31.795 | 29.246 | 25.336 | 21.792 | 19.820 | 17.292 | 15.379 | 13.409 | 12.198 |
| 27 | 46.963 | 44.140 | 40.113 | 36.741 | 32.912 | 30.319 | 26.336 | 22.719 | 20.703 | 18.114 | 16.151 | 14.125 | 12.879 |
| 28 | 48.278 | 45.419 | 41.337 | 37.916 | 34.027 | 31.391 | 27.336 | 23.647 | 21.588 | 19.939 | 16.928 | 14.847 | 13.565 |
| 29 | 49.588 | 46.693 | 42.557 | 39.087 | 35.139 | 32.461 | 28.336 | 24.577 | 22.475 | 19.768 | 17.708 | 15.574 | 14.256 |
| 30 | 50.892 | 47.962 | 43.773 | 40.256 | 36.250 | 33.530 | 29.336 | 25.508 | 23.364 | 20.599 | 18.493 | 16.306 | 14.953 |

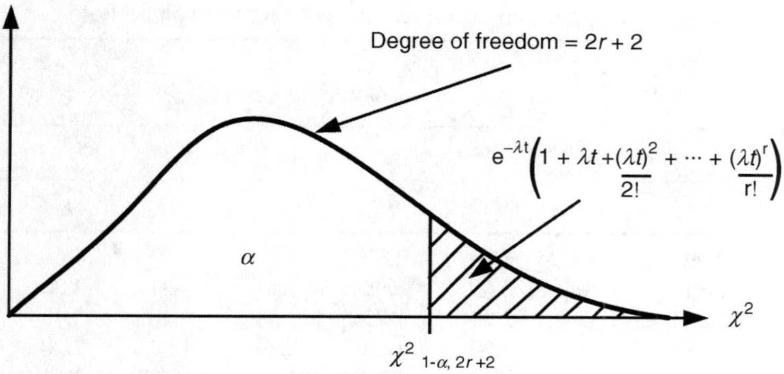


Figure 3.12. Chi-squared distributions for a degree of freedom of $2r+2$.

This factor represents the uncertainty in the estimated failure rate and gets smaller as the amount of data (i.e., $r$) gets greater. Factors for upper and lower confidence limits are given in Table 3.3. Table 3.3 gives the multipliers for various number of failures and confidence levels for both cases of observations.

For equipment in their useful life period, the mean time between failure MTTF $= 1/\lambda$. The lower confidence level limit of MTTF is the reciprocal of the upper confidence limit of $\lambda$, so in calculating the lower limit of MTTF, the estimated MTTF will be divided by the same multiplier used for failure rates.

---

## Example 3.5

Two failures were observed in 10 years.

$$\text{Estimated MTTF} = \frac{10}{2} = 5 \text{ years}$$

$$\text{Lower confidence level} = \frac{3}{3.15} = 1.59 \text{ years}$$

A two-sided confidence interval of failure rates can be established for both types of observations.

For observations of a predetermined time,

$$\text{UCL} = \frac{1}{2r}\chi^2_{\frac{1-\alpha}{2},2r+2\lambda_{\text{est}}}$$

$$\text{LCL} = \frac{1}{2r}\chi^2_{\frac{1+\alpha}{2},2r\lambda_{\text{est}}}$$

TABLE 3.3. Factors for Upper and Lower Confidence Limit of Failure Rate

| Number of Failures Observed ($r$) | Confidence Level 80% $k_1$ | 80% $k_2$ | 90% $k_1$ | 90% $k_2$ | 95% $k_1$ | 95% $k_2$ | 99% $k_1$ | 99% $k_2$ |
|---|---|---|---|---|---|---|---|---|
| 1 | 2.99 | 1.61 | 3.89 | 2.30 | 4.47 | 2.99 | 6.64 | 4.61 |
| 2 | 2.14 | 1.50 | 2.66 | 1.95 | 3.15 | 2.37 | 4.20 | 3.33 |
| 3 | 1.84 | 1.43 | 2.23 | 1.77 | 2.58 | 2.10 | 3.49 | 2.79 |
| 4 | 1.68 | 1.38 | 2.00 | 1.68 | 2.29 | 1.94 | 2.90 | 2.51 |
| 5 | 1.58 | 1.34 | 1.85 | 1.60 | 2.10 | 1.83 | 2.62 | 2.31 |
| 6 | 1.51 | 1.32 | 1.75 | 1.54 | 1.97 | 1.75 | 2.43 | 2.18 |
| 7 | 1.46 | 1.30 | 1.68 | 1.51 | 1.88 | 1.69 | 2.29 | 2.08 |
| 8 | 1.42 | 1.28 | 1.62 | 1.47 | 1.80 | 1.64 | 2.18 | 2.00 |
| 9 | 1.39 | 1.27 | 1.58 | 1.44 | 1.75 | 1.60 | 2.09 | 1.94 |
| 10 | 1.37 | 1.25 | 1.54 | 1.42 | 1.70 | 1.57 | 2.01 | 1.88 |
| 11 | 1.34 | 1.24 | 1.51 | 1.40 | 1.66 | 1.54 | 1.95 | 1.83 |
| 12 | 1.32 | 1.23 | 1.48 | 1.38 | 1.62 | 1.52 | 1.90 | 1.79 |
| 13 | 1.31 | 1.22 | 1.46 | 1.37 | 1.59 | 1.50 | 1.86 | 1.75 |
| 14 | 1.29 | 1.21 | 1.44 | 1.35 | 1.56 | 1.48 | 1.82 | 1.72 |
| 15 | 1.28 | 1.21 | 1.42 | 1.34 | 1.53 | 1.46 | 1.76 | 1.70 |
| 16 | 1.27 | 1.20 | 1.40 | 1.33 | 1.51 | 1.44 | 1.73 | 1.67 |
| 17 | 1.26 | 1.20 | 1.39 | 1.32 | 1.49 | 1.43 | 1.70 | 1.65 |
| 18 | 1.25 | 1.19 | 1.37 | 1.31 | 1.47 | 1.42 | 1.68 | 1.63 |
| 19 | 1.25 | 1.19 | 1.36 | 1.30 | 1.46 | 1.40 | 1.65 | 1.61 |
| 20 | 1.24 | 1.18 | 1.35 | 1.30 | 1.45 | 1.40 | 1.63 | 1.59 |
| 21 | 1.23 | 1.18 | 1.34 | 1.29 | 1.43 | 1.38 | 1.62 | 1.58 |
| 22 | 1.22 | 1.17 | 1.33 | 1.28 | 1.42 | 1.38 | 1.60 | 1.56 |
| 23 | 1.22 | 1.17 | 1.32 | 1.27 | 1.41 | 1.37 | 1.58 | 1.55 |
| 24 | 1.21 | 1.17 | 1.31 | 1.27 | 1.40 | 1.36 | 1.57 | 1.54 |
| 25 | 1.21 | 1.16 | 1.31 | 1.26 | 1.39 | 1.35 | 1.56 | 1.52 |
| 26 | 1.20 | 1.16 | 1.30 | 1.26 | 1.38 | 1.34 | 1.54 | 1.52 |
| 27 | 1.20 | 1.16 | 1.29 | 1.25 | 1.37 | 1.34 | 1.53 | 1.50 |
| 28 | 1.19 | 1.16 | 1.29 | 1.25 | 1.37 | 1.33 | 1.52 | 1.49 |
| 29 | 1.19 | 1.15 | 1.28 | 1.24 | 1.36 | 1.32 | 1.51 | 1.48 |
| 30 | 1.19 | 1.15 | 1.28 | 1.24 | 1.35 | 1.32 | 1.50 | 1.48 |

For observations truncated at predetermined time, $\lambda(\alpha \text{ confidence}) = [\chi^2_{1-\alpha,2r+2}/2r]\lambda_{\text{est}} = k_1\lambda_{\text{est}}$. For obser-vations truncated at $r$th time, $\lambda(\alpha \text{ confidence}) = [\chi^2_{1-\alpha,2r+2}/2r]\lambda_{\text{est}} = k_2\lambda_{\text{est}}$, $r > 30$, $k_1 = (1/4r)\,[z_\alpha + \sqrt{4r+3}]^2$, and $k_2 = (1/4r)\,\left[z_\alpha + \sqrt{4r-1}\right]^2$, where $Z\alpha$ is the $\alpha$-point of the cumulative normal distribution.

For observations terminated at the $r$th failure,

$$\text{UCL} = \frac{1}{2r}\chi^2_{\frac{1-\alpha}{2},2r\lambda_{\text{est}}}$$

$$\text{LCL} = \frac{1}{2r}\chi^2_{\frac{1+\alpha}{2},2r\lambda_{\text{est}}}$$

Two-sided confidence intervals for MTTF can be established similarly. Attention should be drawn to the fact that the upper limits for two-sided and one-sided confidence intervals have different percentage points.

## Problem 3.9

In an assembly line of porcelain insulators, a sample of 20 is drawn every hour to determine the percentage of defective units. The results from the eight samples drawn in a day are used to construct the control chart for the next day. The following is an exhibit of the eight samples drawn today:

| Sample Number, $i$ | Time | Number Inspected, $n_i$ | Number of Defectives | Probability of a Defective, $p_i$ |
|---|---|---|---|---|
| 1 | 8:00 | 20 | 1 | 1/20 |
| 2 | 9:00 | 20 | 0 | 0 |
| 3 | 10:00 | 20 | 2 | 2/20 |
| 4 | 11:00 | 20 | 1 | 1/20 |
| 5 | 12:00 | 20 | 3 | 3/20 |
| 6 | 13:00 | 20 | 5 | 5/20 |
| 7 | 14:00 | 20 | 2 | 2/20 |
| 8 | 15:00 | 20 | 4 | 4/20 |

*Solution*:
The average defective rate for all eight samples is

$$\bar{P} = \frac{\sum n_i p_i}{\sum n_i} = \frac{18}{160} = 0.1125$$

$$\text{UCL} = 0.1125 + \sqrt{0.1125(1 - 0.1125)/20} = 0.1832$$

$$\text{LCL} = 0.1125 - \sqrt{0.1125(1 - 0.1125)/20} = 0.0418$$

## Problem 3.10

We have one hundred 138–69 kV transformers and have experienced five failures in the past 10 years. What is the estimated failure rate? What is the 95% upper confidence limit of the failure rate? A bigger utility with 1000 transformers experiences 50 failures in 10 years (for the same failure rate). What is the 95% upper confidence limit of their failure rate?

*Solution*:

$$\begin{aligned}\text{Estimated failure rate} &= \frac{5}{100 \times 10} = 0.005 \\ 95\%\ \text{confidence limit} &= 0.005 \times k_1\ (5.95\%) \quad \text{(Table 3.3)} \\ &= 0.005 \times 2.10 \\ &= 0.0105\end{aligned}$$

Uncertainty factor for 50 observations (footnote of Table 3.3):

$$\begin{aligned}k_1 &= \frac{1}{4r}\left(Z_{0.95} + \sqrt{4r+3^2}\right) \\ &= \frac{1}{200}(1.645 + 14.248)^2 \\ &= \frac{253}{200} \\ &= 1.27\end{aligned}$$

$$95\%\ \text{confidence limit} = 0.005 \times 1.27 = 0.0064$$

---

## Problem 3.11

In ground testing, a double sampling plan is used. For a station area with 400 grounds, two samples of 35 grounds are selected. The acceptance and rejection numbers are 5 and 9 for the first sample and 12 and 13 for the combined sample (of 70).

In a test, the first sample showed 10 bad grounds. The second sample was tested and five more bad grounds were found. Does that station area pass? Based on the sampling result, what is the 95% confidence interval of bad grounds in that station area?

*Solution*:

$$\text{Number of bad grounds in a sample of } 70 = 15/70 = 0.214$$

$$\text{Proportion of bad grounds in station area} = 0.214 \pm z s_x$$

where

$$s_x = \sqrt{\frac{N-n}{N-1}} \times \frac{S}{\sqrt{n}}$$

$$\begin{aligned}\text{For proportions, } S &= \sqrt{p(1-p)} \\ &= \sqrt{0.214(1-0.214)} = 0.4101\end{aligned}$$

$$
\begin{aligned}
s_x &= \sqrt{\frac{400-70}{400-1}} \times \sqrt{\frac{0.214x(1-0.214)}{70}} \\
&= 0.0446
\end{aligned}
$$

$z$ for 95% confidence interval is 1.96 (see Table 3.3).

Proportion of bad grounds in station area

$$
\begin{aligned}
\bar{P} &= 0.214 \pm 1.96 \times 0.0446 \\
&= 0.1266 \pm 0.3014
\end{aligned}
$$

or 12.66% = 30.14% at 95% confidence.

## 3.8 CONCLUSIONS

This chapter has presented a number of methods for evaluating the reliability of complex problems. A particular method may be more suitable for a specific problem than other methods; however, it is difficult to specify which method is most suitable for a given system problem since most methods can be used for any system problems. Sometimes, it is best to use a mixture of methods to solve a particular problem. The many numerical examples that have been used to illustrate different methods will help the reader to easily apply these techniques in solving real-world problems.

## REFERENCES

1. R. Porkess (ed.), *The Harper Collins Dictionary of Statistics*, Harper Collins Publishers, New York, 1991.
2. D. J. Koosis, *Statistics a Self-Teaching Guide*, John Wiley & Sons, Inc., New York, 1997.
3. S. K. Kachigan, *Multivariate Statistical Analysis: A Conceptual Introduction*, 2nd edition, Radius Press, New York, 1991.
4. R. E. Walpole and R. H. Myers, *Probability and Statistics for Engineers and Scientists*, 3rd edition, Macmillan Publishing Company, New York, 1985.
5. R. Langley, *Practical Statistics*, revised edition, Dover Publications, Inc., New York, 1970.
6. J. Mandel, *The Statistical Analysis of Experimental Data*, Dover Publications, Inc., New York, 1984.
7. R. E. Walpole and R. H. Myers, *Probability and Statistics for Engineers and Scientists*, 3rd edition, Macmillan Publishing Company, New York, 1985.
8. A. M. Breipohl, *Probabilistic Systems Analysis*, John Wiley & Sons, Inc., New York, 1970.
9. A. W. Drake, *Fundamentals of Applied Probability Theory*, McGraw-Hill, New York, 1967.
10. A. M. Neville and J. B. Kennedy, *Basic Statistical Methods for Engineers and Scientists*, International Textbook Co., Scranton, PA, 1964.

# 4

# APPLICATIONS OF SIMPLE RELIABILITY MODELS

## 4.1 EQUIPMENT FAILURE MECHANISM

### 4.1.1 Introduction

This chapter presents simple example applications of the various reliability techniques presented in Chapters 2 and 3 to facilitate the understanding of the reliability techniques discussed in the latter two chapters. The basic expectation is that the reader will be able to grasp the underlying principles of the reliability models and their practical applications in real-world problems. The illustrative example situations considered are simple enough so that almost all calculations can be done by hand. Electric utilities collect and maintain outage databases for the equipment in their systems. Past experience records could be of vital use in predicting system performance and maintaining power systems integrity. A number of real-distribution system problems will be assessed using the various reliability techniques in this chapter.

Virtually all the electric utilities in North America keep statistics on failures of equipment. The record keeping, however, was initially limited to some major equipment and was rather rudimentary. At present, there is an increased emphasis on greater efficiency in operation, which means getting more out of the equipment. Utilities are

*Power Distribution System Reliability.* By Ali A. Chowdhury and Don O. Koval

pushing the system and its equipment to highest possible limits. To do so without risking catastrophic results, utilities have to understand the capabilities of the system and its equipment better. It also requires better preventive maintenance (PM) practices. One thing vital to all this is a better assessment of the failures, including failure rate, failure probability, failure mode, and so on, which in turn means better record keeping on virtually all equipment. With the advent of high-speed computer technology, this is quite achievable in the twenty-first century.

### 4.1.2 Utilization of Forced Outage Statistics

Basically, two types of outage data are most important: mode of failure and frequency of failure. The first is required in the determination of the cause of failure, leading to design and operation improvements, which prevents or reduces recurrences. The second is required in determining the reliability of components leading to realistic operating constraints and maintenance schedules. In the past, not enough attention was given to these aspects of outage data gathering by utilities, especially in the distribution segment of the power system. At present, however, a lot of decisions are routinely made based on reliability considerations.

### 4.1.3 Failure Rate Computation

One of the most important and frequently used statistics on forced outages of equipment is the failure rate. To compute the failure rate statistic, the total unit time of operation of all units must be taken into account. For example, if 20 units were observed in the last 10 years and two failures were recorded, one in Year 2 and one in Year 6, the failure rate would be

$$\lambda = \frac{2}{200} = 0.01 \text{ failures/year}$$

and the mean time to failures would be

$$\text{MTTF} = \frac{1}{\lambda} = 100 \text{ years}$$

Saying that the $\text{MTTF} = \frac{2+6}{2} = 4$ years would be wrong.

The failure rate obtained from a limited amount of data is subject to error. It may be too high or too low. For engineering applications, the worst-case scenario is normally considered. This is accomplished by multiplying the calculated failure rate by an uncertainty factor. The uncertainty factor is derived from the chi-square distribution and depends on the confidence level desired. For the example mentioned above, which has only two data points, the uncertainty factor for 95% confidence is 3.15, so the failure rate could be as high as

$$\lambda = 3.15 \times 0.01 = 0.0315 \text{ failures/year at a } 95\% \text{ confidence level.}$$

In most equipment failure problems, the equipment is assumed to be in its useful lifetime where the failure rate is constant and the exponential distribution of failure applies, that is,
probability of failing in time $t$:

$$Q(t) = 1 - e^{-\lambda t}$$

probability of functioning in time $t$:

$$R(t) = 1 - e^{-\lambda t}$$

The exponential distribution is not always applicable. The probability of failure may follow a normal distribution or a Weibull distribution or some other distributions especially when the equipment is in the wear-out stage of the Bath tub curve; however, for the great majority of engineering applications, the equipment or system is in its useful lifetime and the exponential distribution is assumed.

Various parameters can be derived from the failure rate and population size. Some of the more important of these include mean time to failure, number of failures expected in an interval of time, expected cost of failure, probability of a given number of failures, reliability level, and so on.

## 4.2 AVAILABILITY OF EQUIPMENT

### 4.2.1 Availability Considerations and Requirements

To render continuous service and to cover or minimize the equipment failure, spares are necessary. Having units as standby, however, means tying up a certain amount of capital in a nonrevenue-generating activity. To save costs, such expenses should be kept down, but at the same time an adequate level of reliability has to be provided. Otherwise, an even greater loss of revenue due to service interruptions may be incurred, not to mention the revenue loss resulting from failures in the primary mission of providing reliable service to customers at the reasonable cost. A balance therefore has to be struck between the reliability of having an optimum number of spares and the need to cut down on inventory costs.

Running a system to failure without preventative maintenance can be very costly. Regular preventive maintenance practices can prevent equipment failures from occurring. There are, however, many conflicting interests. Frequent and more maintenance will keep the equipment in better shape and will cut down the chance of failure. However, it also will result in more maintenance costs, more replacement units, and the maintenance procedures will lead to more frequent interruptions to customers.

In both spare provisioning and maintenance scheduling, an optimal compromise thus has to be reached between cutting costs and maintaining adequate service reliability level. The job is usually too complicated to be done by an arbitrary educated guess or gut feeling. Moreover, the decision maker would need to justify the decisions made using hard quantitative and realizable figures. The technical way of accomplishing this is by means of a formal reliability assessment with emphasis on the availability of equipment or systems. To begin with, an acceptable reliability level has to be defined: maximum allowable length of outages, highest allowable probability of not having a spare when needed, and so on. Then alternatives are evaluated to ascertain if they meet the required reliability criteria. Usually, the cheapest way of meeting the set criteria is selected and executed.

### 4.2.2 Availability Model

In equipment availability computations, the most widely used distribution is the Poisson probability distribution. According to the Poisson distribution, the probability of having $x$ failures in time $t$ is given by

$$P(x) = \frac{e^{-\lambda t}(\lambda t)^x}{x!} \tag{4.1}$$

For spare unit requirement problems, the availability is often defined as the probability of having a spare when needed. Since the system can tolerate $k$ failures if there are $k$ spares, the availability problem is often defined as the probability of having up to $k$ failures in a given time period. This is given by the Poisson series:

$$P(x = k) = e^{-\lambda t}\left(1 + \lambda t + \frac{(\lambda t)}{2!} + \frac{(\lambda t)^3}{3!} + \cdots + \frac{(\lambda t)^k}{k!}\right) \tag{4.2}$$

This probability is often called the *availability* although actually it is the probability of availability in that period. Similarly, the probability of unavailability, $P(x > k) = 1 - P(x = k)$, is often called the *unavailability*.

The length of time of interest is usually 1 year or the repair time of the equipment. The choice of 1 year is made because it is the most common and convenient unit of time in reliability problems. The choice of the repair time as the time period is due to the fact that a system with $n$ spares will be short of units if and only if all $n$ spares are used up and the $(n+1)$th failure occurs; so if the $(n+1)$th failure occurs at a time interval longer than the repair time after the first failure, the first failed unit will have been repaired to take care of the $(n+1)$th failure, and the system will remain operative.

The availability obtained from the Poisson series obviously has its reliability significance; however, it should be noted that it is actually the probability of availability and not the percentage of system uptime as availability is traditionally defined. It is important to also note that this availability probability applies to the initial period only. The initial period, whether it is 1 year or the repair time of the equipment, is characterized by the fact that everything is in operating condition initially. This initial condition does not always hold for subsequent periods; therefore, availability for subsequent periods is, in general, different. For example, a system with one operating unit and one spare unit has a first year availability of $R(t) = e^{-\lambda t}(1 + \lambda t)$. At the end of 1 year, however, there may be 0, 1, or 2 units left, depending on how many failures have occurred in the first year. The situation is further complicated by repair activities. The availability of the second year is therefore not given by the above Poisson series.

### 4.2.3 Long-Run Availability

If the availability of the initial period is $A$ and the availability for subsequent periods can be assumed to be the same as that of the initial period, then the availability of $n$ consecutive periods will be

$$A(n) = A^n \tag{4.3}$$

This assumption holds when the failure rate is small or the repair rate is large.

---

## Example 4.1

The following is an example of mobile substation requirements for a different number of substations to backup for different transformer failure rates.

| Number of Stations for Which Mobile is Backup | 1 Spare Mobile | | | 2 Spare Mobiles | | | 3 Spare Mobiles | | |
|---|---|---|---|---|---|---|---|---|---|
| | $\lambda = 0.0093$ | $\lambda = 0.01$ | $\lambda = 0.02$ | $\lambda = 0.0093$ | $\lambda = 0.01$ | $\lambda = 0.02$ | $\lambda = 0.0093$ | $\lambda = 0.01$ | $\lambda = 0.02$ |
| 30 | 0.9933 | 0.9630 | 0.8781 | 0.9997 | 0.9964 | 0.9769 | 0.9999 | 0.9997 | 0.9966 |
| 40 | 0.9885 | 0.9384 | 0.8088 | 0.9994 | 0.9921 | 0.9526 | 0.9999 | 0.9992 | 0.9909 |
| 50 | 0.9825 | 0.9098 | 0.7357 | 0.9988 | 0.9856 | 0.9197 | 0.9999 | 0.9982 | 0.9810 |
| 60 | 0.9754 | 0.8781 | 0.6666 | 0.9981 | 0.9769 | 0.8847 | 0.9998 | 0.9966 | 0.9720 |
| 70 | 0.9674 | 0.8442 | 0.5964 | 0.9970 | 0.9659 | 0.8399 | 0.9998 | 0.9943 | 0.9536 |
| 80 | 0.9585 | 0.8088 | 0.5301 | 0.9957 | 0.9526 | 0.7911 | 0.9997 | 0.9909 | 0.9303 |
| 90 | 0.9488 | 0.7725 | 0.4662 | 0.9941 | 0.9371 | 0.7359 | 0.9995 | 0.9865 | 0.8977 |
| 100 | 0.9384 | 0.7358 | 0.4095 | 0.9921 | 0.9197 | 0.6825 | 0.9992 | 0.9809 | 0.8645 |

$$P = e^{-n\lambda t}\left[1 + n\lambda t + \frac{(n\lambda t)^2}{2!} + \frac{(n\lambda t)^3}{3!} + \cdots \frac{(n\lambda t)^k}{k!}\right] \tag{4.4}$$

where $P$ is the probability of having a spare, $\lambda$ is the failure rate (failures/year), $t = 1$ year, $n$ is the number of transformers in service, and $k$ is the number of spares.

From the above table, the 1-year availability for 1 mobile backing up 30 stations, with $\lambda = 0.00393$ is calculated to be 0.9933. Actually this is the first year probability of availability; but assuming the probability for subsequent years to be the same, the probability of going through, say, 5 years trouble free will be

$$\mathrm{A}(5) = (0.9933)^5 = 0.9669$$

Note that by using 1 year as the time period in computing availability, it is implicitly assumed that the mobile will be tied up for 1 year when a failure occurs. If, for example, the mobile is intended to be used as a replacement for 2 weeks, that is, 0.04 year only, the availability will increase dramatically from 0.9933 to 0.999989.

Using the first-year availability of 0.9933

$$\begin{aligned}\text{Unavailability} &= 1 - 0.9933 = 0.0067\\ \text{Frequency of occurrence} &= \text{once in } 1/0.0067 = 149 \text{ years.}\end{aligned}$$

The parameters in the above example are only approximate as the *availability* is actually the probability of availability for the first year and is an indicator of how likely it is that a system failure (spare is unavailable when demand occurs) will occur and is not an indicator of frequency and duration of occurrence.

To calculate the system parameters for any time instant, a state space analysis using the Markov process (Section 3.6.3) has to be used. The two parameters of particular importance are the steady-state probabilities of being in an operating state and in a failed state and are the long-run availability and unavailability, respectively, since by definition

$$\text{Availability} = \frac{\text{system uptime}}{\text{total time}} \tag{4.5}$$

$$\text{Unavailability} = \frac{\text{system downtime}}{\text{total time}} \tag{4.6}$$

---

## Example 4.2

Again, consider the case of 1 mobile backing up 60 stations:

A state space diagram for the system is set up in Fig. 4.1.

State 3 is the unavailable state

$$\begin{aligned}\text{Unavailability} &= \text{P}(3)\\ &= \frac{(60\lambda)^2}{(60\lambda)^2 + (60\lambda)(2\mu) + 2\mu^2}\\ &= 0.01236\\ \text{Availability} &= 1 - 0.01236\\ &= 0.9876\end{aligned}$$

$$\begin{aligned}\text{Frequency of system unavailability occurrence} &= \text{P}(3) \times 2\mu\\ &= 0.0247/\text{year}\\ &= \text{once every 40.5 years}\\ \text{Duration of each occurrence} &= \frac{1}{2\mu}\\ &= 0.5 \text{ year}\end{aligned}$$

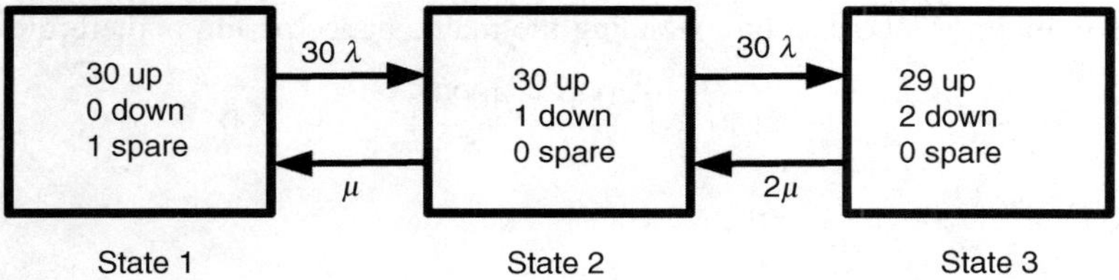


Figure 4.1. State space diagram for 30 stations backed up by 1 mobile station. $\lambda = 0.00393$ and $\mu = 1$.

## 4.3 OIL CIRCUIT RECLOSER (OCR) MAINTENANCE ISSUES

### 4.3.1 Introduction

In the past, a general maintenance procedure for hydraulic oil circuit reclosers was to check the number of operations every 6 months. The normal practice was to replace any unit with a reading exceeding 100 and also to replace any unit that had been in service for 4 years regardless of the number of operations. The reason was that 100 operations or 4 years would degrade the oil and the contacts sufficiently to warrant a complete overhaul.

For some utilities, the usual practice was to use a field-checking period once a year to reduce the workload of the crew centers. This is just an arbitrary decision with no data to back it up. At present, utilities are starting to use statistical techniques.

### 4.3.2 Study Methods

Compute the impact of extending the maintenance period from every 4 to 5 years.

1. The total number of failures for the population of 4000 OCRs (excluding lightning) is four failures in 1 year, for which the failure rate is 0.001 failures/year for each unit.

$$\begin{aligned}\text{Probability of failure in 4 years} &= 1 - e^{-0.001 \times 4} \\ &= 0.0040 \\ \text{Probability of failure in 4 years} &= 1 - e^{-.001 \times 5} \\ &= 0.0050\end{aligned}$$

Number of additional failures incurred by extending the maintenanceperiod from 4 to 5 years:

$$\begin{aligned}N &= (0.005 - 0.004) \times 4000 \\ &= 4.0\end{aligned}$$

The maintenance cost of a unit is \$150 and the loss from scrapping ranges from \$400 for type H to \$900 for type L (\$2010) with the majority being of the cheaper variety.

Savings in maintenance cost by extending the maintenance period is therefore

$$\$150 \times \left(\frac{4000}{4} - \frac{4000}{5}\right) = \$30,000$$

The economics is obvious in this case.

## 4.4 DISTRIBUTION POLE MAINTENANCE PRACTICES

For many years, there has been no formal pole maintenance program in electric utilities. Only recently, some utilities are formally instituting pole maintenance policies due to the fact that in some jurisdictions regulatory agencies require distribution utilities to annually submit reliability improvement programs. In the past, poles were replaced as the need arose. The general population of poles in most utilities is aging and approaching 25–30 years, just the right time to start implementing a regular maintenance program. A generic inspection schedule that could be adopted by a distribution utility is shown in Table 4.1.

There are close to millions of poles on a utility distribution system. To inspect the poles one by one is very time consuming especially when some rigorous tests are involved. To obtain a reasonably accurate estimate of the condition of the poles, a sampling inspection plan can be used. For most inspections, it is the percentage defective that is of concern. The sampling is established such that a statement like this can be made:

$$\text{Percentage defective} = D\ (\text{from sample}) + t \text{ at } 95\% \text{ confidence}$$

Based on the result of the samples, a conclusion is made. If the percentage defective is very small, perhaps no maintenance of the poles is needed. If it is large, then treatment or replacement may be necessary for the entire population without bothering to check every pole individually. The sample size is based on binomial probability distribution.

TABLE 4.1. Pole Inspection Schedule

| Sequence | Plant | Species of Wood | Initial In-Service Inspection | Inspection Period |
|---|---|---|---|---|
| 1 | Pole | Fir larch salt-treated pine | N/A | 2 years |
| 2 | Pole | Treated Western red cedar | 25 years | 10 years |
| 3 | Pole | Jack pine-lodgepole pine pressure treated | 30 years | 10 years |
| 4 | Spare arms and braces | Miscellaneous | 20 years | 5 years |

## 4.5 PROCEDURES FOR GROUND TESTING

### 4.5.1 Concept

Ground testing has traditionally been done on a cyclic basis. In a typical North American integrated utility, for example, in the rural areas, it is normally done once every 10 years. In the suburban areas, it is once every 6 years. Poor grounds are upgraded as they are discovered. Many utilities use statistical methods in ground testing.

### 4.5.2 Statistical Methods for Ground Testing

A practical statistical method for ground checking normally works as follows:

1. For each station area, two equal samples are selected by random number generation in the computer. The sample size depends on the population size and is given by a table.
2. Only the first sample is tested. Mean and standard deviation of the readings are compiled. The number of defective grounds is recorded. If it is below the acceptable number, the sample passes and no more tests are done.
3. If the first sample fails, the second sample is tested. The readings are combined with that of the first sample for computation of mean and standard deviation. An acceptance number of defectives is set for the combined sample (not the second sample).
4. If the first sample or the combined sample passes, the station will be revisited in 4 years. If it fails in both the first and the combined samples, it will be marked as failed. The station area will receive a 100% ground test and defects will be mitigated. They are usually mitigated in the following year.

## 4.6 INSULATORS MAINTENANCE

### 4.6.1 Background

Normally, insulators are left on the line with no maintenance once they are installed. The only kind of maintenance usually performed is to replace units that were seen to be damaged from vandalism or lightning. This practice worked well for many years with little trouble due to the fact that the failure rate of insulators is very low. In addition, a fairly high safety factor is built into the insulator design. As the general population grows older, the electrical and mechanical strength of the insulators will generally deteriorate, and some latent defects such as cement growth also start to show up. A more active maintenance program is therefore required.

### 4.6.2 Inspection Program for Insulators

A typical inspection program presented in the following is divided into four parts based on where the insulators are located.

1. *High-Voltage Transmission Lines*
   The insulators are normally divided into categories based on voltage, type, and usage; for example, ball and socket insulators are in a different category from clevis type units and dead-end insulators are separated from suspension insulators. Within each category, a sample of several towers is selected and tested. Towers already checked are excluded from sample selection in the next 2 years. Sample size is usually chosen to provide a reasonable accuracy at a 95% confidence level. The division of categories and the sample size for each category are determined by crew centers in a distribution utility. The actual sample selection is up to the crew centers.
2. *Subtransmission Lines and Distribution Lines*
   This plan is similar to the high-voltage transmission line insulator testing procedure. Insulators are categorized according to voltage and type of each group, and a given number of insulators are tested.
3. *Station Dead-End Insulators*
   A significant number of failures have been reported in the published literature for dead-end insulators. In some situations, both dead-end units have been reported to be punctured. There are hundreds of stations in a typical utility system. Obviously, a 100% annual inspection program is impossible. The inspection plan therefore divides the stations into three categories, namely, those built before 1980, from 1981 to 2006, and after 2006. For the first group, every station is checked once every 3 years and for the second group, two from each crew center are tested, and the usual statistical inferences are applied. For the third group, no inspections are planned or executed at the current time. Test results are reported periodically. If test results indicate a hazardous condition, then immediate reporting and changeout will be required. The suggested hazardous level could be as follows:

| | |
|---|---|
| Distribution dead-ends: | 6% defective |
| 72 kV: | 7% defective |
| 25 kV: | 3% defective |

4. *Insulators on Power Equipment*
   Insulators on dynamic parts such as power fuses and switches are under serious mechanical stress, and therefore they are more prone to failure than those on stationary parts. This causes additional danger than just a flashover. There have been real instances of the insulator stack falling due to parting of the insulator cap.

### 4.6.3 Voltage Surges on Lines

The basic reason that insulators are required on a line is to separate the line from the supporting structure. Two of the most prevailing transient overvoltage conditions are lightning and switching surges. Lightning is characterized by the surge current rather than voltage, although this current surge does generate a voltage surge across the insulators. Direct lightning strikes are not too frequent when there is a skywire shielding the phase conductors. Generally, voltage withstand criterion of insulators is governed by the switching surge.

Switching as well as lightning surges are statistical quantities depending on the chance combination of different variables. It is impossible to identify what a surge value will be. However, from past experience and line parameters, a prediction can be made to estimate the probability that a surge will fall within a certain range of values. The value of a surge can be described by a normal probability distribution with a mean value and a standard deviation expressed as a percentage of the mean value. From statistical theory, 99.7% of all surges will fall within $\pm 3$ standard deviations from the mean. The values of the mean and standard deviation depend on various factors such as line capacity, line configuration, and so on and are not deterministic. A reasonable estimate can, however, be obtained from calculation, computer simulation, past experience, and literature. Chapter 21 discusses voltage dip and surge issues in detail.

### 4.6.4 Critical Flashover

The strength of insulators is defined in terms of the voltage the insulators can withstand as determined by the voltage at which a flashover occurs. This flashover voltage is, however, not a fixed quantity. It varies from one test to another, so it has to be described in statistical terms. The critical flashover (CFO) voltage is the voltage at which a flashover will occur half of the time. In other words, if an insulator is given 500 shots at the CFO voltage, 250 flashovers will be recorded. If the flashover frequency is plotted against the voltage, a normal probability curve will result. It may therefore be stated that at 3 standard deviations below the CFO voltage, there is only a 0.3% chance of a flashover.

If a line has more than one insulator string in parallel, the probability of a flashover increases due to the fact that there is more than one path from the line to the structure. If the probability of withstand for one string is $p$, the probability of withstand for $n$ strings in parallel will be $p^n$.

---

## Example 4.3

There are 500 towers in 100 miles of a 345 kV line. If the probability of flashover of a string in a year is 0.002, what is the probability of a flashover per 100 mile per year?

$$P(\text{withstand, 1 string}) = 1 - 0.002 = 0.998$$

$$\begin{aligned} P(\text{withstand, 500 strings}) &= (0.998)^{500} \\ &= 0.3675 \end{aligned}$$

$$\begin{aligned} P(\text{F/O, 500}) &= 1 - 3675 \\ &= 0.6325 \end{aligned}$$

In computing the insulation strength of the line, there are two probability curves—(1) the surge voltage distribution and (2) the withstand strength distribution, the latter of which is converted to a *cumulative probability curve*. The probability that the surge voltage will exceed the withstand strength is given by the convolution of the cumulative probability curve with the surge voltage distribution curve.

---

## Example 4.4

A utility conducted a series of flashover tests on an insulator string with 7 defective units on a string of 12. The cumulative probability curve is available. Suppose that for a 230 kV line, the switching surge distribution follows a normal distribution with a mean of 2.325 pu and a standard deviation of 0.12 pu, what is the probability of flashover if one of the insulator strings has seven defective units?

$$\text{For a 230 kV line, } 1\text{ pu} = 230 \times \sqrt{\frac{2}{3}}\text{kV crest} = 187.79\text{kV}$$

By superimposing the cumulative probability of the flashover curve and the switching surge distribution curve, the curve shown in Fig. 4.2 can be obtained (Table 4.2).

This is the probability of flashover for a string with seven defective insulators at line end if a switching surge occurs. Therefore, the actual probability of a hazard is 2.2045% × probability of switching (or lightning) surge.

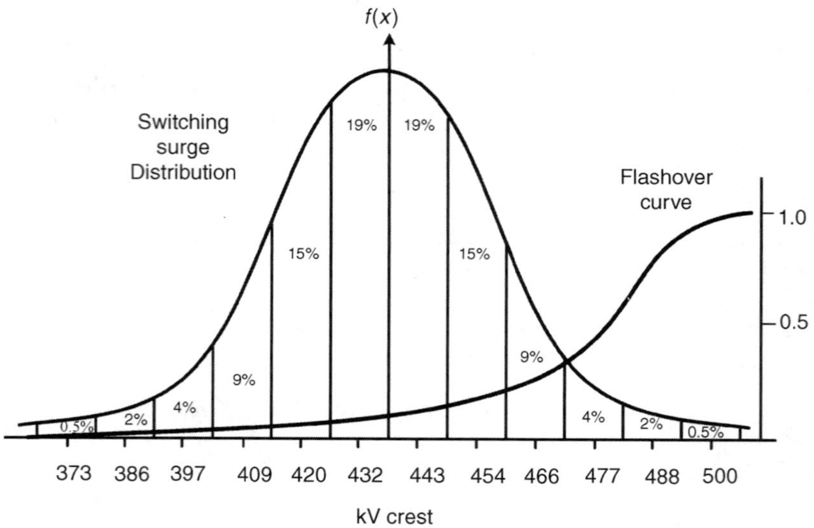


Figure 4.2. Superposition of the cumulative probability of flashover curve and the switching surge distribution curve.

TABLE 4.2. Calculation of the Probability of Flashover

| Class Interval | Class Midpoint | | Calculation | | Probability |
|---|---|---|---|---|---|
| 369–380 kV | 373 | = | 0.5% × 0% | = | 0% |
| 380–392 kV | 386 | = | 2% × 0% | = | 0% |
| 392–403 kV | 397 | = | 4% × 0% | = | 0% |
| 403–414 kV | 409 | = | 9% × 0% | = | 0% |
| 414–426 kV | 420 | = | 15% × 0% | = | 0% |
| 426–437 kV | 432 | = | 19% × 0% | = | 0% |
| 437–448 kV | 443 | = | 19% × 0.05% | = | 0.0095% |
| 448–460 kV | 454 | = | 15% × 0.4% | = | 0.06% |
| 460–471 kV | 466 | = | 9% × 3% | = | 0.27% |
| 471–482 kV | 477 | = | 4% × 15% | = | 0.6% |
| 482–494 kV | 488 | = | 2% × 42% | = | 0.84% |
| 494–505 kV | 500 | = | 0.5% × 85% | = | 0.425% |
| | | | | | 2.2045% |

### 4.6.5 Number of Insulators in a String

High-voltage lines have strings of insulators. The number of insulators in a string is based on the required insulation strength plus a safety factor. In other words, under normal conditions, a number of insulators in the string could be defective while the string still maintains adequate insulation. The condition of each individual string will not be known until it is individually checked. However, the general condition of the population can be estimated with reasonable accuracy from a sampling inspection, and on the basis of the knowledge of the condition of the population, the condition of an individual string can be predicted. Specifically, we want to know the number of defects in a string given the percentage of defects in the general population. The answer comes from a simple binomial probability distribution (strictly speaking, it should be hypergeometric distribution, but the difference is negligible). Given that the defective rate is $p$, probability of having $x$ defectives in $n$ units is given by

$$P(x) - \frac{n!}{(n-x)!x!}p^x(1-p)^{n-x} \tag{4.7}$$

For example, if the defective rate of insulators on a 230 kV line is 5%, the probability of finding three defective units in a suspension string is

$$P(x) = \frac{12!}{3! \times 9!}(0.05)^3(0.95)^9$$
$$= 1.73\%$$

The important question of safety is "How many insulators are required for a given voltage?" The regular $5^3/_4$ in. × 10 in. suspension insulators have a dry flashover rating of 80 kV and wet flashover rating of 50 kV, so by simple division of the line-to-ground crest voltage by 50 kV, the number of insulators required to withstand the nominal power frequency line voltage is known. For 240 kV, this is 4, for 115 kV, 2, and so on.

TABLE 4.3. Typical Number of Insulators on Lines at Various Voltages

| | Suspension | Dead-End |
|---|---|---|
| 500 kV | 26 | 28 |
| 240 kV | 12 | 14 |
| 144 kV | 8 | 10 |
| 115 kV | 7 | 9 |
| 72 kV | 4 | 5 |

If a transient overvoltage should develop, this may not be enough. The typical number of insulators on lines of various voltages is shown in Table 4.3.

The safety factor is very high; however, not as high as it seems. This is due to the following two factors: (1) contamination lowers the strength of the insulators and (2) the voltage distribution along a string is not even. The unit closest to the line bears a much higher stress than the average value. Figure 4.3 gives the probability of power frequency flashover

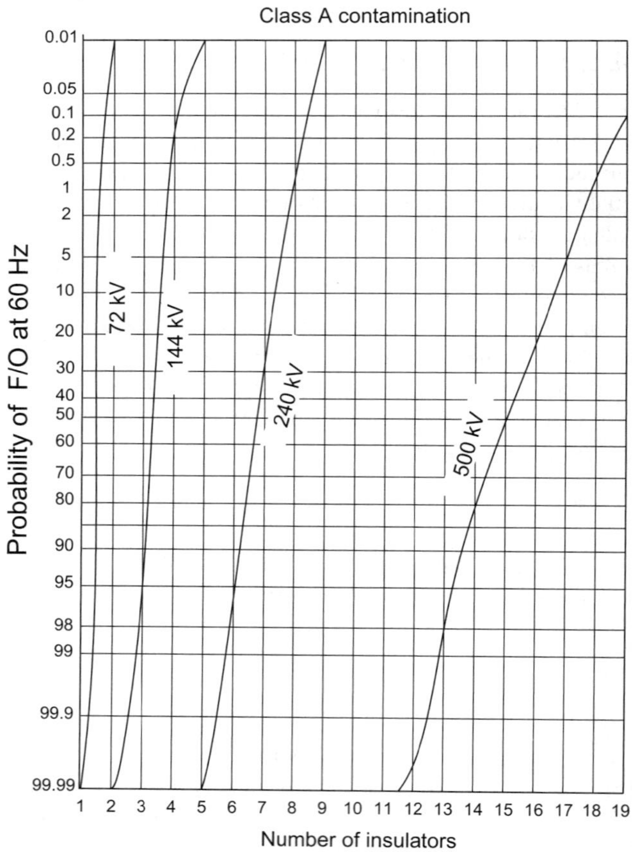


Figure 4.3. Probability of flashover at 60 Hz versus the number of insulators.

TABLE 4.4. Live Line Method

| Nominal Line Voltage | Number of Defective Units in String | Live Line Work Method |
|---|---|---|
| 72 kV | 2 or less | Blocking mandatory |
| 115 and 144 kV | 3 or less | Blocking mandatory |
| 240 kV | 5 or less on dead ends | Blocking mandatory |
| 240 kV | 4 or less on suspension | Blocking mandatory |
| 500 kV | 7 or less | Blocking mandatory |

in a lightly contaminated environment for different numbers of sound insulators at different voltages.

Taking these factors into consideration, some safety limits can be set in the live line methods regarding the number of sound insulators required in a string before live line work can be performed. Table 4.4 shows data taken from a Canadian utility's Live Line Methods Directive issued in the 1990s.

Normally, replacement can be deferred until weather and road conditions improve.

## 4.7 CUSTOMER SERVICE OUTAGES

### 4.7.1 Background

Reliability of a power system is quantified in terms of the number of power supply outages. Generally at the generation level, this signifies capacity inadequacy; at the transmission level, it usually means outage of a line or terminal; and at the distribution level, this means interruption of service to the customer. It is at the distribution level that reliability is most relevant from a customer's viewpoint and a utility's reliability performance is normally quantified at this level in the eyes of both the customers and of the regulatory agencies. Utilities in North America have been collecting distribution system service continuity statistics for many years. Each year data are collected from participating utilities, are compiled, and are analyzed, and the results appear in an annual "Service Continuity Report." In Canada, the Canadian Electricity Association publishes annual service continuity statistics for all Canadian utilities. In the United States, the Edison Electric Institute annually publishes similar statistics for participating utilities. The basic reliability indices included in these reports are defined in Section 4.7.2.

### 4.7.2 Popular Distribution Reliability Indices

The reliability of customer service is indicated by several widely used indices as follows:

- SAIFI is the System Average Interruption Frequency Index. This is a measure of the average number of interruptions in a year and is defined as

$$\text{SAIFI} = \frac{\text{total number of customer interruptions}}{\text{total number of customers}}$$

- SAIDI is the System Average Interruption Duration Index. This is the average outage time in a year for each customer in the system and is defined as

$$\text{SAIDI} = \frac{\text{total number of customer hours of interruptions}}{\text{total number of customers}}$$

- CAIDI is the Customer Average Interruption Duration Index. This is the average duration of an interruption experienced by the customer interrupted and is defined as

$$\text{CAIDI} = \frac{\text{total number of customer hours of interruptions}}{\text{total number of customer interruptions}}$$

- ASAI is the Average System Availability Index. This is also known as the System Reliability Index and is defined as

$$\text{ASAI} = \frac{8760 - \text{SAIDI}}{8760}$$

In compiling the interruption data, some utilities have a policy of not counting outages shorter than 5 min. This does not affect the duration indices much; however, it significantly reduces the frequency index. All these indices have their physical meanings, but the one with most relevance to a customer by far is CAIDI, the average length of an outage, followed by SAIFI, the frequency of customer outages.

These indices represent the average frequency or duration; however, these indices neither give any indication of their limits nor of the manner in which they may vary. For example, CAIDI is the average duration of an outage. Apparently, an outage can be longer or shorter than CAIDI. How likely is it that an outage is longer than a particular value, say, 10 h? The answer will be readily available if there is a probability distribution that describes the variation of CAIDI. This area of pursuit is receiving increasing interest these days, although it is still in its infancy. Even without a theoretically based model, it is possible to use the actual data of previous years to depict the probability distribution of the indices. The IEEE Distribution Subcommittee is working in the area of distribution reliability indices, service continuity report, outage event definition, major event day identification, and so forth.

### 4.7.3 Reliability Criteria

These days more and more emphasis is put on reliability. Reliability in engineering is a quantifiable term, not just an abstract quality. To achieve a certain level of reliability, the main requirement is knowledge of what that level is. In generation, reliability is defined in terms of "loss of load expectation" and is set at 0.1 days/year. In customer service reliability, no such explicit goal is set, although utility management does compare annual reliability indices with the national average and is usually delighted if the comparison is favorable. Regulatory agencies in different jurisdictions in North America are embarking on adopting a performance-based rate (PBR) mechanism to provide an adequate level of service reliability to customers.

### 4.7.4 Cost of Interruption Concept

It is a fairly easy task to set a reliability criterion. There are many factors and issues that need to be taken into consideration in establishing reliability criteria. Chapter 12 identifies many issues and relevant factors in setting reliability performance standards. As indicated earlier, there are many factors to consider; however, the most important one is the cost of interruption. Some surveys have been conducted and cost figures have been published. These cost figures vary from one survey to another because of the difference of time, region, and other factors. It is the methodology that is worth looking into.

The value of loss of load (VOLL) is expressed as $/kW or $/kWh. The derivation of the VOLL depends on many factors; the three most important are time of year, duration, and class of customer. If the time factor is assumed to be the worst, that is, in the middle of a cold winter, a graph can be plotted for the cost of interruption versus the duration of interruption for each class of customer. So on the graph, there are different curves representing residential, commercial, industrial customers, and so on.

The customer outage cost survey result is, of course, subject to variance, and there are many refining techniques available to refine the collected data. In setting reliability criteria, the cost of reliability should be considered. It is always possible to make the system a little more reliable, but the cost incurred should bring in a worthwhile benefit. In the past, the benefit of improvement in reliability was an abstract quality; even the improvement in reliability itself is hard to measure; however nowadays, with reliability quantified into indices and parameters, and with a cost attached to an outage, it is possible to obtain some cost/benefit comparison. Obviously, there are many other considerations in reliability besides economic ones, such as utility image, political, technical, and even health hazards (e.g., a long outage in winter); however, the cost/benefit analysis gives an indication of the marginal improvement rate. This is an added advantage in proper system planning and design.

## 4.8 CONCLUSIONS

This chapter has been concerned with practical applications of different reliability techniques in distribution systems reliability and maintainability assessments. Simple numerical examples have been used to illustrate the utilization of reliability models in solving practical system problems. The chapter dealt with past and predictive reliability assessment metrics. The past and predictive assessments of distribution system reliability performance metrics and methods will be further discussed in later chapters. The concept of cost of service interruptions to customers is briefly introduced in this chapter. The utilization reliability cost–reliability benefit concept will be further enhanced and used in later chapters.

## REFERENCES

1. IEEE Standard 1366™-2003, IEEE Guide for Electric Power Distribution Reliability Indices, 2003.
2. 2001 Annual Service Continuity Report on Distribution System Performance in Electrical Utilities, Canadian Electricity Association, Montreal, Canada.
3. REA Bulletin 161-1, March 1972, Interruption Reporting and Service Continuity Standards for Electric Distribution Systems.
4. R. Porkess, *The Harper Collins Dictionary of Statistics*, Harper Perennial, 1991.
5. R. E. Walpole and R. H. Myers, *Probability and Statistics for Engineers and Scientists*, 3rd edition, Macmillan Publishing Company, 1985.
6. A. Papoulis, *Probability, Random Variables and Stochastic Processes*, McGraw-Hill. 1965.
7. W. Feller, *An Introduction to Probability Theory and Its Applications*, Vol. 1, John Wiley, 1968.
8. P. L. Meyer, *Introductory Probability and Statistical Applications*, Addison-Wesley, 1965.

# 5

# ENGINEERING ECONOMICS

## 5.1 INTRODUCTION

Engineers do routinely make technical and nontechnical business decisions. There are many factors and issues that can influence the decision-making processes such as technical, social, and political considerations. However, for the most part, decisions are made on the basis of economics. Decisions based on economics are considered to be the most unbiased and the least controversial since everybody understands in terms of dollars and cents. Technical people, such as engineers, try to make most decisions on an economic basis. That means assigning dollar value to all items, even to some intangible ones. This is more relevant in today's electric utility business environment as the electricity market is deregulated and competition is introduced.

The fundamental approach to economic planning is making engineering economic decisions by first choosing from several options and then obtaining a dollar value for each as best one can and compare, normally choosing the least costly one. It is important to note that, because of the complexity of the options under considerations, care must be taken that the comparisons are done on equal basis. A few very basic economics terms are provided in this chapter. The intent is to facilitate general understanding of economic planning in simple distribution system situations.

*Power Distribution System Reliability.* By Ali A. Chowdhury and Don O. Koval

## 5.2 CONCEPT OF INTEREST AND EQUIVALENT

*Interest* may be thought of as the return obtainable by the productive investment of capital. The rate of interest is the ratio between the interest chargeable or payable at the end of a period and the money owed or invested at the beginning of that period. If the period of investment is longer than the period for interest computation, the interest becomes compounded, that is, the interest earned is automatically reinvested to earn more interest.

Since money can "grow," the time of payment becomes significant. Payments that differ in total magnitude but are made at different dates may be equivalent to one another; for example, a payment of \$100 now is equivalent to a payment of \$110 1 year later if the annual interest rate is 10%, because if we have the \$100 now, we can earn \$10 in a year at 10% interest rate for a total asset of \$110 at the end of the year.

## 5.3 COMMON TERMS

There are many terminologies representing the various amounts of money and payments at various points in time in the lifetime of a system. Some of the most common terms used are the following:

*Present Value*: the value at a reference date of a future payment or receipt. The present worth of a sum of money is of course just the face value of the sum.

*Future Value*: the value of a sum of money at a future date. That sum of money could be the result of investment/payments from earlier times, or could be just a payment or value at that future date.

*Interest Period*: the period of time at the end of which interest is calculated and compounded. While the interest period used most often is a year, it could be a half-year, a month, or anything.

*Annuity*: a series of equal cash payments or receipts made at the end of each period (usually a year) in a uniform series.

*Sinking Fund*: a fund established to produce a desired amount at the end of a given period by means of a series of payments throughout the period. It is a form of annuity.

## 5.4 FORMULAS FOR COMPUTING INTEREST

In typical engineering economics studies, all the quantities such as present value, future value, annual payment, and so on are all related. An investment may "grow" to a certain value in the future; a series of payments for $n$ years is established to pay off a debt incurred now, or to meet an expense at a future date; different financial plans are converted to their equivalent annual costs for comparison; and so on. The relationship between these quantities is given by the following interest formulas:

$i$: interest rate per interest period

$n$: number of interest periods

$P$: present sum of money or present value of future expenses/payments
$S$: sum of money at the end of $n$ periods or future value of investment payments made during the $n$ periods
$R$: end-of-period payment or receipt in a uniform series continuing for the coming $n$ periods, the entire series equivalent to $P$ or $S$

Given $P$, to find $S$

$$S = P(1+i)^n \tag{5.1}$$

Given $S$, to find $P$

$$P = S\left[\frac{1}{(1+i)^n}\right] \tag{5.2}$$

Given $S$, to find $R$

$$R = S\left[\frac{i}{(1+i)^n - 1}\right] \tag{5.3}$$

$$= S \times \text{sinking fund deposit factor.}$$

Given $P$, to find $R$

$$R = P\left[\frac{i(1+i)^n}{(1+i)^n - 1}\right] \tag{5.4}$$

$$= P \times \text{capital recovery factor.}$$

Given $R$, to find $S$

$$S = R\left[\frac{(1+i)^n - 1}{i}\right] \tag{5.5}$$

Given $R$, to find $P$

$$P = R\left[\frac{(1+i)^n - 1}{i(1+i)^n}\right] \tag{5.6}$$

Capital recovery factor = sinking fund deposit factor + interest rate. This process is illustrated in Fig. 5.1.

With the advent of high-speed computers and several commercial spreadsheet packages, the computations involved in these formulas are not difficult.

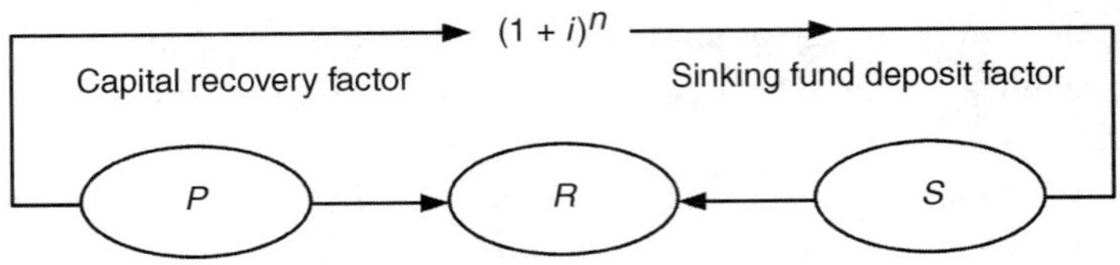


Figure 5.1. Capital recovery process.

## Example 5.1

If a person deposits \$5000 in the bank at an interest rate of 10% per year, how much will it be worth in 20 years?

$$\begin{aligned} S &= P(1+i)^n \\ &= \$5000(1+0.10)^{20} \\ &= \$33,637.50 \end{aligned}$$

## Example 5.2

A person wins a lottery of \$50 million and decides to be paid by a 15-year annuity at 5% per year. How much will he get per year?

$$\begin{aligned} R &= P\,[\text{capital recovery factor, } 5\%, 15 \text{ years}] \\ &= P \times \frac{i(1+i)^n}{(1+i)^n - 1} \\ &= \$50,000,000 \times \frac{0.05(1+0.05)^{15}}{(1+0.05)^{15} - 1} \\ &= \$4,814,963.68 \end{aligned}$$

## Example 5.3

A person wants to buy a house worth \$500,000 in 5 years. He saved the amount by depositing a fixed sum at the end of each year. How much should that annual deposit be if interest rate is 10% per year?

$$\begin{aligned} R &= S\,[\text{sinking fund deposit factor, } 10\%, 5 \text{ year}] \\ &= S \times \frac{i}{(1+i)^n - 1} \\ &= \$500,000 \times \frac{0.1}{(1+0.1)^5 - 1} \\ &= \$81,898.74 \end{aligned}$$

## Problem 5.1

A man sets up a fund for his newborn son's college education. He figures that his son will go to the college at the age of 18 years for a cost of $200,000. How much should be put into the fund if the interest rate is 5%.

*Solution*:
The formula to use is

$$\begin{aligned} P &= S\left[\frac{1}{(1+i)^n}\right] \\ &= \$200,000 \times \frac{1}{(1.05)^{18}} \\ &= \$83,104.13 \end{aligned}$$

## Problem 5.2

A man retires with $3,000,000 at the age of 65. He wants to convert it to an annuity that will look after him till he is 100 years. If the interest rate is 7%, what will be his annual retirement income?

*Solution*:
The formula to use is

$$\begin{aligned} R &= P\left[\frac{i(1+i)^n}{(1+i)^n - 1}\right] \\ &= \$3,000,000 \times \frac{0.07(1.07)^{35}}{(1.07)^{35} - 1} \\ &= \$231,600.00 \end{aligned}$$

## 5.5 ANNUAL COST

### 5.5.1 Concept of Annual Cost

Projects normally span a period of time with monetary transactions made at the beginning, end, or anywhere in between that period. The comparison of different options is meaningless unless the costs are brought to the same common basis. One common ground that is usually used is the *annual cost*. The cost of a project is converted to a uniform series of annual cost as if the project were financed by equal annual sums. Different options are meaningless if the option costs are not brought to the same basis as stated earlier. Different options can then be compared on this basis.

## Example 5.4

A company will spend \$5000 every year for a service for the next 20 years. If it buys its own machine there will be a maintenance cost of \$1000 and an insurance premium of \$200. The machine costs \$50,000, and in 20 years it will still have a 10% salvage value. If the interest rate is 10% per year, is it worthwhile to buy a new machine?

Everything is already an annual cost except the purchase and salvage price of the new machine. The initial expense of \$50,000 is equivalent to an annual expense of a certain amount for 20 years. That amount is given by

$$\begin{aligned} R &= P \times [\text{capital recovery factor}, 10\%, 20\text{ year}] \\ &= P \times \frac{i(1+i)^n}{(1+i)^n - 1} \\ &= \$50,000 \times \frac{0.1(1+0.1)^{20}}{(1+0.1)^{20} - 1} \\ &= \$5872.98 \end{aligned}$$

The salvage value = \$50,000 × 10% = \$5000 but that is the money 20 years down the road. To express that as an annual credit for the preceding 20 years,

$$\begin{aligned} R &= S \times [\text{sinking fund deposit factor}, 10\%, 20\text{ year}] \\ &= S \times \frac{i}{(1+i)^n - 1} \\ &= \$5000 \times \frac{0.1}{(1+0.1)^{20} - 1} \\ &= \$87.30 \end{aligned}$$

$$\begin{aligned} \text{Annual cost with new machine} &= \$(5872.98 - 87.30 + 1000 + 200) \\ &= \$6985.68 \\ \text{Annual cost at present} &= \$5000.00 \end{aligned}$$

So it is not worthwhile to buy a new machine.

### 5.5.2 Alternatives with Different Life Times

In many situations, the alternatives under considerations for comparison purposes will have different lifetimes. The method used earlier is still valid if it can be assumed that at the end of the shorter lived alternative, the alternative can be repeated. To compare the alternatives, a time span equal to the least common multiple of the life spans of the alternatives will have to be tabulated. For example, if one alternative has a life span of 25 years and another has a life span of 10 years, a 50-year period will have to be used.

In most practical cases, the conditions of an alternative will not repeat themselves at the end of its life. Costs may go up. There may be new technologies and new constraints. If price increases are predicted, the longer lived alternative will have an advantage. If technological improvements and cost reductions are forecast, the shorter lived alternative will be favored. Economic comparisons become very complicated in these circumstances.

## 5.6 PRESENT VALUE (PV) CONCEPT

The concept of present value is frequently used in planning and designing power systems. In lieu of converting various expansion plans to annual costs, present values for expansion projects are computed for comparison purposes. Normally a complex project spanning a long period, with disbursements and receipts at various stages of the project, is converted to a single value at present. Actually, it does not have to be the present date; it could be any reference date. This is called *capitalization*. It is only one step away from the annual cost because of the following formula:

$$\begin{aligned} R &= P \times \text{capital recovery factor} \\ &= P \times \frac{i(1+i)^n}{(1+i)^n - 1} \end{aligned} \tag{5.7}$$

In comparing alternatives of different lifetimes, a time span equal to the least common multiple of the lives is normally used. Interest rates are assumed to be constant for the entire period of comparison. The method is very simple. Each payment or receipt can be converted to an equivalent value at a reference date, that is, present value if the interest rate and time of transaction are known. Summing these up, the present value of the entire project is obtained.

---

## Example 5.5

Rework the Example 5.4 in Section 5.5.1 using present value comparison.

Alternative 1*:* Buying a new machine

$$\begin{aligned} \text{Purchase price} &= \$50,000 \\ \text{Salvage price} &= \$50,000 \times 10\% \\ &= \$5000 \text{ (in year 20)} \\ \text{PV of salvage price} &= \$5000 \times \frac{1}{(1.1)^{20}} \\ &= \$743.22 \end{aligned}$$

Annual cost = $(1000 + 200) = $1200, for 20 years

$$\begin{aligned} \text{PV} &= \$1200 \times \frac{(1.1)^{20} - 1}{0.1(1.1)^{20}} \\ &= \$10,216.27 \end{aligned}$$

$$\begin{aligned}\text{Total present value} &= \$(50,000 - 743.22 + 10,216.27)\\ &= \$59,473.05\end{aligned}$$

Alternative 2*:* Pay for the service for 20 years

$$\begin{aligned}\text{Annual service cost} &= \$5000\\ \text{PV of service cost} &= \$7500 \times \frac{(1.1)^{20} - 1}{0.1(1.1)^{20}}\\ &= \$42,567.82\end{aligned}$$

Alternative 2 is a lot cheaper.

---

## Problem 5.3

Power plant expenses for a 5-year project are as follows:

Year 0: $260 million dam and site construction
Year 3: $60 million powerhouse construction
Year 5: $220 million equipment
Plus $30 million annually for operating expenses

A contractor offers to provide a turnkey project for $700 million, half to be paid at the beginning and half to be paid at the end of the 5-year project. If interest rate is 8%, is that offer worth accepting?

*Solution*:
Everything is capitalized to present value

$$\begin{aligned}\$260\text{ million at Year }0 &= \$260\text{ million PV}\\ \$60\text{ million at Year }3 &= \$60\text{ million } \times \frac{1}{(1+0.08)^3}\text{ PV}\\ &= \$47.62\text{ million}\\ \$220\text{ million at Year }5 &= \$220\text{ million } \div (1.08)^5\text{ PV}\\ &= \$149.72\text{ million}\\ \$30\text{ million for 5 years} &= \$30\text{ million} \times \frac{(1.08)^5 - 1}{0.08(1.08)^5}\\ &= \$119.78\text{ million}\end{aligned}$$

Total present value of project = $577.12 million
Contractor's proposal:

$$\begin{aligned}\$350\text{ million at Year }0 &= \$350\text{ million PV}\\ \$350\text{ million at Year }5 &= \$350 \div (1.08)^5\text{ PV}\\ &= \$238.2\text{ million}\end{aligned}$$

Total cost of proposal = $588.20 million
The contractor's proposal is expensive.

TABLE 5.1. Five-Year Cash Flow of the $10,000 Investment

| Year | Disbursements | Receipts | Cash Flow |
|---|---|---|---|
| 0 | 10,000 | | – 10,000 |
| 1 | 1000 | 5000 | 4000 |
| 2 | 800 | 5000 | 4200 |
| 3 | 1200 | 4800 | 3600 |
| 4 | 1300 | 4500 | 3200 |
| 5 | 1500 | 4000 | 2500 |

## 5.7 THEORY OF RATE OF RETURN

In a bank account, interest is paid as a percentage of the principal and then compounded every interest period. A return on an investment is equivalent to the return from an annual interest rate of $i$ and is said to have a rate of return of $i$. For example, if a person invests \$20,000 and at the end of 5 years has \$46,007.75, the rate of return will be 18.13%, because the person would have received the same amount of money by putting it in a bank account with an interest rate of 18.13%:

$$\$20,000(1+0.1813)^5 = \$46,007.75$$

Most financial investments, however, consist of a series of disbursements and receipts during that period. For example, the 5-year cash flow of a \$10,000 investment may turn out to be like the results shown in Table 5.1.

To calculate the rate of return on this investment, one has to find an interest rate such that the net present value of the cash flow is zero. This has to be done by trial and error using the following formula:

$$P = \frac{S}{(1+i)^n} \tag{5.8}$$

For this example,
If $i = 15\%$

$$\begin{aligned} \text{PV} &= \$4000/(1.15) + \$4200/(1.15)^2 + \$3600/(1.15)^3 + \$3200/(1.15)^4 \\ &\quad + \$2500/(1.15)^5 - \$10,000 \\ &= \$2093.68 \end{aligned}$$

$i$ is not high enough.
If $i = 24\%$

$$\text{PV} = \$51.78$$

$i$ is not high enough.
If $i = 24.27\%$

$$\text{PV} = 0$$

So the rate of return $= 24.27\%$.

As an illustration, in the previous example, the present value of the cash flow is zero at $i = 0.1813$

$$\$46,007/(1.1813)^5 - \$20,000 = 0$$

therefore, the rate of return is 18.13%.

## 5.8 COST–BENEFIT ANALYSIS APPROACH

Before a project is considered, it is to be proved that the project is worthwhile and is not a waste of money; or when there are several alternatives to be chosen from, it is necessary to determine the alternative that is most beneficial, and in both situations, the conclusion is made from a cost–benefit analysis.

Cost–benefit assessment is an economic approach where all the economic principles apply. It is also applicable to all economic activities, from financial investment decisions to the making of public policy. In the case of investment problems, the costs are the capital and the benefits are the profits. The applications to other projects can be more complicated. In many situations, there are a lot of intangibles. If those intangibles can be quantified and measured, they need to be included in the analysis. Otherwise, the analysis will include only the economic factors and the decision makers will have to include other noneconomic factors in their considerations.

### Example 5.6 (Substation Reliability)

A very simplified representation of the investment decision is adopted in this example. The basic intent behind this is to focus on the framework of cost–benefit analysis in similar situations and the very essence of the trade-off involved. The example, for simplicity, assumed the alternative considered would provide perfect reliability. The cost figures used for the customer interruption and for the facility in the example are in Canadian funds. The economic parameters assumed are 2% inflation rate and 9.71% discount rate. An aggregate outage cost value of $14/kWh was used in computing impact of customer interruptions.

Assume that conventional power flow and dynamic performance analyses reveal substandard reliability performance for the load point served by the substation in question according to the utility reliability planning criteria. Planning engineers would normally perform conventional technical and economic studies to investigate different transmission upgrade projects for the area network served by the radial substation. Let the least-cost transmission project recommended by the planning engineers taking into account cost, technical superiority, long-term fit, flexibility, environment, safety, and all other pros and cons is to add another transformer to the substation to improve the service reliability to the area customers.

If implemented, the proposed second transformer is expected to eliminate the customer outage costs. The next step would be to perform a cost–benefit analysis for the proposed addition of a second transformer to the substation to ascertain whether the capital investment would be justified from customers' point of interest in terms of reduced customer interruption cost rendered by the transformer addition. Table 5.2 describes steps to calculate the cumulative present value of

TABLE 5.2. An Additional Transformer to be Added to a Substation

| Economic Life of Transformer | 30 years |
|---|---|
| Average frequency of transformer outage | 0.2709 occurrences/year |
| Average length of an outage | 6 h |
| Expected outage hours avoided | 1.63 h/year |
| Unserved peak load | 25 MW |
| Load factor | 85% |
| Avoidable expected unserved energy | 35 MWh |
| Customer interruption cost | $14/kWh |
| Expected reliability benefits | $490,000 per year |
| The cumulative present value of reliability benefits over 30 years | $6,189,000 |
| The capital cost of a transformer | $1,000,000 |
| The cumulative present value of the cost of a transformer over 30 years | $1,529,000 |

reliability benefits, that is, avoided customer interruption costs, should a second transformer be provided to the substation with substandard performance at present.

In this illustrative example, an additional transformer to be added at a substation to serve an area load of 25 MW under contingency operation. The method for the cost–benefit analysis is illustrated in Table 5.2.

Assume that the expected frequency of a transformer outage is 0.2709 occurrence/year. The average length of an outage is 6 h. The expected outage hours avoided due to the installation of a second transformer at the substation is 1.63 h/year. The unserved peak load is 25 MW with a load factor of 0.85. Therefore, the avoidable expected unserved energy is 35 MWh/year. Assuming the customer interruption cost for the mix of customers served by that substation to be $14/kWh yields a cumulative present value of reliability benefits of $6,189,000 over 30 years. The cumulative present value of the capital cost to add a second transformer over 30 years is $1,529,000. For this example, the reliability benefit of adding a second transformer to the substation exceeds the cost of adding a second transformer and may therefore be justified from the customers' perspective.

## 5.9 FINANCIAL RISK ASSESSMENT

### 5.9.1 Basic Concept

Many times the decision maker does not have certainty in the outcome of things and yet has to make a decision. In that situation, the person is taking a risk. Sometimes, this cannot be avoided, as any action including inaction would also involve a risk. Therefore, there should be some understanding of dealing with uncertainties, which is basically a matter of probability.

### 5.9.2 Principles

The underlying principle in risk assessment is one of the expected value. The consequences of all possible outcomes of an action are weighted according to the probabilities of the outcomes and summed together. The result is known as the expected value of the action. This expected value is a representation of the average outcome of a decision made under uncertainty. A typical example in power system application is the loss of load expectation value computed in generation adequacy assessments.

---

## Example 5.7

An investment in a growth stock has a 45% chance of doubling its value, 35% chance of keeping its value and 20% chance of losing out completely. Is it worthwhile to invest?

$$\text{The expected value, } E = 2 \times 45\% + 1 \times 35\% + 0 \times 20\% = 125\%$$

In this case, the expected value is higher than the original investment, so it is profitable.

In a lot of risk assessment situations, the outcomes are much more complex than the above simple illustrative example; however, the principle stays the same.

### 5.9.3 Concept of Risk Aversion

The above-mentioned principle is sound conceptually; however, in real life, there is another factor that needs to be considered: the tendency to avoid risks. For example, an investment with a 55% chance of doubling in value and a 45% chance of losing out completely is mathematically the same as one with a sure gain, that is, 100% chance of 10%; however, just about every investor will prefer the 10% sure gain. In fact, an investor may prefer a 5% sure gain over the 10% expected gains.

The basic reason for this kind of behavior is the diminishing marginal utility of money, that is, additional investments of money becoming increasingly less useful to the investor. This could be due to the cash flow requirements, less actual use for the additional money, or simply psychology; it is, however, widely prevalent. For example, given a choice between a sure $1 and a 50/50 chance of $2 or nothing, a person will choose the sure $1 if the second dollar the person may gain half the time is of less use than the first dollar the person may lose half the time. So to take more of a risk, the person would want a higher rate of return to compensate for it. This kind of attitude is true not only in investment situations but also in other operations decisions. To take this behavior into account, it is necessary to include a risk adjustment factor in the assessment of alternative actions.

## 5.10 CONCLUSIONS

This chapter has presented basic models used in engineering economic analysis. Different principles such as the concepts of interest, rate of return, present value, risk assessment, risk aversion, and so on are discussed with examples. The primary objective is to make engineering economic decisions by choosing from alternatives. The normal approach is to specify each alternative and obtain the best dollar value one can for each and compare, normally choosing the least costly alternative. No attempt was made in this book to provide a chapter with comprehensive economic theories and principles. The readers are encouraged to consult various textbooks on the subject matter.

## REFERENCES


1. M. Kurtz, *Engineering Economics for Professional Engineers' Examinations*, 2nd edition, McGraw-Hill Book Company, Inc., 1975.
2. H. G. Stoll, *Least Cost Electric Utility Planning*, John Wiley & Sons, 1989
3. C. V. Jones, *Financial Risk Analysis of Infrastructure Debt: The Cost of Water and Power Investments*, Quorum Books, 1991.
4. G. W. Smith, *Engineering Economy: Analysis of Capital Expenditures*, 2nd edition, The Iowa State University Press, 1973.
5. T. W. Berrie, *Electricity Economics and Planning*, Peter Peregrinus Ltd., 1992.
6. D. G. Newman, *Engineering Economic Analysis*, Engineering Press, Inc., 1983.
7. IEEE Tutorial Course, Engineering Economic Analysis: Overview and Current Applications, 91EH0345-9-PWR, 1991.
8. N. G. Mankiw, *Principles of Economics*, 2nd edition, Harcourt College Publishers, 2001.
9. R. L. Miller, *Economics Today*, 9th edition, Addison-Wesley Educational Publishers, Inc., 1997.

# 6

# RELIABILITY ANALYSIS OF COMPLEX NETWORK CONFIGURATIONS

## 6.1 INTRODUCTION

The techniques described in Chapters 3 and 4 are limited in their application to systems and networks that have a series and parallel type of design structure. The majority of the systems in the real world either do not have simple structures or operate by complex operational logic. In solving complex networks or systems, additional modeling and evaluation techniques are required to evaluate the reliability of such networks. This chapter will introduce such models for assessing the reliability of complex network configurations. The basic models for complex network solutions will be illustrated using numerical examples.

The reliability analysis of series and parallel system configurations is generally not complicated and depends upon the application of a set of basic equations. As the network configurations become more complicated, these basic series–parallel methodologies become compromised. There are many reliability analytical techniques for evaluating the reliability of complex network configurations. Some of the common methodologies used in practice are (1) state enumeration methods (event space methods), (2) network reduction methods, and (3) path enumeration methods.

*Power Distribution System Reliability.* By Ali A. Chowdhury and Don O. Koval

TABLE 6.1. Number of Possible States for a Given Number of Components

| Number of Components | Number of Possible States |
|---|---:|
| 1 | 2 |
| 2 | 4 |
| 4 | 16 |
| 5 | 32 |
| 10 | 1024 |
| 20 | 1,048,576 |
| 25 | 33,554,432 |
| 50 | 1,125,899,906,842,620 |
| 60 | 1,152,921,504,606,850,000 |
| 75 | 37,778,931,862,957,200,000,000 |
| 90 | 1,237,940,039,285,380,000,000,000,000 |
| 100 | 1,267,650,600,228,230,000,000,000,000,000 |

## 6.2 STATE ENUMERATION METHODOLOGIES

These methods involve defining all possible mutually exclusive states of a system based on the states of its components. A state is defined by listing the successful and failed elements in a system. For a system with $n$ elements or components, there are $2^n$ possible states, so that a system of 5 components would have 32 states. The number of possible states quickly becomes computationally unfeasible for systems with a large number of components, as shown in Table 6.1.

The states that result in successful network system operation are identified, and the probability of occurrence of each successful state is calculated. The reliability of the system is the sum of all the successful state probabilities. The event tree technique is a typical method that uses the state enumeration methodology and is computationally efficient for systems containing a small number of components (e.g., five or fewer).

## Example 6.1 (Event Tree Technique Applied to a Distribution Network)

A utility distribution system single-line diagram is shown in Fig. 6.1.

### 6.2.1 Basic Assumptions: Criteria for System Success—Power is Delivered to All Loads

A utility using the state enumeration methodology for evaluating system reliability, an event tree can be constructed showing all possible combinations of component states in the system as shown in Fig. 6.2. Each cable branch can reside in two mutually exclusive states, either operational (up) or failure (down). There are 32 possible branches in the tree diagram. Exhaustive enumeration of all possible states is, however, often not necessary.

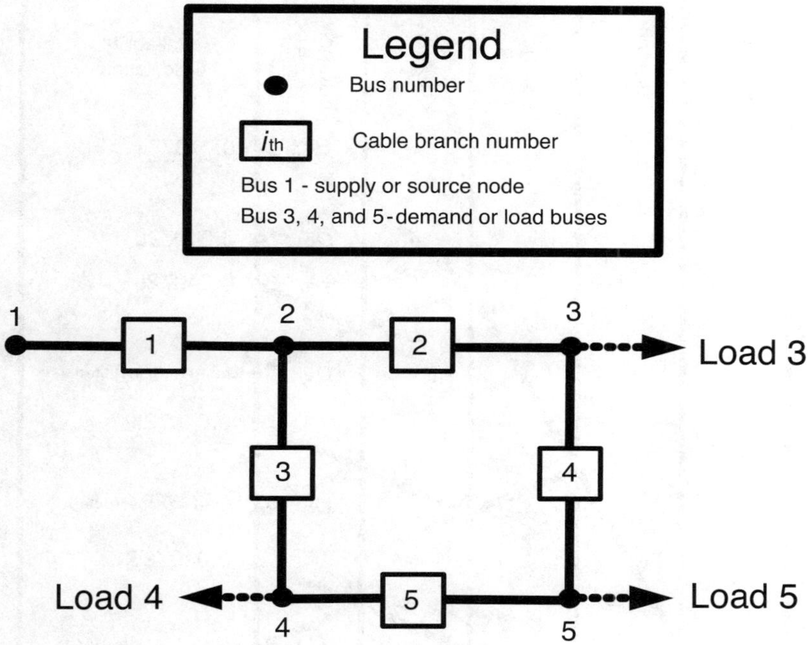


Figure 6.1. An illustrative utility distribution system.

System reliability can be obtained by summing up the probabilities associated with all the operating branches. If the reliability of each cable branch is assumed to be 0.95 then the reliability of the system $R_s$ is

$$R_s = \sum_{i=1}^{5} R(\text{branch}_i) = 0.936682$$

where $n = 5$.

$$R(\text{branch}(1)) = R1 \times R2 \times R3 \times R4 \times R5 = 0.95 \times 0.95 \times 0.95 \times 0.95 \times 0.95 = 0.773781$$
$$R(\text{branch}(2)) = R1 \times R2 \times R3 \times R4 \times Q5 = 0.95 \times 0.95 \times 0.95 \times 0.95 \times 0.05 = 0.040725$$
$$R(\text{branch}(3)) = R1 \times R2 \times R3 \times Q4 \times R5 = 0.95 \times 0.95 \times 0.95 \times 0.05 \times 0.95 = 0.040725$$
$$R(\text{branch}(4)) = R1 \times R2 \times Q3 \times R4 \times R5 = 0.95 \times 0.95 \times 0.05 \times 0.95 \times 0.95 = 0.040725$$
$$R(\text{branch}(5)) = R1 \times Q2 \times R3 \times R4 \times R5 = 0.95 \times 0.05 \times 0.95 \times 0.95 \times 0.95 = 0.040725$$

## Example 6.2 (Event Tree Technique Applied to a Bridge Network)

A utility distribution system single line diagram is shown in Fig. 6.3.

$$R_s = \sum_{i=1}^{5} R(\text{branch}_i) = 0.994781$$

where $n$ is the number of successful branches $= 5$.

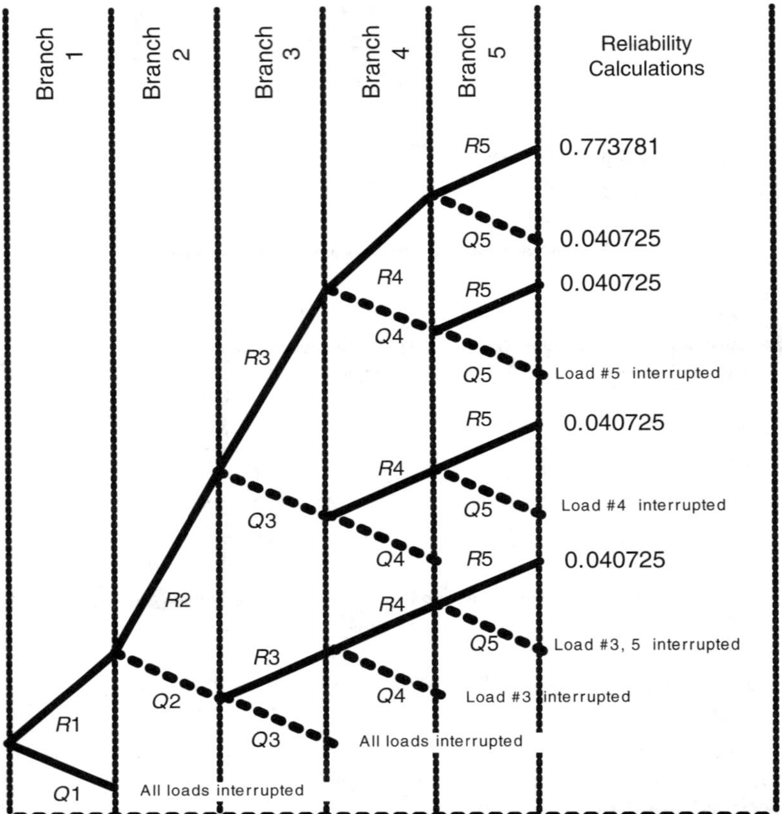


$$R_S = 0.773781 + 0.040725 + 0.040725 + 0.040725 + 0.040725$$
$$= 0.936682$$

Figure 6.2. Event tree diagram for state enumeration of a distribution network.

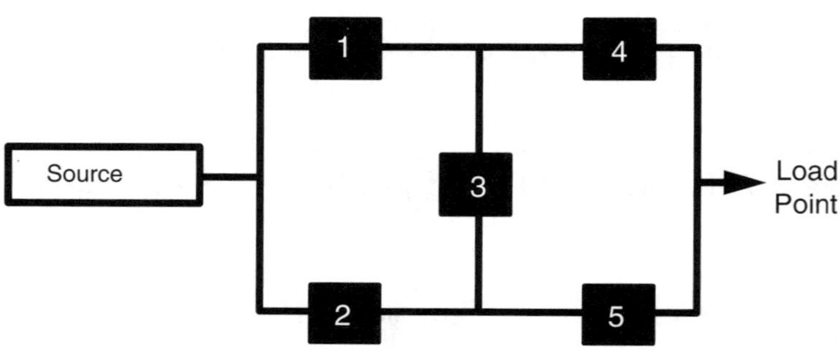


Figure 6.3. An illustrative bridge network.

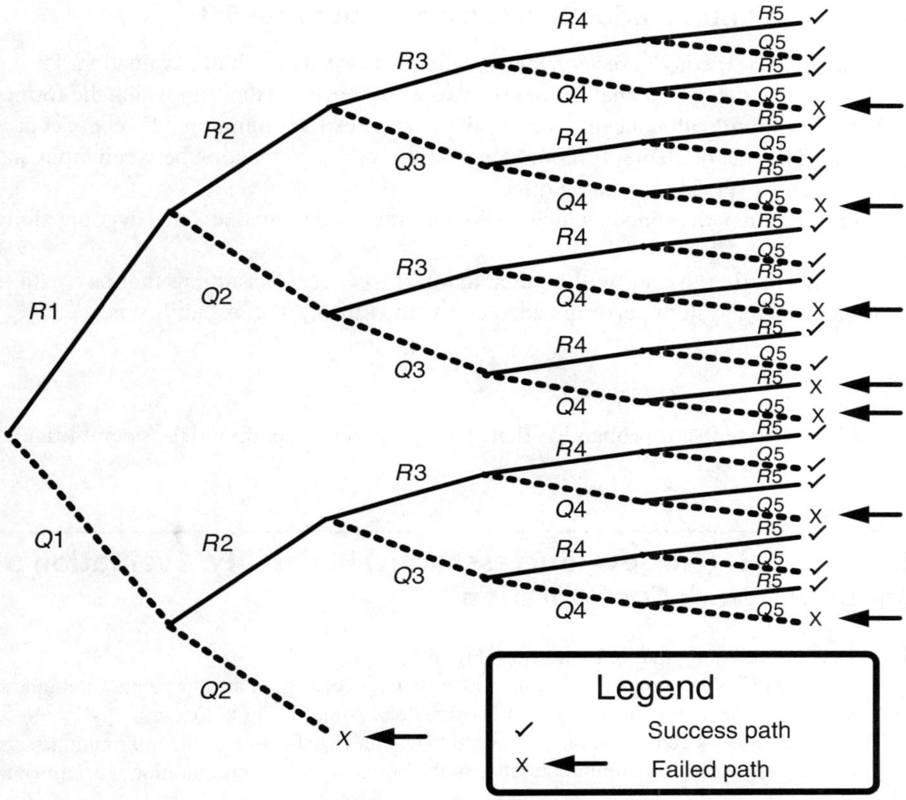


Figure 6.4. Event tree diagram for state enumeration of bridge network.

The reliability of each component is assumed to be 0.95 and $R1 = R2 = R3 = R4 = R5 = R$ and $Q_i = 1 - R_i$ where $i$ is the $i$th component number. $Q = 1 - R$.

The system reliability is given as

$$R_s = R^5 + 5R^4Q + 8R^3Q^2 + 2R^2Q^3 = 2R^2 + 2R^3 - 5R^4 + 2R^5$$
$$R_s = 0.994781$$

The event tree for Fig. 6.3 is illustrated in Fig. 6.4.

## 6.3 NETWORK REDUCTION METHODS

Network reduction methods combine series, parallel, and series–parallel subsystems until a nonseries–parallel system results, which cannot be further reduced. Factoring theorems are then used to obtain the reliability of the system. In general, network reduction methods are useful if the network system under investigation consists of a single-source node and a single-sink node. Multiple sink and source nodes cannot be readily solved by network reduction methodologies.

### 6.3.1 Path Enumeration Methods: Minimum Tie Set

Path enumeration methods are very valuable tools for system reliability evaluation. The tie set analysis and the cut set analysis are the two well-known methods in which the former uses the minimum path concept whereas the latter uses the minimum cut set concept.

A *path* is a set of elements (components) that form a connection between input and output when traversed in a stated direction.

A minimum path is one in which no node is traversed more than once in going along the path.

The *i*th minimum path will be denoted as $T_i$, $i = 1, n$. Assuming that any path is operable and the system performs adequately, then the system reliability is

$$R_s = P\left[\bigcup_{i=1}^{n} T_i\right]$$

where $P$ [ ] represents the probability that at least one of the $n$ paths will be operable and ∪ denotes the union.

---

## Example 6.3 (Tie Sets (Success Paths) Reliability: Evaluation of Bridge Network Configuration)

A bridge network configuration is shown in Fig. 6.3.

A tie set identifies the components within a given network configuration that forms a continuous path that links the source node(s) with the network load point(s) (Fig. 6.5).

The bridge network configuration is repeated to enable visualization of the minimum tie sets.

The success paths (i.e., minimum tie sets) for the bridge network configuration are defined as follows:

T1 = [1,4]—path 1

T2 = [2,5]—path 2

T3 = [1,3,5]—path 3

T4 = [2,3,4]—path 4

A visual description of defining the above minimum tie sets is shown in Fig. 6.6.

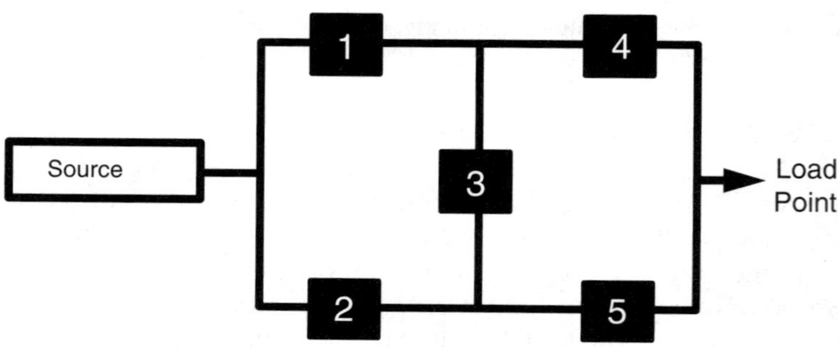


Figure 6.5. Power system bridge network.

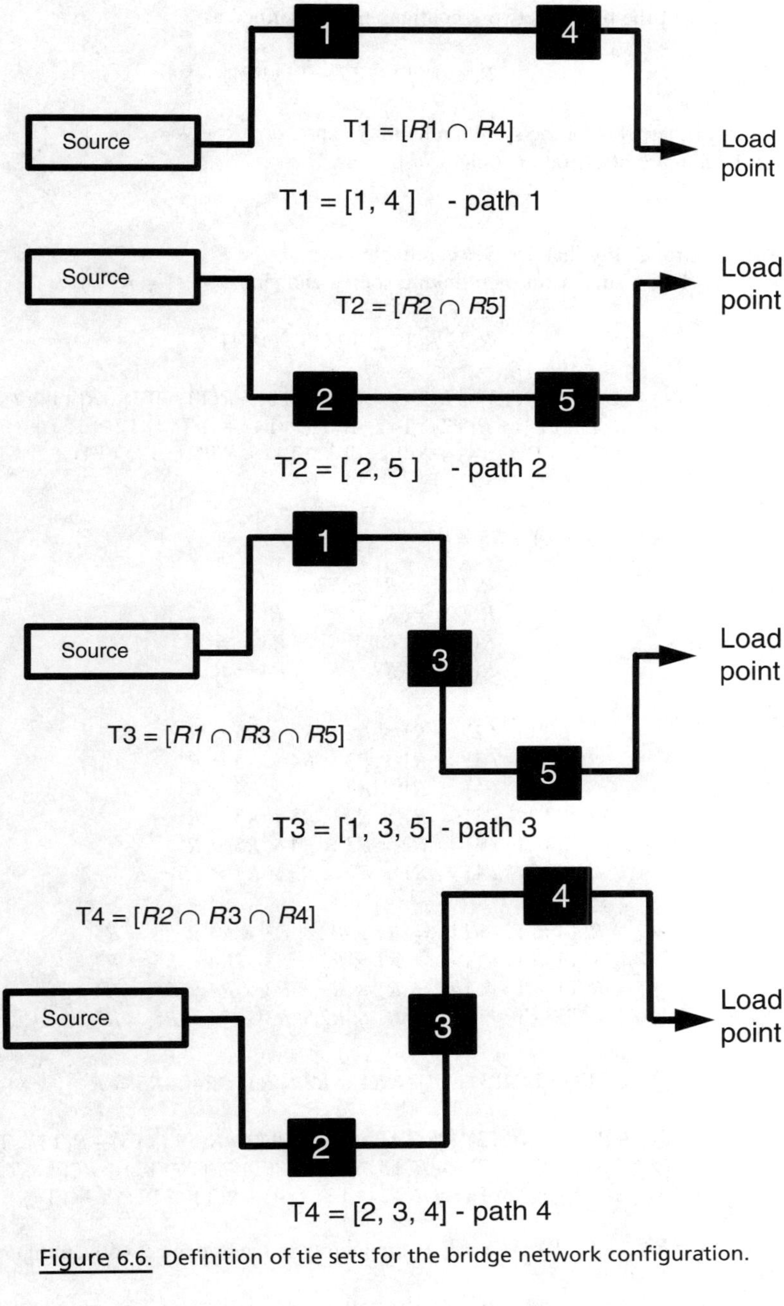


Figure 6.6. Definition of tie sets for the bridge network configuration.

The reliability of the bridge network configuration is defined as

$$R_s = P(\mathrm{T1} \cup \mathrm{T2} \cup \mathrm{T3} \cup \mathrm{T4})$$

The number of terms $N$ for the system reliability expression $R_s$ is $N = 2^n - 1 = 2^4 - 1 = 15$, where $n$ is the number of operational paths or tie sets.

*Note*:

$R(\mathrm{T}i)$ = probability that Tie Set is reliable (i.e., a successful path linking source and sink nodes) = $R(\mathrm{T}i)$

$$R_s = P(\mathrm{T1} \cup \mathrm{T2} \cup \mathrm{T3} \cup \mathrm{T4})$$

$$\begin{aligned} R_s = {} & R(\mathrm{T1}) + R(\mathrm{T2}) + R(\mathrm{T3}) + R(\mathrm{T4}) - R(\mathrm{T1} \cap \mathrm{T2}) - R(\mathrm{T1} \cap \mathrm{T3}) - R(\mathrm{T1} \cap \mathrm{T4}) + \cdots \\ & + \cdots - R(\mathrm{T2} \cap \mathrm{T3}) - R(\mathrm{T2} \cap \mathrm{T4}) - R(\mathrm{T3} \cap \mathrm{T4}) + R(\mathrm{T1} \cap \mathrm{T2} \cap \mathrm{T3}) + \cdots \\ & + \cdots + R(\mathrm{T1} \cap \mathrm{T2} \cap \mathrm{T4}) + R(\mathrm{T1} \cap \mathrm{T3} \cap \mathrm{T4}) + R(\mathrm{T2} \cap \mathrm{T3} \cap \mathrm{T4}) + \cdots \\ & + \cdots - R(\mathrm{T1} \cap \mathrm{T2} \cap \mathrm{T3} \cap \mathrm{T4}) \end{aligned}$$

Assuming $R1 = R2 = R3 = \mathrm{R}4 = R5 = R$

$$\begin{aligned} R(T1) &= R1 \times R2 = R^2 \\ R(T2) &= R2 \times R5 = R^2 \\ R(T3) &= R1 \times R3 \times R5 = R^3 \\ R(T4) &= R2 \times R3 \times R4 = R^3 \end{aligned}$$

$$\begin{aligned} R(\mathrm{T1} \cap T2) &= R1 \times R2 \times R4 \times R5 = R^4 \\ R(\mathrm{T1} \cap T3) &= R1 \times R3 \times R4 \times R5 = R^4 \\ R(\mathrm{T1} \cap T4) &= R1 \times R2 \times R3 \times R4 = R^4 \\ R(\mathrm{T2} \cap T3) &= R1 \times R2 \times R3 \times R5 = R^4 \\ R(\mathrm{T2} \cap T4) &= R2 \times R3 \times R4 \times R5 = R^4 \\ R(\mathrm{T3} \cap T4) &= R1 \times R2 \times R3 \times R4 \times R5 = R^5 \end{aligned}$$

$$\begin{aligned} R(\mathrm{T1} \cap \mathrm{T2} \cap \mathrm{T3}) &= R1 \times R2 \times R3 \times R4 \times R5 = R^5 \\ R(\mathrm{T1} \cap \mathrm{T2} \cap \mathrm{T4}) &= R1 \times R2 \times R3 \times R4 \times R5 = R^5 \\ R(\mathrm{T1} \cap \mathrm{T3} \cap \mathrm{T4}) &= R1 \times R2 \times R3 \times R4 \times R5 = R^5 \\ R(\mathrm{T2} \cap \mathrm{T3} \cap \mathrm{T4}) &= R1 \times R2 \times R3 \times R4 \times R5 = R^5 \end{aligned}$$

$$R(\mathrm{T1} \cap \mathrm{T2} \cap \mathrm{T3} \cap \mathrm{T4}) = R1 \times R2 \times R3 \times R4 \times R5 = R^5$$

$$\begin{aligned} R_s = {} & R(\mathrm{T1}) + R(\mathrm{T2}) + R(\mathrm{T3}) + R(\mathrm{T4}) - R(\mathrm{T1} \cap \mathrm{T2}) - R(\mathrm{T1} \cap \mathrm{T3}) - R(\mathrm{T1} \cap \mathrm{T4}) \cdots \\ & R(\mathrm{T2} \cap \mathrm{T3}) - R(\mathrm{T2} \cap \mathrm{T4}) - R(\mathrm{T3} \cap \mathrm{T4}) + R(\mathrm{T1} \cap \mathrm{T2} \cap \mathrm{T3}) + R(\mathrm{T1} \cap \mathrm{T2} \\ & \cap \mathrm{T4}) + R(\mathrm{T1} \cap \mathrm{T3} \cap \mathrm{T4}) + R(\mathrm{T2} \cap \mathrm{T3} \cap \mathrm{T4}) - R(\mathrm{T1} \cap \mathrm{T2} \cap \mathrm{T3} \cap \mathrm{T4}) \end{aligned}$$

Assuming $R1 = R2 = R3 = R4 = R5 = R$, the reliability of the bridge network configuration is

$$R_s = 2R^2 + 2R^3 - 5R^4 + 2R^5$$

If $R = 0.95$, then $R_s = 0.994781$.

## Example 6.4 (Tie Sets (Success Paths) Reliability Methodology Applied to the Distribution Network)

A utility distribution system single-line diagram, as shown in Fig. 6.1, is repeated below and the tie sets reliability methodology is illustrated using this network (Fig. 6.7).

As was the case for the state enumeration approach presented in Section 6.1, the criterion for system success is power delivered to all loads.

The minimum tie set (or paths) for the above network is defined as follows:

T1 = [1, 2, 4, 5]—path 1: component 3 failed

T2 = [1, 3, 4, 5]—path 2: component 2 failed

T3 = [1, 2, 3, 4]—path 3: component 5 failed

T4 = [1, 2, 3, 5]—path 4: component 4 failed

A visual description of defining the above minimum tie sets is shown in Figs. 6.8–6.11.

$$R_s = R(\mathrm{T1} \cup \mathrm{T2} \cup \mathrm{T3} \cup \mathrm{T4})$$

$$\begin{aligned} R_s = {} & R(\mathrm{T1}) + R(\mathrm{T2}) + R(\mathrm{T3}) + R(\mathrm{T4}) - R(\mathrm{T1} \cap \mathrm{T2}) - R(\mathrm{T1} \cap \mathrm{T3}) - R(\mathrm{T1} \cap \mathrm{T4}) + \cdots \\ & + \cdots - R(\mathrm{T2} \cap \mathrm{T3}) - R(\mathrm{T2} \cap \mathrm{T4}) - R(\mathrm{T3} \cap \mathrm{T4}) + R(\mathrm{T1} \cap \mathrm{T2} \cap \mathrm{T3}) + \cdots \\ & + \cdots + R(\mathrm{T1} \cap \mathrm{T2} \cap \mathrm{T4}) + R(\mathrm{T1} \cap \mathrm{T3} \cap \mathrm{T4}) + R(\mathrm{T2} \cap \mathrm{T3} \cap \mathrm{T4}) + \cdots \\ & + \cdots - R(\mathrm{T1} \cap \mathrm{T2} \cap \mathrm{T3} \cap \mathrm{T4}) \end{aligned}$$

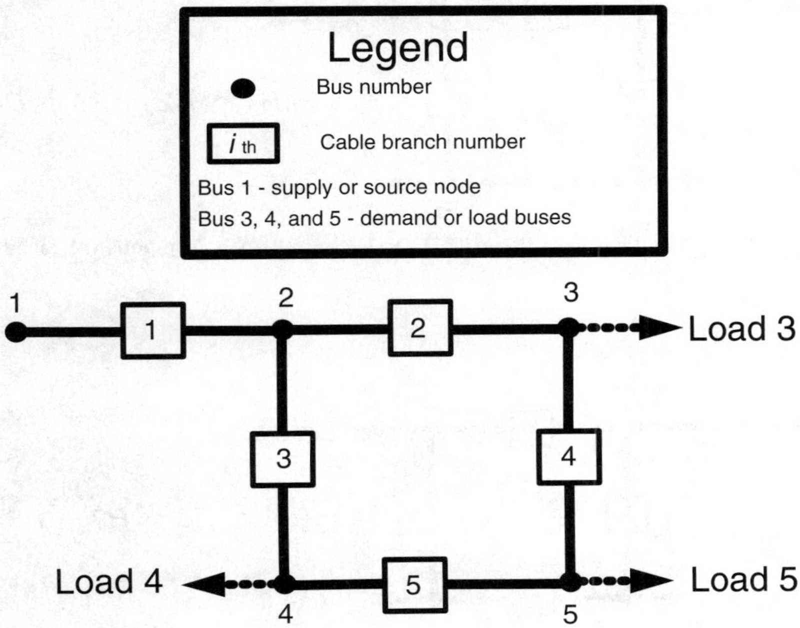


Figure 6.7. A bridge network configuration.

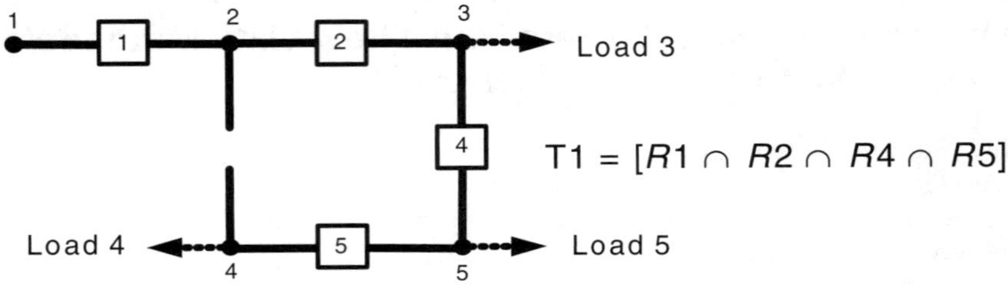


Figure 6.8. Definition of tie set T1. T1 = [1,2,4,5]—Path 1...component 3 failed.

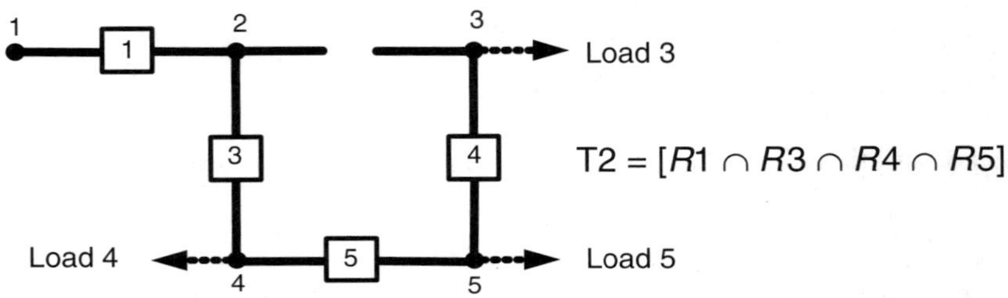


Figure 6.9. Definition of tie set T2. T2 = [1,3,4,5]—Path 2...component 2 failed.

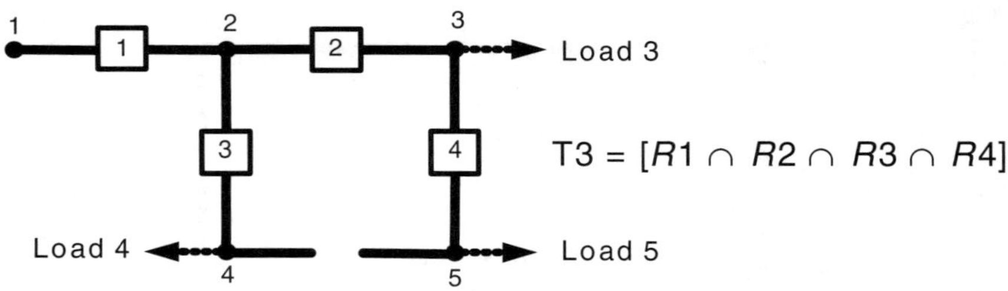


Figure 6.10. Definition of tie set T3. T3 = [1,2,3,4]—Path 3...component 5 failed.

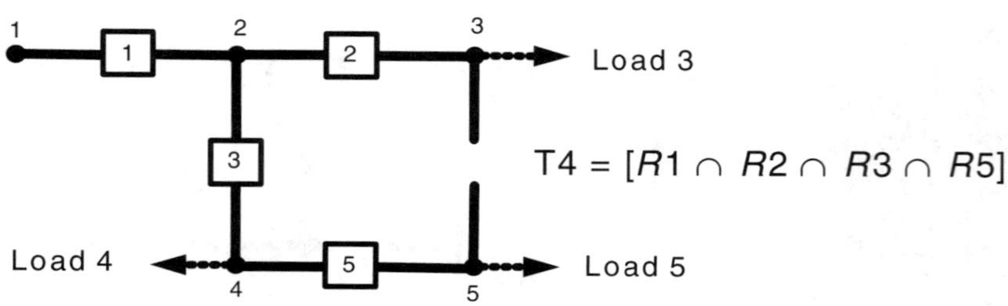


Figure 6.11. Definition of tie set T4. T4 = [1,2,3,5]—Path 4...component 4 failed.

Assuming $R1 = R2 = R3 = R4 = R5 = R$

$$\begin{aligned}
R(\mathrm{T1}) &= R1 \times R2 \times R4 \times R5 = R^4 \\
R(\mathrm{T2}) &= R1 \times R3 \times R4 \times R5 = R^4 \\
R(\mathrm{T3}) &= R1 \times R2 \times R3 \times R4 = R^4 \\
R(\mathrm{T4}) &= R1 \times R2 \times R3 \times R5 = R^4
\end{aligned}$$

$$\begin{aligned}
R(\mathrm{T1} \cap \mathrm{T2}) &= R1 \times R2 \times R3 \times R4 \times R5 = R^5 \\
R(\mathrm{T1} \cap \mathrm{T3}) &= R1 \times R2 \times R3 \times R4 \times R5 = R^5 \\
R(\mathrm{T1} \cap \mathrm{T4}) &= R1 \times R2 \times R3 \times R4 \times R5 = R^5 \\
R(\mathrm{T2} \cap \mathrm{T3}) &= R1 \times R2 \times R3 \times R4 \times R5 = R^5 \\
R(\mathrm{T2} \cap \mathrm{T4}) &= R1 \times R2 \times R3 \times R4 \times R5 = R^5 \\
R(\mathrm{T3} \cap \mathrm{T4}) &= R1 \times R2 \times R3 \times R4 \times R5 = R^5
\end{aligned}$$

$$\begin{aligned}
R(\mathrm{T1} \cap \mathrm{T2} \cap \mathrm{T3}) &= R1 \times R2 \times R3 \times R4 \times R5 = R^5 \\
R(\mathrm{T1} \cap \mathrm{T2} \cap \mathrm{T4}) &= R1 \times R2 \times R3 \times R4 \times R5 = R^5 \\
R(\mathrm{T1} \cap \mathrm{T3} \cap \mathrm{T4}) &= R1 \times R2 \times R3 \times R4 \times R5 = R^5 \\
R(\mathrm{T2} \cap \mathrm{T3} \cap \mathrm{T4}) &= R1 \times R2 \times R3 \times R4 \times R5 = R^5 \\
R(\mathrm{T1} \cap \mathrm{T2} \cap \mathrm{T3} \cap \mathrm{T4}) &= R1 \times R2 \times R3 \times R4 \times R5 = R^5
\end{aligned}$$

$$R_s = 4R^4 - 6R^5 + 4R^5 - R^5 = 4R^4 - 3R^5$$

If $R = 0.95$, then

$$\begin{aligned}
R_s &= 4(0.814506) - 3(0.773781) \\
&= 0.936682
\end{aligned}$$

### 6.3.2 Path Enumeration Methods: Minimum Cut Set

A *cut set* is defined as a set of elements that, if fails, causes the system to fail regardless of the condition of the other elements in the system. A *minimum cut set* is one in which there is no proper subset of elements whose failure alone will cause the system to fail.

A minimal cut set is such that if any component is removed from the set, the remaining elements collectively are no longer a cut set. The $i$th minimum path of possible $n$ paths will be denoted as $C_i$. Assuming that any path is operable and the system performs adequately, then the system reliability is

$$R_s = 1 - P\left[\bigcup_{i=1}^{n} C_i\right] \quad \text{or} \quad Q_s = P\left[\bigcup_{i=1}^{n} C_i\right]$$

where $P[\,]$ represents the probability that at least one of the $n$ paths will have failed and $\cup$ denotes the union.

## Example 6.5 (Cut Sets (Failure Paths) Reliability Methodology Applied to the Distribution Network)

The cut sets reliability methodology is illustrated using the utility distribution system as shown in Fig. 6.12.

The criterion for system success is power delivered to all loads.

Based on the system reliability criteria, the minimum cut sets for the above network are

$C_1 = [1]$—component 1 failed

$C_2 = [2, 3]$—components 2 and 3 failed

$C_3 = [2, 4]$—components 2 and 4 failed

$C_4 = [2, 5]$—components 2 and 5 failed

$C_5 = [3, 4]$—components 3 and 4 failed

$C_6 = [3, 5]$—components 3 and 5 failed

$C_7 = [4, 5]$—components 4 and 5 failed

A visual description of defining the above cut sets is shown in Figs. 6.13–6.20.

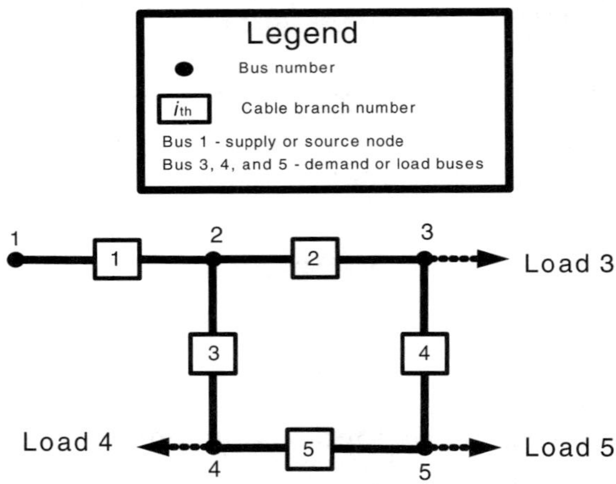


Figure 6.12. A bridge network configuration.

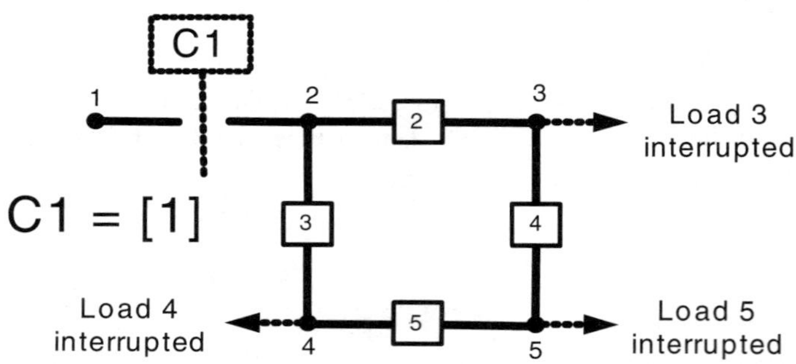


Figure 6.13. Definition of cut set $C_1$. $C_1 = [1]$...component 1 failed.

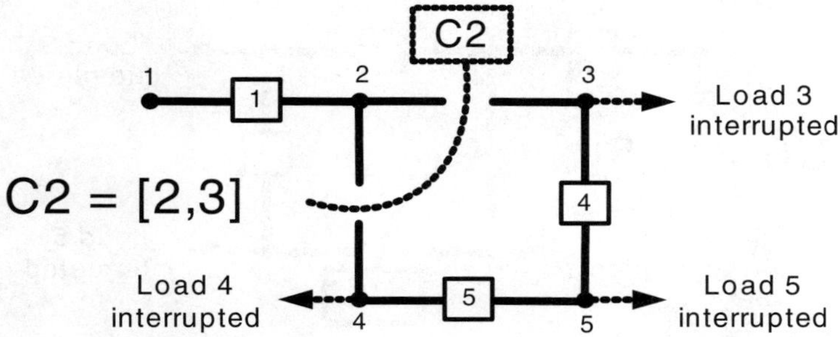


Figure 6.14. Definition of cut set $C_2$. $C_2 = [2,3]$...components 2 and 3 failed.

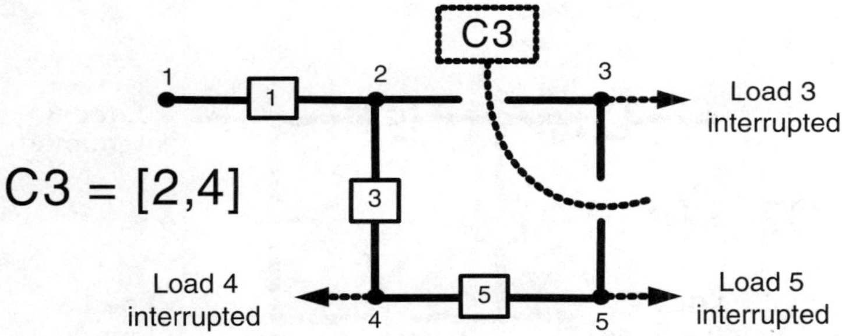


Figure 6.15. Definition of cut set $C_3$. $C_3 = [2,4]$...components 2 and 4 failed.

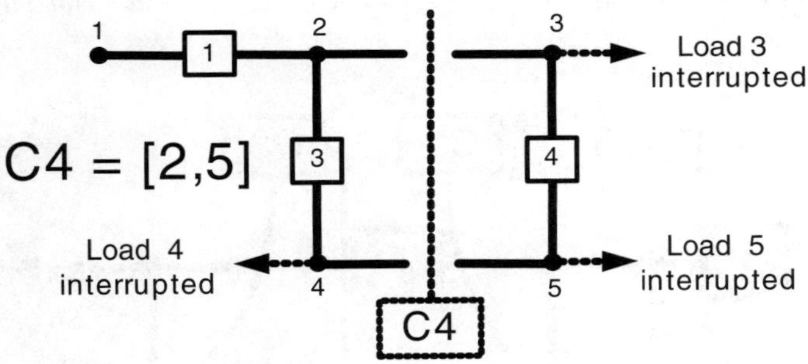


Figure 6.16. Definition of cut set $C_4$. $C_4 = [2,5]$...components 2 and 5 failed.

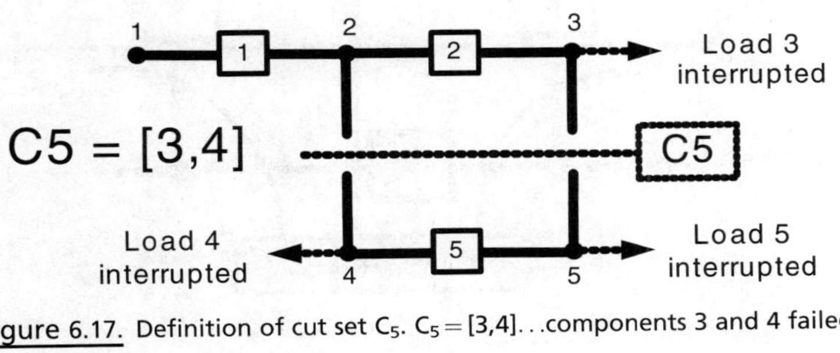


Figure 6.17. Definition of cut set $C_5$. $C_5 = [3,4]$...components 3 and 4 failed.

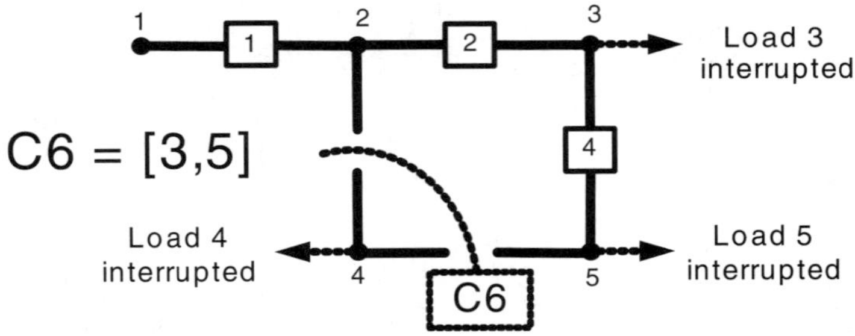


Figure 6.18. Definition of cut set $C_6$. $C_6 = [3,5]$...components 3 and 5 failed.

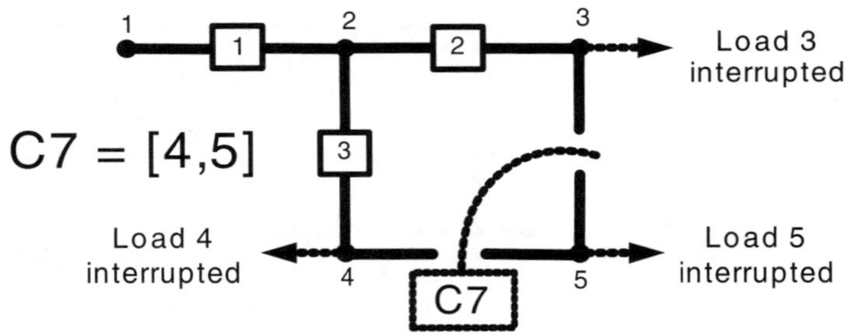


Figure 6.19. Definition of cut set $C_7$. $C_7 = [4,5]$...components 4 and 5 failed.

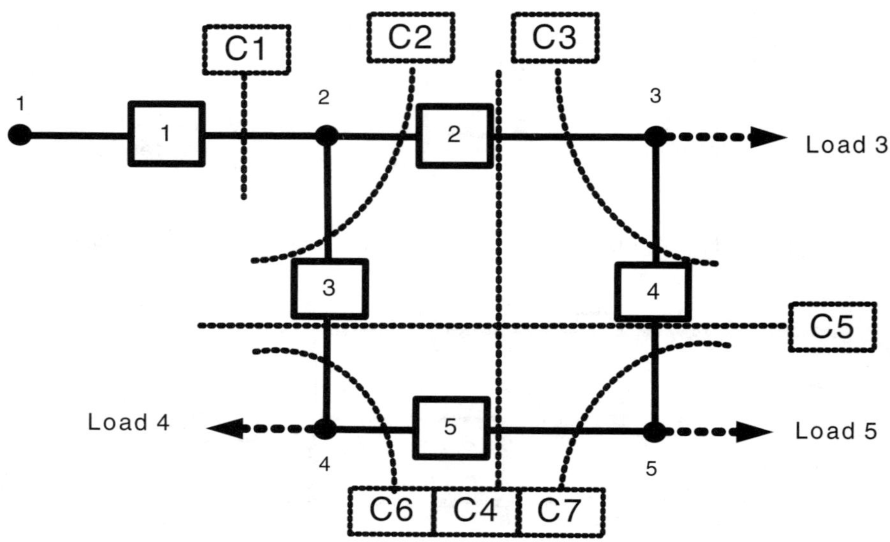


Figure 6.20. Definition of all cuts.

Given

$C_1 = [1]$—component 1 failed

$C_2 = [2, 3]$—components 2 and 3 failed

$C_3 = [2, 4]$—components 2 and 4 failed

$C_4 = [2, 5]$—components 2 and 5 failed

$C_5 = [3, 4]$—components 3 and 4 failed

$C_6 = [3, 5]$—components 3 and 5 failed

$C_7 = [4, 5]$—components 4 and 5 failed

$$R_s = 1 - P\left[\bigcup_{i=1}^{n} C_i\right] \quad \text{or} \quad Q_s = P\left[\bigcup_{i=1}^{n} C_i\right]$$

The above cut set can be summarized in Fig. 6.20.

Given

$C_1 = [1]$—component 1 failed

$C_2 = [2, 3]$—components 2 and 3 failed

$C_3 = [2, 4]$—components 2 and 4 failed

$C_4 = [2, 5]$—components 2 and 5 failed

$C_5 = [3, 4]$—components 3 and 4 failed

$C_6 = [3, 5]$—components 3 and 5 failed

$C_7 = [4, 5]$—components 4 and 5 failed

$$Q_s = P\left[\bigcup_{i=1}^{n} C_i\right]$$

### Seven First-Order Cut Combinations and Their Probability Expressions

| $Q(C_1)$ | $Q(C_2)$ | $Q(C_3)$ | $Q(C_4)$ | $Q(C_5)$ | $Q(C_6)$ | $Q(C_7)$ |
|---|---|---|---|---|---|---|
| $Q_1$ | $Q_2 \times Q_3$ | $Q_2 \times Q_4$ | $Q_2 \times Q_5$ | $Q_3 \times Q_4$ | $Q_3 \times Q_5$ | $Q_4 \times 5$ |

### Twenty-One Second-Order Cut Combinations and Their Probability Expressions

| | | | | | |
|---|---|---|---|---|---|
| $Q(C_1 \cap C_2)$<br>$Q_1 \times Q_2 \times Q_3$ | $Q(C_1 \cap C_3)$<br>$Q_1 \times Q_2 \times Q_4$ | $Q(C_1 \cap C_4)$<br>$Q_1 \times Q_2 \times Q_5$ | $Q(C_1 \cap C_5)$<br>$Q_1 \times Q_3 \times Q_4$ | $Q(C_1 \cap C_6)$<br>$Q_1 \times Q_3 \times Q_5$ | $Q(C_1 \cap C_7)$<br>$Q_1 \times Q_4 \times Q_5$ |
| | $Q(C_2 \cap C_3)$<br>$Q_2 \times Q_3 \times Q_4$ | $Q(C_2 \cap C_4)$<br>$Q_2 \times Q_3 \times Q_5$ | $Q(C_2 \cap C_5)$<br>$Q_2 \times Q_3 \times Q_4$ | $Q(C_2 \cap C_6)$<br>$Q_2 \times Q_3 \times Q_5$ | $Q(C_2 \cap C_7)$<br>$Q_2 \times Q_3 \times Q_4 \times Q_5$ |
| | | $Q(C_3 \cap C_4)$<br>$Q_2 \times Q_4 \times Q_5$ | $Q(C_3 \cap C_5)$<br>$Q_2 \times Q_3 \times Q_5$ | $Q(C_3 \cap C_6)$<br>$Q_2 \times Q_3 \times Q_4 \times Q_5$ | $Q(C_3 \cap C_7)$<br>$Q_2 \times Q_4 \times Q_5$ |
| | | | $Q(C_4 \cap C_5)$<br>$Q_2 \times Q_3 \times Q_4 \times Q_5$ | $Q(C_4 \cap C_6)$<br>$Q_2 \times Q_3 \times Q_5$ | $Q(C_4 \cap C_7)$<br>$Q_2 \times Q_4 \times Q_5$ |
| | | | | $Q(C_5 \cap C_6)$<br>$Q_3 \times Q_4 \times Q_5$ | $Q(C_5 \cap C_7)$<br>$Q_3 \times Q_4 \times Q_5$ |
| | | | | | $Q(C_6 \cap C_7)$<br>$Q_3 \times Q_4 \times Q_5$ |

### Thirty-Five Third-Order Cut Combinations and Their Probability Expressions

| | | | | |
|---|---|---|---|---|
| $Q(C_1 \cap C_2 \cap C_3)$ | $Q(C_1 \cap C_2 \cap C_4)$ | $Q(C_1 \cap C_2 \cap C_5)$ | $Q(C_1 \cap C_2 \cap C_6)$ | $Q(C_1 \cap C_2 \cap C_7)$ |
| $Q_1 \times Q_2$ | $Q_1 \times Q_2 \times Q_3 \times Q_5$ | $Q_1 \times Q_2 \times Q_3 \times Q_4$ | $Q_1 \times Q_2 \times Q_3 \times Q_5$ | $Q_1 \times Q_2 \times Q_3$ |
| $\times Q_3 \times Q_4$ | | | | $\times Q_4 \times Q_5$ |
| $\times$ | $Q(C_1 \cap C_3 \cap C_4)$ | $Q(C_1 \cap C_3 \cap C_5)$ | $Q(C_1 \cap C_3 \cap C_6)$ | $Q(C_1 \cap C_3 \cap C_7)$ |
| | $Q_1 \times Q_2 \times Q_4 \times Q_5$ | $Q_1 \times Q_2 \times Q_3 \times Q_4$ | $Q_1 \times Q_2 \times Q_3$ | $Q_1 \times Q_2 \times Q_4 \times Q_5$ |
| | | | $\times Q_4 \times Q_5$ | |
| | | $Q(C_1 \cap C_4 \cap C_5)$ | $Q(C_1 \cap C_4 \cap C_6)$ | $Q(C_1 \cap C_4 \cap C_7)$ |
| | | $Q_1 \times Q_2 \times Q_3$ | $Q_1 \times Q_2 \times Q_3 \times Q_5$ | $Q_1 \times Q_2 \times Q_4 \times Q_5$ |
| | | $\times Q_4 \times Q_5$ | | |
| | | | $Q(C_1 \cap C_5 \cap C_6)$ | $Q(C_1 \cap C_5 \cap C_7)$ |
| | | | $Q_1 \times Q_3 \times Q_4 \times Q_5$ | $Q_1 \times Q_3 \times Q_4 \times Q_5$ |
| | | | | $Q(C_1 \cap C_6 \cap C_7)$ |
| | | | | $Q_1 \times Q_3 \times Q_4 \times Q_5$ |
| | $Q(C_2 \cap C_3 \cap C_4)$ | $Q(C_2 \cap C_3 \cap C_5)$ | $Q(C_2 \cap C_3 \cap C_6)$ | $Q(C_2 \cap C_3 \cap C_7)$ |
| | $Q_2 \times Q_3 \times Q_4 \times Q_5$ | $Q_2 \times Q_3 \times Q_4$ | $Q_2 \times Q_3 \times Q_4 \times Q_5$ | $Q_2 \times Q_3 \times Q_4 \times Q_5$ |
| | | $Q(C_2 \cap C_4 \cap C_5)$ | $Q(C_2 \cap C_4 \cap C_6)$ | $Q(C_2 \cap C_4 \cap C_7)$ |
| | | $Q_2 \times Q_3 \times Q_4 \times Q_5$ | $Q_2 \times Q_3 \times Q_5$ | $Q_2 \times Q_3 \times Q_4 \times Q_5$ |
| | | | $Q(C_2 \cap C_5 \cap C_6)$ | $Q(C_2 \cap C_5 \cap C_7)$ |
| | | | $Q_2 \times Q_3 \times Q_4 \times Q_5$ | $Q_2 \times Q_3 \times Q_4 \times Q_5$ |
| | | | | $Q(C_2 \cap C_6 \cap C_7)$ |
| | | | | $Q_2 \times Q_3 \times Q_4 \times Q_5$ |
| | | $Q(C_3 \cap C_4 \cap C_5)$ | $Q(C_3 \cap C_4 \cap C_6)$ | $Q(C_3 \cap C_4 \cap C_7)$ |
| | | $Q_2 \times Q_3 \times Q_4 \times Q_5$ | $Q_2 \times Q_3 \times Q_4 \times Q_5$ | $Q_2 \times Q_4 \times Q_5$ |
| | | | $Q(C_3 \cap C_5 \cap C_6)$ | $Q(C_3 \cap C_5 \cap C_7)$ |
| | | | $Q_2 \times Q_3 \times Q_4 \times Q_5$ | $Q_2 \times Q_3 \times Q_4 \times Q_5$ |
| | | | | $Q(C_3 \cap C_6 \cap C_7)$ |
| | | | | $Q_2 \times Q_3 \times Q_4 \times Q_5$ |
| | | | $Q(C_4 \cap C_5 \cap C_6)$ | $Q(C_4 \cap C_5 \cap C_7)$ |
| | | | $Q_2 \times Q_3 \times Q_4 \times Q_5$ | $Q_2 \times Q_3 \times Q_4 \times Q_5$ |
| | | | | $Q(C_4 \cap C_6 \cap C_7)$ |
| | | | | $Q_2 \times Q_3 \times Q_4 \times Q_5$ |
| | | | | $Q(C_5 \cap C_6 \cap C_7)$ |
| | | | | $Q_3 \times Q_4 \times Q_5$ |

### Thirty-Five Fourth-Order Cut Combinations and Their Probability Expressions

| | | | |
|---|---|---|---|
| $Q(C_1 \cap C_2 \cap C_3 \cap C_4)$ | $Q(C_1 \cap C_2 \cap C_3 \cap C_5)$ | $Q(C_1 \cap C_2 \cap C_3 \cap C_6)$ | $Q(C_1 \cap C_2 \cap C_3 \cap C_7)$ |
| $Q_1 \times Q_2 \times Q_3 \times Q_4 \times Q_5$ | $Q_1 \times Q_2 \times Q_3 \times Q_4$ | $Q_1 \times Q_2 \times Q_3 \times Q_4 \times Q_5$ | $Q_1 \times Q_2 \times Q_3 \times Q_4 \times Q_5$ |
| | $Q(C_1 \cap C_2 \cap C_4 \cap C_5)$ | $Q(C_1 \cap C_2 \cap C_4 \cap C_6)$ | $Q(C_1 \cap C_2 \cap C_4 \cap C_7)$ |
| | $Q_1 \times Q_2 \times Q_3 \times Q_4 \times Q_5$ | $Q_1 \times Q_2 \times Q_3 \times Q_5$ | $Q_1 \times Q_2 \times Q_3 \times Q_4 \times Q_5$ |
| | | $Q(C_1 \cap C_2 \cap C_5 \cap C_6)$ | $Q(C_1 \cap C_2 \cap C_5 \cap C_7)$ |
| | | $Q_1 \times Q_2 \times Q_3 \times Q_4 \times Q_5$ | $Q_1 \times Q_2 \times Q_3 \times Q_4 \times Q_5$ |
| | | | $Q(C_1 \cap C_2 \cap C_6 \cap C_7)$ |
| | | | $Q_1 \times Q_2 \times Q_3 \times Q_4 \times Q_5$ |
| | $Q(C_1 \cap C_3 \cap C_4 \cap C_5)$ | $Q(C_1 \cap C_3 \cap C_4 \cap C_6)$ | $Q(C_1 \cap C_3 \cap C_4 \cap C_7)$ |
| | $Q_1 \times Q_2 \times Q_3 \times Q_4 \times Q_5$ | $Q_1 \times Q_2 \times Q_3 \times Q_4 \times Q_5$ | $Q_1 \times Q_2 \times Q_4 \times Q_5$ |
| | | $Q(C_1 \cap C_3 \cap C_5 \cap C_6)$ | $Q(C_1 \cap C_3 \cap C_5 \cap C_7)$ |
| | | $Q_1 \times Q_2 \times Q_3 \times Q_4 \times Q_5$ | $Q_1 \times Q_2 \times Q_3 \times Q_4 \times Q_5$ |
| | | | $Q(C_1 \cap C_3 \cap C_6 \cap C_7)$ |
| | | | $Q_1 \times Q_2 \times Q_3 \times Q_4 \times Q_5$ |

*(continued)*

| | | |
|---|---|---|
| | $Q(C_1 \cap C_4 \cap C_5 \cap C_6)$ | $Q(C_1 \cap C_4 \cap C_5 \cap C_7)$ |
| | $Q_1 \times Q_2 \times Q_3 \times Q_4 \times Q_5$ | $Q_1 \times Q_2 \times Q_3 \times Q_4 \times Q_5$ |
| | | $Q(C_1 \cap C_4 \cap C_6 \cap C_7)$ |
| | | $Q_1 \times Q_2 \times Q_3 \times Q_4 \times Q_5$ |
| | | $Q(C_1 \cap C_5 \cap C_6 \cap C_7)$ |
| | | $Q_1 \times Q_3 \times Q_4 \times Q_5$ |
| $Q(C_2 \cap C_3 \cap C_4 \cap C_5)$ | $Q(C_2 \cap C_3 \cap C_4 \cap C_6)$ | $Q(C_2 \cap C_3 \cap C_4 \cap C_7)$ |
| $Q_2 \times Q_3 \times Q_4 \times Q_5$ | $Q_2 \times Q_3 \times Q_4 \times Q_5$ | $Q_2 \times Q_3 \times Q_4 \times Q_5$ |
| | $Q(C_2 \cap C_3 \cap C_5 \cap C_6)$ | $Q(C_2 \cap C_3 \cap C_5 \cap C_7)$ |
| | $Q_2 \times Q_3 \times Q_4 \times Q_5$ | $Q_2 \times Q_3 \times Q_4 \times Q_5$ |
| | | $Q(C_2 \cap C_3 \cap C_6 \cap C_7)$ |
| | | $Q_2 \times Q_3 \times Q_4 \times Q_5$ |
| | $Q(C_2 \cap C_4 \cap C_5 \cap C_6)$ | $Q(C_2 \cap C_4 \cap C_5 \cap C_7)$ |
| | $Q_2 \times Q_3 \times Q_4 \times Q_5$ | $Q_2 \times Q_3 \times Q_4 \times Q_5$ |
| | | $Q(C_2 \cap C_4 \cap C_6 \cap C_7)$ |
| | | $Q_2 \times Q_3 \times Q_4 \times Q_5$ |
| | | $Q(C_2 \cap C_5 \cap C_6 \cap C_7)$ |
| | | $Q_2 \times Q_3 \times Q_4 \times Q_5$ |
| | $Q(C_3 \cap C_4 \cap C_5 \cap C_6)$ | $Q(C_3 \cap C_4 \cap C_5 \cap C_7)$ |
| | $Q_2 \times Q_3 \times Q_4 \times Q_5$ | $Q_2 \times Q_3 \times Q_4 \times Q_5$ |
| | | $Q(C_3 \cap C_4 \cap C_6 \cap C_7)$ |
| | | $Q_2 \times Q_3 \times Q_4 \times Q_5$ |
| | | $Q(C_3 \cap C_5 \cap C_6 \cap C_7)$ |
| | | $Q_2 \times Q_3 \times Q_4 \times Q_5$ |
| | | $Q(C_4 \cap C_5 \cap C_6 \cap C_7)$ |
| | | $Q_2 \times Q_3 \times Q_4 \times Q_5$ |

## Twenty-One Fifth-Order Cut Combinations and Their Probability Expressions

| | | |
|---|---|---|
| $Q(C_1 \cap C_2 \cap C_3 \cap C_4 \cap C_5)$ | $Q(C_1 \cap C_2 \cap C_3 \cap C_4 \cap C_6)$ | $Q(C_1 \cap C_2 \cap C_3 \cap C_4 \cap C_7)$ |
| $Q_1 \times Q_2 \times Q_3 \times Q_4 \times Q_5$ | $Q_1 \times Q_2 \times Q_3 \times Q_4 \times Q_5$ | $Q_1 \times Q_2 \times Q_3 \times Q_4 \times Q_5$ |
| | $Q(C_1 \cap C_2 \cap C_3 \cap C_5 \cap C_6)$ | $Q(C_1 \cap C_2 \cap C_3 \cap C_5 \cap C_7)$ |
| | $Q_1 \times Q_2 \times Q_3 \times Q_4 \times Q_5$ | $Q_1 \times Q_2 \times Q_3 \times Q_4 \times Q_5$ |
| | | $Q(C_1 \cap C_2 \cap C_3 \cap C_6 \cap C_7)$ |
| | | $Q_1 \times Q_2 \times Q_3 \times Q_4 \times Q_5$ |
| | $Q(C_1 \cap C_2 \cap C_4 \cap C_5 \cap C_6)$ | $Q(C_1 \cap C_2 \cap C_4 \cap C_5 \cap C_7)$ |
| | $Q_1 \times Q_2 \times Q_3 \times Q_4 \times Q_5$ | $Q_1 \times Q_2 \times Q_3 \times Q_4 \times Q_5$ |
| | | $Q(C_1 \cap C_2 \cap C_4 \cap C_6 \cap C_7)$ |
| | | $Q_1 \times Q_2 \times Q_3 \times Q_4 \times Q_5$ |
| | | $Q(C_1 \cap C_2 \cap C_5 \cap C_6 \cap C_7)$ |
| | | $Q_1 \times Q_2 \times Q_3 \times Q_4 \times Q_5$ |
| | $Q(C_1 \cap C_3 \cap C_4 \cap C_5 \cap C_6)$ | $Q(C_1 \cap C_3 \cap C_4 \cap C_5 \cap C_7)$ |
| | $Q_1 \times Q_2 \times Q_3 \times Q_4 \times Q_5$ | $Q_1 \times Q_2 \times Q_3 \times Q_4 \times Q_5$ |
| | $Q(C_1 \cap C_3 \cap C_4 \cap C_6 \cap C_7)$ | $Q(C_1 \cap C_3 \cap C_5 \cap C_6 \cap C_7)$ |
| | $Q_1 \times Q_2 \times Q_3 \times Q_4 \times Q_5$ | $Q_1 \times Q_2 \times Q_3 \times Q_4 \times Q_5$ |
| | | $Q(C_1 \cap C_4 \cap C_5 \cap C_6 \cap C_7)$ |
| | | $Q_1 \times Q_2 \times Q_3 \times Q_4 \times Q_5$ |
| | $Q(C_2 \cap C_3 \cap C_4 \cap C_5 \cap C_6)$ | $Q(C_2 \cap C_3 \cap C_4 \cap C_5 \cap C_7)$ |
| | $Q_2 \times Q_3 \times Q_4 \times Q_5$ | $Q_2 \times Q_3 \times Q_4 \times Q_5$ |
| | $Q(C_2 \cap C_3 \cap C_4 \cap C_6 \cap C_7)$ | $Q(C_2 \cap C_3 \cap C_5 \cap C_6 \cap C_7)$ |

| | |
|---|---|
| $Q_2 \times Q_3 \times Q_4 \times Q_5$ | $Q_2 \times Q_3 \times Q_4 \times Q_5$ |
| $Q(C_2 \cap C_4 \cap C_5 \cap C_6 \cap C_7)$ | $Q(C_3 \cap C_4 \cap C_5 \cap C_6 \cap C_7)$ |
| $Q_2 \times Q_3 \times Q_4 \times Q_5$ | $Q_2 \times Q_3 \times Q_4 \times Q_5$ |

### Seven Sixth-Order Cut Combinations and Their Probability Expressions

| | |
|---|---|
| $Q(C_1 \cap C_2 \cap C_3 \cap C_4 \cap C_5 \cap C_6)$ | $Q(C_1 \cap C_2 \cap C_3 \cap C_4 \cap C_5 \cap C_7)$ |
| $Q_1 \times Q_2 \times Q_3 \times Q_4 \times Q_5$ | $Q_1 \times Q_2 \times Q_3 \times Q_4 \times Q_5$ |
| $Q(C_1 \cap C_2 \cap C_3 \cap C_4 \cap C_6 \cap C_7)$ | $Q(C_1 \cap C_2 \cap C_3 \cap C_5 \cap C_6 \cap C_7)$ |
| $Q_1 \times Q_2 \times Q_3 \times Q_4 \times Q_5$ | $Q_1 \times Q_2 \times Q_3 \times Q_4 \times Q_5$ |
| $Q(C_1 \cap C_2 \cap C_4 \cap C_5 \cap C_6 \cap C_7)$ | $Q(C_1 \cap C_3 \cap C_4 \cap C_5 \cap C_6 \cap C_7)$ |
| $Q_1 \times Q_2 \times Q_3 \times Q_4 \times Q_5$ | $Q_1 \times Q_2 \times Q_3 \times Q_4 \times Q_5$ |
| $Q(C_2 \cap C_3 \cap C_4 \cap C_5 \cap C_6 \cap C_7)$ | |
| $Q_2 \times Q_3 \times Q_4 \times Q_5$ | |

### One Seventh-Order Cut Combination and Its Probability Expressions

| |
|---|
| $Q(C_1 \cap C_2 \cap C_3 \cap C_4 \cap C_5 \cap C_6 \cap C_7)$ |
| $Q_1 \times Q_2 \times Q_3 \times Q_4 \times Q_5$ |

$$Q_s = P\left[\bigcup_{i=1}^{7} C_i\right]$$

### Assuming $Q_1 = Q_2 = Q_3 = Q_4 = Q_5 = Q$ (component)

| Combinations (*i* at a time) | Expression | Sign |
|---|---|---|
| 1 | $Q + 6Q^2$ | + |
| 2 | $18Q^3 + 3Q^4$ | − |
| 3 | $4Q^3 + 28Q^4 + 3Q^5$ | + |
| 4 | $19Q^4 + 16Q^5$ | − |
| 5 | $6Q^4 + 15Q^5$ | + |
| 6 | $Q^4 + 6Q^5$ | − |
| 7 | $Q^5$ | + |

$$Q_s = Q + 6Q^2 - 14Q^3 + 11Q^4 - 3Q^5$$

Deriving an expression for the distribution system in terms of $R$ Substituting $Q = 1 - R$ into the equation for $Q_s$ results in

*Note*:

$$Q = 1 - R$$
$$Q^2 = 1 - 2R + R^2$$

$$Q^3 = 1 - 3R + 3R^2 - R^3$$
$$Q^4 = 1 - 4R + 6R^2 - 4R^3 + R^4$$
$$Q^5 = 1 - 5R + 10R^2 - 10R^3 + 5R^4 - R^5$$

$$\begin{aligned} Q_s &= [1-R] + [6-12R+6R^2] + [-14+42R-42R^2+14R^3] + \cdots \\ &\quad + [11-44R+66R^2-44R^3+11R^4] + [-3+15R-30R^2 \\ &\quad + 30R^3 - 15R^4 + 3R^5] \\ Q_s &= 1-4R^4+3R^5 \\ R_s &= 1-Q_s = 4R-3R^5 \end{aligned}$$

If $Q = 0.05$ and $R = 0.95$ then:

$$\begin{aligned} Q_s &= 0.063318 \\ R_s &= 1-Qs = 0.063318 \end{aligned}$$

## 6.4 BAYES' THEOREM IN RELIABILITY

Not all network configurations can be reduced to either a series or a parallel system or a combination of both systems. In many cases, components are interconnected within a network configuration such that they cannot be classified as being connected in parallel or in series. Bayes' theorem is based on conditional probability and is commonly applied when two mutually exclusive events are associated with each component of a network. The theorem states the following:

If $A$ is an event that depends on one of the two mutually exclusive events, $B_i$ and $B_j$, of which one must necessarily occur, then the probability of the occurrence of $A$ is

$$P(A) = P(A \mid B_i)P(B_i) + P(A \mid B_j)P(B_j)$$

This is the same expression we used previously for conditional probability problems. Now, we will refer to the conditional probability methodology as Bayes' theorem when applied to evaluating the reliability of network configurations. Bayes' theorem stated in reliability terms is

$$R_s = P(\text{SS} \mid \text{component } i \text{ is up})\, R_i + P(\text{SS} \mid \text{component } i \text{ is down})\, Q_i$$

where SS stands for "system success."

As the network configurations become more complex, it is imperative that the system successful paths between the input and the output terminals of the reliability block diagram be clearly defined.

*Note*: The system operates if there is a successful path between the input and the output terminals of the reliability block diagram.

## Example 6.6 (Bayes' Theorem)

The reliability block diagram of a network configuration is given in Fig. 6.5. Calculate the reliability of the network configuration given in Fig. 6.5 by using Bayes' theorem (Fig. 6.21).

System success operational paths: A–D, B–D, B–E, C–E

$$R_s = P(\text{SS} \mid \text{component B is up})R_B + P(\text{SS} \mid \text{component B is down})Q_B$$

*Given: component B is up*

The original network configuration given that component B is up can be redrawn as follows (Fig. 6.22):

The probability evaluation of *P*(SS|component B is up) reduces to two components in parallel (i.e., D and E).

$$P(\text{SS} \mid \text{component B is up}) = [1-(1-R_D)(1-R_E)] = R_D + R_E - R_D R_E$$

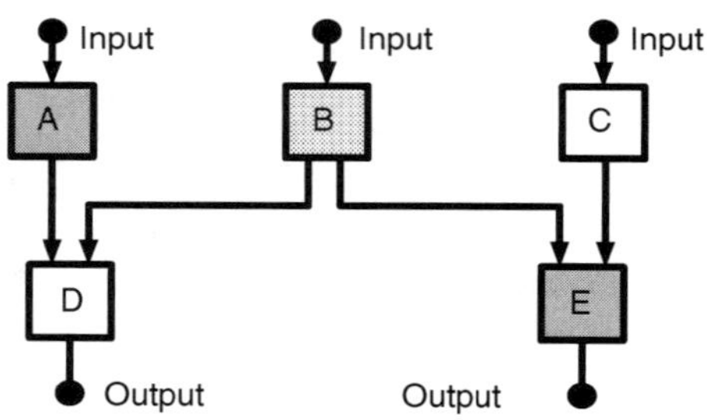

Figure 6.21. Example system configuration.

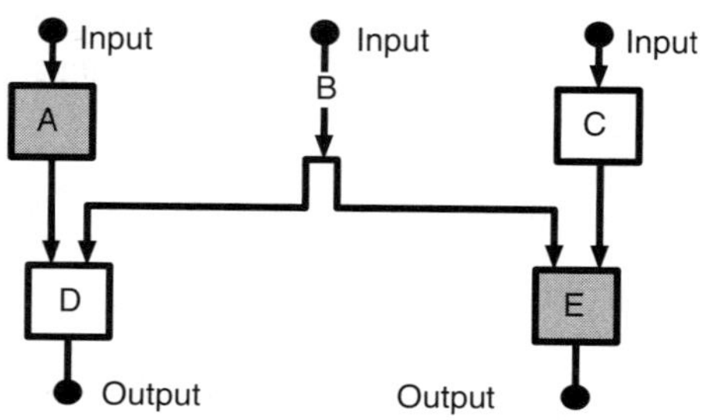

Figure 6.22. Reliability block diagram: given component B is UP.

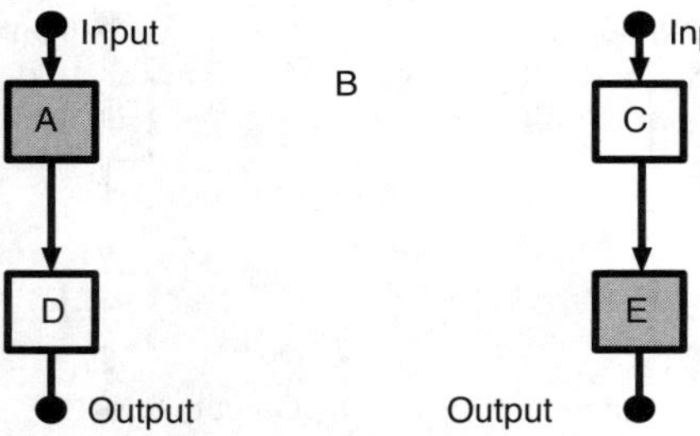


Figure 6.23. Reliability block diagram: given component B is DOWN.

*Given: component B is down*

The original network configuration given that component B is down can be redrawn as follows (Fig. 6.23):

The probability evaluation of $P$(SS|component B is down) reduces to two *series paths* in parallel (i.e., A–D and C–E).

$$P(\text{SS}\mid \text{component B is down}) = [1-(1-R_A R_D)(1-R_C R_E)] = R_A R_D + R_C R_E - R_A R_D R_C R_E$$

$$R_s = P(\text{SS}\mid \text{component B is up})R_B + P(\text{SS}\mid \text{component B is down})Q_B$$

$$\begin{aligned} R_s &= [1-(1-R_D)(1-R_E)]R_B + [1-(1-R_A R_D)(1-R_C R_E)](1-R_B) \\ &= R_D R_B + R_E R_B - R_D R_E R_B + R_A R_D + R_C R_E - R_A R_D R_B - R_C R_E R_B - R_A R_D R_C R_E \\ &\quad + R_A R_D R_C R_E R_B \end{aligned}$$

*An Alternative Approach to the Problem*

The reliability of the previous network configuration was solved from a *system success viewpoint*; however, it can also be solved from a *system failure viewpoint* as follows:

$$Q_s = P(\text{SF}\mid \text{component B is up})R_B + P(\text{SF}\mid \text{component B is down})Q_B$$

where SF stands for system failure.

*Given: component B is up*

The original network configuration given that component B is up can be redrawn as follows (Fig. 6.24):

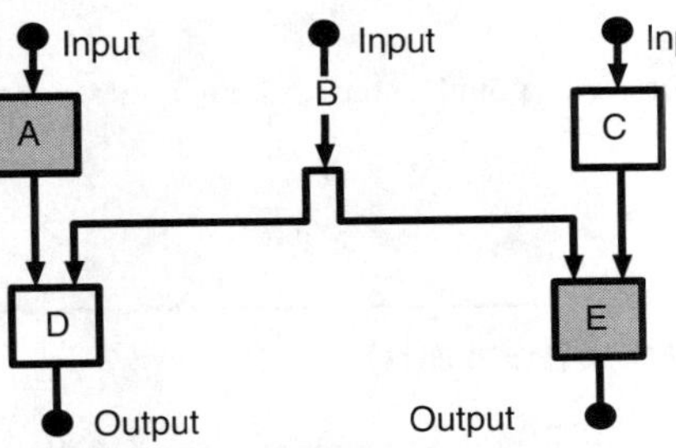


Figure 6.24. Reliability block diagram: given component B is UP.

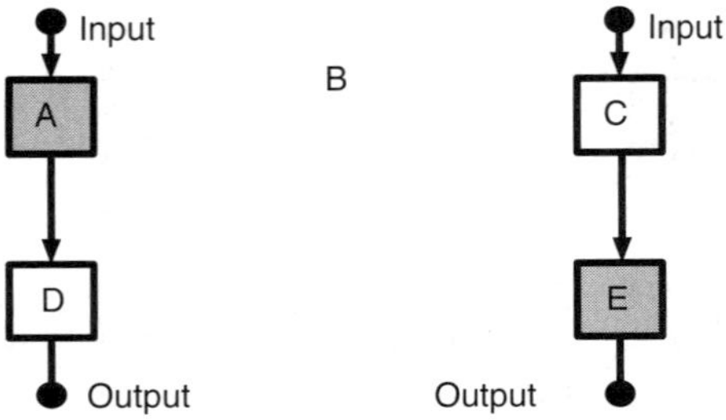


Figure 6.25. Reliability block diagram: given component B is DOWN.

The probability evaluation of $P$(SF|component B is up) reduces to two components in parallel (i.e., D and E). Both components D and E must fail to cause system failure.

$$P(\text{SF}|\text{component B is up}) = Q_D Q_E = [(1-R_D)(1-R_E)] = 1-R_D-R_E-R_DR_E$$

*Given: component B is down*

The original network configuration given that component B is down can be redrawn as follows (Fig. 6.25):

The probability evaluation of $P$(SF|component B is down) reduces to two series paths in parallel (i.e., A–D and C–E). Both paths must fail to cause system failure.

$$\begin{aligned} P(\text{SF}|\text{component B is down}) = Q_{AD}Q_{CE} &= [(1-R_AR_D)(1-R_CR_E)] \\ &= 1-R_AR_D-R_CR_E+R_AR_DR_CR_E \end{aligned}$$

$$Q_s = P(\text{SF}|\text{component B is up})R_B + P(\text{SF}|\text{component B is down})Q_B$$

$$\begin{aligned} Q_s &= [1-R_D-R_E-R_DR_E]R_B + [1-R_AR_D-R_CR_E+R_AR_DR_CR_E](1-R_B) \\ &= 1-R_DR_B-R_ER_B+R_DR_ER_B-R_AR_D-R_CR_E+R_AR_DR_B+R_CR_ER_B+R_AR_DR_CR_E \\ &\quad -R_AR_DR_CR_ER_B \end{aligned}$$

$$\begin{aligned} R_s &= 1-Q_s \\ &= R_DR_B+R_ER_B-R_DR_ER_B+R_AR_D+R_CR_E-R_AR_DR_B-R_CR_ER_B-R_AR_DR_CR_E \\ &\quad +R_AR_DR_CR_ER_B \end{aligned}$$

*Note*: Either the "system success viewpoint" or the "system failure viewpoint" approach will result in the same answer.

## Problem 6.1 (Bayes' Theorem)

The reliability block diagram of a network configuration is shown in the figure below. Calculate the reliability of the network configuration below by using Bayes' theorem (Fig. 6.26).

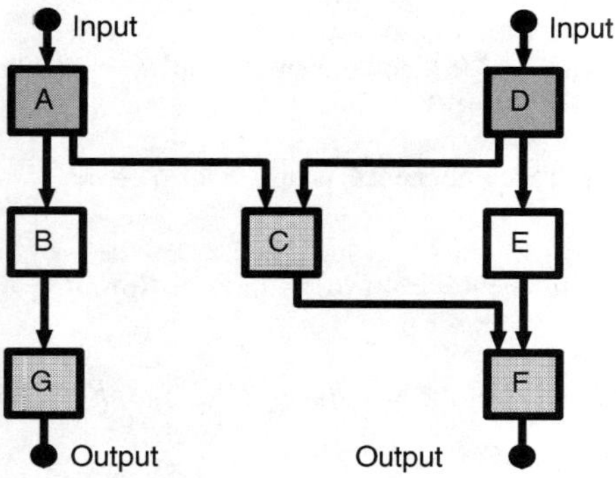


Figure 6.26. Example system configuration.

System success operational paths: A–B–G, A–C–F, D–C–F, D–E–F

$$R_s = P(\text{SS}|\text{component A is up})R_A + P(\text{SS}|\text{component A is down})Q_A$$

*Given: component A is up*

$$P(\text{SS}|\text{component A is up}) = P(\text{SS}|\text{component C is up})R_C + P(\text{SS}|\text{component C is down})\ Q_C$$

Given component A is up, the resulting network configurations, subject to component C being up and down, are shown in Fig. 6.27.

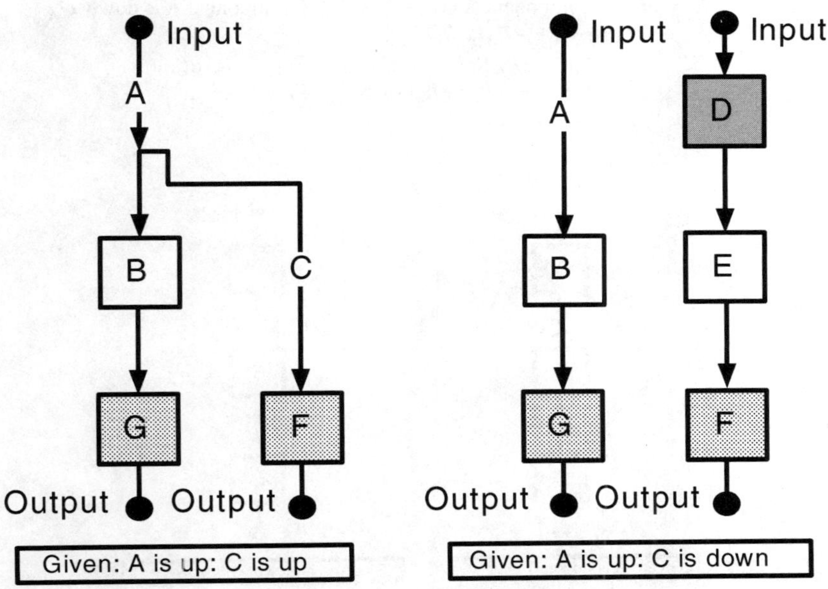


Figure 6.27. System configuration: given component A is up.

$$P(\text{SS}|\text{component A is up}) = P(\text{SS}|\text{component C is up})R_C + P(\text{SS}|\text{component C is down})Q_C$$

$$P(\text{SS}|\text{component C is up}) = R_B R_G + R_F - R_B R_G R_F$$

$$P(\text{SS}|\text{component C is down}) = R_B R_G + R_D R_E R_F - R_B R_G R_D R_E R_F$$

$$P(\text{SS}|\text{component A is up}) = [R_B R_G + R_F - R_B R_G R_F]R_C + [R_B R_G + R_D R_E R_F - R_B R_G R_D R_E R_F]Q_C$$

*Given: component A is down*

$$P(\text{SS}|\text{component A is down}) = P(\text{SS}|\text{component C is up})R_C + P(\text{SS}|\text{component C is down})Q_C$$

Given component A is down, the resulting network configurations, subject to component C being up and down, are shown in Fig. 6.28.

$$P(\text{SS}|\text{component A is down}) = P(\text{SS}|\text{component C is up})R_C + P(\text{SS}|\text{component C is down})Q_C$$

$$P(\text{SS}|\text{component C is up}) = R_D R_F$$

$$P(\text{SS}|\text{component C is down}) = R_D R_E R_F$$

$$P(\text{SS}|\text{component A is down}) = [R_D R_F]R_C + [R_D R_E R_F]Q_C$$

$$\begin{aligned} R_s &= P(\text{SS}|\text{component A is up})R_A + P(\text{SS}|\text{component A is down})Q_A \\ &= [R_B R_G + R_F - R_B R_G R_F]R_C R_A + \cdots \\ &\quad + [R_B R_G + R_D R_E R_F - R_B R_G R_D R_E R_F](1 - R_C)R_A + \cdots \\ &\quad + [R_D R_F R_C + R_D R_E R_F(1 - R_C](1 - R_A) \end{aligned}$$

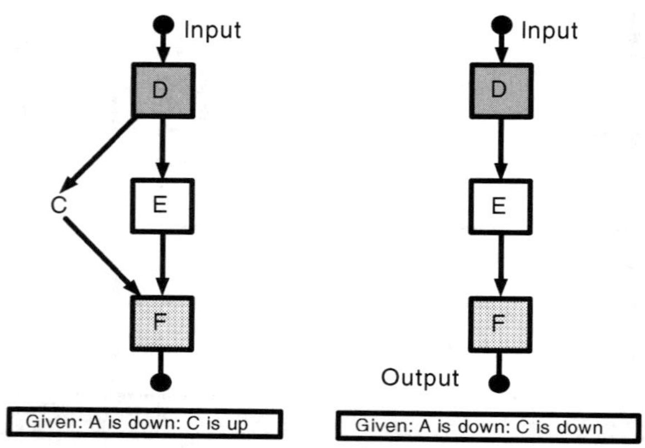


Figure 6.28. System configuration: given component A is down.

If all the components are identical (i.e., $R_A = R_B = R_C = R_D = R_E = R_F = R_G = 0.90$), then the system reliability would be $R_s = 0.960165$.

## Problem 6.2 (Bayes' Theorem)

The reliability block diagram of a network configuration is shown in Fig. 6.29 (i.e., the same configuration as Problem 6.1). Calculate the reliability of the network configuration below selecting component C as the starting component in the reliability analysis.

System success operational paths: A–B–G, A–C–F, D–C–F, D–E–F

$$R_s = P(\text{SS}|\text{component C is up})R_C + P(\text{SS}|\text{component C is down})Q_C$$

*Given: component C is up*

$$\begin{aligned} P(\text{SS}|\text{component C is up}) &= P(\text{SS}|\text{component A is up})R_A + P(\text{SS}|\text{component A is down})Q_A \\ &= [R_F + R_B R_G - R_F R_B R_G]R_A + [R_D R_F](1 - R_A) \end{aligned}$$

*Given: component C is down*

$$\begin{aligned} P(\text{SS}|\text{component C is down}) &= P(\text{SS}|\text{component A is up})R_A + P(\text{SS}|\text{component A is down} \\ &= [R_B R_G + R_D R_E R_F - R_B R_G R_D R_E R_F]R_A + R_D R_E R_F(1 - R_A) \end{aligned}$$

$$\begin{aligned} R_s &= P(\text{SS}|\text{component C is up})R_C + P(\text{SS}|\text{component C is down})Q_C \\ &= [[R_F + R_B R_G - R_F R_B R_G]R_A + [R_D R_F](1 - R_A)]R_C \\ &\quad + [[R_B R_G + R_D R_E R_F - R_B R_G R_D R_E R_F]R_A + R_D R_E R_F(1 - R_A)](1 - R_C) \end{aligned}$$

*Note*: The same solution for reliability will be obtained irrespective of which component is selected in the first expression for the reliability of the system. Usually, if the component with the most interconnections is selected initially, the analysis is simplified.

In some publications, $P$(SSlcomponent $i$ is up) is written $R_s$ (if component $i$ is up).

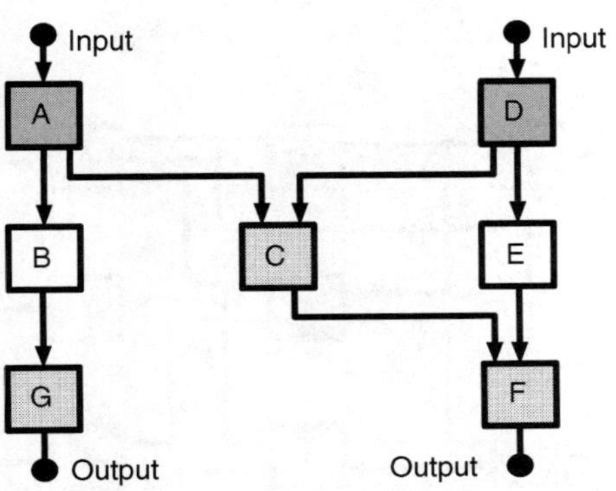


Figure 6.29. System configuration.

## Problem 6.3 (Bayes' Theorem)

A network configuration is shown below. All components in the network have the same mission reliability equal to "$p$" (Fig. 6.30).

Develop an expression for the reliability of the network configuration shown above.

*Note*: The final reliability expression *must* be in the form of a polynomial in terms of "$p$" only.

$$R_s = R_s(\text{if component 2 is up})\,R_2 + R_s(\text{if component 2 is down})\,Q_2$$

*Given: component 2 is up*

$$\begin{aligned} R_s(\text{If component 2 is up}) &= R_s(\text{if component 4 is up})R_4 + R_s(\text{if component 4 is down})\,Q_4 \\ &= (R_6 + R_8 + R_5R_7 - R_6R_8 - R_5R_6R_7 - R_5R_7R_8 + R_5R_6R_7R_8)R_4 \\ &\quad + [(R_6 + R_5(R_7 + R_8 - R_7R_8) - R_5R_6(R_7 + R_8 - R_7R_8)]Q_4 \end{aligned}$$

*Given: component 2 is down*

$$\begin{aligned} R_s(\text{if component 2 is down}) &= R_s(\text{if component 4 is up})R_4 + R_s(\text{if component 4 is down})Q_4 \\ &= R_1(R_6 + R_8 - R_6R_8)R_4 + [R_1R_3R_6]Q_4 \end{aligned}$$

$$\begin{aligned} R_s &= R_s\ (\text{if component 2 is up})R_2 + R_s\ (\text{if component 2 is down})Q_2 \\ &= R_2\{(R_6 + R_8 + R_5R_7 - R_6R_8 - R_5R_6R_7 - R_5R_7R_8 + R_5R_6R_7R_8)R_4\} \\ &\quad + R_2[(R_6 + R_5(R_7 + R_8 - R_7R_8) - R_5R_6(R_7 + R_8 - R_7R_8)]Q_4 \\ &\quad + Q_2[R_1(R_6 + R_8 - R_6R_8)R_4 + [R_1R_3R_6]Q_4] \\ &= R_2R_4R_6 + R_2R_4R_8 + R_2R_4R_5R_7 - R_2R_4R_6R_8 - R_2R_4R_5R_6R_7 - R_2R_4R_5R_7R_8 \\ &\quad + R_2R_4R_5R_6R_7R_8 + R_2R_6 + R_2R_5R_7 + R_2R_5R_8 - R_2R_5R_7R_8 - R_2R_5R_6R_7 \\ &\quad - R_2R_5R_6R8 + R_2R_5R_6R_7R_8 - R_2R_4R_6 - R_2R_4R_5R_7 - R_2R_4R_5R_8 + R_2R_4R_5R_7R_8 \\ &\quad + R_2R_4R_5R_6R_7 + R_2R_4R_5R_6R_8 - R_2R_4R_5R_6R_7R_8 + R_1R_4R_6 + R_1R_4R_8 \\ &\quad - R_1R_4R_6R_8 - R_1R_2R_4R_6 - R_1R_2R_4R_8 + R_1R_2R_4R_6R_8 + R_1R_3R_6 - R_1R_4R_3R_6 \\ &\quad - R_1R_2R_3R_6 + R_1R_2R_4R_3R_6 \end{aligned}$$

If $R_1 = R_2 = R_3 = R_4 = R_5 = R_6 = R_7 = R_8 = p$

Then, $R_s = p^2 + 6p^3 - 10p^4 + 4p^5$

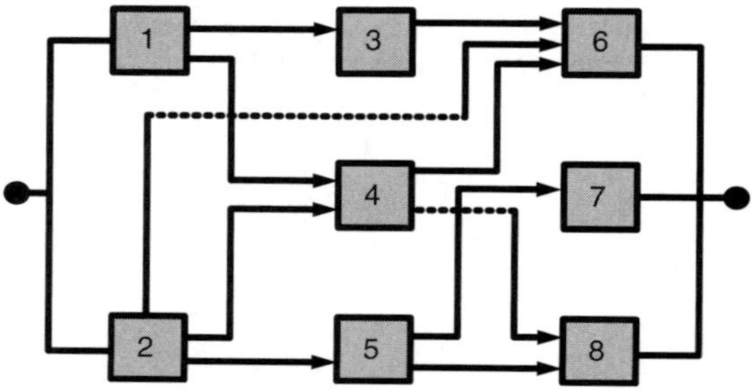


Figure 6.30. System configuration.

Calculate the system network reliability if $p = 0.90$.

$$\begin{aligned} R_s &= p^2 + 6p^3 - 10p^4 + 4p^5 \\ &= 0.984960 \end{aligned}$$

## Problem 6.4 (Bayes' Theorem)

A network configuration is shown below. All components in the network have the same mission reliability equal to "$p$" (Figs. 6.31–6.35).

Develop an expression for the reliability of the network configuration shown above.

*Note*: The final reliability expression *must* be in the form of a polynomial *in terms of* "$p$" *only*.

$$R_s = R_s(\text{if component 2 is up})R_2 + R_s(\text{if component 2 is down})Q_2$$

*Given: component 2 is up*

$$R_s(\text{if component 2 is up}) = R_s(\text{if component 4 is up})R_4 + R_s(\text{if component 4 is down})Q_4$$

$$R_s(\text{if component 4 is up}) = (R_6 + R_8 + R_5R_7 - R_6R_8 - R_5R_6R_7 - R_5R_7R_8 + R_5R_6R_7R_8)$$

$$R_s(\text{if component 4 is down}) = [(R_6 + R_5(R_7 + R_8 - R_7R_8) - R_5R_6(R_7 + R_8 - R_7R_8)]$$

*Given: component 2 is down*

$$R_s(\text{if component 2 is down}) = R_s(\text{if component 4 is up})\ R_4 + R_s(\text{if component 4 is down})Q_4$$

$$R_s(\text{if component 4 is up}) = R_1(R_6 + R_8 - R_6R_8)$$

$$R_s(\text{if component 4 is down}) = [R_1R_3R_6]$$

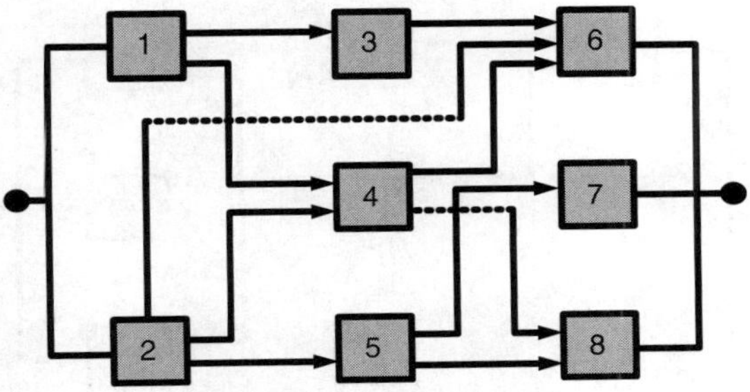


Figure 6.31. System configuration.

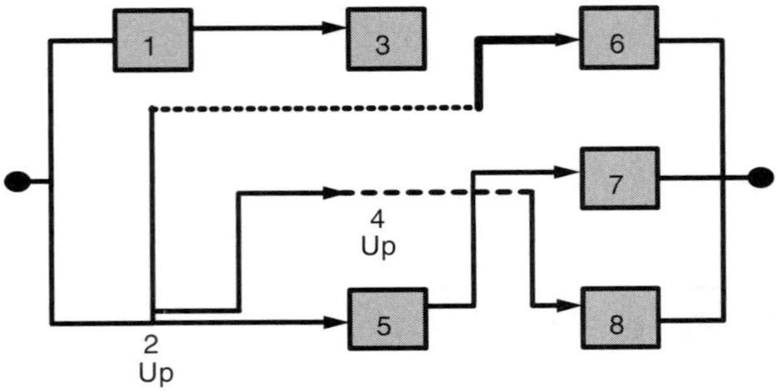


Figure 6.32. System configuration: given component 2 is UP and component 4 is UP.

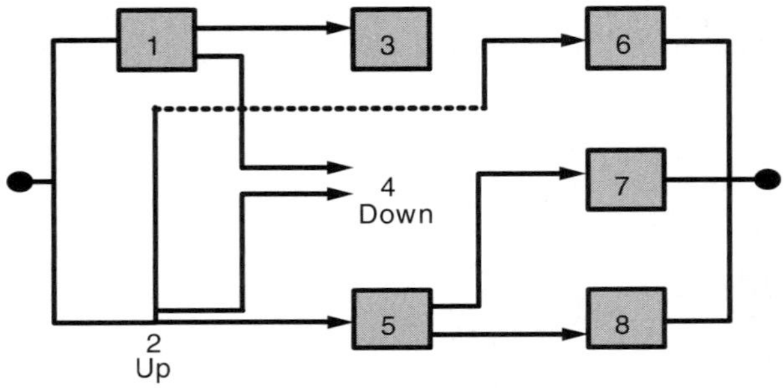


Figure 6.33. System configuration: given component 2 is UP and component 4 is Down.

$$
\begin{aligned}
R_s &= R_s(\text{if component 2 is up})R_2 + R_s(\text{if component 2 is down})Q_2 \\
&= R_2\{(R_6 + R_8 + R_5R_7 - R_6R_8 - R_5R_6R_7 - R_5R_7R_8 + R_5R_6R_7R_8)R_4\} \\
&\quad + R_2[(R_6 + R_5(R_7 + R_8 - R_7R_8) - R_5R_6(R_7 + R_8 - R_7R_8)]Q_4 \\
&\quad + Q_2[R_1(R_6 + R_8 - R_6R_8)R_4 + [R_1R_3R_6]Q_4] \\
&= R_2R_4R_6 + R_2R_4R_8 + R_2R_4R_5R_7 - R_2R_4R_6R_8 - R_2R_4R_5R_6R_7 - R_2R_4R_5R_7R_8 \\
&\quad + R_2R_4R_5R_6R_7R_8 + R_2R_6 + R_2R_5R_7 + R_2R_5R_8 - R_2R_5R_7R_8 - R_2R_5R_6R_7 \\
&\quad - R_2R_5R_6R_8 + R_2R_5R_6R_7R_8 - R_2R_4R_6 - R_2R_4R_5R_7 - R_2R_4R_5R_8 + R_2R_4R_5R_7R_8 \\
&\quad + R_2R_4R_5R_6R_7 + R_2R_4R_5R_6R_8 - R_2R_4R_5R_6R_7R_8 + R_1R_4R_6 + R_1R_4R_8 - R_4R_6R_8 \\
&\quad - R_1R_2R_4R_6 - R_1R_2R_4R_8 + R_1R_2R_4R_6R_8 + R_1R_3R_6 - R_1R_4R_3R_6 - R_1R_2R_3R_6 \\
&\quad + R_1R_2R_4R_3R_6
\end{aligned}
$$

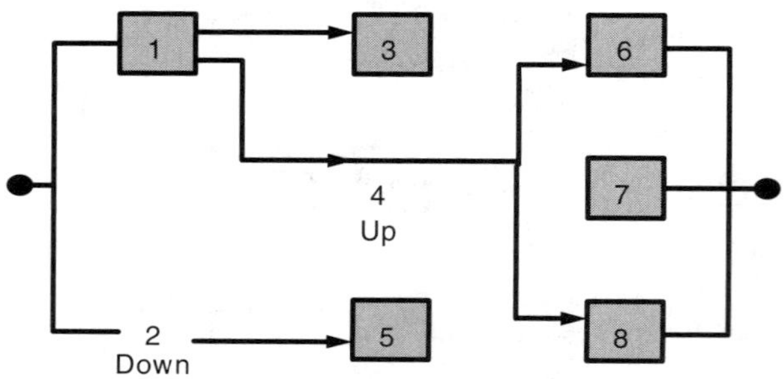


Figure 6.34. System configuration: given component 2 is Down and component 4 is UP.

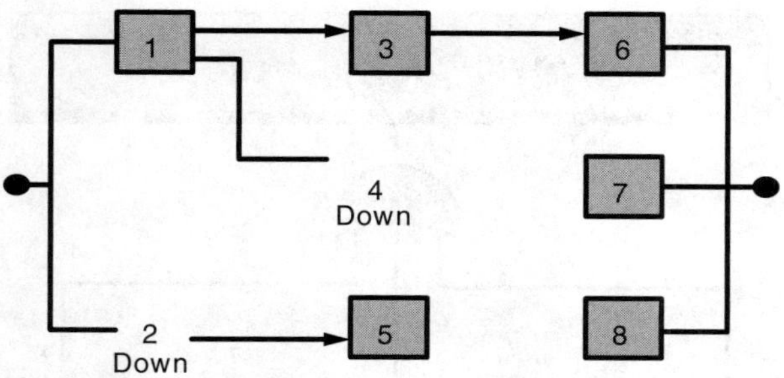


Figure 6.35. System configuration: given component 2 is Down and component 4 is Down.

If $R_1 = R_2 = R_3 = R_4 = R_5 = R_6 = R_7 = R_8 = p$

Then, $R_s = p^2 + 6p^3 - 10p^4 + 4p^5$

Calculate the system network reliability if $p = 0.90$.

$$\begin{aligned} R_s &= p^2 + 6p^3 - 10p^4 + 4p^5 \\ &= 0.984960 \end{aligned}$$

## 6.5 CONSTRUCTION OF FAULT TREE DIAGRAM

First identify all the major failure events associated with the components and/or subsystems of the system or portion of the system to be considered for analysis. Then connect the various failure events by appropriate OR gates or AND gates. The methodology will be illustrated with several examples.

---

## Problem 6.5 (Fault Tree Analysis)

A TV monitor will not operate. It is known that this major failure event can be the result of a combination of the following single failures (i.e., minor events):

1. No electrical power
2. Cable television malfunction
3. Defective TV

The criterion for system (i.e., TV) failure is that if *any* of the minor failure events occur, then the TV will not operate (i.e., the condition for an OR gate). The following fault tree diagram can be used to represent the failure process of the TV monitor (Fig. 6.36).

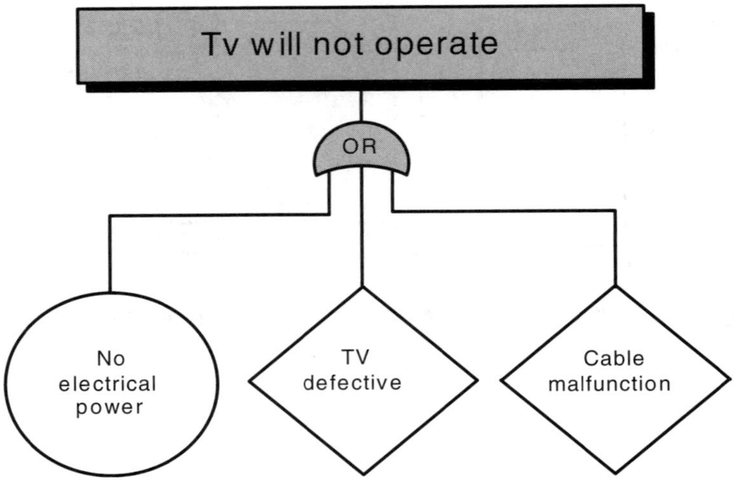


Figure 6.36. Fault tree diagram: nonoperating TV monitor.

---

## Problem 6.6 (Fault Tree Analysis)

Failure of the electrical supply to a critical life support system results in a system failure. This major failure event depends upon the following single-failure events:

1. Utility power system failure
2. Uninterruptible power system (UPS) failure.

The criterion for system (i.e., critical life support) failure is that if all of the minor failure events occur, then the critical life support system will fail (i.e., the condition for an AND gate). The following fault tree diagram can be used to represent the failure process of the critical life support system (Fig. 6.37).

### 6.5.1 Basic Rules for Combining the Probability of Independent Input Failure Events to Evaluate the Probability of a Single-Output Failure Event

It will be assumed that events A and B are two independent failure events. Various logical combinations of these two event failures will cause a single output failure event (OFE) to occur. If the probabilities of the input failure events A and B are known, then the probability of the OFE can be evaluated depending upon the logical relationship between the input failure events and the resulting output failure event.

***6.5.1.1 AND Gate.*** A fault tree diagram for a general OFE with two input failure events interconnected by an AND gate is shown in Fig. 6.38.

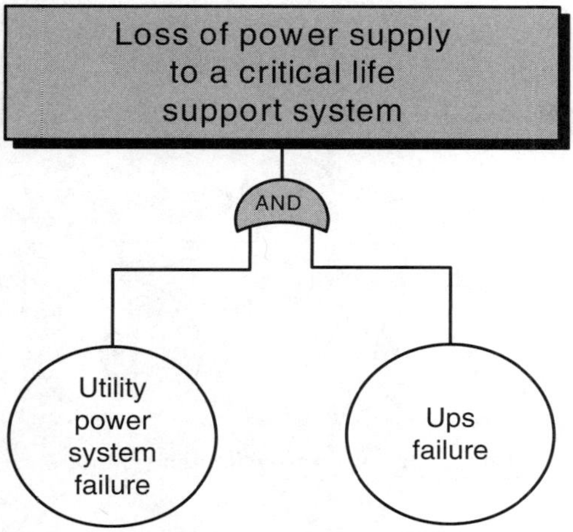


Figure 6.37. Fault tree diagram: critical life support system.

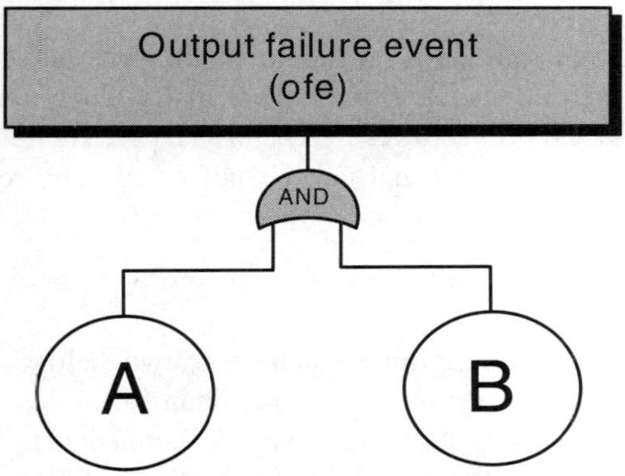


Figure 6.38. Fault tree diagram: AND gate.

If $P(\text{A})$ is equal to the probability of failure event A and $P(\text{B})$ is equal to the probability of failure event B, then $P(\text{OFE}) = P(\text{A} \cap \text{B}) = P(\text{A})\ P(\text{B})$.

In general, if a set of input failure events $\text{A}_1$ to $\text{A}_n$ connected to an AND gate causes the failure event (OFE) to occur, the probability of the output failure event is

$$P(\text{OFE}) = P(\text{A}_1 \cap \text{A}_2 \cap \text{A}_3 \cdots \cap \text{A}_n) = P(\text{A}_1)P(\text{A}_2)P(\text{A}_3) \cdots P(\text{A}_n)$$

In general, for AND gate logical connections, the probability of the output failure event is the product of the input failure event probabilities.

***6.5.1.2 OR Gate.*** A fault tree diagram for a general output failure event with two input failure events interconnected by an OR gate is given in Fig. 6.39.

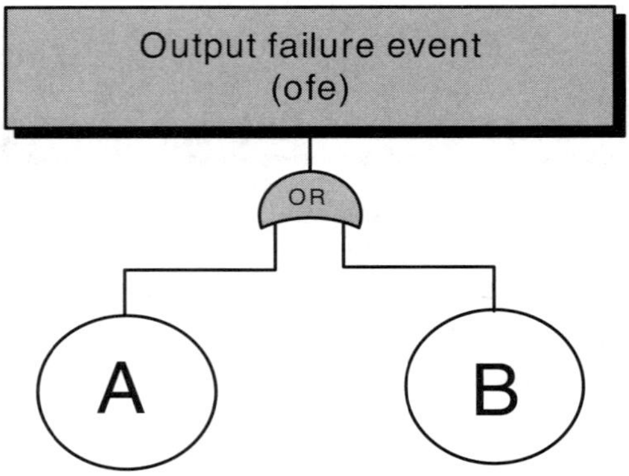


Figure 6.39. Fault tree diagram: OR gate.

If $P(\mathrm{A})$ is equal to the probability of failure event A and $P(\mathrm{B})$ is equal to the probability of failure event B, then

$$P(\mathrm{OFE}) = P(\mathrm{A} \cup \mathrm{B}) = P(\mathrm{A}) + P(\mathrm{B}) - P(\mathrm{A})P(\mathrm{B})$$

Depending upon the magnitude of the input probability failure events for an OR gate, the equation for $P(\mathrm{OFE})$ can be simplified, particularly when there are three or more input failure events. If the probabilities $P(\mathrm{A})$ and $P(\mathrm{B})$ are each less than 0.1, then the product $P(\mathrm{A})$ $P(\mathrm{B})$ is very small and can be neglected. The resulting approximate expression for $P(\mathrm{OFE})$ becomes

$$P(\mathrm{OFE}) = P(\mathrm{A} \cup \mathrm{B}) = P(\mathrm{A}) + P(\mathrm{B})$$

*Note*: Prior to applying the simplified equation shown below, it is imperative that probabilities of input failure events are each less than 0.10.

In general, if a set of input failure events $\mathrm{A}_1$ to A$n$ connected to an OR gate causes the failure event (OFE) to occur, the probability of the output failure event is

$$P(\mathrm{OFE}) = P(\mathrm{A}_1 \cup \mathrm{A}_2 \cup \mathrm{A}_3 \cdots \cup \mathrm{A}_n) = P(\mathrm{A}_1) + P(\mathrm{A}_2) + P(\mathrm{A}_3) \cdots + P(\mathrm{A}_n)$$

In general, for OR gate logical connections, the probability of the output failure event is the *sum of the input failure event probabilities.*

## Problem 6.7 (Fault Tree Analysis)

An automobile in perfect mechanical condition will not start in cold weather. Five major failure events affect this, namely,

1. no fuel,
2. engine does not turn over,

3. blocked exhaust,
4. no ignition was identified, and
5. cold engine block.

If any of the first four major events occurs, then the automobile engine will not start in cold weather. One possible fault tree diagram describing the failure events contributing to the failure event of the engine not starting in cold weather is provided in Fig. 6.40. The minor failure events and their associated probabilities are defined in the tables after the fault tree diagram for each major event.

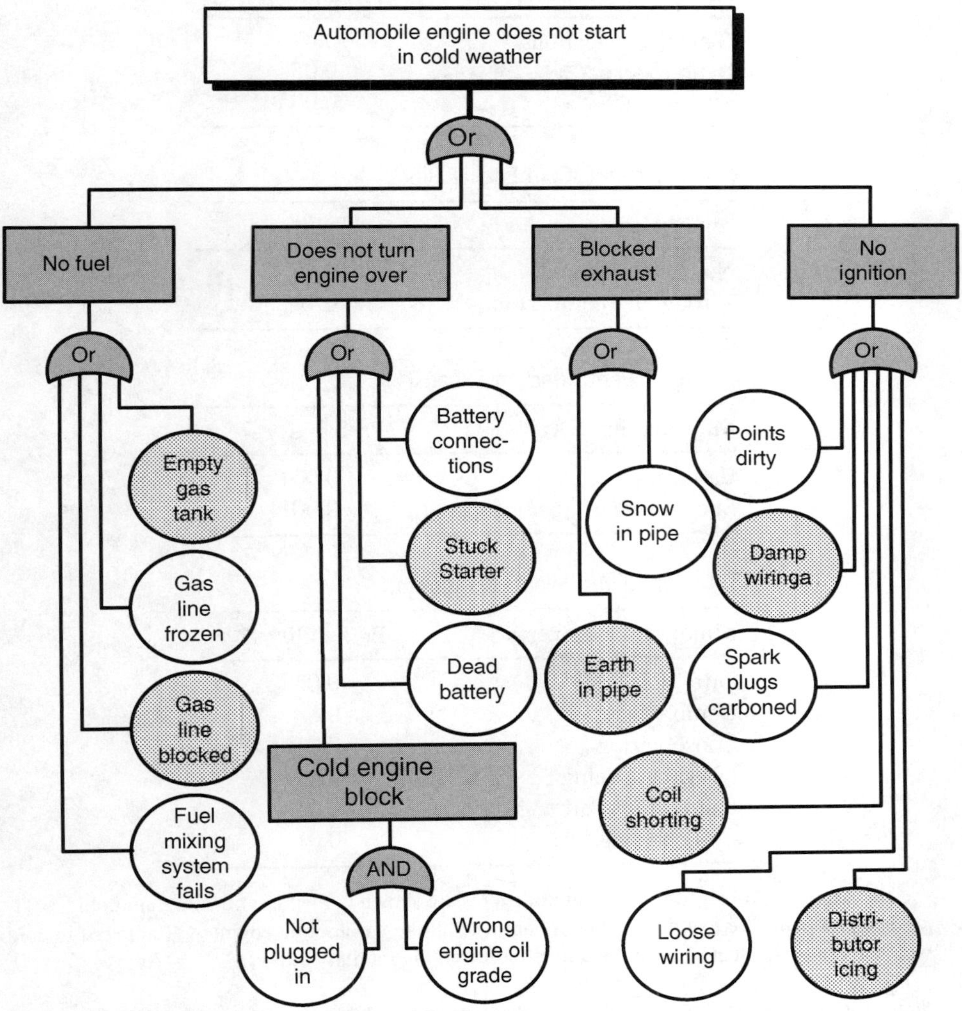


Figure 6.40. Fault tree diagram: automobile engine not starting in cold weather.

1. Major Failure Event: No Fuel

| Minor Failure Events | Probability |
|---|---|
| Empty gas tank | 0.05 |
| Gas line frozen | 0.02 |
| Gas line blocked | 0.01 |
| Fuel mixing system fails | 0.01 |

2. Major Failure Event: Does Not Turn Engine Over

| Minor Failure Events | Probability |
|---|---|
| Battery connections | 0.03 |
| Battery dead | 0.05 |
| Starter stuck | 0.005 |

3. Major Event: Cold Engine Block

| Minor Failure Events | Probability |
|---|---|
| Not plugged in | 0.10 |
| Wrong engine oil grade | 0.10 |

4. Major Event: Blocked Exhaust

| Minor Failure Events | Probability |
|---|---|
| Earth in pipe | 0.001 |
| Snow in pipe | 0.001 |

5. Major Failure Event: No Ignition

| Minor Failure Events | Probability |
|---|---|
| Dirty points | 0.03 |
| Damp wiring | 0.03 |
| Loose wiring | 0.01 |
| Distributor icing | 0.01 |
| Spark plugs carboned | 0.01 |
| Coil shorting | 0.01 |

The assigned probabilities of the minor events are shown in the fault tree diagram in Fig. 6.41.

The procedure for evaluating the probability that the automobile engine will not start in cold weather can be evaluated from the above fault tree diagram as follows:

*Step 1*: Starting from the bottom of the tree diagram and working your way to the top, evaluate each major failure event probability by noting the input minor failure event probabilities and the

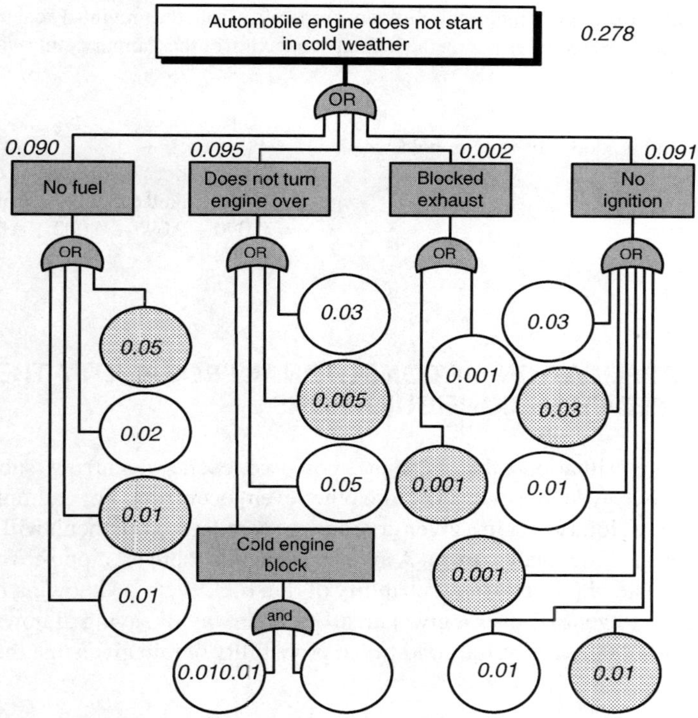


Figure 6.41. Fault tree diagram: automobile engine not starting in cold weather—probabilities defined.

logical connection between the desired output major event and the input minor events, and then *add* the input minor failure event probabilities for OR gate interconnections to evaluate the major failure event probability.

$$\text{In this case } P\text{ (blocked exhaust)} = P(\text{earth in pipe}) + P(\text{snow in pipe}) = 0.001 + 0.001 = 0.002$$

Multiply the input minor failure event probabilities for AND gate interconnections to evaluate the major failure event probability.

For instance, $P$ (cold engine block) $= P$ (not plugged in) $\times P$ (wrong engine oil grade) $= 0.1(0.1) = 0.01$.

*Step 2*: Evaluate each major output failure event probability from the previously calculated major and minor input failure probabilities.

$$\begin{aligned}\text{For example, } P(\text{does not turn engine over}) &= P(\text{battery connections}) \\ &\quad + P(\text{battery dead}) \\ &\quad + P(\text{starter stuck}) + \cdots \\ &\quad + P(\text{cold engine block}) \\ &= 0.03 + 0.05 + 0.005 + 0.01 = 0.095\end{aligned}$$

*Step 3* : Evaluate the single major failure event probability from the previously calculated major event failure probabilities noting the logical connection of these input events with the major failure event.

$$\begin{aligned} P(\text{automobile engine does not start in cold weather}) &= P(\text{No fuel}) + \cdots \\ &\quad \cdots + P(\text{does not turn engine over}) \\ &\quad + P(\text{blocked exhaust}) + P(\text{no ignition}) \\ &= 0.090 + 0.095 + 0.002 + 0.091 \\ &= 0.278 \end{aligned}$$

## 6.6 THE APPLICATION OF CONDITIONAL PROBABILITY THEORY TO SYSTEM OPERATING CONFIGURATIONS

There are many situations that depend upon two events occurring such that the occurrence of one event depends upon the other event occurring. For example, suppose we are observing cloud cover in a given area and are wondering whether it will rain or not. If the event "rain" is denoted as event A and the event that there is "cloud cover" as event B, we are interested in what is the probability of rain (i.e., event A) knowing that there is cloud cover (i.e., event B) over a given area. Another way of saying "knowing that" is "given that." Mathematically, the conditional probability of rain given that there is cloud cover is defined as

$$P(\text{A}|\text{B})$$

where the vertical bar (i.e., |) between A and B is read "given."

Suppose we have observed that 10% of the time it is "rainy" and there is cloud cover, or mathematically

$$P(\text{A} \cap \text{B}) = 0.10$$

Suppose we have also observed that 40% of the time there is cloud cover, or mathematically

$$P(\text{B}) = 0.40$$

Therefore, in a 100-day period, cloud cover will be experienced in 40 days and it will be raining with cloud cover in 10 days. Think of the probabilities of each event as representing a specified percentage of a known population (e.g., 100 days). In this particular example, the known subpopulation is the fact that there is cloud cover (i.e., event B) for 40 days and we are interested in the portion or subpopulation or the number of days in which there will be rain (i.e., event A) given that there is cloud cover.

$$P(\text{A}|\text{B}) = \frac{\text{number of days with rain and cloud cover}}{\text{number } of \text{ days there is cloud cover}} = \frac{P(\text{A} \cap \text{B})}{P(\text{B})} \times \frac{100}{100} = \frac{10}{40} = 0.25$$

Therefore, we conclude that 25% of the time it will rain if there is cloud cover.

$$P(\mathrm{A}|\mathrm{B}) = \frac{P(\mathrm{A} \cap \mathrm{B})}{P(\mathrm{B})}$$

The probability of the simultaneous occurrence of two events A and B is equal to the product of the probability of B and the conditional probability of A given that B has occurred.

$$P(\mathrm{A} \cap \mathrm{B}) = P(\mathrm{B})P(\mathrm{A}|\mathrm{B})$$

or the probability of the simultaneous occurrence of the two events A and B is also equal to the product of the probability of A and the conditional probability of B given that A has occurred.

$$P(\mathrm{A} \cap \mathrm{B}) = P(\mathrm{A})P(\mathrm{B}|\mathrm{A})$$

## Problem 6.8 (Conditional Probabilities)

A warehouse contains eight transformers, two of which are made by a local manufacturer in Alberta. If two transformers are selected at random (i.e., with equal probability) one at a time from the warehouse, what is the probability that the two locally made transformers will be selected for the job?

Let A be the event that the first locally made transformer is selected and B the event that the second locally made transformer is selected.

The probability of A is simply 2/8 or 1/4. Once the first transformer is selected, then the probability of selecting the second transformer is altered and depends upon the condition that the first transformer has been selected. With seven remaining transformers in the warehouse after the first locally made transformer has been selected, the probability of selecting the second locally made transformer is now 1/7.

$$P(\mathrm{A} \cap \mathrm{B}) = P(\mathrm{A})P(\mathrm{B}|\mathrm{A}) = (1/4)(1/7) = 1/28 = 0.03571428571$$

*Mutually Exclusive Conditional Events*

There are many situations in which a single event A is conditional or depends upon the occurrence of a number "$n$" of individual events $\mathrm{B}_i$, which are mutually exclusive. The probability of a single event A can be expressed as follows:

$$P(\mathrm{A}) = \sum_{i=1}^{n} P(\mathrm{A}|\mathrm{B}_i)P(\mathrm{B}_i)$$

In reliability engineering, the single event A is often defined as system successful operation (i.e., labeled SS). The probability of system success (i.e., $P(\mathrm{A}) = P(\mathrm{SS}) = R_\mathrm{s}$) is defined as the reliability of the system. If a single component (e.g., labeled component 1) is assumed to reside in one of the two (i.e., $n = 2$) mutually exclusive states, that is, an operating state and a failed state, let

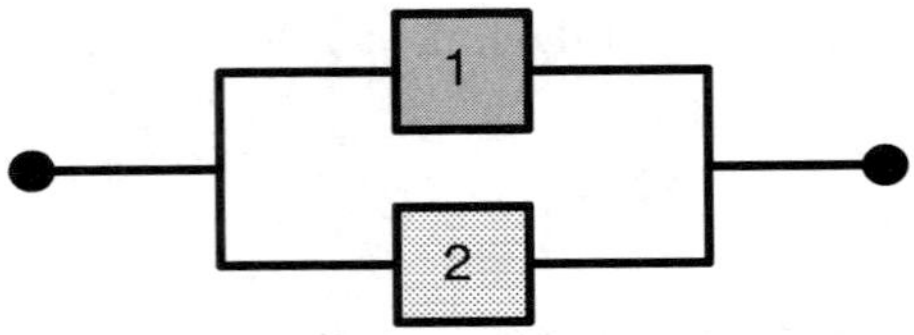


Figure 6.42. System configuration: two blocks connected in parallel.

the event $B_1$ represent the component's operating state (e.g., a good state, an upstate, etc.) and event $B_2$ represent the component's failed state (e.g., a bad state, a downstate, etc.). Then

$$P(A) = P(A|B_1)P(B_1) + P(A|B_2)P(B_2)$$

$$R_s = P(\text{SS}|\text{component 1 is up})R_1 + P(\text{SS}|\text{component 1 is down})Q_1$$

where $R_1$ is the probability of component 1 being up and $Q_1$ is the probability of component 1 being down.

*Note*: $R_1 + Q_1 = 1.0$.

---

## Problem 6.9 (System Success Criteria)

A system consisting of two components whose block diagrams are connected in parallel is shown in Fig. 6.42. Calculate the reliability of the network configuration.

System success criterion: *at least* one component must operate for system success.

*Note*: At least means one or more (i.e., one or more components operating, for example, component 1 operating or component 2 operating or both components 1 and 2 operating).

$$R_s = P(\text{SS}|\text{component 1 is } up)R_1 + P(\text{SS}|\text{component 1 is } down)Q_1$$

$$P(\text{SS}|\text{component 1 is } up) = 1.0$$
$$P(\text{SS}|\text{component 1 is } down) = R2$$
$$R_s = R_1 + R_2Q_1 = R_1 + R_2(1 - R2) = R_1 + R_2 - R_1R_2$$

*Note*: $R_1 + Q_1 = 1.0 = R_2 + Q_2$.

---

## Problem 6.10 (System Success Criteria)

A system consisting of two components whose block diagrams are connected in series is shown in Fig. 6.43. Calculate the reliability of the network configuration.

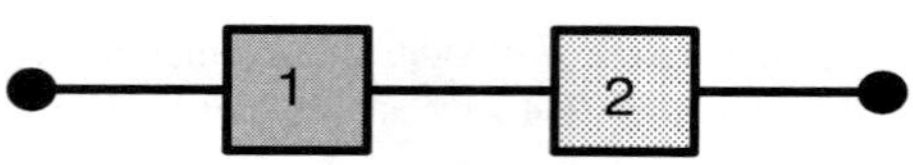


Figure 6.43. System configuration: two block diagrams connected in series.

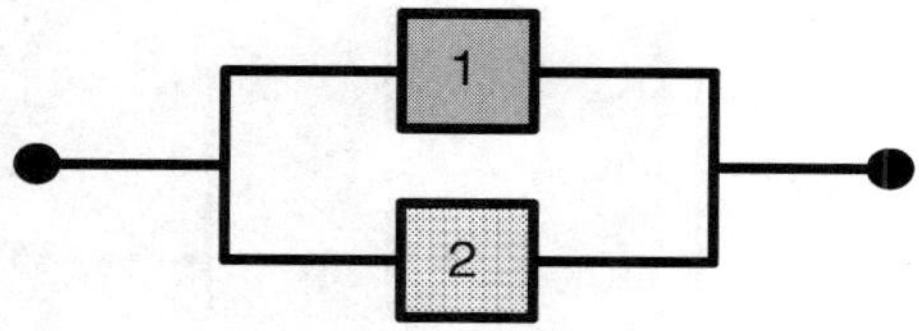


Figure 6.44. System configuration: two block diagrams connected in parallel.

System success criterion: *both* components must operate for system success.

$$R_s = P(\text{SS}|\text{component 1 is up})R_1 + P(\text{SS}|\text{component 1 is down})Q_1$$

$$P(\text{SS}|\text{component 1 is up}) = R_2$$
$$P(\text{SS}|\text{component 1 is down}) = 0.0$$
$$R_s = R_1R_2 + (0.0)Q_1 = R_1R_2$$

The reliability approach applied in the two previous problems was based on a system success criterion. Alternatively, the same problems can be evaluated by solving the conditional probabilities based on system failure criteria. System failure will be labeled SF, and the probability of system failure will be defined as $Q_s$ (i.e., $R_s + Q_s = 1.0$).

## Problem 6.11 (System Failure Criteria)

A system consisting of two components whose block diagrams are connected in parallel is shown in Fig. 6.44. Calculate the reliability of the network configuration based on the system's failure criterion.

System failure criterion: both components must fail to cause system failure.

$$Q_s = P(\text{SF}|\text{component 1 is up})R_1 + P(\text{SF}|\text{component 1 is down})Q_1$$

$$P(\text{SF}|\text{component 1 is up}) = 0.0$$
$$P(\text{SF}|\text{component 1 is down}) = Q_2$$
$$Q_s = (0.0)R_1 + Q_2Q_1 = Q_1Q_2$$
$$R_s = 1 - Q_s = 1 - (1 - R_1)(1 - R_2) = R_1 + R_2 - R_1R_2$$

## Problem 6.12 (System Failure Criteria)

A system consisting of two components whose block diagrams are connected in series is shown in Fig. 6.45. Calculate the reliability of the network configuration based on the system's failure criteria.

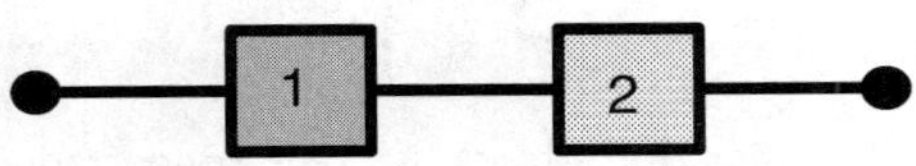


Figure 6.45. System configuration: two block diagrams connected in series.

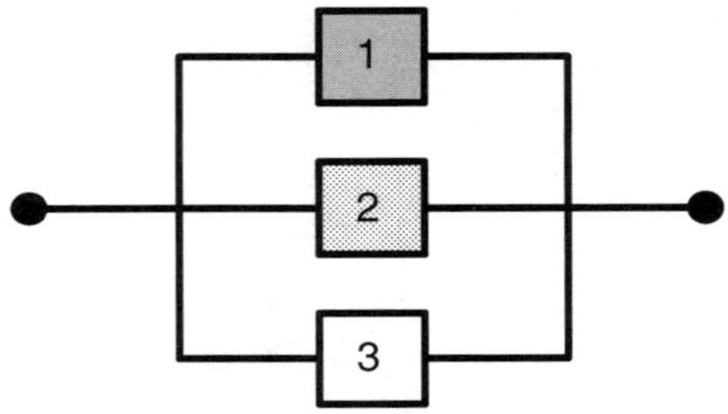


Figure 6.46. System configuration: three block diagrams connected in parallel.

System failure criterion: either one or both components must fail to cause system failure.

$$Q_s = P(\text{SF}|\text{component 1 is up})R_1 + P(\text{SF}|\text{component 1 is down})Q_1$$
$$P(\text{SF}|\text{component 1 is up}) = Q_2$$
$$P(\text{SF}|\text{component 1 is down}) = 1.0$$
$$Q_s = Q_2R_1 + (1.0)Q_1 = (1 - R_2)R_1 + (1 - R_1) = R_1 - R_1R_2 + 1 - R_1 = 1 - R_1R_2$$

Therefore, $R_s = 1 - Q_s = R_1R_2$.

## Problem 6.13 (System Success Criteria)

A system consisting of three components whose block diagrams are connected in parallel is shown in Fig. 6.46. Calculate the reliability of the network configuration based on its system success criteria.

System success criterion: *at least* one component must operate to cause system success.

$$R_s = P(\text{SS}|\text{component 1 is up})R_1 + P(\text{SS}|\text{component 1 is down})Q_1$$

$$P(\text{SS}|\text{component 1 is up}) = 1.0$$

$P(\text{SS}|\text{component 1 is down}) = ?$ that is, given component 1 is down, the network configuration is reduced to the following network configuration (i.e., subsystem) (Fig. 6.47):

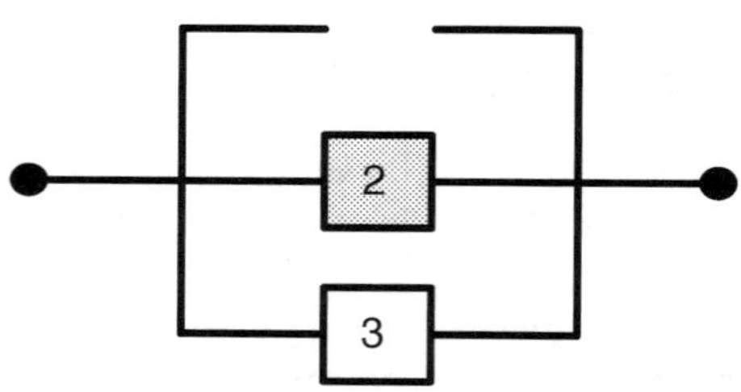


Figure 6.47. System configuration: three block diagrams connected in parallel, given component 1 is down.

$$P(\text{SS}|\text{component 1 is down}) = P(\text{SS}|\text{component 2 is up})R_2 + P(\text{SS}|\text{component 2 is down})Q_2$$
$$P(\text{SS}|\text{component 2 is up}) = 1.0$$
$$P(\text{SS}|\text{component 2 is down}) = R_3$$
$$P(\text{SS}|\text{component 1 is down}) = (1.0)R_2 + R_3Q_2$$

Substituting these results into the above equation yields

$$\begin{aligned} R_s &= P(\text{SS}|\text{component 1 is up})R_1 + P(\text{SS}|\text{component 1 is down})Q_1 \\ R_s &= (1.0)R_1 + [(1.0)R_2 + R_3Q_2]Q_1 \\ &= R_1 + [R_2 + R_3(1 - R_2)][1 - R_1] \\ R_s &= R_1 + R_2 + R_3 - R_1R_2 - R_2R_3 - R_3R_1 - R_1R_2R_3 \end{aligned}$$

## 6.7 CONCLUSIONS

This chapter has discussed a variety of methods for evaluating the reliability of complex systems or network configurations. The relationships and similarities between different techniques have been illustrated. The methods are used in real-life situations in one form or another for a variety of reasons and purposes. For example, the fault tree analysis is extensively used in nuclear power plant safety and reliability assessments. For an analyst, the first task is to assess the particular problem at hand and judge which method would render the required reliability assessment when applied.

## REFERENCES

1. P. A. Jensen and M. Bellemore, An algorithm to determine the reliability of a complex network, *IEEE Transactions on Reliability*, **R-25**, 1976, 169–174.
2. J. B. Fussell and W. E. Vessly, A new methodology for obtaining cut sets for fault trees, *Transactions of the American Nuclear Society*, **15**, 1972.
3. K. C. Kapur and L. R. Lamberson, *Reliability in Engineering Design*, John Wiley, Inc., 1977.
4. S. Garriba, P. Mussion, F. Naldi, G. Reina, and G. Volta, Efficient construction of minimal cut sets from fault trees, *IEEE Transactions on Reliability*, **R-26**, 1977, 88–93.
5. M. L. Shooman, *Probabilistic Reliability, An Engineering Approach*, McGraw-Hill, 1968.
6. G. J. Hahn and S. S. Shapiro, *Statistical Models in Engineering*, John Wiley, Inc., 1967.
7. E. Hansler, A fast recursive algorithm to calculate the reliability of a communication network, *IEEE Transactions on Communications*, **COM-20**, 1972, 637–640.
8. C. Singh and R. Billinton, A new method to determine the failure frequency of a complex system, *Microelectronics and Reliability*, **12**, 1973, 459–465.
9. A. C. Nelson, J. R. Batts, and R. L. Beadles, A computer program for approximating system reliability, *IEEE Transactions on Reliability*, **R-19**, 1970, 61–65.
10. R. N. Allan, R. Billinton, and M. F. De Olivera, An efficient algorithm for deducing the minimal cuts and reliability indices of a general network configuration, *IEEE Transactions on Reliability*, **R-25**, 1976, 226–233.

# 7

# DESIGNING RELIABILITY INTO INDUSTRIAL AND COMMERCIAL POWER SYSTEMS

## 7.1 INTRODUCTION

A description is given of how to make quantitative reliability and availability predictions for proposed new configurations of industrial power distribution systems. Seven examples are worked out, including a simple radial system, a primary selective system, and a secondary selective system. A brief tabulation of pertinent reliability data needed to make the reliability and availability predictions is also given. The simple radial system that was analyzed had an average number of forced hours of downtime per year that was 19 times larger than a secondary selective system; the failure rate was 6 times larger. The importance of having two separate power supply sources from the electric utility provider has been identified and analyzed. This approach could be used to assist in cost–reliability trade-off decisions in the design of the power distribution system.

An industrial power distribution system may receive power at 13.8 kV from an electric utility and then distribute that power throughout the plant for use at various locations. One of the questions often raised during the designing of the power distribution system is whether there is a way of making a quantitative comparison of the failure rate and the forced hours downtime per year for a secondary selective system with a primary

*Power Distribution System Reliability.* By Ali A. Chowdhury and Don O. Koval

selective system and a simple radial system. This comparison can then be used in cost–reliability and cost–availability trade-off decisions in the design of the power distribution system. The estimated cost of power outages at the various plant locations may also be factored into the decision regarding the type of power distribution system to use. The decisions may therefore be based upon "total owning cost over the useful life of the equipment" rather than on "first cost."

Only forced outages of the electrical equipment are considered in the seven examples to follow. It is assumed that scheduled maintenance will be performed at those times when 480 V power output is not needed. The frequency of scheduled outages and the average duration can be estimated, and, if necessary, these can be added to the forced outages given in the seven examples.

When doing a reliability study, it is necessary to define what a failure of the 480 V power is. Some of the failure definitions for 480 V power that are often used are as follows:

1. Complete loss for more than one cycle
2. Complete loss for more than 10 cycles
3. Complete loss for more than 5 s
4. Complete loss for more than 2 min

Definition 3 will be used in the seven examples studied in this chapter. This definition of failure can have an effect on the determination of the necessary speed of automatic throwover equipment used in primary selective or secondary selective systems. In some cases when conducting reliability studies, it might be necessary to further define what is "complete loss of incoming power," for example, "voltage drops below 70%."

## 7.2 EXAMPLE 1: SIMPLE RADIAL DISTRIBUTION SYSTEM

One of the main benefits of a reliability and availability analysis is that a disciplined look is taken at the alternative choices in the design of the power distribution system. By using published reliability data collected from industrial plants by a technical society, the best possible attempt is made to use historical experience to aid in the design of the new system.

The following seven examples of common low-voltage industrial power distribution systems are analyzed in this chapter:

1. Example 1—Simple radial
2. Example 2—Primary selective to 13.8 kV utility supply
3. Example 3—Primary selective to load side of 13.8 kV circuit breaker
4. Example 4—Primary selective to primary of transformer
5. Example 5—Secondary selective
6. Example 6—Simple radial with spares
7. Example 7—Simple radial system with cogeneration

### 7.2.1 Description of a Simple Radial System

A simple radial system is shown in Fig. 7.1. Power is received from the electric utility at 13.8 kV. It goes through a 13.8 kV circuit breaker inside the industrial plant, 600 ft of cable in underground conduit, an enclosed disconnect switch, to a transformer that reduces the voltage to 480 V, and then through a 480 V main circuit breaker, a second 480 V circuit breaker, 300 ft of cable in an aboveground conduit, to the point where the power is used in the industrial plant.

### 7.2.2 Results: Simple Radial System Example 1

The results from the reliability and availability calculations are given in Table 7.1. The failure and repair rates are obtained from IEEE Standard 493–2007.

### 7.2.3 Conclusions: Simple Radial System Example 1

The electric utility supply is the largest contributor to both the failure rate and the forced hours of downtime per year at the 480 V point of use. A significant improvement can be

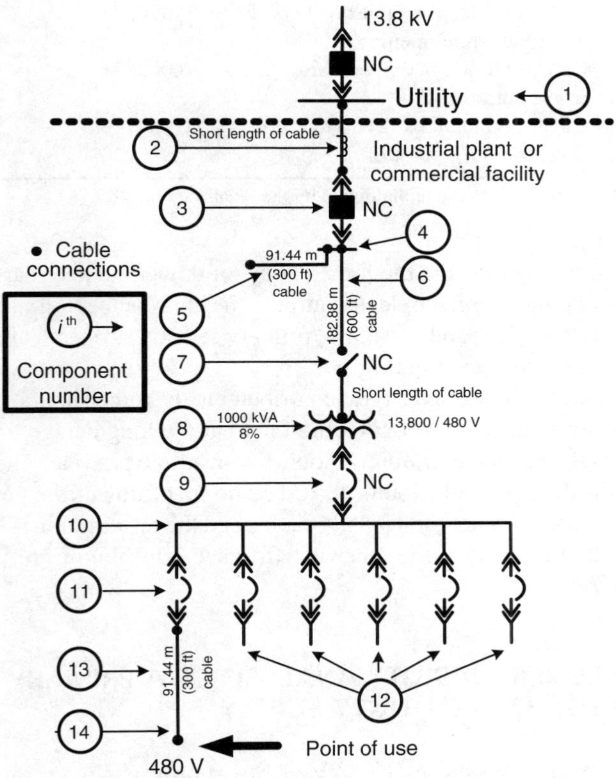


Figure 7.1. Simple radial system—Example 1.

TABLE 7.1. A Simple Radial System: Calculation of Failure Rate and Forced Hours Downtime per Year at 480 V Point of Use—Example 1

| Component Number | Component | $\lambda$ (failures/year) | $\lambda r$ (forced hours of downtime per year) |
|---|---|---|---|
| 1 | 13.8 kV power source from electric utility | 1.956000 | 2.582000 |
| 2 | Primary protection and control system | 0.000600 | 0.003000 |
| 3 | 13.8 kV metalclad circuit breaker | 0.001850 | 0.000925 |
| 4 | 13.8 kV switchgear bus—insulated | 0.004100 | 0.153053 |
| 5 | Cable (13.8 kV); 900 ft, conduit below ground | 0.002124 | 0.033347 |
| 6 | Cable terminations (8) at 13.8 kV | 0.002960 | 0.002220 |
| 7 | Disconnect switch (enclosed) | 0.001740 | 0.001740 |
| 8 | Transformer | 0.010800 | 1.430244 |
| 9 | 480 V metalclad circuit breaker | 0.000210 | 0.001260 |
| 10 | 480 V switchgear bus—bare | 0.009490 | 0.069182 |
| 11 | 480 V metalclad circuit breaker | 0.000210 | 0.001260 |
| 12 | 480 V metalclad circuit breakers (5) (failed while opening) | 0.000095 | 0.000378 |
| 13 | Cable (480 V); 300 ft, conduit aboveground | 0.000021 | 0.000168 |
| 14 | Cable terminations (2) at 480 V | 0.000740 | 0.000555 |
| | Total at 480 V point of use | 1.990940 | 4.279332 |

*Note*: $\lambda r$ is the product of the failure rate multiplied by the repair duration "*r*."

made in both the failure rate and the forced hours of downtime per year by having two 13.8 kV sources of power from the electric utility. The improvements that can be obtained are shown in Examples 2, 3, and 4 using a "primary selective system" and in Example 5 using a "secondary selective system."

The transformer is the second largest contributor to the forced hours of downtime per year. The transformer has a very low failure rate, but the long outage time of 132.43 h after a failure results in a large $\lambda r$, forced hours of downtime per year. The 3.8 kV circuit breaker is the third largest contributor to forced hours of downtime per year, and the fourth largest contributors are the 13.8 kV cables and the terminations. The failure rate and the forced hours of downtime per year for each contributor are summarized in Tables 7.2 and 7.3.

## 7.3 EXAMPLE 2: RELIABILITY ANALYSIS OF A PRIMARY SELECTIVE SYSTEM TO THE 13.8 kV UTILITY SUPPLY

The primary selective system to 13.8 kV utility supply is shown in Fig. 7.2. Two subexamples will be considered.

TABLE 7.2. A Simple Radial System—Example 1—Relative Ranking of Failure Rates

| Ranking | Component | λ (failures/year) |
|---|---|---|
| 1 | 13.8 kV power source from electric utility | 1.956000 |
| 8 | Transformer | 0.010800 |
| 10 | 480 V switchgear bus—bare | 0.009490 |
| 4 | 13.8 kV switchgear bus—insulated | 0.004100 |
| 6 | Cable terminations (8) at 13.8 kV | 0.002960 |
| 5 | Cable (13.8 kV); 900 ft, conduit below ground | 0.002124 |
| 3 | 13.8 kV metalclad circuit breaker | 0.001850 |
| 7 | Disconnect switch (enclosed) | 0.001740 |
| 14 | Cable terminations (2) at 480 V | 0.000740 |
| 2 | Primary protection control system | 0.000600 |
| 9 | 480 V metalclad circuit breaker | 0.000210 |
| 11 | 480 V metalclad circuit breaker | 0.000210 |
| | Total at 480 V point of use | 1.990940 |

*Example 2a*: After a power failure of utility power source number 1, assume a 9 min "*manual switchover time*" to utility power source number 2.

*Example 2b*: Assume an "*automatic switchover time*" of less than 5 s after a failure is assumed. (*Note*: loss of 480 V power for less than 5 s is not counted as a failure.)

### 7.3.1 Description: Primary Selective System to the 13.8 kV Utility Supply

This is a simple radial system with the addition of a second 13.8 kV power source from the electric utility; the second power source is normally disconnected. In the event of

TABLE 7.3. Simple Radial System—Example 1—Relative Ranking of Forced Hours of Downtime per Year

| Ranking | Component | λ (failures/year) |
|---|---|---|
| 1 | 13.8 kV power source from electric utility | 2.582000 |
| 8 | Transformer | 1.430244 |
| 4 | Switchgear bus—insulated | 0.153053 |
| 10 | Switchgear bus—bare | 0.069182 |
| 5 | Cable connections (8) at 13.8 kV | 0.033347 |
| 2 | 480 V metalclad circuit breakers (5) (failed while opening) | 0.003823 |
| 6 | Primary protection and control system | 0.003000 |
| 7 | Cable connections (8) at 13.8 kV | 0.002220 |
| 12 | Disconnect switch (enclosed) | 0.001740 |
| 9 | 480 V metalclad circuit breaker | 0.001260 |
| 11 | 480 V metalclad circuit breaker | 0.001260 |
| 3 | 13.8 kV metalclad circuit breaker | 0.000925 |
| | Total at 480 V output | 4.279332 |

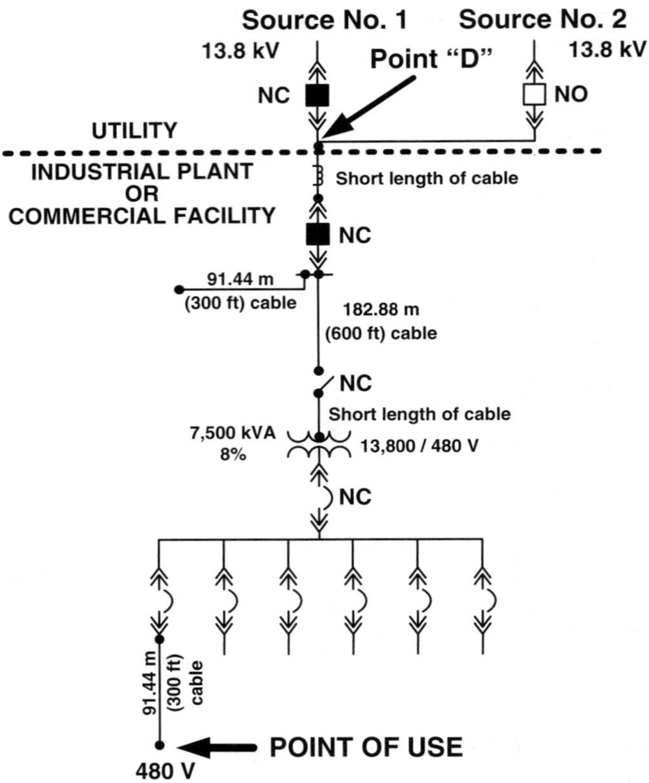


Figure 7.2. Example 2: A primary selective system to the 13.8 kV utility supply.

a failure in the first 13.8 kV utility power source, the second 13.8 kV utility power source is switched on to replace the failed power source. Assume that the two utility power sources are synchronized.

### 7.3.2 Results: A Primary Selective System to the 13.8 kV Utility Supply

*Example 2a*: If the time to switch to a second utility power source takes 9 min after a failure of the first source, then there would be a power supply failure of 9 min duration (Table 7.4). Using the data from *IEEE Standard 493-2007*, for double-circuit utility supplies, this would occur *1.644 times per year* (1.956–0.312). This is in addition to losing both power sources simultaneously *0.312 times per year*, for an average outage time of 0.52 h. Adding these utility supply data together, for the simple radial system, would result in a reduction of the forced hours of downtime per year at the 480 V point of use from 4.3033 to 2.1301. *The failure rate, however, would stay the same at 1.9896 failures/year.*

*Example 2b*: If the time to switch to a second utility power source takes less than 5 s after a failure of the first source, then there would be *no failure* of the electric utility power

TABLE 7.4. A Primary Selective System to the 13.8 kV Utility Supply; Calculation of Failure Rate and Forced Hours Downtime per Year at 480 V Point of Use—Example 2a (9 min Manual Switchover Time)

| Component Number | Component | λ (failures/year) | λr (forced hours of downtime per year) |
|---|---|---|---|
| 1 | 13.8 kV power source from electric utility | 1.644000 | |
| 2 | Primary protection and control system | 0.000600 | |
| 3 | 13.8 kV metalclad circuit breaker | 0.001850 | |
| | | 1.646450 | 0.246968 |
| | Loss of both 13.8 kV power sources simultaneously | 0.312000 | 0.162240 |
| Total to point D | | 1.958450 | 0.409208 |
| 4 | 13.8 kV switchgear bus—insulated | 0.004100 | 0.153053 |
| 5 | Cable (13.8 kV); 900 ft, conduit belowground | 0.002124 | 0.033347 |
| 6 | Cable terminations (8) at 13.8 kV | 0.002960 | 0.002220 |
| 7 | Disconnect switch (enclosed) | 0.001740 | 0.001740 |
| 8 | Transformer | 0.010800 | 1.430244 |
| 9 | 480 V metalclad circuit breaker | 0.000210 | 0.001260 |
| 10 | 480 V switchgear bus—bare | 0.009490 | 0.069182 |
| 11 | 480 V metalclad circuit breaker | 0.000210 | 0.001260 |
| 12 | 480 V metalclad circuit breakers (5) (failed while opening) | 0.000095 | 0.000378 |
| 13 | Cable (480 V); 300 ft, conduit aboveground | 0.000021 | 0.000168 |
| 14 | Cable terminations (2) at 480 V | 0.000740 | 0.000555 |
| | Total at 480 V point of use | 1.990940 | 2.102614 |

supply (Table 7.5). The only time a failure of the utility power source would occur is when both sources fail simultaneously. It will be assumed that the data shown in the Gold Book are applicable to the loss of both power supply circuits simultaneously. This is 0.312 failures/year with an average outage time of 0.52 h (i.e., $\lambda r \approx 0.1622$).

A comparison of the results for the simple radial system and the primary selective system to a 13.8 kV utility supply is shown in Table 7.6.

### 7.3.3 Conclusions: Primary Selective System to 13.8 kV Utility Supply

The use of a primary selective to the 13.8 kV utility supply with a 9 min manual switchover time reduces the forced hours downtime per year at the 480 V point of use by about 50%, but the failure rate is the same as for a simple radial system.

TABLE 7.5. A Primary Selective System to the 13.8 kV Utility Supply: Calculation of the Frequency and Duration of Interruptions to Point of Use (480 V Load)—Example 2b

| Number | Component | λ (failures/year) | λr (forced hours of downtime per year) |
|---|---|---|---|
| 1 | 13.8 kV power source from electric utility | 1.644000 | |
| 2 | Primary protection and control system | 0.000600 | |
| 3 | 13.8 kV metalclad circuit breaker | 0.001850 | |
| | | 0.0 | 0.0 |
| | Loss of both 13.8 kV power sources simultaneously | 0.312000 | 0.162240 |
| Total to point D | | 0.312000 | 0.162240 |
| 4 | 13.8 kV switchgear bus—insulated | 0.004100 | 0.153053 |
| 5 | Cable (13.8 kV); 900 ft, conduit belowground | 0.002124 | 0.033347 |
| 6 | Cable terminations (8) at 13.8 kV | 0.002960 | 0.002220 |
| 7 | Disconnect switch (enclosed) | 0.001740 | 0.001740 |
| 8 | Transformer | 0.010800 | 1.430244 |
| 9 | 480 V metalclad circuit breaker | 0.000210 | 0.001260 |
| 10 | 480 V switchgear bus—bare | 0.009490 | 0.069182 |
| 11 | 480 V metalclad circuit breaker | 0.000210 | 0.001260 |
| 12 | 480 V metalclad circuit breakers (5) (failed while opening) | 0.000095 | 0.000378 |
| 13 | Cable (480 V); 300 ft, conduit aboveground | 0.000021 | 0.000168 |
| | Total at 480 V point of use | 0.344490 | 1.855647 |

Automatic switchover can be accomplished in less than 5 s after an outage of utility source 1 assuming that the loss of 480 V power for less than 5 s is not counted as an outage.

TABLE 7.6. Comparison of a Simple Radial System and a Primary Selective System to 13.8 kV Utility Supply

| System Configuration | λ (failures/year) | λr (forced hours of downtime per year) |
|---|---|---|
| Example 1 | | |
| Simple radial system | 1.990940 | 4.279332 |
| Example 2a | | |
| Primary selective system to 13.8 kV utility supply (with 9 min switchover after a supply failure) | 1.990940 | 2.102614 |
| Example 2b | | |
| Primary selective system to 13.8 kV utility supply (with switchover less than 5 s after a supply failure) | 0.344490 | 1.855647 |

The use of automatic throwover equipment that can sense a failure of one 13.8 kV utility supply and switchover to the second supply in less than 5 s would give a 6–1 improvement in the failure rate at the 480 V point of use (a loss of 480 V power for less than 5 s is not counted as a failure).

## 7.4 EXAMPLE 3: A PRIMARY SELECTIVE SYSTEM TO THE LOAD SIDE OF A 13.8 kV CIRCUIT BREAKER

### 7.4.1 Description of a Primary Selective System to the Load Side of a 13.8 kV Circuit Breaker

Figure 7.3 shows a one-line diagram of the power distribution system for a primary selective to primary of transformer. What are the failure rate and the forced hours downtime per year at the 480 V point of use?

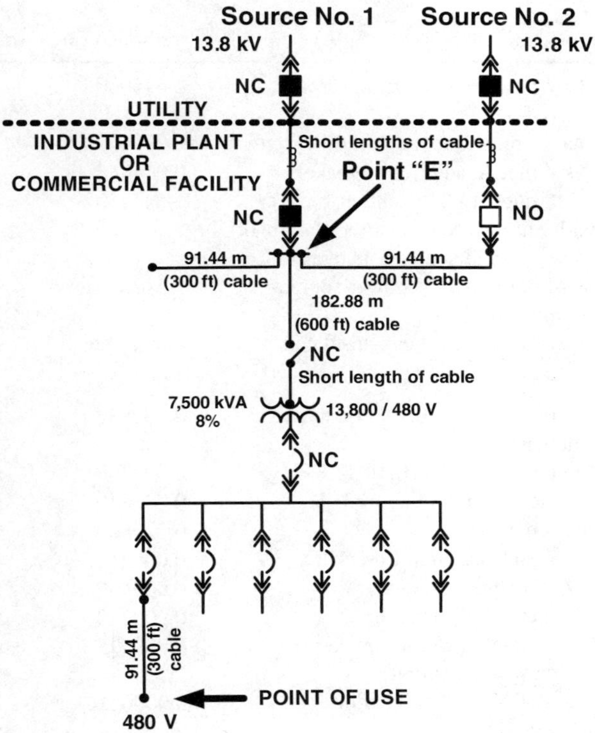


Figure 7.3. Example 3: A primary selective system to the load side of a 13.8 kV circuit breaker.

Example 3a—utility switchover time 9.0 min
Example 3b—utility switchover time 5 s

(Assuming that the 5 s switchover time will not result in a load failure at the 480 V point of use.)

### 7.4.2 Results: Primary Selective System to Load Side of 13.8 kV Circuit Breaker

The results from the reliability and availability calculations for a 9.0 min utility switchover time are given in Table 7.7 and for a 5 s utility switchover time are given in Table 7.8.

TABLE 7.7. A Primary Selective System to the Load Side of a 13.8 kV Circuit Breaker: Calculation of the Frequency and Duration of Interruptions to Point of Use (480 V Load) Assuming a 9 min "Manual Switchover Time" to Utility Power Source 2

| Component Number | Component | λ (failures/year) | λr (forced hours of downtime per year) |
|---|---|---|---|
| 1 | 13.8 kV power source from electric utility | 1.644000 | |
| 2 | Primary protection and control system | 0.000600 | |
| 3 | 13.8 kV metalclad circuit breaker | 0.001850 | |
| | Total through 13.8 kV circuit breaker with 9 min switchover after a failure of source 1 (and source 2 is okay) | 1.646450 | 0.246968 |
| | Loss of both 13.8 kV power sources simultaneously | 0.312000 | 0.162240 |
| 4 | 13.8 kV switchgear bus—insulated | 0.004100 | 0.153053 |
| Total to point E | | 1.962550 | 0.562261 |
| 5 | Cable (13.8 kV); 1200 ft, conduit belowground | 0.002832 | 0.044462 |
| 6 | Cable terminations (10) at 13.8 kV | 0.003700 | 0.002775 |
| 7 | Disconnect switch (enclosed) | 0.001740 | 0.001740 |
| 8 | Transformer | 0.010800 | 1.430244 |
| 9 | 480 V metalclad circuit breaker | 0.000210 | 0.001260 |
| 10 | 480 V switchgear bus—bare | 0.009490 | 0.069182 |
| 11 | 480 V metalclad circuit breaker | 0.000210 | 0.001260 |
| 12 | 480 V metalclad circuit breakers (5) (failed while opening) | 0.000095 | 0.000378 |
| 13 | Cable (480 V); 300 ft, conduit aboveground | 0.000021 | 0.000168 |
| 14 | Cable terminations (2) at 480 V | 0.000740 | 0.000555 |
| Total at 480 V point of use | | 1.992388 | 2.114285 |

TABLE 7.8. A Primary Selective System to the Load Side of 13.8 kV Circuit Breaker: Calculation of the Frequency and Duration of Interruptions to Point of Use (480 V Load) Assuming 5 s "Automatic Transfer" to Utility Power Source 2

| Component Number | Component | $\lambda$ (failures/year) | $\lambda r$ (forced hours of downtime per year) |
|---|---|---|---|
| 1 | 13.8 kV power source from electric utility | | |
| 2 | Primary protection and control system | | |
| 3 | 13.8 kV metalclad circuit breaker | | |
| | Total through 13.8 kV circuit breaker with 9 s switchover after a failure of source 1 (and source 2 is okay) | 0.0 | 0.0 |
| | Loss of both 13.8 kV power sources simultaneously | 0.312000 | 0.162240 |
| 4 | 13.8 kV switchgear bus—insulated | 0.004100 | 0.153053 |
| Total to point E | | 0.316100 | 0.315293 |
| 5 | Cable (13.8 kV); 1200 ft, conduit below ground | 0.002832 | 0.044462 |
| 6 | Cable terminations (10) at 13.8 kV | 0.003700 | 0.002775 |
| 7 | Disconnect switch (enclosed) | 0.001740 | 0.001740 |
| 8 | Transformer | 0.010800 | 1.430244 |
| 9 | 480 V metalclad circuit breaker | 0.000210 | 0.001260 |
| 10 | 480 V switchgear bus—bare | 0.009490 | 0.069182 |
| 11 | 480 V metalclad circuit breaker | 0.000210 | 0.001260 |
| 12 | 480 V metalclad circuit breakers (5) (failed while opening) | 0.000095 | 0.000378 |
| 13 | Cable (480 V); 300 ft, conduit aboveground | 0.000021 | 0.000168 |
| 14 | Cable terminations (2) at 480 V | 0.000740 | 0.000555 |
| Total at 480 V point of use | | 0.345938 | 1.867318 |

### 7.4.3 Conclusions: A Primary Selective System to the Load Side of a 13.8 kV Circuit Breaker

The forced hours downtime per year at the 480 V point of use in Example 3 (primary selective to load side of 13.8 kV circuit breaker) is about 10% lower than in Example 2 (primary selective to 13.8 kV utility supply). The failure rate is about the same.

## 7.5 EXAMPLE 4: A PRIMARY SELECTIVE SYSTEM TO THE PRIMARY OF THE TRANSFORMER

### 7.5.1 Description of a Primary Selective System to the Primary of the Transformer

Figure 7.4 shows a one-line diagram of the power distribution system for a primary selective system to the primary of transformer. What are the failure rate and the forced hours downtime per year at the 480 V point of use?

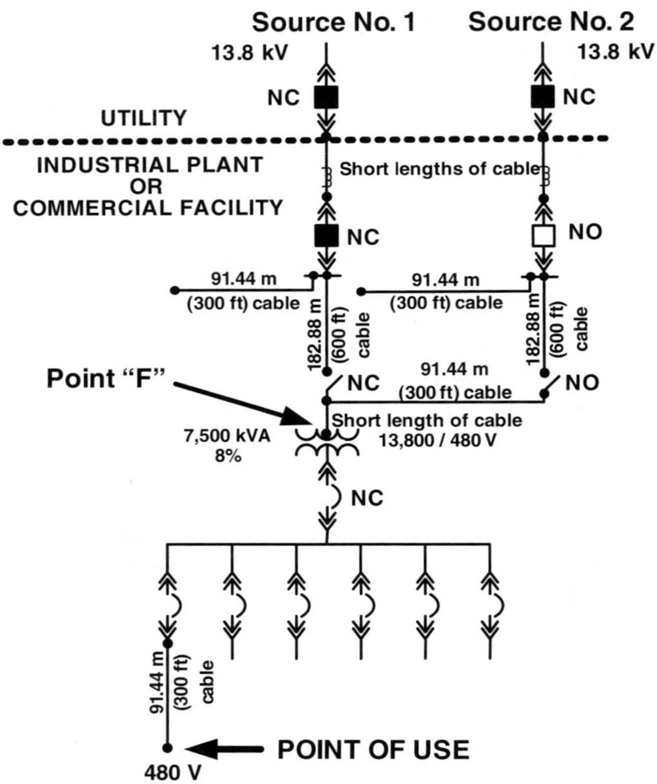


Figure 7.4. Example 4: A primary selective system to the primary of a transformer.

## 7.5.2 Results: A Primary Selective System to the Primary of the Transformer

The results from the reliability and availability calculations are given in Tables 7.9 and 7.10.

## 7.5.3 Conclusions: Primary Selective System to Primary of Transformer

The forced hours downtime per year at the 480 V point of use in Example 4 (primary selective to primary of transformer) is about 32% lower than for the simple radial system in Example 1. The failure rate is the same in Examples 4 and 1.

# 7.6 EXAMPLE 5: A SECONDARY SELECTIVE SYSTEM

## 7.6.1 Description of a Secondary Selective System

A one-line diagram of the power distribution system for a secondary selective system is provided in Fig. 7.5. What are the failure rate and forced hours of downtime per year at the 480 V point of use?

TABLE 7.9. A Primary Selective System to the Primary of the Transformer: Calculation of the Frequency and Duration of Interruptions to Point of Use (480 V Load) Assuming a 9 min "Manual Switchover Time" to Utility Power Source 2

| Component Number | Component | λ (failures/year) | λr (forced hours of downtime per year) |
|---|---|---|---|
| 1 | 13.8 kV power source from electric utility | 1.644000 | |
| 2 | Primary protection and control system | 0.000600 | |
| 3 | 13.8 kV metalclad circuit breaker | 0.001850 | |
| 4 | 13.8 kV switchgear bus—insulated | 0.004100 | 0.153053 |
| 5 | Cable (13.8 kV); 1200 ft, conduit below ground | 0.002832 | 0.044462 |
| 6 | Cable terminations (9) at 13.8 kV | 0.003330 | 0.002498 |
| 7 | Disconnect switch (enclosed) | 0.001740 | 0.001740 |
| | Total through 13.8 kV circuit breaker with 9 min switchover after a failure of source 1 (and source 2 is okay) | 1.658452 | 0.248768 |
| | Loss of both 13.8 kV power sources simultaneously | 0.312000 | 0.162240 |
| Total to point F | | 1.970452 | 0.411008 |
| 8 | Transformer | 0.010800 | 1.430244 |
| 9 | 480 V metalclad circuit breaker | 0.000210 | 0.001260 |
| 10 | 480 V switchgear bus—bare | 0.009490 | 0.069182 |
| 11 | 480 V metalclad circuit breaker | 0.000210 | 0.001260 |
| 12 | 480 V metalclad circuit breakers (5) (failed while opening) | 0.000095 | 0.000378 |
| 13 | Cable (480 V); 300 ft, conduit aboveground | 0.000021 | 0.000168 |
| 14 | Cable terminations (2) at 480 V | 0.000740 | 0.000555 |
| Total at 480 V point of use | | 1.992018 | 1.914055 |

### 7.6.2 Results: A Secondary Selective System

The results from the reliability and availability calculations are given in Tables 7.11 and 7.12.

### 7.6.3 Conclusions: A Secondary Selective System

The simple radial system in Example 1 had an average forced hours downtime per year that was 18 times larger than the secondary selective system in Example 5b with automatic throwover in less than 5 s. The failure rate of the simple radial system was six times larger than the secondary selective system in Example 5b with automatic switchover in less than 5 s.

TABLE 7.10. A Primary Selective System to the Primary of the Transformer: Calculation of the Frequency and Duration of Interruptions to Point of Use (480 V Load) Assuming a 5 s "Automatic Transfer" to Utility Power Source 2

| Component Number | Component | λ (failures/year) | λr (forced hours of downtime per year) |
|---|---|---|---|
| 1 | 13.8 kV power source from electric utility | 1.644000 | |
| 2 | Primary protection and control system | 0.000600 | |
| 3 | 13.8 kV metalclad circuit breaker | 0.001850 | |
| 4 | 13.8 kV switchgear bus—insulated | 0.004100 | 0.153053 |
| 5 | Cable (13.8 kV); 1200 ft, conduit belowground | 0.002832 | 0.044462 |
| 6 | Cable terminations (9) at 13.8 kV | 0.003330 | 0.002498 |
| 7 | Disconnect switch (enclosed) | 0.001740 | 0.001740 |
| | Total through 13.8 kV circuit breaker with 9 min switchover after a failure of source 1 (and source 2 is okay) | 0.0 | 0.0 |
| | Loss of both 13.8 kV power sources simultaneously | 0.312000 | 0.162240 |
| Total to point F | | 0.312000 | 0.162240 |
| 8 | Transformer | 0.010800 | 1.430244 |
| 9 | 480 V metalclad circuit breaker | 0.000210 | 0.001260 |
| 10 | 480 V switchgear bus—bare | 0.009490 | 0.069182 |
| 11 | 480 V metalclad circuit breaker | 0.000210 | 0.001260 |
| 12 | 480 V metalclad circuit breakers (5) (failed while opening) | 0.000095 | 0.000378 |
| 13 | Cable (480 V); 300 ft, conduit aboveground | 0.000021 | 0.000168 |
| 14 | Cable terminations (2) at 480 V | 0.000740 | 0.000555 |
| Total at 480 V point of use | | 0.333566 | 1.665287 |

## 7.7 EXAMPLE 6: A SIMPLE RADIAL SYSTEM WITH SPARES

### 7.7.1 Description of a Simple Radial System with Spares

What are the failure rate and forced hours of downtime per year of the 480 V point of use if all of the following spare parts are available and can be installed as a replacement in these average times?

1. 13.8 kV circuit breaker (inside plant only)—2.1 h
2. 900 ft of cable (13.8 kV)—19 h
3. 1000 kVA transformer—130 h

The above three "*replace with spare*" times were calculated from the actual values obtained from the IEEE Reliability Survey of Industrial Plants (IEEE Standard

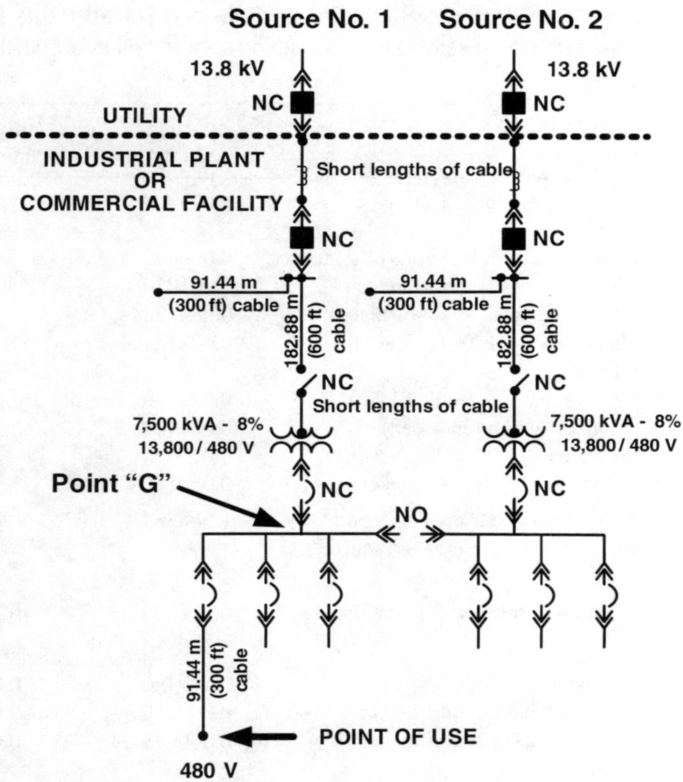


Figure 7.5. Example 5: Secondary selective system.

493-2007). The times are much lower than the "repair failed unit" times that were used in Examples 1–5.

## 7.7.2 Results: A Simple Radial System with Spares

The results of the reliability and availability calculations are given in Table 7.13. They are compared with those of the simple radial system in Example 1 using average outage times based upon "*repair failed unit.*"

## 7.7.3 Conclusions: Simple Radial System with Spares

The simple radial system with spares in Example 6 had a forced hours downtime per year that was 21% lower than the simple radial system in Example 1. The failure rate is the same in both examples.

TABLE 7.11. Secondary Selective System: Failure Rate and Forced Hours Downtime per Year at 480 V Point of Use—Example 5a—Assuming a 9 min "Manual Switchover Time" to Utility Power Source 2

| Component Number | Component | λ (failures/year) | λr (forced hours of downtime per year) |
|---|---|---|---|
| 1 | 13.8 kV power source from electric utility | 1.644000 | |
| 2 | Primary protection and control system | 0.000600 | |
| 3 | 13.8 kV metalclad circuit breaker | 0.001850 | |
| 4 | 13.8 kV switchgear bus—insulated | 0.004100 | |
| 5 | Cable (13.8 kV); 1200 ft, conduit belowground | 0.002124 | |
| 6 | Cable terminations (9) at 13.8 kV | 0.002960 | |
| 7 | Disconnect switch (enclosed) | 0.001740 | |
| 8 | Transformer | 0.010800 | |
| 9 | 480 V metalclad circuit breaker | 0.000210 | |
| Total through 13.8 kV circuit breaker with 9 min switchover after a failure of source 1 (and source 2 is okay) | | 1.668384 | 0.250258 |
| Loss of both 13.8 kV power sources simultaneously | | 0.312000 | 0.162240 |
| Total to point G | | 1.980384 | 0.412498 |
| 10 | 480 V switchgear bus—bare | 0.009490 | 0.069182 |
| 11 | 480 V metalclad circuit breaker | 0.000210 | 0.001260 |
| 12 | 480 V metalclad circuit breakers (5) (failed while opening) | 0.000038 | 0.000151 |
| 13 | Cable (480 V); 300 ft, conduit aboveground | 0.000021 | 0.000168 |
| 14 | Cable terminations (2) at 480 V | 0.000740 | 0.000555 |
| Total at 480 V point of use | | 1.990883 | 0.483814 |

## 7.8 EXAMPLE 7: A SIMPLE RADIAL SYSTEM WITH COGENERATION

### 7.8.1 Description of a Simple Radial System with Cogeneration

A single-line diagram of the power distribution system for a simple radial system with cogeneration is shown in Fig. 7.6. What are the failure rates and forced hours of downtime per year at the 480 V point of use, assuming that the utility and cogeneration sources are operated in parallel?

### 7.8.2 Results: Simple Radial System with Cogeneration

The results from the reliability and availability calculations are given in Table 7.14.

TABLE 7.12. A Secondary Selective System: Failure Rate and Forced Hours Downtime per Year at 480 V Point of Use—Example 5b—Assuming a 5 s "Automatic Transfer" to Utility Power Source 2

| Component Number | Component | λ (failures/year) | λr (forced hours of downtime per year) |
|---|---|---|---|
| 1 | 13.8 kV power source from electric utility | 1.644000 | |
| 2 | Primary protection and control system | 0.000600 | |
| 3 | 13.8 kV metalclad circuit breaker | 0.001850 | |
| 4 | 13.8 kV switchgear bus—insulated | 0.004100 | |
| 5 | Cable (13.8 kV); 1200 ft, conduit belowground | 0.002124 | |
| 6 | Cable terminations (9) at 13.8 kV | 0.002960 | |
| 7 | Disconnect switch (enclosed) | 0.001740 | |
| 8 | Transformer | 0.010800 | |
| 9 | 480 V metalclad circuit breaker | 0.000210 | |
| Total through 13.8 kV circuit breaker with 9 min switchover after a failure of source 1 (and source 2 is okay) | | 0.00 | 0.0 |
| Loss of both 13.8 kV power sources simultaneously | | 0.312000 | 0.162240 |
| Total to point G | | 0.312000 | 0.162240 |
| 10 | 480 V switchgear bus—bare | 0.009490 | 0.069182 |
| 11 | 480 V metalclad circuit breaker | 0.000210 | 0.001260 |
| 12 | 480 V metalclad circuit breakers (5) (failed while opening) | 0.000038 | 0.000151 |
| 13 | Cable (480 V); 300 ft, conduit aboveground | 0.000021 | 0.000168 |
| 14 | Cable terminations (2) at 480 V | 0.000740 | 0.000555 |
| Total at 480 V point of use | | 0.322499 | 0.233556 |

### 7.8.3 Conclusions: A Simple Radial System with Cogeneration

The simple radial system in Example 1 yielded an average forced hours downtime per year that was about twice as large as the radial system with cogeneration in Example 7. The largest contributor to the average forced hours of downtime per year is the transformer; for example, if the transformer was replaced with a spare in 48 h, the downtime per year would be 0.781933 h compared to 1.741527 h and 0.522733 h compared to 1.741527 for a 24 h spare changeout. The failure rate of the simple radial system was about 37 times larger than the radial system with cogeneration in Example 7.

TABLE 7.13. Simple Radial System With and Without a Spare Transformer: Failure Rate and Forced Hours Downtime per Year at 480 V Point of Use—Example 6

| Component Number | Component | λ (failures/year) | λr (forced hours of downtime per year) |
|---|---|---|---|
| 1 | 13.8 kV power source from electric utility | 1.956000 | 2.582000 |
| 2 | Primary protection and control system | 0.000600 | 0.003000 |
| 3 | 13.8 kV metalclad circuit breaker | 0.001850 | 0.000925 |
| 4 | 13.8 kV switchgear bus—insulated | 0.004100 | 0.153053 |
| 5 | Cable (13.8 kV); 900 ft, conduit belowground | 0.002124 | 0.033347 |
| 6 | Cable terminations (8) at 13.8 kV | 0.002960 | 0.002220 |
| 7 | Disconnect switch (enclosed) | 0.001740 | 0.001740 |
| 8 | Transformer—replace with spare when it fails—48 h | 0.010800 | 0.518400 |
| 9 | 480 V metalclad circuit breaker | 0.000210 | 0.001260 |
| 10 | 480 V switchgear bus—bare | 0.009490 | 0.069182 |
| 11 | 480 V metalclad circuit breaker | 0.000210 | 0.001260 |
| 12 | 480 V metalclad circuit breakers (5) (failed while opening) | 0.000095 | 0.000378 |
| 13 | Cable (480 V); 300 ft, conduit aboveground | 0.000021 | 0.000168 |
| 14 | Cable terminations (2) at 480 V | 0.000740 | 0.000555 |
| Total at 480 V point of use (with a spare transformer) | | 1.990940 | 3.367488 |
| Total at 480 V point of use (without a spare transformer) | | 1.990940 | 4.279332 |

## 7.9 RELIABILITY EVALUATION OF MISCELLANEOUS SYSTEM CONFIGURATIONS

### Example 7.1

An industrial power system configuration supplies power to an industrial process from an on-site cogenerator unit (G1) as shown in Fig. 7.7. Cable 1 path and Cable 2 path are fully redundant.

The reliability data for the industrial power system electrical components are shown in Table 7.15.

(a) Calculate the load point reliability of the industrial process for the existing configuration shown in the above figure.
Block 1: G1–CB1–T1–CB2
Block 2: CB3–cable 1–CB4

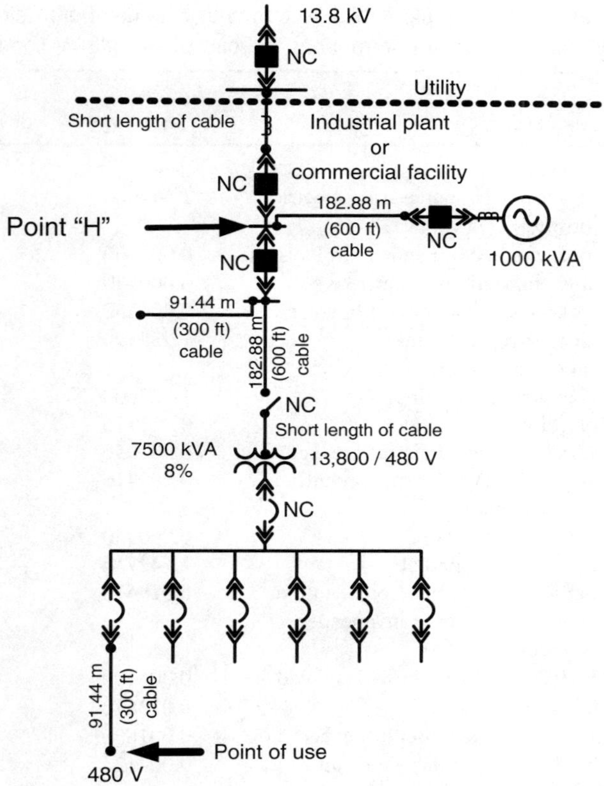


Figure 7.6. A simple radial system with cogeneration.

Block 3: CB5–cable 2–CB6
Block 4: cable 3–CB7

$$\begin{aligned} R(\text{block 1}) &= R(\text{G1}) \times R(\text{CB1}) \times R(\text{CB2}) \times R(\text{T1}) \\ &= 0.9500 \times (0.9985) \times 0.9985 \times (0.9950) = 0.94241637681250 \end{aligned}$$

$$\begin{aligned} R(\text{block 2}) &= R(\text{CB3}) \times R(\text{cable 1}) \times R(\text{CB4}) = 0.9975 \times (0.9900) \times 0.9975 \\ &= 0.98505618750000 \end{aligned}$$

$$\begin{aligned} R(\text{block 3}) &= R(\text{CB5}) \times R(\text{cable 2}) \times R(\text{CB6}) = 0.9975 \times (0.9900) \times 0.9975 \\ &= 0.98505618750000 \end{aligned}$$

$$\begin{aligned} R(\text{block 4}) &= R(\text{cable 3}) \times R(\text{CB7}) = (0.9900) \times 0.9975 \\ &= 0.98752500000000 \end{aligned}$$

$$\begin{aligned} R(\text{block 2//block 3}) &= R(\text{block 2}) + R(\text{block 3}) - R(\text{block 2}) \times R(\text{block 3}) \\ &= 0.99977668246796 \end{aligned}$$

$$\begin{aligned} R(\text{load point}) &= R(\text{block 1}) \times R(\text{block 2//block 3}) \times R(\text{block 4}) \\ &= 0.93045189987714 \end{aligned}$$

TABLE 7.14. Example 7: A Simple Radial System with Cogeneration: Calculation of the Frequency and Duration of Interruptions to Point of Use (480 V Load)

| Component Number | Component | λ (failures/year) | λr (forced hours of downtime per year) |
|---|---|---|---|
| | Utility supply | | |
| 1 | 13.8 kV power source from electric utility | 1.644000 | 2.582000 |
| 2 | Primary protection and control system | 0.000600 | 0.003000 |
| | Cable connections (2) at 13.8 kV | 0.000740 | 0.000555 |
| 3 | 13.8 kV metalclad circuit breaker | 0.001850 | 0.000925 |
| | **Utility source subtotal** | **1.959190** | **2.586480** |
| | Local cogeneration | | |
| | Generator (gas turbine) | 1.727600 | 47.318964 |
| | Control panel generator | 0.011110 | 0.023442 |
| | 13.8 kV metalclad circuit breaker | 0.001850 | 0.000925 |
| | Cable (13.8 kV); 600 ft, conduit belowground | 0.001416 | 0.022231 |
| | Cable connections (2) at 13.8 kV | 0.000740 | 0.000555 |
| | Cogeneration subtotal | **1.742716** | **47.366117** |
| | Combined utility and cogeneration sources (assuming independent sources) | **0.019470** | **0.047750** |
| | 13.8 kV switchgear bus—insulated | 0.004100 | 0.153053 |
| | **Total to point H** | **0.023570** | **0.200803** |
| 4 | 13.8 kV metalclad circuit breaker | 0.001850 | 0.000925 |
| 5 | Cable (13.8 kV); 900 ft, conduit belowground | 0.002124 | 0.033347 |
| 6 | Cable connections (6) at 13.8 kV | 0.002220 | 0.001665 |
| 7 | Disconnect switch (enclosed) | 0.001740 | 0.001740 |
| 8 | Transformer | 0.010800 | 1.430244 |
| 9 | 480 V metalclad circuit breaker | 0.000210 | 0.001260 |
| 10 | 480 V switchgear bus—bare | 0.009490 | 0.069182 |
| 11 | 480 V metalclad circuit breaker | 0.000210 | 0.001260 |
| 12 | 480 V metalclad circuit breakers (5) (failed while opening) | 0.000095 | 0.000378 |
| 13 | Cable (480 V); 300 ft, conduit above ground | 0.000021 | 0.000168 |
| 14 | Cable connections (2) at 480 V | 0.000740 | 0.000555 |
| **Total at 480 V point of use** | | 0.053069 | 1.741527 |

Data for hours of downtime per failure are based on *repair failed unit*.

(b) Calculate the load point reliability of the industrial process if the cable 1 feed is removed from service for maintenance (i.e., CB3 and CB4 are opened).

$$\begin{aligned} R(\text{load point}) &= R(\text{block } 1) \times R(\text{block } 3) \times R(\text{block } 4) \\ &= 0.94241637681250 \times (0.98505618750000) \times 0.98752500000000 \\ &= 0.91675212796781 \end{aligned}$$

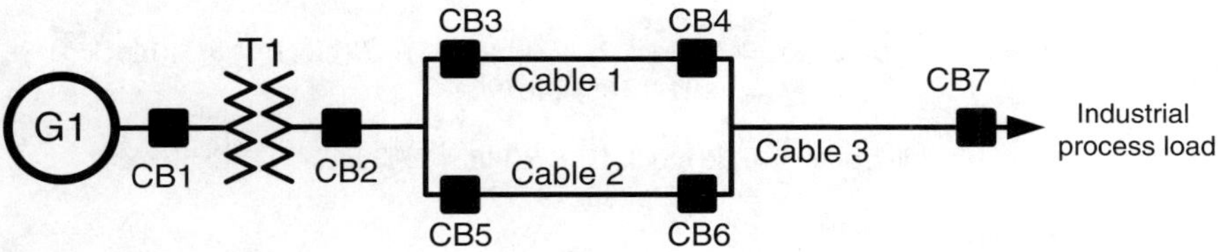


Figure 7.7. An industrial power system configuration.

(c) Calculate the load point reliability of the industrial process if the cable paths are *not* fully redundant.

$$\begin{aligned} R(\text{load point}) &= R(\text{block }1) \times R(\text{block }2) \times R(\text{block }3) \times R(\text{block }4) \\ &= 0.94241637681250 \times (0.98505618750000) \\ &\quad \times 0.98505618750000 \times (0.98752500000000) \\ &= 0.90305235605848 \end{aligned}$$

(d) If the industrial cogenerating unit's reliability is increased from 0.95 to 0.99, calculate the load point reliability of the industrial process for the existing configuration shown in Fig. 7.7 above.

$$\begin{aligned} R(\text{block }1) &= R(\text{G1}) \times R(\text{CB1}) \times R(\text{CB2}) \times R(\text{T1}) \\ &= 0.9900 \times (0.9985) \times 0.9985 \times (0.9950) \\ &= 0.98209706636250 \end{aligned}$$

$$\begin{aligned} R(\text{block }2) &= R(\text{CB3}) \times R(\text{cable }1) \times R(\text{CB4}) = 0.9975 \times (0.9900) \times 0.9975 \\ &= 0.98505618750000 \end{aligned}$$

$$\begin{aligned} R(\text{block }3) &= R(\text{CB5}) \times R(\text{cable }2) \times R(\text{CB6}) = 0.9975 \times (0.9900) \times 0.9975 \\ &= 0.98505618750000 \end{aligned}$$

$$\begin{aligned} R(\text{block }4) &= R(\text{cable }3) \times R(\text{CB7}) = (0.9900) \times 0.9975 \\ &= 0.98752500000000 \end{aligned}$$

TABLE 7.15. Reliability Data for the Electrical Equipment in the Industrial Power System of Fig. 7.7

| Component | Reliability | Component | Reliability |
|---|---|---|---|
| G1: generator 1 | 0.9500 | CB5: circuit breaker 5 | 0.9975 |
| CB1: circuit breaker 1 | 0.9985 | CB6: circuit breaker 6 | 0.9975 |
| CB2: circuit breaker 2 | 0.9985 | CB7: circuit breaker 7 | 0.9975 |
| T1: transformer 1 | 0.9950 | Feeder cable 1 | 0.9900 |
| CB3: circuit breaker 3 | 0.9975 | Feeder cable 1 | 0.9900 |
| CB4: circuit breaker 4 | 0.9975 | Feeder cable 1 | 0.9900 |

$$\begin{aligned} R(\text{block } 2//\text{block } 3) &= R(\text{block } 2) + R(\text{block } 3) - R(\text{block } 2) \times R(\text{block } 3) \\ &= 0.99977668246796 \end{aligned}$$

$$\begin{aligned} R(\text{load point}) &= R(\text{block } 1) \times R(\text{block } 2//\text{block } 3) \times R(\text{block } 4) \\ &= 0.96962882197723 \end{aligned}$$

---

## Example 7.2

Two *independent* transmission lines, 1 and 2, serve a large industrial plant with a reliability of 0.999700070. A failure of any or both of the transmission lines results in a plant outage. Both lines have the same repair rate. It is known that the failure rate of transmission line 1 is 1.0 failure/year and of transmission line 2 is 2.0 failures/year.

(a) Calculate the average time to repair "$r$", a transmission line expressed in minutes per outage.

$$R_s = \frac{[\mu]}{\mu+\lambda}\frac{\mu}{\mu+\lambda} = \frac{\mu^2}{(1+\mu)(2+\mu)} = \frac{\mu^2}{2+3\mu+\mu^2}$$

$$R_s = \frac{1}{2r^2+3r+1}$$

$$2r^2+3r+(1-(1/R_s)) = 0.0$$

$$2r^2+3r+(1-(1/0.999700070)) = 0.0 = 2r^2+3r-0.000300020$$

$$\text{Solving for } r = 0.000010000 \text{ years/outage}$$

$$\text{Therefore, } r = 0.000010000 \text{ years/outage} \times 8760 \text{ h/year} = 0.8760 \text{ h/failure}$$

$$\text{Therefore, } r = 0.8760 \text{ h/outage} \times 60 \text{ min/h} = 52.56 \text{ min/outage}$$

---

## Example 7.3

A single-line diagram of a power system is shown in Fig. 7.8.

The reliability data for the overhead lines and utility supplies are shown in Table 7.16. Note that all other component failure rates are assumed to be zero (e.g., overhead line termination, splices, and bus connections).

(a) *Calculate the frequency* (i.e., failures/year) *and the average duration of interruptions per outage* (i.e., h/interruption) *at the load point.*

$$\begin{aligned} \lambda(\text{path } 1) &= \lambda_{\text{ut1}} + \lambda_{\text{OH1}} \times \lambda_{\text{OH2}}(r_{\text{OH1}} + r_{\text{OH2}}) + \lambda_{\text{OH3}} \\ &= 0.50 + 1(2.0)3/8760 + 3.0 \\ &= 3.50068493 \text{ failures/year} \end{aligned}$$

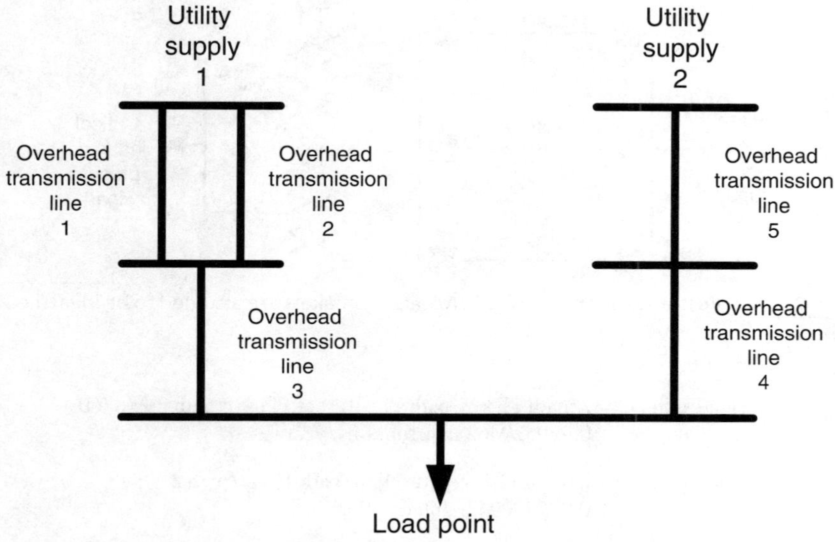


Figure 7.8. A simple power system example.

$$\begin{aligned}\lambda(\text{path }2) &= \lambda_{ut2} + \lambda_{OH5} + \lambda_{OH4}\\ &= 1.0 + 5.0 + 4.0\\ &= 10.0\text{ failures/year}\end{aligned}$$

$$\begin{aligned}\lambda r(\text{path }1) &= \lambda_{ut1} \times r_{ut1} + \lambda_{OH1} \times \lambda_{OH2}(r_{OH1} \times r_{OH2}) + \lambda_{OH3} \times r_{OH3}\\ &= 0.5(0.5) + 1(2)(1 \times 2) + 3(3.0) = 13.25\text{ h/year}\end{aligned}$$

$$\begin{aligned}\lambda r(\text{path }2) &= \lambda_{ut2} \times r_{ut2} + \lambda_{OH5} \times r_{OH5} + \lambda_{OH4} \times r_{OH4}\\ &= 0.5(0.5) + 5.0(1.0) + 4.0(2.0) = 13.5\text{ h/year}\end{aligned}$$

$$r(\text{path }1) = \lambda r(\text{path }1)/\lambda(\text{path }1) = 13.25/3.50068493 = 3.784973586\text{ h/interruption}$$

$$r(\text{path }2) = \lambda r(\text{path }2)/\lambda(\text{path }2) = 13.5/10.0 = 1.35\text{ h/interruption}$$

TABLE 7.16. Typical Equipment Reliability Data for Components of the System of Fig. 7.8

| Component | λ (failures/year) | Average repair time (r) (h/failure) |
|---|---|---|
| Utility supply 1 | 0.5 | 0.5 |
| Utility supply 2 | 1.0 | 0.5 |
| Overhead transmission line 1 | 1.0 | 1.0 |
| Overhead transmission line 2 | 2.0 | 2.0 |
| Overhead transmission line 3 | 3.0 | 3.0 |
| Overhead transmission line 4 | 4.0 | 2.0 |
| Overhead transmission line 5 | 5.0 | 1.0 |

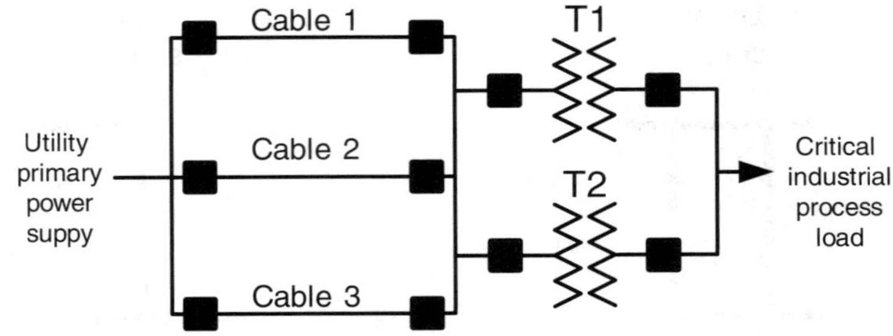


Figure 7.9. An illustrative supply network. *Note*: All breakers are assumed to be ideal (i.e., zero failure rate).

$$\lambda(\text{load point}) = \lambda(\text{path 1}) \times \lambda(\text{path 2}) \times (r(\text{path 1}) + r(\text{path 2}))/8760$$
$$= 0.020520462 \text{ interruptions/year}$$

$$r(\text{load point}) = [r(\text{path 1}) \times r(\text{path 2})]/[r(\text{path 1}) + r(\text{path 2})]$$
$$= 0.99508094 \text{ h/interruption.}$$

## Example 7.4

Three *independent identical* cable sections 1, 2, and 3 are connected to two transformers operating in parallel that are connected to an industrial process load shown in Fig. 7.9.

| | System Operating Criteria |
|---|---|
| 1. | One out of three cables are required to maintain continuity of service to the load |
| 2. | Both transformers are required (i.e., nonredundant components) |

The reliability data for the components in the above system configuration are shown in Table 7.17.

(a) Calculate the reliability of the power system configuration.

(b) Calculate the frequency (i.e., failures per year) and the average duration of interruptions per outage (i.e., hours per interruption) at the load point (i.e., industrial process load).

TABLE 7.17. Reliability Data for the Components of the System of Fig. 7.9

| Component | λ (failures/year) | Average Repair Time (*r*) (h/failure) |
|---|---|---|
| Utility primary power supply | 0.50 | 0.5 |
| Cable 1 | 1.0 | 2.0 |
| Cable 2 | 1.0 | 2.0 |
| Cable 3 | 1.0 | 2.0 |
| Transformer 1 | 0.5 | 10.0 |
| Transformer 2 | 1.0 | 10.0 |

(c) If transformers T1 and T2 are upgraded so that they are redundant, what is the frequency and average duration of interruptions per outage at the load point?
(d) Calculate the reliability of the power system configuration if the system operating criteria for the cable supply are changed to the following: *all cables are required to maintain continuity of service to the load, transformers not redundant.*

*Solution:*

(a) Calculate the reliability of the power system configuration.

$$\mu_s = 1/rs = 8760/0.5 = 17520.0 \text{ repairs/year}$$

$$r_{\text{utility}} = \mu_s/(\lambda_s + \mu_s) = 17520.0/(1 + 17520.0) = 0.99997146200166$$

$$\mu_{c1} = 1/r_{c1} = 8760/2.0 = 4380.0 \text{ repairs/year}$$

$$r_{\text{cable 1}} = \mu_{c1}/(\lambda_{c1} + \mu_{c1}) = 4380.0/(1.0 + 4380.0) = 0.99977174161150$$

$$R_{\text{cable 1}} = R_{\text{cable 2}} = R_{\text{cable 3}} = 0.99977174161150$$

$$\begin{aligned} Q_{\text{cable 1}} &= Q_{\text{cable 2}} = Q_{\text{cable 3}} = 1.0 - R_{\text{cable 1}} = 1 - 0.99977174161150 \\ &= 0.00022825838849 58278 \end{aligned}$$

$$R_{\text{cable 1-cable 2-cable 3}} = 1 - Q_{\text{cable 1}} \times Q_{\text{cable 2}} \times Q_{\text{cable 3}} = 0.99999999998811$$

$$\mu_{T1} = 1/r_{T1} = 8760.0/10.0 = 876.0 \text{ repairs/year}$$

$$\lambda_{T1} = 0.5 \text{ failures/year}$$

$$R_{\text{transformer 1}} = \mu_{T1}/(\lambda_{T1} + \mu_{T1}) = 0.99942954934398$$

$$\mu_{T2} = 1/r_{T2} = 8760.0/10.0 = 876.0 \text{ repairs/year}$$

$$\lambda_{T2} = 1.0 \text{ failures/year}$$

$$R_{\text{transformer 2}} = \mu_{T2}/(\lambda_{T2} + \mu_{T2}) = 0.99828994894564$$

$$\begin{aligned} R_{\text{system}} &= R_{\text{utility}} \times R_{\text{cable 1-cable 2-cable 3}} \times R_{\text{transformer 1}} \times R_{\text{transformer 2}} \\ &= 0.999971462001668 \times (0.99999999998811) \times 0.99942954934398 \\ &\quad \times (0.99828994894564) \\ &= 0.99826145973686 \end{aligned}$$

(b) Calculate the frequency (i.e., failures per year) and the average duration of interruptions per outage (i.e., hours per interruption) at the load point (i.e., industrial process load). Given

$$\mu_s = 1/r_s = 8760/0.5 = 17520.0 \text{ repairs/year}$$

$$\lambda_s = 0.5 \text{ failures/year}$$

$$\mu_{c1} = 1/r_{c1} = 8760/2.0 = 4380.0\ \text{repairs/year}$$

$$\lambda_{c1} = 1.0\ \text{failures/year}$$

$$\mu_{c2} = 1/r_{c2} = 8760/2.0 = 4380.0\ \text{repairs/year}$$

$$\lambda_{c2} = 1.0\ \text{failures/year}$$

$$\mu_{c3} = 1/r_{c3} = 8760/2.0 = 4380.0\ \text{repairs/year}$$

$$\lambda_{c3} = 1.0\ \text{failures/year}$$

$$\begin{aligned}\lambda_{c1c2c3} &= \lambda_{c1} \times \lambda_{c1} \times \lambda_{c1}[r_{c1} \times r_{c2} + r_{c1} \times r_{c3} + r_{c1} \times r_{c2}] \\ &= 1.0 \times (1.0) \times 1.0 \times (2 \times 2 + 2 \times 2 + 2 \times 2)/(8760 \times 8760) \\ &= 1.563770563582911 \times 10^{-7}\ \text{failures/year}\end{aligned}$$

$$r_{c1c2c3} = r_{c1} \times r_{c2} \times r_{c3}/[r_{c1} \times r_{c2} + r_{c1} \times r_{c3} + r_{c1} \times r_{c2}] = 8/12 = 0.6666667\ \text{h/failure}$$

$$\lambda_{c1c2c3} \times r_{c1c2c3} = 1.563770563582911 \times 10^{-7}(0.6666667) = 0.000000104251\ \text{h/year}$$

$$\mu_{T1} = 1/r_{T1} = 8760.0/10.0 = 876.0\ \text{repairs/year}$$

$$\lambda_{T1} = 0.5\ \text{failures/year}$$

$$\mu_{T2} = 1/r_{T2} = 8760.0/10.0 = 876.0\ \text{repairs/year}$$

$$\lambda_{T2} = 1.0\ \text{failures/year}$$

$$\begin{aligned}\lambda_{\text{load point}} &= \lambda_s + \lambda_{c1c2c3} + \lambda_{T1} + \lambda_{T2} = 0.5 + 1.563770563582911 \times 10^{-7} + 0.5 + 1.0 \\ &= 2.00000015637706\ \text{failures/year}\end{aligned}$$

$$\begin{aligned}U_{\text{load point}} &= \lambda_s \times r_s + \lambda_{c1c2c3} \times r_{c1c2c3} + \lambda_{T1} \times r_{T1} + \lambda_{T2} \times r_{T1} \\ &= 0.5(0.5) + (0.000000104251) + 0.5(10.0) + 1.0 \times (10.0) \\ &= 15.250000104251\ \text{h/year}\end{aligned}$$

$$\begin{aligned}r_{\text{load point}} &= U_{\text{load point}}/\lambda_{\text{load point}} = 15.250000104251/2.00000015637706 \\ &= 7.62499945593820\ \text{h/interruption}\end{aligned}$$

(c) If transformers T1 and T2 are upgraded so that they are redundant, what is the frequency and average duration of interruptions per outage at the load point?

$$\mu_s = 1/r_s = 8760/0.5 = 17520.0\ \text{repairs/year}$$

$$\lambda_s = 0.5\ \text{failures/year}$$

$$\mu_{c1} = 1/r_{c1} = 8760/2.0 = 4380.0\ \text{repairs/year}$$

$$\lambda_{c1} = 1.0\ \text{failures/year}$$

$$\mu_{c2} = 1/r_{c2} = 8760/2.0 = 4380.0\ \text{repairs/year}$$

$$\lambda_{c2} = 1.0\ \text{failures/year}$$

$$\mu_{c3} = 1/r_{c3} = 8760/2.0 = 4380.0\ \text{repairs/year}$$

$$\lambda_{c3} = 1.0\ \text{failures/year}$$

$$\begin{aligned}\lambda_{c1c2c3} &= \lambda_{c1} \times \lambda_{c1} \times \lambda_{c1}[r_{c1} \times r_{c2} + r_{c1} \times r_{c3} + r_{c1} \times r_{c2}] \\ &= 1.0 \times (1.0) \times 1.0 \times (2 \times 2 + 2 \times 2 + 2 \times 2)/(8760 \times 8760) \\ &= 1.563770563582911 \times 10^{-7} \text{ failures/year}\end{aligned}$$

$$r_{c1c2c3} = r_{c1} \times r_{c2} \times r_{c3}/[r_{c1} \times r_{c2} + r_{c1} \times r_{c3} + r_{c1} \times r_{c2}] = 8/12 = 0.6666667 \text{ h/failure}$$

$$\lambda_{c1c2c3} \times r_{c1c2c3} = 1.563770563582911 \times 10^{-7}(0.6666667) = 0.000000104251 \text{ h/year}$$

$$\mu_{T1} = 1/r_{T1} = 8760.0/10.0 = 876.0 \text{ repairs/year}$$

$$\lambda_{T1} = 0.5 \text{ failures/year}$$

$$\mu_{T2} = 1/r_{T2} = 8760.0/10.0 = 876.0 \text{ repairs/year}$$

$$\lambda_{T2} = 1.0 \text{ failures/year}$$

$$\begin{aligned}\lambda_{T1 \text{ and } T2} &= \lambda_{T1}\lambda_{T2}(r_{T1} + r_{T2})/8760 \\ &= 0.50(1.0)8(10.0 + 10.0)/8760.0 \\ &= 0.00114155251142 \text{ failures/year}\end{aligned}$$

$$r_{T1 \text{ and } T2} = r_{T1} \times r_{T2}/(r_{T1} + r_{T2}) = 10.0 \times (10.0/(10.0 + 10.0) = 5.0 \text{ h/repair}$$

$$\begin{aligned}\lambda_{\text{load point}} &= \lambda_s + \lambda_{c1c2c3} + \lambda_{T1} + \lambda_{T2} = 0.5 + 1.563770563582911 \times 10^{-7} \\ &\quad + 0.00114155251142 = 0.50114170888847 \text{ failures/year}\end{aligned}$$

$$\begin{aligned}U_{\text{load point}} &= \lambda_s \times r_s + \lambda_{c1c2c3} \times r_{c1c2c3} + \lambda_{T1} \times r_{T1} + \lambda_{T2} \times r_{T1} \\ &= 0.5(0.5) + (0.000000104251) + 0.5(10.0) + 1.0 \times (10.0) \\ &= 15.250000104251 \text{ h/year}\end{aligned}$$

$$\begin{aligned}r_{\text{load point}} &= U_{\text{load point}}/\lambda_{\text{load point}} = 15.250000104251/2.00000015637706 \\ &= 7.62499945593820 \text{ h/interruption}\end{aligned}$$

(d) Calculate the reliability of the power system configuration if the system operating criteria for the cable supply are changed to the following: *all cables are required to maintain continuity of service to the load, transformers not redundant.*

$$\mu_s = 1/r_s = 8760/0.5 = 17520.0 \text{ repairs/year}$$

$$R_{\text{utility}} = \mu_s/(\lambda_s + \mu_s) = 17520.0/(1 + 17520.0) = 0.99997146200166$$

$$\mu_{c1} = 1/r_{c1} = 8760/2.0 = 4380.0 \text{ repairs/year}$$

$$R_{\text{cablel}} = \mu_{c1}/(\lambda_{c1} + \mu_{c1}) = 4380.0/(1.0 + 4380.0) = 0.99977174161150$$

$$R_{\text{cablel}} = R_{\text{cable2}} = R_{\text{cable3}} = 0.99977174161150$$

$$\begin{aligned}Q_{\text{cablel}} &= Q_{\text{cable2}} = Q_{\text{cable3}} = 1.0 - R_{\text{cable1}} = 1 - 0.99977174161150 \\ &= 0.00022825838884958278\end{aligned}$$

$$\begin{aligned}R_{\text{cable1-cable2-cable3}} &= R_{\text{cable1}} \times R_{\text{cable2}} \times R_{\text{cable3}} \\ &= 0.99977174161150 \times (0.99977174161150) \\ &\quad \times 0.99977174161150 \\ &= 0.99931538112830\end{aligned}$$

$$\mu_{T1} = 1/r_{T1} = 8760.0/10.0 = 876.0 \text{ repairs/year}$$

$$\lambda_{T1} = 0.5 \text{ failures/year}$$

$$R_{\text{transformer1}} = \mu_{T1}/(\lambda_{T1} + \mu_{T1}) = 0.99942954934398$$

$$\mu_{T2} = 1/r_{T2} = 8760.0/10.0 = 876.0 \text{ repairs/year}$$

$$\lambda_{T2} = 1.0 \text{ failures/year}$$

$$R_{\text{transformer2}} = \mu_{T2}/(\lambda_{T2} + \mu_{T2}) = 0.99828994894564$$

$$\begin{aligned} R_{\text{system}} &= R_{\text{utility}} \times R_{\text{cable1-cable2-cable3}} \times \mathrm{R}_{\text{transformer1 and transformer2}} \\ &= 0.999971462001668(0.99931538112830) \times 0.99942954934398 \\ &\quad \times (0.99828994894564) \\ &= 0.99757803111449 \end{aligned}$$

---

## Example 7.5

An independent set of utility sources supplies an industrial plant as shown in Fig. 7.10.

The reliability data for the components in the above system configuration are shown in Table 7.18.

(a) Calculate the frequency (i.e., failures per year) and the average duration of interruptions per interruption (i.e., hours per interruption) at load points 1 and 2 (Tables 7.19 and 7.20).

*Solution:*

$$\lambda(\text{utility supply 1}) = \lambda_{u1} = 0.5 \text{ failures/year}$$

$$r(\text{utility supply 1}) = r_{u1} = 0.5 \text{ h/failure}$$

$$\lambda(\text{utility supply 2}) = \lambda_{u2} = 1.0 \text{ failures/year}$$

$$r(\text{utility supply 2}) = r_{u2} = 0.5 \text{ h per failure}$$

$$\begin{aligned} \lambda(\text{utility supply 1 and 2}) &= \lambda_{u12} = \lambda_{u1} \times \lambda_{u2} \times (r_{u1} + r_{u2})/8760.0 \\ &= (0.5)1.0 \times (0.5 + 0.5)/8760.0 \\ &= 57.07762557077625 \times 10^{-6} \text{ failures/year} \end{aligned}$$

$$\begin{aligned} r(\text{utility supply 1 and 1}) &= r_{u12} = r_{u1} \times r_{u2}/(r_{u1} + r_{u2}) = 0.50.5/(0.5 + 0.5) \\ &= 0.25 \text{ h/failure} \end{aligned}$$

$$\lambda_{u12} r_{u12} = 57.07762557077625 \times 10^{-6}(0.25)$$

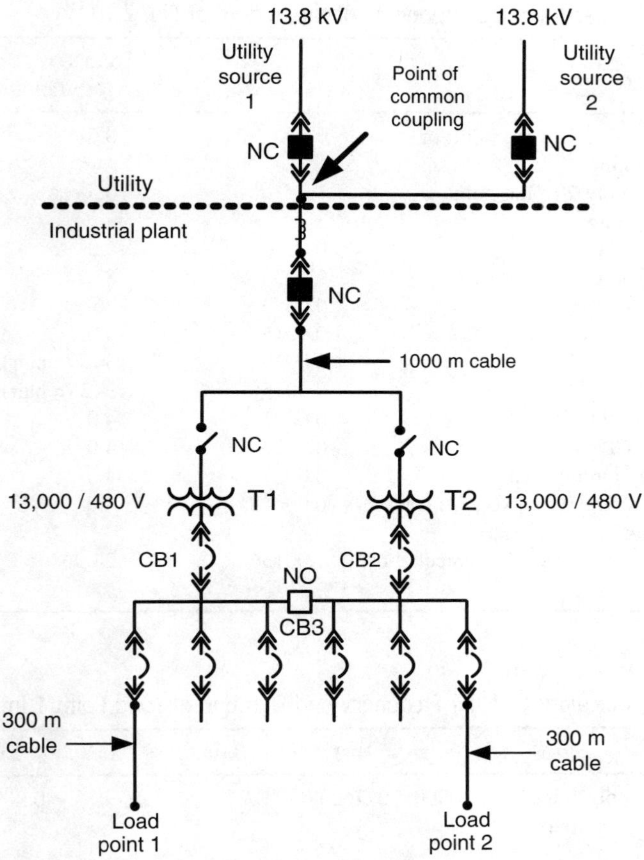


Figure 7.10. An industrial plant supplied by two independent utility lines. *Note*: Circuit breaker CB3 is assumed to be ideal (i.e., zero failure rate); all disconnect switches are assumed to be ideal; all cable terminations are assumed to be ideal; assume that a transformer is repaired (no spares are available); failure of either T2 or CB2 are added to load point 1 reliability indices; failure of either T1 or CB1 are added to load point 2 reliability indices; the T1 and T2 transformers are not redundant; loss of both utility supplies causes a plant outage; loss of either utility supply does not cause a plant outage; switching time for manual disconnect switches is 30.0 min.

$$\lambda(\text{load point 1}) = 0.02829707762557 \text{ interruptions/year}$$

$$r(\text{load point 1}) = 53.95731282253195 \text{ h/interruption}$$

$$\lambda(\text{load point 2}) = 0.02829707762557 \text{ interruptions/year}$$

$$r(\text{load point 2}) = 64.64922963469772 \text{ h/interruption}$$

TABLE 7.18. Reliability for Components of the System of Fig. 7.10

| Component | λ (failures/year) | Average repair time (*r*) (h/failure) |
|---|---|---|
| Utility power supply 1 to the point of common coupling | 0.50 | 0.5 |
| Utility power supply 2 to the point of common coupling | 1.0 | 0.5 |
| Protective relays (3) | 0.0006 | 5.0 |
| 13.8 kV metalclad circuit breaker | 0.0036 | 83.1 |
| 1000 m cable | 0.0050 | 26.5 |
| 300 m cable | 0.0016 | 4.0 |
| Transformer 1 | 0.0030 | 342 (repair); 130.0 (spare) |
| Transformer 2 | 0.0045 | 342 (repair); 130.0 (spare) |
| Circuit breaker CB1 | 0.0027 | 4.0 |
| Circuit breaker CB2 | 0.0027 | 4.0 |
| 480 V metalclad circuit breaker | 0.0027 | 4.0 |
| 480 V metalclad circuit breakers (2) (failed while opening) | 0.00048 | 4.0 |
| Switchgear bus—bare 480 V (connected to four breakers) | 0.00136 | 24.0 |

TABLE 7.19. Calculation of the Frequency and Duration of Load Point 1 Interruptions

| Component | λ (failures/year) | (*r*) (h/failure) | λ*r* (h/year) |
|---|---|---|---|
| Utility power supply 1 and 2 to the point of common coupling | 0.000057077625 | 0.25 | 0.00001426940639 |
| Protective relays (3) | 0.0006 | 5.0 | 0.003000 |
| 13.8 kV metalclad circuit breaker | 0.0036 | 83.1 | 0.2991600 |
| 1000 m cable | 0.0050 | 26.5 | 0.1325000 |
| Transformer 1 | 0.0030 | 342 (repair) | 1.0260000 |
| Transformer 2 | 0.0045 | 0.5 | 0.0022500 |
| Circuit breaker CB1 | 0.0027 | 4.0 | 0.0108000 |
| Circuit breaker CB2 | 0.0027 | 0.5 | 0.0013500 |
| Switchgear bus—bare 480 V (connected to four breakers) | 0.00136 | 24.0 | 0.0326400 |
| 480 V metalclad circuit breakers (2) (failed while opening) | 0.00048 | 4.0 | 0.0019200 |
| 480 V metalclad circuit breaker | 0.0027 | 4.0 | 0.0108000 |
| 300 m cable | 0.0016 | 4.0 | 0.0064000 |
| | 0.028297077626 | 53.957312822532 | 1.52683426940639 |

TABLE 7.20. Calculation of the Frequency and Duration of Load Point 2 Interruptions

| Component | $\lambda$ (failures/year) | $r$ (h/failure) | $\lambda r$ (h/year) |
|---|---|---|---|
| Utility power supply 1 and 2 to the point of common coupling | 0.000057077625 | 0.25 | 0.00001426940639 |
| Protective relays (3) | 0.0006 | 5.0 | 0.003000 |
| 13.8 kV metalclad circuit breaker | 0.0036 | 83.1 | 0.2991600 |
| 1000 m cable | 0.0050 | 26.5 | 0.1325000 |
| Transformer 1 | 0.0030 | 0.50 | 0.0015000 |
| Transformer 2 | 0.0045 | 295.4 | 1.3293000 |
| Circuit breaker CB1 | 0.0027 | 0.5 | 0.00135000 |
| Circuit breaker CB2 | 0.0027 | 4.0 | 0.0108000 |
| Switchgear bus—bare 480 V (connected to four breakers) | 0.00136 | 24.0 | 0.0326400 |
| 480 V metalclad circuit breakers (2) (failed while opening) | 0.00048 | 4.0 | 0.0019200 |
| 480 V metalclad circuit breaker | 0.0027 | 4.0 | 0.0108000 |
| 300 m cable | 0.0016 | 4.0 | 0.0064000 |
| | 0.028297077626 | 64.649229634698 | 1.82938426940639 |

## Example 7.6

An independent set of utility sources supplies an industrial plant as shown in Fig. 7.11.

The reliability data for the components in the above system configuration are shown in Table 7.21.

(a) Calculate the frequency (i.e., failures/year) and the average duration of interruptions per interruption (i.e., h/interruption) at the load points 1 and 2 (Tables 7.22 and 7.23).

*Solution:*

$$\lambda(\text{utility supply 1}) = \lambda_{\text{u1}} = 0.5 \text{ failures/year}$$

$$r(\text{utility supply 1}) = r_{\text{u1}} = 0.5 \text{ h/failure}$$

$$\lambda(\text{utility supply 2}) = \lambda_{\text{u1}} = 1.0 \text{ failures/year}$$

$$r(\text{utility supply 2}) = r_{\text{u1}} = 0.5 \text{ h/failure}$$

$$\begin{aligned} \lambda(\text{utility supply 1 and 2}) &= \lambda_{\text{u12}} = \lambda_{\text{u1}} \times \lambda_{\text{u2}} \times (r_{\text{u1}} + r_{\text{u2}})/8760 \\ &= (0.5)1.0 \times (0.5 + 0.5)/8760.0 \\ &= 57.07762557077625 \times 10^{-6} \text{ failures/year} \end{aligned}$$

$$\begin{aligned} r(\text{utility supply 1 and 2}) &= r_{\text{u12}} = r_{\text{u1}} \times r_{\text{u2}}/(r_{\text{u1}} + r_{\text{u2}}) = 0.5 \times 0.5/(0.5 + 0.5) \\ &= 0.25 \text{ h/failure} \end{aligned}$$

$$\lambda_{\text{u12}} r_{\text{u12}} = 57.07762557077625 \times 10^{-6}(0.25)$$

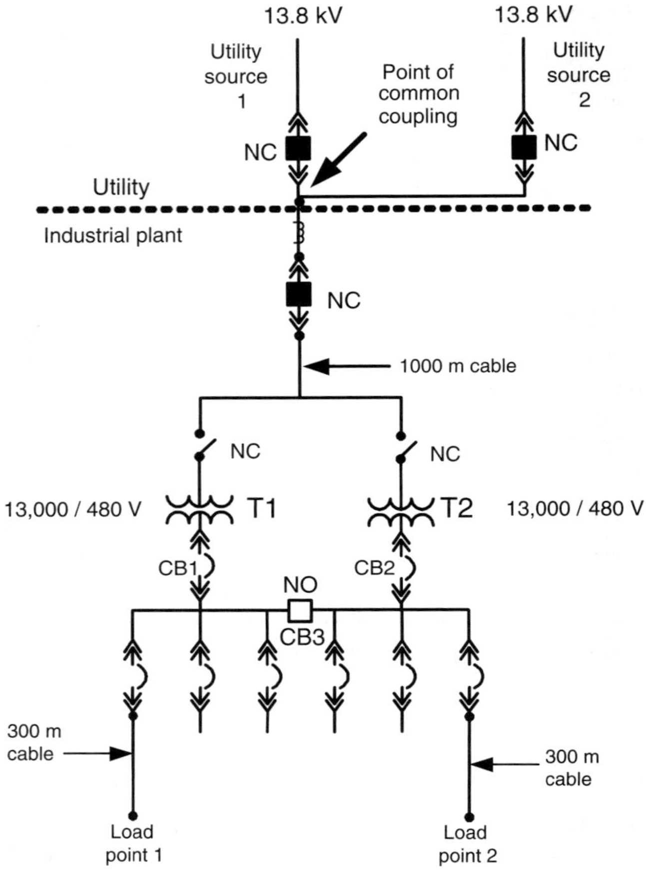


Figure 7.11. An industrial plant supply system. *Note*: Circuit breaker CB3 is assumed to be ideal (i.e., zero failure rate); all disconnects switches are assumed to be ideal; all cable terminations are assumed to be ideal; assume that *a transformer is not repaired but replaced with a spare*; failure of either T2 or CB2 is added to load point 1 reliability indices; failure of either T1 or CB1 is added to load point 2 reliability indices; T1 and T2 transformers are not redundant; loss of both utility supplies causes a plant outage; loss of either utility supply *does not* cause a plant outage; the switching time for manual disconnect switches is 30.0 min.

$$\lambda(\text{load point 1}) = 0.028297077626 \text{ interruptions/year}$$

$$r(\text{load point 1}) = 31.481493643760 \text{ h/interruption}$$

$$\lambda(\text{load point 2}) = 0.028297077626 \text{ interruptions/year}$$

$$r(\text{load point 2}) = 38.346160114635 \text{ h/interruption}$$

TABLE 7.21. Reliability Data for the Components of the System of Fig. 7.11

| Component | λ (failures/year) | Average repair time ($r$) (h/failure) |
|---|---|---|
| Utility power supply 1 to the point of common coupling | 0.50 | 0.5 |
| Utility power supply 2 to the point of common coupling | 1.0 | 0.5 |
| Protective relays (3) | 0.0006 | 5.0 |
| 13.8 kV metalclad circuit breaker | 0.0036 | 83.1 |
| 1000 m cable | 0.0050 | 26.5 |
| 300 m cable | 0.0016 | 4.0 |
| Transformer 1 | 0.0030 | 342 (repair); 130.0 (spare) |
| Transformer 2 | 0.0045 | 342 (repair); 130.0 (spare) |
| Circuit breaker CB1 | 0.0027 | 4.0 |
| Circuit breaker CB2 | 0.0027 | 4.0 |
| 480 V metalclad circuit breaker | 0.0027 | 4.0 |
| 480 V metalclad circuit breakers (2) (failed while opening) | 0.00048 | 4.0 |
| Switchgear bus—bare 480 V (connected to four breakers) | 0.00136 | 24.0 |

TABLE 7.22. Calculation of the Frequency and Duration of Load Point 1 Interruptions

| Component | λ (failures/year) | ($r$) (h/failure) | $\lambda r$ (h/year) |
|---|---|---|---|
| Utility power supply 1 and 2 to the point of common coupling | 0.000057077625 | 0.25 | 0.00001426940639 |
| Protective relays (3) | 0.0006 | 5.0 | 0.003000 |
| 13.8 kV metalclad circuit breaker | 0.0036 | 83.1 | 0.2991600 |
| 1000 m cable | 0.0050 | 26.5 | 0.1325000 |
| Transformer 1 | 0.0030 | 130.0 (spare) | 0.3900000 |
| Transformer 2 | 0.0045 | 0.5 | 0.0022500 |
| Circuit breaker CB1 | 0.0027 | 4.0 | 0.0108000 |
| Circuit breaker CB2 | 0.0027 | 0.5 | 0.0013500 |
| Switchgear bus—bare 480 V (connected to four breakers) | 0.00136 | 24.0 | 0.0326400 |
| 480 V metalclad circuit breakers (2) (failed while opening) | 0.00048 | 4.0 | 0.0019200 |
| 480 V metalclad circuit breaker | 0.0027 | 4.0 | 0.0108000 |
| 300 m cable | 0.0016 | 4.0 | 0.0064000 |
| | 0.028297077626 | 31.481493643760 | 0.890834269406 |

TABLE 7.23. Calculation of the Frequency and Duration of Load Point 2 Interruptions

| Component | λ (failures/year) | (r) (h/failure) | λr (h/year) |
|---|---|---|---|
| Utility power supply 1 and 2 to the point of common coupling | 0.000057077625 | 0.25 | 0.00001426940639 |
| Protective relays (3) | 0.0006 | 5.0 | 0.003000 |
| 13.8 kV metalclad circuit breaker | 0.0036 | 83.1 | 0.2991600 |
| 1000 m cable | 0.0050 | 26.5 | 0.1325000 |
| Transformer 1 | 0.0030 | 0.50 | 0.0015000 |
| Transformer 2 | 0.0045 | 130.0 (spare) | 0.585000 |
| Circuit breaker CB1 | 0.0027 | 0.5 | 0.00135000 |
| Circuit breaker CB2 | 0.0027 | 4.0 | 0.0108000 |
| Switchgear bus—bare 480 V (connected to four breakers) | 0.00136 | 24.0 | 0.0326400 |
| 480 V metalclad circuit breakers (2) (failed while opening) | 0.00048 | 4.0 | 0.0019200 |
| 480 V metalclad circuit breaker | 0.0027 | 4.0 | 0.0108000 |
| 300 m cable | 0.0016 | 4.0 | 0.0064000 |
| | 0.028297077626 | 38.346160114635 | 1.08508426940639 |

## Example 7.7

A distribution looped radial system is shown in Fig. 7.12. The following reliability and load data are defined:

$$\lambda(\text{feeder section 1}) = 1.0\,\text{failures/year};\ \text{load (customer A)} = 1000.0\,\text{kW}$$

$$\lambda(\text{feeder section 2}) = 2.0\,\text{failures/year};\ \text{load (customer B)} = 2000.0\,\text{kW}$$

$$\lambda(\text{feeder section 3}) = 3.0\,\text{failures/year};\ \text{load (customer C)} = 3000.0\,\text{kW}$$

$$\lambda(\text{feeder section 4}) = 4.0\,\text{failures/year};\ \text{load (customer D)} = 4000.0\,\text{kW}$$

$$\text{Average time to repair each line section} = 2.0\,\text{h}$$

$$\text{Average switching and isolation time } r(\text{switching}) = 1.0\,\text{h}$$

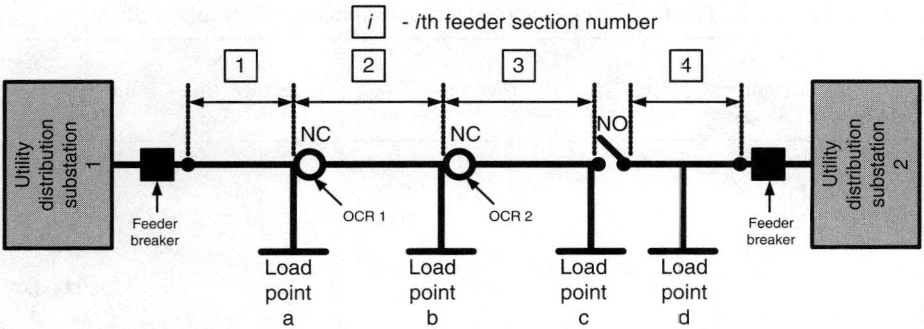


Figure 7.12. A looped radial distribution system.

TABLE 7.24. Load Point "A" Reliability Indices and Annual Cost of Interruptions

| Section Number | λ (failures/ year) | r (h) | U (h/ year) | Load (kW) | Cost of Interruptions ($/kW) | Annual Cost of Interruptions ($/year) |
|---|---|---|---|---|---|---|
| 1 | 1.0 | 2.0 | 2.0 | 1000.0 | 20.00 | 20,000.00 |
| 2 | | | | | | |
| 3 | | | | | | |
| 4 | | | | | | |
| Total | 1.0 | $R_{av} = U/\lambda = 2.0$ | 2.0 | | | $20,000.00 |

TABLE 7.25. Load Point "B" Reliability Indices and Annual Cost of Interruptions

| Section Number | λ (failures/ year) | r (h) | U (h/year) | Load (kW) | Cost of Interruptions ($/kW) | Annual Cost of Interruptions ($/year) |
|---|---|---|---|---|---|---|
| 1 | 1.0 | 1.0 | 2.0 | 2000.00 | 20.00 | 20,000.00 |
| 2 | 2.0 | 2.0 | 4.0 | 2000.00 | 20.00 | 80,000.00 |
| 3 | | | | | | |
| 4 | | | | | | |
| Total | 3.0 | $R_{av} = U/\lambda = 1.66$ | 5.0 | | | $100,000.00 |

The OCR and feeder breakers are assumed to be ideal. The cost of interruptions is $20.00/kW load interrupted.

(a) Calculate the reliability indices and cost of interruptions for load points A, B, C, and D. The calculated results are shown in Tables 7.24–7.27.

TABLE 7.26. Load Point "C" Reliability Indices and Annual Cost of Interruptions

| Section Number | λ (failures/ year) | r (h) | U (h/ year) | Load (kW) | Cost of Interruptions ($/kW) | Annual Cost of Interruptions ($/year) |
|---|---|---|---|---|---|---|
| 1 | 1.0 | 1.0 | 2.0 | 3000.0 | 20.00 | 60,000.00 |
| 2 | 2.0 | 1.0 | 2.0 | 3000.0 | 20.00 | 120,000.00 |
| 3 | 3.0 | 2.0 | 6.0 | 3000.0 | 20.00 | 180,000.00 |
| 4 | | | | | | |
| Total | 6.0 | $R_{av} = U/\lambda = 1.5$ | 9.0 | | | $360,000.00 |

TABLE 7.27. Load Point "D" Reliability Indices and Annual Cost of Interruptions

| Section Number | λ (failures/ year) | r (h) | U (h/year) | Load (kW) | Cost of Interruptions ($/kW) | Annual Cost of Interruptions ($/year) |
|---|---|---|---|---|---|---|
| 1 | | | | | | |
| 2 | | | | | | |
| 3 | | | | | | |
| 4 | 4.0 | 2.0 | 8.0 | 4000.0 | 20.00 | 320,000.00 |
| Total | 1.0 | $R_{av} = U/\lambda = 2.0$ | 2.0 | | | $320,000.00 |

## 7.10 CONCLUSIONS

This chapter has attempted to provide the basic reliability methodology to evaluate the frequency and duration of load point interruptions being served by various industrial and commercial power system configurations. The cost of load point interruptions is included in the reliability methodology to assess the annual cost of interruptions for a given operating configuration. The reliability cost–reliability worth methodology can be applied to assess whether economic improvements can be made to the existing or future power system configuration. The simple reliability index computation techniques illustrated using a practical distribution system will prove very useful to practicing distribution engineers and undergraduate and graduate students pursuing a power program.

## REFERENCES

1. R. J. Ringlee, P. Albrecht, R. N. Allan, et al., Bulk power system criteria and indices: trends and future needs, *IEEE Transactions on Power Systems*, **9** (1), 1994, 181–190.
2. G. Toleffson, G. Wacker, and R. Billinton, NSERC cost of interruption study preliminary results, *CEA Transactions*, **31**, 1992.

3. M. J. Sullivan, B. N. Suddeth, T. Vardell, and A. Vojdani, Interruption costs, customer satisfaction and expectations for service reliability, *IEEE Transactions on Power Systems*, **11** (2), 1996, 989–995.
4. B. L. Silverstein and D. M. Porter, Contingency ranking for bulk system reliability criteria, *IEEE/PES Transaction on Power Systems*, **7** (3), 1992, 956–964.
5. A. A. Chowdhury, J. Voss, K. Berg, and D. Frehlich, Development of consistent transmission outage data reporting and analysis procedures by Alberta utilities, *CEA Transactions*, **33**, 1994.
6. A. A. Chowdhury, C. F. Fuchshuber, and R. Billinton, Bulk system delivery point performance reporting at Alberta Power Limited, *CEA Transactions*, **32**, 1993.
7. R. Billinton and L. Goel, Electric service reliability worth evaluation in power systems, *CEA Transactions*, **31**, 1992.
8. A. P. Sanghvi, Economic Penalties, Including Customer Costs, for Loss of Service Continuity, Report for CEA, SD-273, July 1991.
9. E. G. Neudorf, et al., Cost–benefit analysis of power system reliability: two utility case studies, *IEEE/PES Transactions on Power Systems*, **10** (13), 1995, 1667–1675.
10. Electric Power Research Institute, Cost–Benefit Analysis of Power System Reliability: Determination of Interruption Costs—Vol. 1: Measurement Methods and Potential Applications in Reliability Cost–Benefit Analysis, EPRI IL-6791, Palo Alto, CA, April 1990.

# 8

# ZONE BRANCH RELIABILITY METHODOLOGY

## 8.1 INTRODUCTION

There are many methods available for evaluating the frequency and duration of load point interruptions within a given industrial power system configuration. As systems become larger and more interconnected, these existing methods can become computationally bound and limited in their ability to assess the impact of unreliable protective equipment and protection–coordination schemes on individual load point reliability indices within a given plant configuration. These methods can also often not account for complex isolation and restoration procedures within an industrial plant configuration. This chapter presents a zone branch methodology that overcomes many of these limitations and applies the methodology to a large industrial plant power system configuration. The primary advantage of the zone branch methodology is that it can readily identify faulty protection schemes involving all the components of an industrial power system and can evaluate load point reliability indices.

Designing reliable industrial and commercial power systems is important because of the high costs associated with interruptions to these facilities. To design and operate these facilities reliably, it is necessary to be able to readily estimate the frequency and duration of load point interruptions within these distribution power systems. There are many

*Power Distribution System Reliability.* By Ali A. Chowdhury and Don O. Koval

methods that have evolved over the past 40 years for evaluating the reliability of power system networks. The minimal cut set and the series–parallel reliability methodologies are recommended in IEEE Standard 493-1997 (*IEEE Gold Book*). These methods are systematic and lend themselves to either manual or computer computation. However, the impact of faulty protection–coordination schemes on load point reliability indices can be particularly difficult to evaluate for large industrial and commercial power systems involving complex isolation–restoration procedures. This chapter presents a zone branch methodology that can readily be used to evaluate the impact of protection–coordination schemes on individual load point reliability indices within a large industrial or commercial power system.

## 8.2 ZONE BRANCH CONCEPTS

The basic element in a graphical or digital representation of an industrial power system is a link (i.e., a homogeneous connection between any two nodes or buses in the system). A link may be a piece of electrical equipment connecting two points in the circuit such as a transformer or regulator, or a length of overhead line or cable or bus duct. Protective equipment is normally installed at the beginning of a link or branch or feeder section to protect subsequent equipment from faults within that link or branch or feeder section. Some of the basic protective equipment used in industrial systems is

1. Breakers and relays
2. Fuses
3. Reclosers
4. Sectionalizers
5. Automatic and manual isolating switches.

It is important to note that operation or nonoperation of protective devices directly affects the reliability of an industrial power system; that is, they are dependent variables. To evaluate the protection–coordination and reliability characteristics of a given industrial power system, it is necessary to divide the industrial power system into protective zones. Essentially, a protective zone is part of the power system that can isolate or detach itself automatically or manually from the remaining power system if a fault occurs in any of its links. Evaluation of the protective equipment operating characteristics establishes whether the protective equipment can isolate faults in the branches of the affected portion of the power system from the remaining power system.

In this chapter, the concept and formation of protective zone branches are based upon the following assumptions:

1. All faults are permanent;
2. The protective equipment perfectly isolates all permanent faults instantaneously;
3. The protective equipment is perfectly coordinated, that is, the device closest to the fault operates first.

Generally, each industrial power system is connected to a source of normal supply, which has protective equipment to isolate industrial plant outages from the rest of the utility power system. The *first step* in defining the first zone is to identify all branches, transformers, and related equipment in which a permanent fault of this equipment would result in only the normal supply protective equipment recognizing and isolating the permanent fault. These branches and transformers are labeled as zone 1 equipment. These are often unprotected branch lines or feeder sections and transformers connected radially to the source or main of the industrial power system.

When two protective devices are connected in series, the protective equipment nearest to the permanent fault is assumed to isolate the fault first. The *second step* in identifying protective zones is to identify permanent faults in links that would result in some protective device, other than the normal supply, isolating the fault of the link with its connection with the zone 1 link. These links are labeled as zone 2, branch $i$, where "$i$" is the branch number. The zone number represents the number of protective components between the source and the individual links that sense the fault. This procedure of classifying links into their respective zones is continued until all the links have been labeled. An example of an industrial circuit configuration and its respective zone branches is illustrated in Fig. 8.1.

A zone branch single-line diagram of the industrial power system shown in Fig. 8.1 can be drawn by visual inspection and is shown in Fig. 8.2.

The symbol $\lambda(i, j)$, for example, $\lambda(2, 1)$ is the failure rate of zone 2, branch 1. Associated with each zone branch is an isolating device labeled $S(i, j)$, where "$i$" is the zone number and "$j$" the branch number. These isolating devices can be manual or automatic switches, fuses, reclosers, sectionalizers, or a relay breaker combination, and so on. Associated with each isolating device is a probability $q(i, j)$ that the devices will *not*

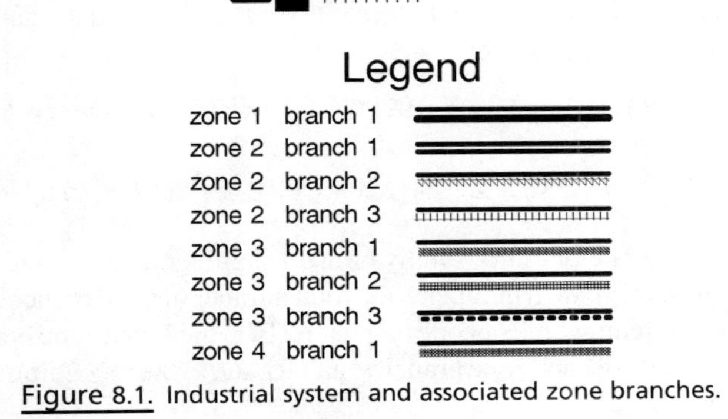


Figure 8.1. Industrial system and associated zone branches.

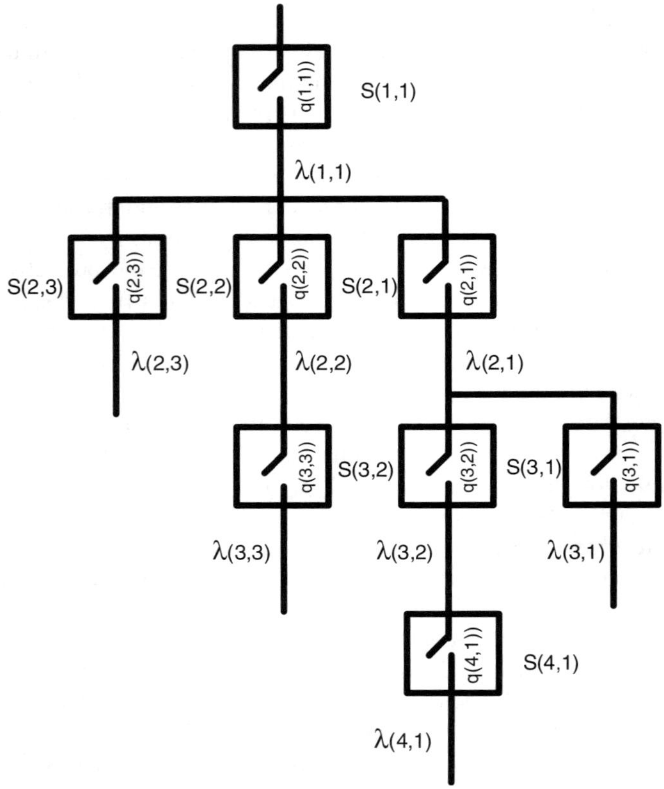


Figure 8.2. Industrial power system zone branch single-line diagram.

recognize and isolate any permanent faults of the equipment within its zone. Note: if the isolating device is a manual switch, then $q(i, j) = 1.0$; if the device is an ideal breaker–relay scheme, then $q(i, j) = 0.0$, nonideal scheme $q(i, j) > 0.0 < 1.0$.

The failure rate $\lambda(i, j)$ of any zone $i$, branch $j$ is the sum of all the equipment failure rates whose failure will result in only the operation of the isolating device of zone $I$, branch $j$. It can be shown that the total failure rate (i.e., $\lambda_T(i, j)$) and the annual downtime (i.e., $\lambda r(i, j)$) for any zone $i$, branch $j$ are

$$\lambda_T(i,j) = \lambda_s + \Sigma\,\text{RIA}(z,k) \times \text{FZB}(k)^T \text{ failures/year} \tag{8.1}$$

$$\lambda r(i,j) = \lambda_s \times rs + \Sigma\,\text{RIA}(z,k) \times \text{FZB}(k)^T \times \text{R}(z,k) \text{ h/year} \tag{8.2}$$

where $\lambda_s$ is the failure rate of utility supply or plant supply, $r_s$ the restoration duration of utility supply, $z$ the zone branch number, $k$ the total number of zone branches in system, $R(z, k)$ the repair or switching time of zone branch, FZB($k$) the failed zone branch array that contains the failure rate of each zone branch $k$, and RIA($z$, $k$) the recognition and isolation array coefficients.

Each zone branch in the industrial power system is given a unique number (i.e., 1–8). The transposed failed zone branch array FZB($k$) is defined below:

| $\lambda(1, 1)$ | $\lambda(2, 1)$ | $\lambda(2, 2)$ | $\lambda(2, 3)$ | $\lambda(3, 1)$ | $\lambda(3, 2)$ | $\lambda(3, 3)$ | $\lambda(4, 1)$ |
|---|---|---|---|---|---|---|---|
| 1 | 2 | 3 | 4 | 5 | 6 | 7 | 8 |

Failed zone branch array FZB($k$)

*Note:* FZB($k$) contains the failure rate of each zone branch $k$.

The transposed failed zone branch array, FZB($k$), and the recognition and isolation array, RIA($z$, $k$), are defined below.

| $z$ | | | | |
|---|---|---|---|---|
| 1 | 1 | $q(2, 1)$ | $q(2, 2)$ | $q(2, 3)$ |
| 2 | 1 | 1 | $q(2, 2)$ | $q(2, 3)$ |
| 3 | 1 | | 1 | $q(2, 3)$ |
| 4 | 1 | | $q(2, 2)$ | 1 |
| 5 | 1 | 1 | $q(2, 2)$ | $q(2, 3)$ |
| 6 | 1 | 1 | $q(2, 2)$ | $q(2, 3)$ |
| 7 | 1 | | 1 | $q(2, 3)$ |
| 8 | 1 | 1 | $q(2, 2)$ | $q(2, 3)$ |
| | 1 | 2 | 3 | 4 |
| | | | $k$ | |

RIA($z$, $k$) (columns 1–4)

| $z$ | | | | |
|---|---|---|---|---|
| 1 | $q(3, 1) \times q(2, 1)$ | $q(3, 2) \times q(2, 1)$ | $q(3, 3) \times q(2, 2)$ | $q(4, 1) \times q(3, 2) \times q(2, 1)$ |
| 2 | $q(3, 1)$ | $q(3, 2)$ | $q(3, 3) \times q(2, 2)$ | $q(4, 1) \times q(3, 2)$ |
| 3 | $q(3, 1) \times q(2, 1)$ | $q(3, 2) \times q(2, 1)$ | $q(3, 3)$ | $q(4, 1) \times q(3, 2) \times q(2, 1)$ |
| 4 | $q(3, 1) \times q(2, 1)$ | $q(3, 2) \times q(2, 1)$ | $q(3, 3) \times q(2, 2)$ | $q(4, 1) \times q(3, 2) \times q(2, 1)$ |
| 5 | 1 | $q(3, 2)$ | $q(3, 3) \times q(2, 2)$ | $q(4, 1) \times q(3, 2)$ |
| 6 | $q(3, 1)$ | 1 | $q(3, 3) \times q(2, 2)$ | $q(4, 1)$ |
| 7 | $q(3, 1) \times q(2, 1)$ | $q(3, 2) \times q(2, 1)$ | 1 | $q(4, 1) \times q(3, 2) \times q(2, 1)$ |
| 8 | $q(3, 1)$ | 1 | $q(3, 3) \times q(2, 2)$ | 1 |
| | 5 | 6 | 7 | 8 |
| | | | $k$ | |

RIA($z$, $k$) (columns 5–8)

The total failure rate of any zone branch depends upon the failure rates of all zone branches in the industrial power system and the probability of the isolation devices detecting a permanent fault within their respective zone branches. With reference to Fig. 8.2, the total failure rate of zone 4, branch 1 is

$$\begin{aligned}\lambda_T(4,1) \quad = \quad & [1] \times \lambda(1,1) + [1] \times \lambda(2,1) + [q(2,2)] \times \lambda(2,2) + \\ & \cdots + [q(2,3)] \times \lambda(2,3) + [q(3,1)] \times \lambda(3,1) + \\ & \cdots + [1] \times \lambda(3,2) + [q(3,3)q(2,2)] \times \lambda(3,3) + \\ & \cdots + [1] \times \lambda(4,1) + \lambda_s\end{aligned}$$

*Note:* The elements in square brackets correspond to the elements of RIA($z$, $k$) or can be obtained by a visual inspection of Fig. 8.2.

The annual outage duration [$\lambda r(4, 1)$] of equipment connected to zone 4, branch 1 is

$$\begin{aligned}\lambda r(4,1) = {} & [1]\times\lambda(1,1)\times r(1,1)+[1]\times\lambda(2,1)\times r(2,1)+\\ & \cdots+q(2,2)]\times\lambda(2,2)\times \mathrm{Rsw}(2,2)+q(2,3)]\times\lambda(2,3)\times \mathrm{Rsw}(2,3)+\\ & \cdots+q(3,1)]\times\lambda(3,1)\times \mathrm{Rsw}(3,1)+[1]\times\lambda(3,2)\times r(3,2)+\\ & \cdots+q(3,3)q(2,2)]\times\lambda(3,3))\times \mathrm{Rsw}(3,1)+[1]\times\lambda(4,1)\times r(4,1)+\lambda_s\times rs\end{aligned}$$

where $r(i,j)$ is the repair or restoration of zone $i$, branch $j$ and Rsw($i,j$) is the switching and isolation time of zone $i$, branch $j$.

*Note:* Any zone branch in the direct path to a particular zone branch requires repair activities while those zone branches off the "herring bone" configuration require only switching and isolation duration activities.

## 8.3 INDUSTRIAL SYSTEM STUDY

A single-line diagram of an *industrial power system* is shown in Fig. 8.3. The reliability data for the circuit equipment are listed in Tables 8.1–8.7. *Note*: The tables are truncated (i.e., to comply with the page restrictions of this publication, for example, 15 more pages would be required, and to illustrate what reliability data are required for the case study presented). The data not shown in these tables are available in the *IEEE Gold Book* (i.e., Standard 493-1997, Chapter 3).

The zone branch single-line diagram for the industrial power system shown in Fig. 8.3 is provided in Fig. 8.4.

Based on the equipment reliability data (i.e., Tables 8.1–8.7), the failure rate and average repair duration for each zone branch can be calculated and are listed in Table 8.8. The frequency and the duration of interruptions for each motor group being served by the industrial power system shown in Fig. 8.3 are listed in Table 8.9. The protective equipment is assumed to be ideal (i.e., $q(i, j) = 0.0$) and the average switching and isolation time was assumed at 15 min (Fig. 8.4).

Once the failure rate and the average repair duration of each zone branch (i.e., Table 8.8) in the industrial power system have been calculated from the reliability data (i.e., IEEE Standard 493-1997, *IEEE Gold Book*), the frequency and the duration of interruptions at each load point (i.e., motor groups) can be calculated. With reference to Table 8.9, the failure rate of each motor group varied from a minimum of 1.164940 to a high of 1.322572 failures/year. The small difference in the frequency of interruptions was due to the protection–coordination scheme adopted for this particular industrial power system configuration. Other less efficient schemes would result in a significantly larger difference in the frequency of motor group interruptions. For example, if all the protective devices in the industrial power system were manual switches or completely faulty (i.e., $q(i, j) = 1.0$), the frequency of interruptions for all motor groups would jump to 2.512917 interruptions/year.

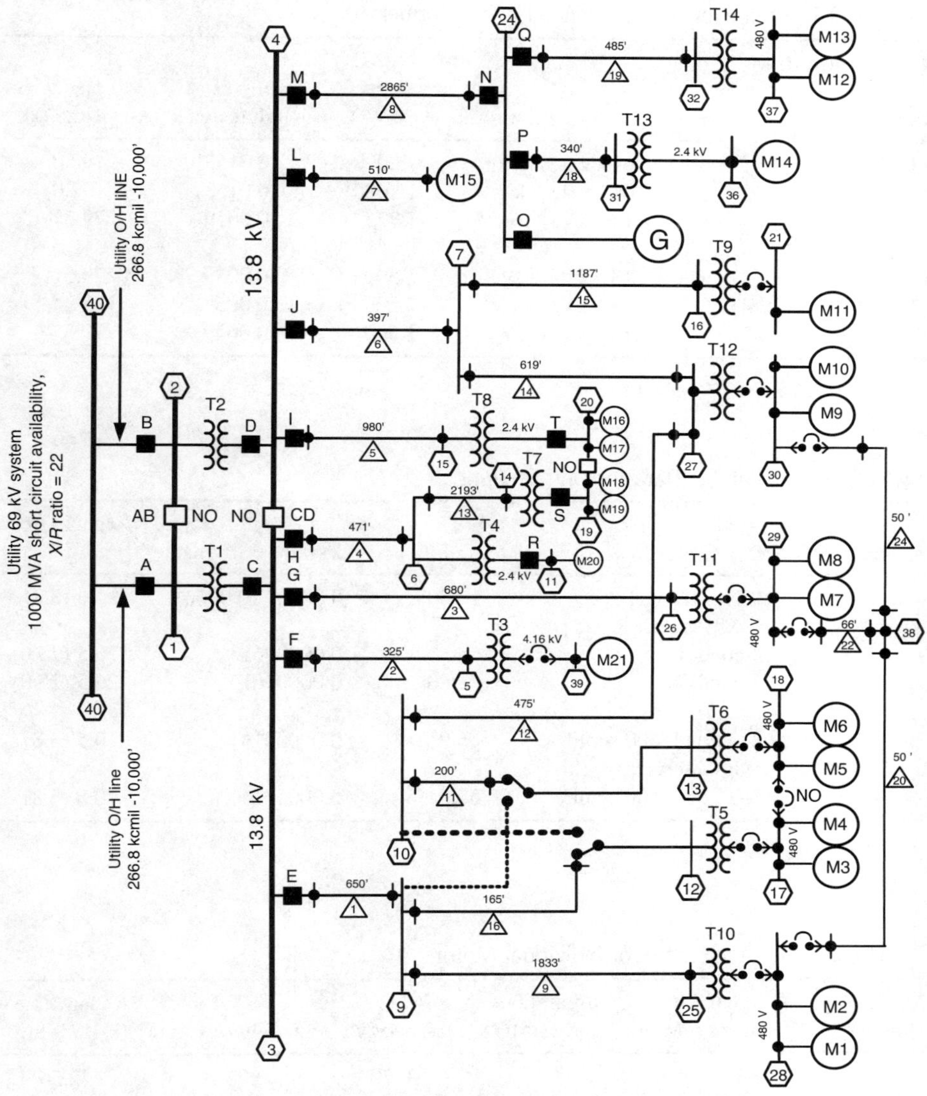

Figure 8.3. Single-line diagram of an industrial power system.

The annual interruption duration for each motor group was varied from a low of 4.009801 to a high of 15.1454204 h per year. The impact of lengthy repair times (i.e., 11.451205 h for motor groups 18 and 19) on the industrial process was identified (not obvious from the single-line diagram). The repair and isolation activities were altered to lower the duration of this motor group to approximately 7 h per interruption. For example, if all the protective devices in the industrial power system were manual

TABLE 8.1. Characteristics of Individual Transformers

| Transformer Number (h/failure) | Bus–Bus | Transformation Ratio | Size (MVA) | λ (failures/year) | Replacement Time (*r*) |
|---|---|---|---|---|---|
| (T1) | 1–3 | 69 kV/13.8 kV | 15.000 | 0.0153 | 192.0 |
| (T2) | 2–4 | 69 kV/13.8 kV | 15.000 | 0.0153 | 192.0 |
| (T3) | 5–39 | 13.8 kV/4.16 kV | 1.725 | 0.0059 | 79.3 |
| ... | ... | ... | ... | ... | ... |
| (T12) | 27–30 | 13.8 kV/480V | 1.500 | 0.0059 | 79.3 |
| (T13) | 31–36 | 13.8 kV/2.4 kV | 3.500 | 0.0059 | 79.3 |
| (T14) | 32–37 | 13.8 kV/480V | 1.500 | 0.0059 | 79.3 |

TABLE 8.2. Reliability Data for Line Sections

| Section Number | Description | Length (ft) | λ (failures/ft/year) | Repair Time (*r*) (h/failure) |
|---|---|---|---|---|
| – | UTILITY OH 266.8 MCM ACSR | 10,000 | 0.00001 (0.0003) | 2.00 (3.8) |
| 1 | 3 cond. #6 | 650.0 | 0.00000613 | 26.5 (25.0) |
| 2 | 3 cond. #6 | 325.0 | 0.00000613 | 26.5 (25.0) |
| ... | ... | ... | ... | ... |
| 22 | 2–3 cond. 400 kcmil | 66.0 | 0.00000141 | 10.5 (3.8) |
| 23 | Out of service | – | – | – |
| 24 | 2–3 cond. 400 kcmil | 50.0 | 0.00000141 | 10.5 (3.8) |

TABLE 8.3. Characteristics of Individual Motors

| Motor Number | Type of Motor Configuration Induction Motor Group (IMG) | Size (MVA) | λ (failures/year) | Repair time (*r*) (h/failure) |
|---|---|---|---|---|
| (M1) | IMG (>50 hp)-480 V | 0.70 | 0.0824 | 42.5 |
| (M2) | IMG (<50 hp)-480 V | 0.50 | 0.0109 | 18.3 |
| (M3) | IMG (>50 hp)-480 V | 1.00 | 0.0824 | 42.5 |
| (M4) | IMG (<50 hp)-480 V | 0.50 | 0.0109 | 18.3 |
| (M5) | IMG (>50 hp)-480 V | 1.00 | 0.0824 | 42.5 |
| ... | ... | ... | ... | ... |
| (M20) | IMG (>50 hp) 2.4 kV | 1.50 | 0.0714 | 75.1 |
| (M21) | 1750 hp induction, motor—1800 rpm, 4.16 kV | 1.75 | 0.0404 | 76.0 |
| G | Generator—0.8 PF, standby and emergency | 10.00 | 0.00536 | 478.0 |

TABLE 8.4. Reliability Data for Circuit Breakers (Includes Connections and Terminations)

| Circuit Breaker Number | Description | $\lambda$ (failures/year) | Repair Time ($r$) (h/failure) |
|---|---|---|---|
| A | 69 kV ($q$A = 0.0) | 0.0036 | 109.0 |
| B | 69 kV ($q$B = 0.0) | 0.0036 | 109.0 |
| C | 13.8 kV MC ($q$C = 0.0) | 0.0036 | 83.1 |
| ... | ... | ... | ... |
| T | 13.8 kV MC ($q$T = 0.0) | 0.0036 | 83.1 |
| ALL | 480 V MC ($q$ = 0.0 | 0.0027 | 4.0 |
| ALL | 2.4 and 4.16 kV metalclad ($q$ = 0.0) | 0.0036 | 83.1 |

TABLE 8.5. Reliability Data for Insulated Switchgear Bus and Bus Duct and Terminations

| Switchgear Bus No. | Description | $\lambda$ (failures/year) | Repair Time ($r$) (h/failure) |
|---|---|---|---|
| 1 | Switchgear bus insulated (connected to 1 breaker) | 0.0034 | 26.8 |
| 2 | Switchgear bus insulated (connected to 1 breaker) | 0.0034 | 26.8 |
| 3 | Switchgear bus insulated (connected to 5 breakers) | 0.0170 | 26.8 |
| ... | ... | ... | ... |
| 10 | Switchgear bus insulated | 0.001050 | 28.0 |
| 27 | Switchgear bus insulated | 0.001050 | 28.0 |
| – | Disconnect switch enclosed | 0.006100 | 1.6 |

TABLE 8.6. Reliability Data for Insulated Motor Bus Connections and Terminations

| Motor Bus No. | Description | $\lambda$ (failures/year) | Repair Time ($r$) (h/failure) |
|---|---|---|---|
| 28 | Motor bus for M1 and M2 (including four connections) | 0.000127 | 4.0 |
| 17 | Motor bus for M3 and M4 (including four connections) | 0.000127 | 4.0 |
| 18 | Motor bus for M5 and M6 (including four connections) | 0.000127 | 4.0 |
| ... | ... | ... | ... |
| 19 | Motor bus for M18 and M19 (including four connections) | 0.004192 | 11.5 |
| 11 | Motor bus for M20 (including 1 connections) | 0.001048 | 11.5 |
| 39 | Motor bus for M21 (including 1 connections) | 0.001048 | 11.5 |

TABLE 8.7. Characteristics of Source Elements

| 69 kV Utility System | | |
|---|---|---|
| 1000 MVA available | $\lambda$ (failures/year) = 0.843 | $r$ (h/failure) = 1.00 |

switches or all the protective devices were completely faulty (i.e., $q(i, j) = 1.0$), the annual interruption duration for all motor groups would be a 91.340941 h with an average repair duration of 36.348571 h per interruption.

Reliable protective schemes and equipment for industrial power systems are critical for minimizing the frequency, duration, and cost of interruptions. If the protection equipment (e.g., breakers, relays, etc.) are not adequately maintained or inspected, then the reliability performance of an industrial power system will deteriorate significantly as illustrated in this chapter. One potential problem with computerized protective schemes

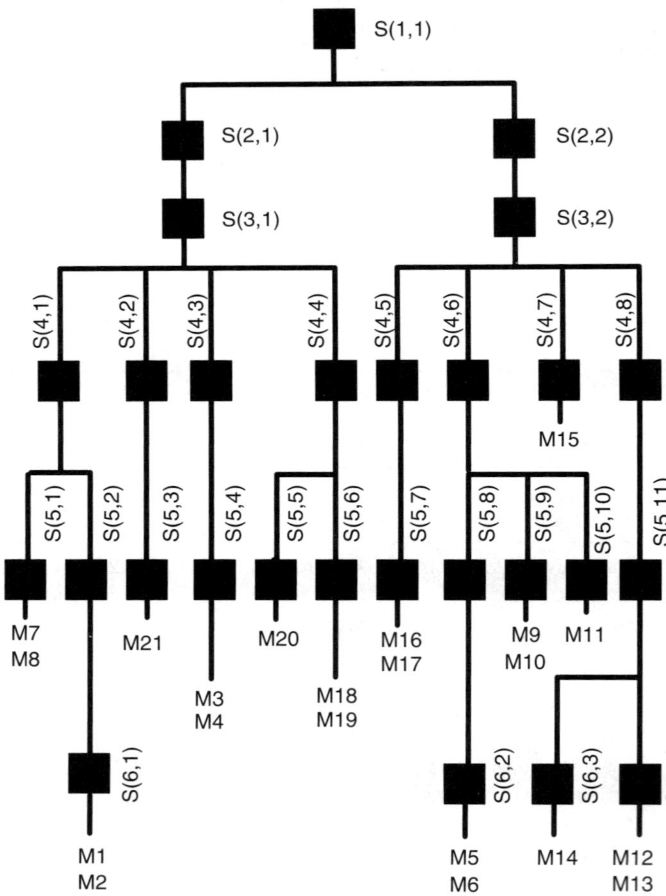


Figure 8.4. Zone branch single-line diagram.

TABLE 8.8. Failure Rate and Duration of Zone Branch Outages Computed from Previous Data

| Zone Branch | Failure Rate (failures/year) | Interruption Duration (h/year) | Average Repair Duration (h/failure) |
|---|---|---|---|
| (1, 1) | 1.058300 | 2.161330 | 2.042265 |
| (2, 1) | 0.022300 | 3.327880 | 149.232286 |
| (2, 2) | 0.022300 | 3.327880 | 149.232286 |
| (3, 1) | 0.031400 | 1.652240 | 52.619108 |
| (3, 2) | 0.031400 | 1.652240 | 52.619108 |
| (4, 1) | 0.034682 | 1.015484 | 29.279664 |
| (4, 2) | 0.012392 | 0.553964 | 48.203321 |
| (4, 3) | 0.013668 | 0.842324 | 44.747929 |
| (4, 4) | 0.038180 | 2.041213 | 67.971887 |
| (4, 5) | 0.016407 | 0.948726 | 57.823061 |
| (4, 6) | 0.048592 | 1.635316 | 33.653934 |
| (4, 7) | 0.036874 | 5.682398 | 154.101879 |
| (4, 8) | 0.022062 | 0.787064 | 35.674411 |
| (5, 1) | 0.093427 | 3.701978 | 39.624284 |
| (5, 2) | 0.008600 | 0.478670 | 55.659302 |
| (5, 3) | 0.041448 | 3.082452 | 74.369137 |
| (5, 4) | 0.093427 | 3.701978 | 39.624284 |
| (5, 5) | 0.072448 | 5.374192 | 55.659302 |
| (5, 6) | 0.172392 | 12.96074 | 75.181841 |
| (5, 7) | 0.107392 | 10.97534 | 102.198934 |
| (5, 8) | 0.008600 | 0.478670 | 55.659302 |
| (5, 9) | 0.093427 | 3.701978 | 39.624284 |
| (5, 10) | 0.010965 | 0.199730 | 18.215230 |
| (5, 11) | 0.024400 | 1.261960 | 51.719672 |
| (6, 1) | 0.093427 | 3.701978 | 39.624284 |
| (6, 2) | 0.093427 | 3.701978 | 39.624284 |
| (6, 3) | 0.107780 | 8.120105 | 75.339489 |
| (6, 4) | 0.103200 | 4.271133 | 41.386935 |

is their susceptibility to power supply anomalies caused by switching and operational activities within an industrial facility. To minimize the impact and cost of these anomalies, proper maintenance and power supply immunity of the protective computerized systems should be undertaken.

## 8.4 APPLICATION OF ZONE BRANCH METHODOLOGY: CASE STUDIES

Applications of zone branch methodology are further illustrated in the following sections using several case studies.

TABLE 8.9. Frequency and Duration of Motor Group Interruptions

| Motor Number | Failure Rate (failures/year) | Interruption Duration (h/year) | Average Repair Duration (h/failure) |
|---|---|---|---|
| M1 and M2 | 1.247659 | 5.891031 | 4.718669 |
| M3 and M4 | 1.247659 | 5.880150 | 4.718669 |
| M5 and M6 | 1.262619 | 5.891031 | 4.665722 |
| M7 and M8 | 1.219095 | 5.880150 | 4.823371 |
| M9 and M10 | 1.262619 | 5.891031 | 4.665722 |
| M11 | 1.180157 | 4.009801 | 3.397684 |
| M14 | 1.261662 | 8.494913 | 6.733111 |
| M15 | 1.148874 | 7.857153 | 6.839002 |
| M16 and M17 | 1.235799 | 13.154204 | 10.644288 |
| M18 and M19 | 1.322572 | 15.145048 | 11.451205 |
| M20 | 1.222628 | 9.590160 | 7.843888 |
| M21 | 1.164940 | 5.811171 | 4.988385 |

### 8.4.1 Case 1: Design "A"—Simple Radial Substation Configuration

This is a simple radial system with only one supply source and no switching options in case of any fault. In case of any fault, the supply to the load point will remain interrupted unless faulty equipment is repaired or replaced. The single-line diagram of substation Design "A"—simple radial configuration—is given in Fig. 8.5 and its zone branch diagram is provided in Fig. 8.6.

***8.4.1.1 Zone Branch Calculations: Design "A".*** The zone branch calculations are illustrated as follows.

Zone 1—Branch 1

$$\begin{aligned}
\lambda(1,1) &= \lambda_s + \lambda_{T1} + 0.50 \times \lambda_{CB1} \\
&= 1.956 + 0.0062 + 0.002 \\
&= 1.9642 \text{ outages/year} \\
\lambda r(1,1) &= \lambda_s r_s + \lambda_{T1} r_{T1} + 0.50 \times \lambda_{CB1} r_{CB1} \\
&= 5.050392 + 2.20782 + 0.2 \\
&= 7.458212 \text{ h/year} \\
r(1,1) &= \lambda r(1,1)/l(1,1) \\
&= 3.797073618 \text{ h/interruption}
\end{aligned}$$

Zone 2—Branch 1

$$\begin{aligned}
\lambda(2,1) &= 0.50 \times \lambda_{CB1} + \lambda_{SGB1} + 0.50 \times \{\lambda_{CB2} + \lambda_{CB3} + \lambda_{CB4} + \lambda_{CB5} + \lambda_{CB6} + \lambda_{CB7}\} \\
&= 0.002 + 0.000802 + 0.002 \times 6 \\
&= 0.014802 \text{ outages/year} \\
\lambda r(2,1) &= 0.50 \times \lambda_{CB1} r_{CB1} + \lambda_{SGB1} r_{SGB1} + 0.50 \\
&\quad \times \{\lambda_{CB2} r_{CB2} + \lambda_{CB3} r_{CB3} + \lambda_{CB4} r_{CB4} + \lambda_{CB5} r_{CB5} + \lambda_{CB6} r_{CB6} + \lambda_{CB7} r_{CB7}\} \\
&= 0.2 + 0.4411 + 0.2 \times 6
\end{aligned}$$

$$= 1.8411\ \text{h/year}$$

$$r(2,1) = \lambda r(2,1)/\lambda(2,1)$$
$$= 124.3818403\ \text{h/outage}$$

Zone 3—Branch 1

$$\lambda(3,1) = 0.50 \times \lambda_{CB2} + \lambda_{SGB2} + 0.50 \times \{\lambda_{DS1} + \lambda_{DS2} + \lambda_{DS3}\} + \lambda_{C1} + 2\lambda_{CT}$$
$$= 0.002 + 0.000802 + 0.00305 \times 3 + 0.00613 + 0.0001 \times 2$$
$$= 0.018282\ \text{outages/year}$$

$$\lambda r(3,1) = 0.50 \times \lambda_{CB2} r_{CB2} + \lambda_{SGB3} r_{SGB3} + 0.50$$
$$\times \{\lambda_{DS1} r_{DS1} + \lambda_{DS2} r_{DS2} + \lambda_{DS3} r_{DS3}\} + \lambda_{C1} r_{C1} + 2\lambda_{CT} r_{CT}$$
$$= 0.2 + 0.4411 + 0.01098 \times 3 + 0.162445 + 0.0025 \times 2$$
$$= 0.841485\ \text{h/year}$$

$$r(3,1) = \lambda r(3,1)/\lambda(3,1)$$
$$= 46.02806039\ \text{h/outage}$$

Zone 3—Branch 4

$$\lambda(3,4) = 0.50 \times \lambda_{CB5} r_{CB5} + \lambda_{SGB4} + 0.50 \times \{\lambda_{DS4} + \lambda_{DS5} + \lambda_{DS6}\} + \lambda_{C2} + 2\lambda_{CT}$$
$$= 0.002 + 0.000802 + 0.00305 \times 3 + 0.00613 + 0.0001 \times 2$$
$$= 0.018282\ \text{outages/year}$$

$$\lambda r(3,4) = 0.50 \times \lambda_{CB5} r_{CB5} + \lambda_{SGB4} r_{SGB4} + 0.50 \times \{\lambda_{DS4} r_{DS4} + \lambda_{DS5} r_{DS5} + \lambda_{DS6} r_{DS6}\}$$
$$+ \lambda_{C2} r_{C2} + 2\lambda_{CT} r_{CT}$$
$$= 0.2 + 0.4411 + 0.01098 \times 3 + 0.162445 + 0.0025 \times 2$$
$$= 0.841485\ \text{h/year}$$

$$r(3,4) = \lambda r(3,4)/\lambda(3,4)$$
$$= 46.02806039\ \text{h/outage}$$

Zone 4—Branch 1

$$\lambda(4,1) = 0.50 \times \{\lambda_{DS1} + \lambda_{F1}\}$$
$$= 0.00305 + 0.00095$$
$$= 0.004\ \text{outages/year}$$
$$\lambda r(4,1) = 0.50 \times \{\lambda_{DS1} r_{DS1} + \lambda_{F1} r_{F1}\}$$
$$= 0.01098 + 0.005225$$
$$= 0.016205\ \text{h/year}$$
$$r(4,1) = \lambda r(4,1)/\lambda(4,1)$$
$$= 4.05125\ \text{h/outage}$$

Zone 4—Branch 2

$$
\begin{aligned}
\lambda(4,2) &= 0.50 \times \{\lambda_{DS2} + \lambda_{F2}\} \\
&= 0.00305 + 0.00095 \\
&= 0.004 \text{ outages/year} \\
\lambda r(4,2) &= 0.50 \times \{\lambda_{DS2} r_{DS2} + \lambda_{F2} r_{F2}\} \\
&= 0.01098 + 0.005225 \\
&= 0.016205 \text{ h/year} \\
r(4,2) &= \lambda r(4,2)/\lambda(4,2) \\
&= 4.05125 \text{ h/outage}
\end{aligned}
$$

Zone 4—Branch 3

$$
\begin{aligned}
\lambda(4,3) &= 0.50 \times \{\lambda_{DS4} + \lambda_{F4}\} \\
&= 0.00305 + 0.00095 \\
&= 0.004 \text{ outages/year} \\
\lambda r(4,3) &= 0.50 \times \{\lambda_{DS3} r_{DS3} + \lambda_{F3} r_{F3}\} \\
&= 0.01098 + 0.005225 \\
&= 0.016205 \text{ h/year} \\
r(4,3) &= \lambda r(4,3)/\lambda(4,3) \\
&= 4.05125 \text{ h/outage}
\end{aligned}
$$

Zone 4—Branch 4

$$
\begin{aligned}
\lambda(4,4) &= 0.50 \times \{\lambda_{DS4} + \lambda_{F4}\} \\
&= 0.00305 + 0.00095 \\
&= 0.004 \text{ outages/year} \\
\lambda r(4,4) &= 0.50 \times \{\lambda_{DS4} r_{DS4} + \lambda_{F4} r_{F4}\} \\
&= 0.01098 + 0.005225 \\
&= 0.016205 \text{ h/year} \\
r(4,4) &= \lambda r(4,4)/\lambda(4,4) \\
&= 4.05125 \text{ h/outage}
\end{aligned}
$$

Zone 4—Branch 5

$$
\begin{aligned}
\lambda(4,5) &= 0.50 \times \{\lambda_{DS5} + \lambda_{F5}\} \\
&= 0.00305 + 0.00095 \\
&= 0.004 \text{ outages/year} \\
\lambda r(4,5) &= 0.50 \times \{\lambda_{DS5} r_{DS5} + \lambda_{F5} r_{F5}\} \\
&= 0.01098 + 0.005225 \\
&= 0.016205 \text{ h/year} \\
r(4,5) &= \lambda r(4,5)/\lambda(4,5) \\
&= 4.05125 \text{ h/outage}
\end{aligned}
$$

Zone 4—Branch 6

$$
\begin{aligned}
\lambda(4,6) &= 0.50 \times \{\lambda_{DS6} + \lambda_{F6}\} \\
&= 0.00305 + 0.00095 \\
&= 0.004 \text{ outages/year} \\
\lambda r(4,6) &= 0.50 \times \{\lambda_{DS6} r_{DS6} + \lambda_{F6} r_{F6}\} \\
&= 0.01098 + 0.005225 \\
&= 0.016205 \text{ h/year} \\
r(4,6) &= \lambda r(4,6)/\lambda(4,6) \\
&= 4.05125 \text{ h/outage}
\end{aligned}
$$

Zone 5—Branch 1

$$\begin{aligned}
\lambda(5,1) &= 0.50 \times \lambda_{F1} + \lambda_{T2} \\
&= 0.00095 + 0.0062 \\
&= 0.00715 \text{ outages/year} \\
\lambda r(5,1) &= 0.50 \times \lambda_{F1} r_{F1} + \lambda_{T2} r_{T2} \\
&= 0.005225 + 2.20782 \\
&= 2.213045 \text{ h/year} \\
r(5,1) &= \lambda r(5,1)/\lambda(5,1) \\
&= 309.5167832 \text{ h/outage}
\end{aligned}$$

Zone 5—Branch 2

$$\begin{aligned}
\lambda(5,2) &= 0.50 \times \lambda_{F2} + \lambda_{T3} \\
&= 0.00095 + 0.0062 \\
&= 0.00715 \text{ outages/year} \\
\lambda r(5,2) &= 0.50 \times \lambda_{F2} r_{F2} + \lambda_{T3} r_{T3} \\
&= 0.005225 + 2.20782 \\
&= 2.213045 \text{ h/year} \\
r(5,2) &= \lambda r(5,2)/\lambda(5,2) \\
&= 309.5167832 \text{ h/outage}
\end{aligned}$$

Zone 5—Branch 3

$$\begin{aligned}
\lambda(5,3) &= 0.50 \times \lambda_{F3} + \lambda_{T4} \\
&= 0.00095 + 0.0062 \\
&= 0.00715 \text{ outages/year} \\
\lambda r(5,3) &= 0.50 \times \lambda_{F3} r_{F3} + \lambda_{T4} r_{T4} \\
&= 0.005225 + 2.20782 \\
&= 2.213045 \text{ h/year} \\
r(5,3) &= \lambda r(6,3)/\lambda(6,3) \\
&= 309.5167832 \text{ h/outage}
\end{aligned}$$

Zone 5—Branch 4

$$\begin{aligned}
\lambda(5,4) &= 0.50 \times \lambda_{F4} + \lambda_{T5} \\
&= 0.00095 + 0.0062 \\
&= 0.00715 \text{ outages/year} \\
\lambda r(5,4) &= 0.50 \times \lambda_{F4} r_{F4} + \lambda_{T5} r_{T5} \\
&= 0.005225 + 2.20782 \\
&= 2.213045 \text{ h/year} \\
r(5,4) &= \lambda r(5,4)/\lambda(5,4) \\
&= 309.5167832\, h\text{/outage}
\end{aligned}$$

Zone 5—Branch 5

$$\begin{aligned}
\lambda(5,5) &= 0.50 \times \lambda_{F5} + \lambda_{T6} \\
&= 0.00095 + 0.0062 \\
&= 0.00715 \text{ outages/year} \\
\lambda r(5,5) &= 0.50 \times \lambda_{F5} r_{F5} + \lambda_{T6} r_{T6} \\
&= 0.005225 + 2.20782 \\
&= 2.213045 \text{ h/year} \\
r(5,5) &= \lambda r(5,1)/\lambda(5,1) \\
&= 309.5167832 \text{ h/outage}
\end{aligned}$$

Zone 5—Branch 6

$$\begin{aligned}
\lambda(5,6) &= 0.50 \times \lambda_{F6} + \lambda_{T7} \\
&= 0.00095 + 0.0062 \\
&= 0.00715 \text{ outages/year} \\
\lambda r(5,6) &= 0.50 \times \lambda_{F6} r_{F6} + \lambda_{T7} r_{T7} \\
&= 0.005225 + 2.20782 \\
&= 2.213045 \text{ h/year}
\end{aligned}$$

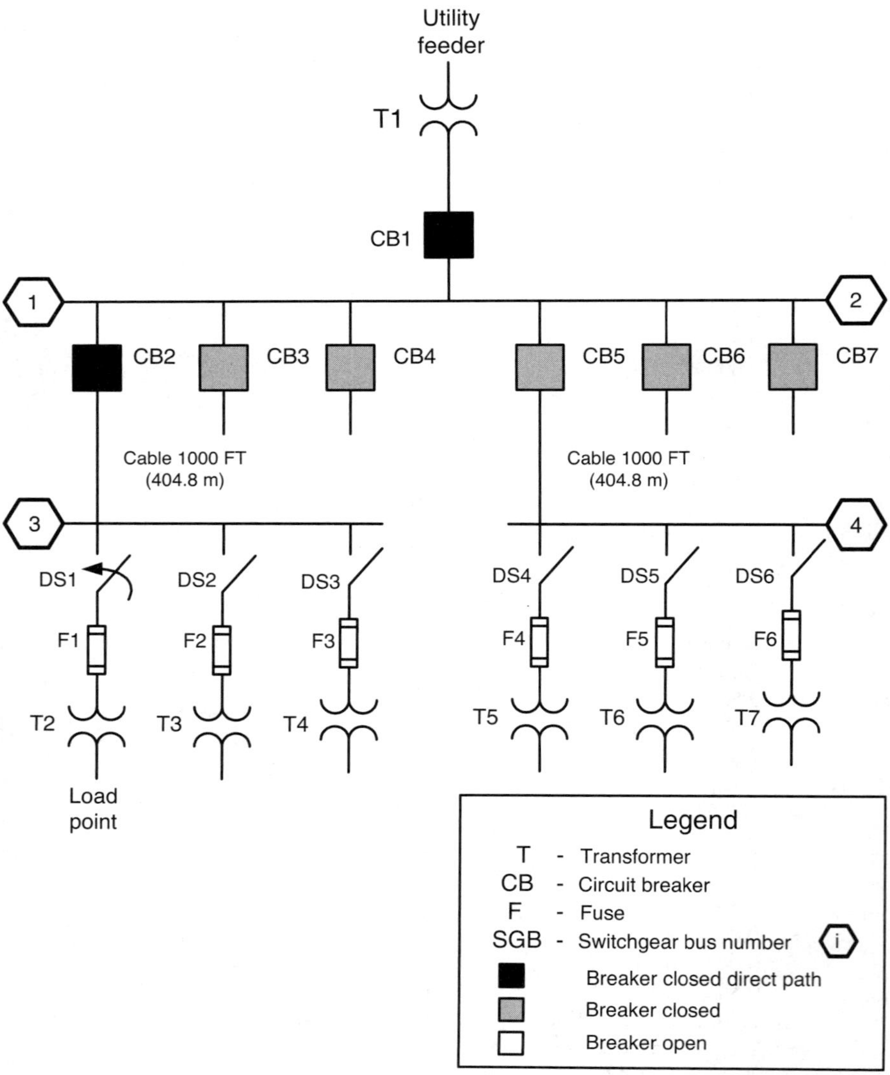


Figure 8.5. Single-line diagram of substation Design "A".

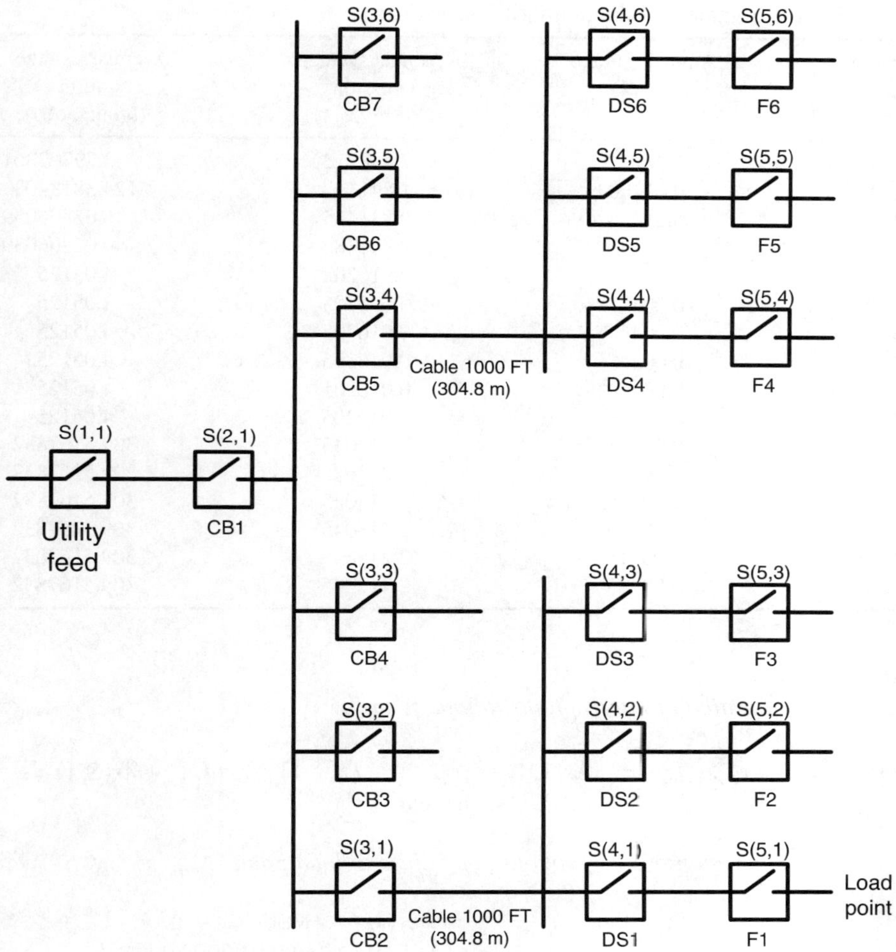


Figure 8.6. Zone branch diagram of substation Design "A".

The failure rate, average outage duration, and annual outage time are summarized in Table 8.10.

***8.4.1.2 Calculating the Load Point Failure Rate and Average Repair Duration of Interruptions: Design "A".*** Assuming the herringbone network for load point zone branch. The herringbone network for a given zone branch has the zone branches in the direct path as the main trunk and has all the zone branches immediately connected to the direct path as laterals.

*Frequency of interruptions at load point*

$$\begin{aligned}\lambda_{\text{load point}} &= \lambda(1,1)+\lambda(2,1)+\lambda(3,1)+\lambda(4,1)+\lambda(5,1)\\ &= 2.008434 \text{ interruptions/year}\end{aligned}$$

TABLE 8.10. Summary of Zone Branch Calculations

| Zone Branch | Annual Outage Rate ($\lambda$) (outages/year) | Annual Outage Duration ($\lambda r$) (h/year) | Average Outage Duration ($r$) (h/outage) |
|---|---|---|---|
| (1, 1) | 1.9642 | 7.458212 | 3.797073618 |
| (2, 1) | 0.014802 | 1.8411 | 124.3818403 |
| (3, 1) | 0.018282 | 0.841485 | 46.02806039 |
| (3, 4) | 0.018282 | 0.841485 | 46.02806039 |
| (4, 1) | 0.004 | 0.016205 | 4.05125 |
| (4, 2) | 0.004 | 0.016205 | 4.05125 |
| (4, 3) | 0.004 | 0.016205 | 4.05125 |
| (4, 4) | 0.004 | 0.016205 | 4.05125 |
| (4, 5) | 0.004 | 0.016205 | 4.05125 |
| (4, 6) | 0.004 | 0.016205 | 4.05125 |
| (5, 1) | 0.00715 | 2.213045 | 309.5167832 |
| (5, 2) | 0.00715 | 2.213045 | 309.5167832 |
| (5, 3) | 0.00715 | 2.213045 | 309.5167832 |
| (5, 4) | 0.00715 | 2.213045 | 309.5167832 |
| (5, 5) | 0.00715 | 2.213045 | 309.5167832 |
| (5, 6) | 0.00715 | 2.213045 | 309.5167832 |

*Annual duration of interruptions to load point*

$$\begin{aligned} \lambda r_{\text{load point}} &= \lambda r(1,1) + \lambda r(2,1) + \lambda r(3,1) + \lambda r(4,1) + \lambda r(5,1) \\ &= 12.370047\ \text{h/year} \end{aligned}$$

*Average duration of interruptions per failure to load point*

$$\begin{aligned} r_{\text{load } point} &= \lambda r_{\text{load point}} / \lambda_{\text{load point}} \\ &= 6.159050783\ \text{h/interruption} \end{aligned}$$

## 8.4.2 Case 2: Design "B"—Dual Supply Radial—Single Bus

This is a single bus dual supply radial system with two supply sources in parallel. Both supply sources are assumed to be capable of feeding the load independently. In case any supply needs repair or maintenance, the other supply can feed the load. The single-line diagram of substation Design "B"—dual supply radial—single bus is given in Fig. 8.7 and its zone branch diagram is provided in Fig. 8.8.

This configuration of Fig. 8.7 can be analyzed in two ways.

1. It is assumed that utility supply no. 1 is capable of supplying power to the whole network and utility supply no. 2 is redundant. In this case, CB2 is normally open. The second supply is only switched in when supply no. 1 has failed.

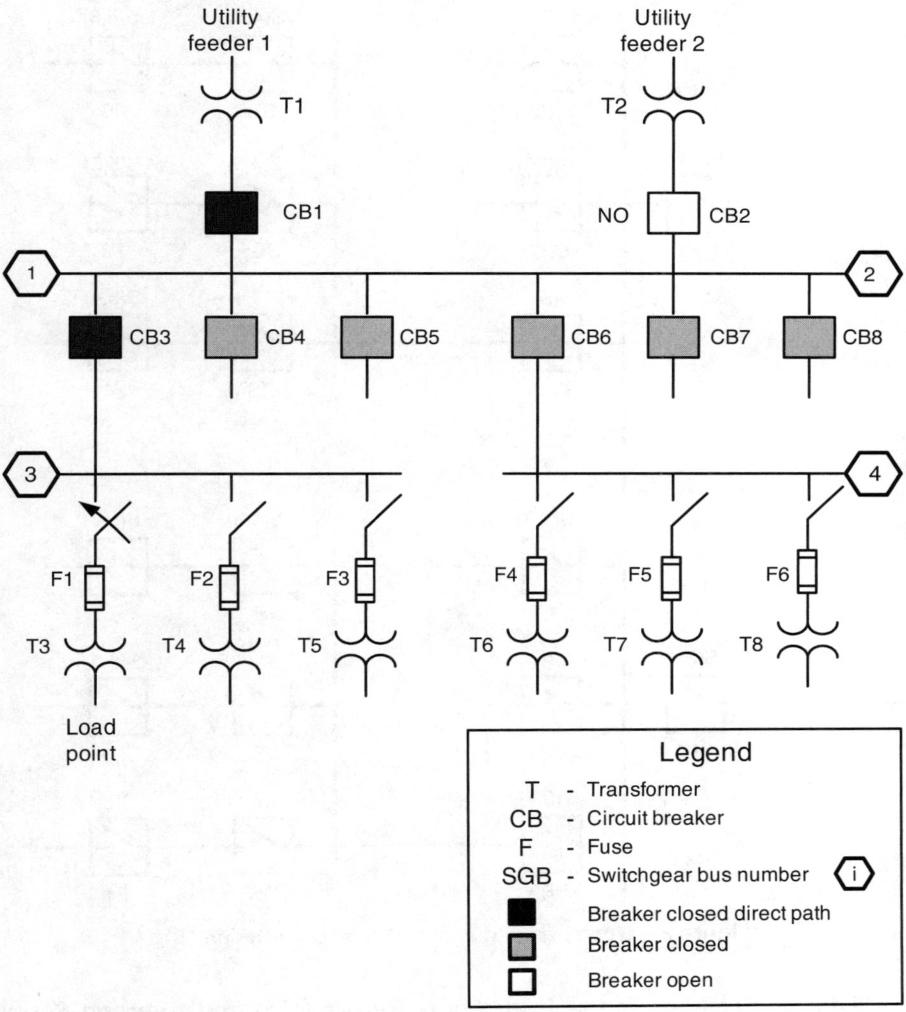


Figure 8.7. Single-line diagram of substation Design "B".

2. It is assumed that both supplies are always connected to the network and CB2 is normally closed.

***8.4.2.1 Zone Branch Calculations—Design "B"—Only One Supply is Connected at a Time.*** A summary of the zone branch parameters is provided in Table 8.11.

Assume the herringbone network for the load point zone branch. The herringbone network for a given zone branch has the zone branches in the direct path as the main trunk and has all the zone branches immediately connected to the direct path as laterals.

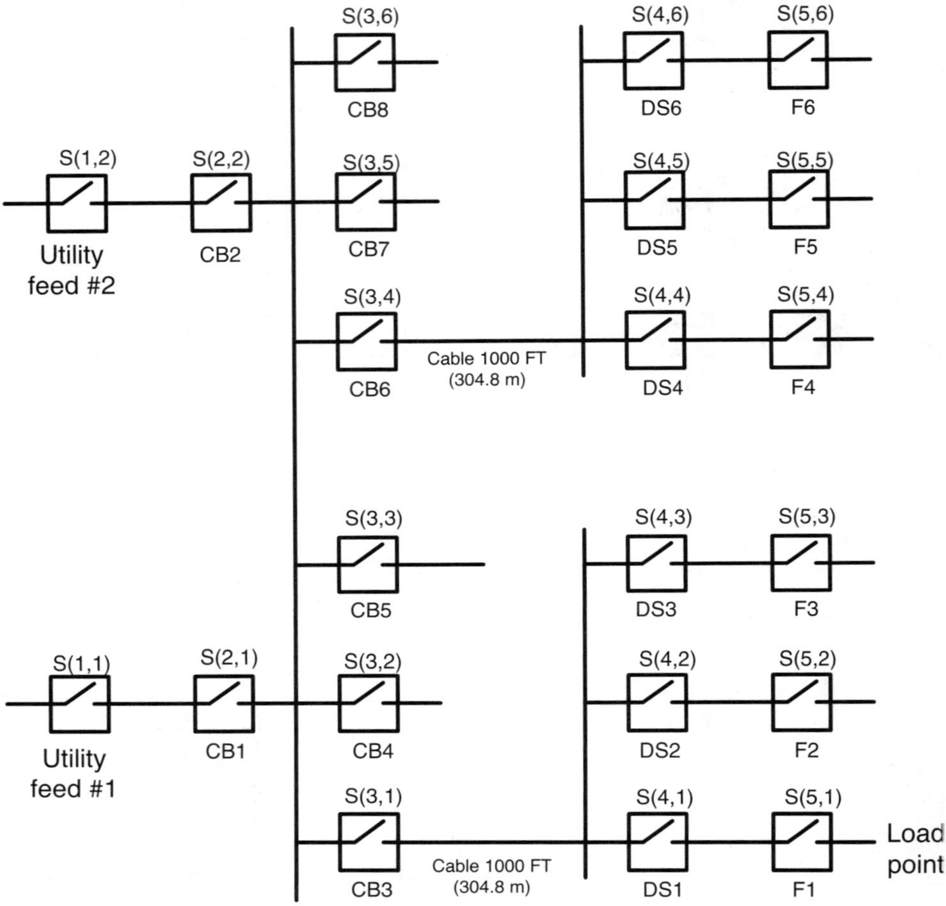


Figure 8.8. Zone branch diagram of substation Design "B".

#### *8.4.2.2 Calculating the Load Point Failure Rate and Average Repair Duration of Interruptions—Design "B"—When CB2 is not Switched in.*

*Frequency of interruptions at load point*

$$\begin{aligned}\lambda_{\text{load } point} &= \lambda(1,1)+\lambda(2,1)+\lambda(3,1)+\lambda(4,1)+\lambda(5,1)\\ &= 2.008434 \text{ interruptions/year}\end{aligned}$$

*Annual duration of interruptions to the load point*

$$\begin{aligned}\lambda r_{\text{load } point} &= \lambda r(1,1)+\lambda r(2,1)+\lambda r(3,1)+\lambda r(4,1)+\lambda r(5,1)\\ &= 12.370047 \text{ h/year}\end{aligned}$$

*Average duration of interruptions per failure to the load point*

$$\begin{aligned}r_{\text{load point}} &= \lambda r_{\text{load point}}/\lambda_{\text{load point}}\\ &= 6.159050783 \text{ h/interruption}\end{aligned}$$

TABLE 8.11. Summary of Zone Branch Calculations for Failure Rate, Average Repair Duration, and Annual Outage Time of the Load

| Zone Branch | Annual Outage Rate ($\lambda$) (outages/year) | Annual Outage Duration ($\lambda r$) (h/year) | Average Outage Duration ($r$) (h/outage) |
|---|---|---|---|
| (1, 1) | 1.9642 | 7.458212 | 3.797073618 |
| (1, 2) | 1.9642 | 7.458212 | 3.797073618 |
| (2, 1) | 0.014802 | 1.8411 | 124.3818403 |
| (2, 2) | 0.014802 | 1.8411 | 124.3818403 |
| (3, 1) | 0.018282 | 0.841485 | 46.02806039 |
| (3, 4) | 0.018282 | 0.841485 | 46.02806039 |
| (4, 1) | 0.004 | 0.016205 | 4.05125 |
| (4, 2) | 0.004 | 0.016205 | 4.05125 |
| (4, 3) | 0.004 | 0.016205 | 4.05125 |
| (4, 4) | 0.004 | 0.016205 | 4.05125 |
| (4, 5) | 0.004 | 0.016205 | 4.05125 |
| (4, 6) | 0.004 | 0.016205 | 4.05125 |
| (5, 1) | 0.00715 | 2.213045 | 309.5167832 |
| (5, 2) | 0.00715 | 2.213045 | 309.5167832 |
| (5, 3) | 0.00715 | 2.213045 | 309.5167832 |
| (5, 4) | 0.00715 | 2.213045 | 309.5167832 |

#### 8.4.2.3 Calculating the Load Point Failure Rate and the Average Repair Duration of Interruptions—Design "B"—When CB2 is Switched in After Failure of Utility Supply No. 1.

*Frequency of interruptions at the load point*

$$\begin{aligned}\lambda_{\text{load point}} &= \lambda(1,1)+\lambda(2,1)+\lambda(3,1)+\lambda(4,1)+\lambda(5,1)\\ &= 2.008434 \text{ interruptions/year}\end{aligned}$$

*Annual duration of interruptions to the load point*

$$\begin{aligned}\lambda r_{\text{load point}} &= \lambda r(1,1)+\lambda r(2,1)+\lambda r(3,1)+\lambda r(4,1)+\lambda r(5,1)\\ &= 12.370047 \text{ h/year}\end{aligned}$$

*Average duration of interruptions per failure to the load point*

$$\begin{aligned}r_{\text{load point}} &= \lambda r_{\text{load point}}/\lambda_{\text{load point}}\\ &= 6.159050783 \text{ h/interruption}\end{aligned}$$

**8.4.2.4 Calculating the Load Point Failure Rate and Average Repair Duration of Interruptions—Design "B"—Both Supplies are Connected at the Same Time.** In this case, there is no switching option and in case of any fault in the switchgear bus or any of the circuit breakers below it, the circuit breakers of both supplies will trip simultaneously and there will be no supply to the load. Also, since there is no switching option, in case of a fault in or below the switchgear bus, the supply to the load point will remain interrupted unless the faulty equipment is repaired or replaced (Figs. 8.9 and 8.10).

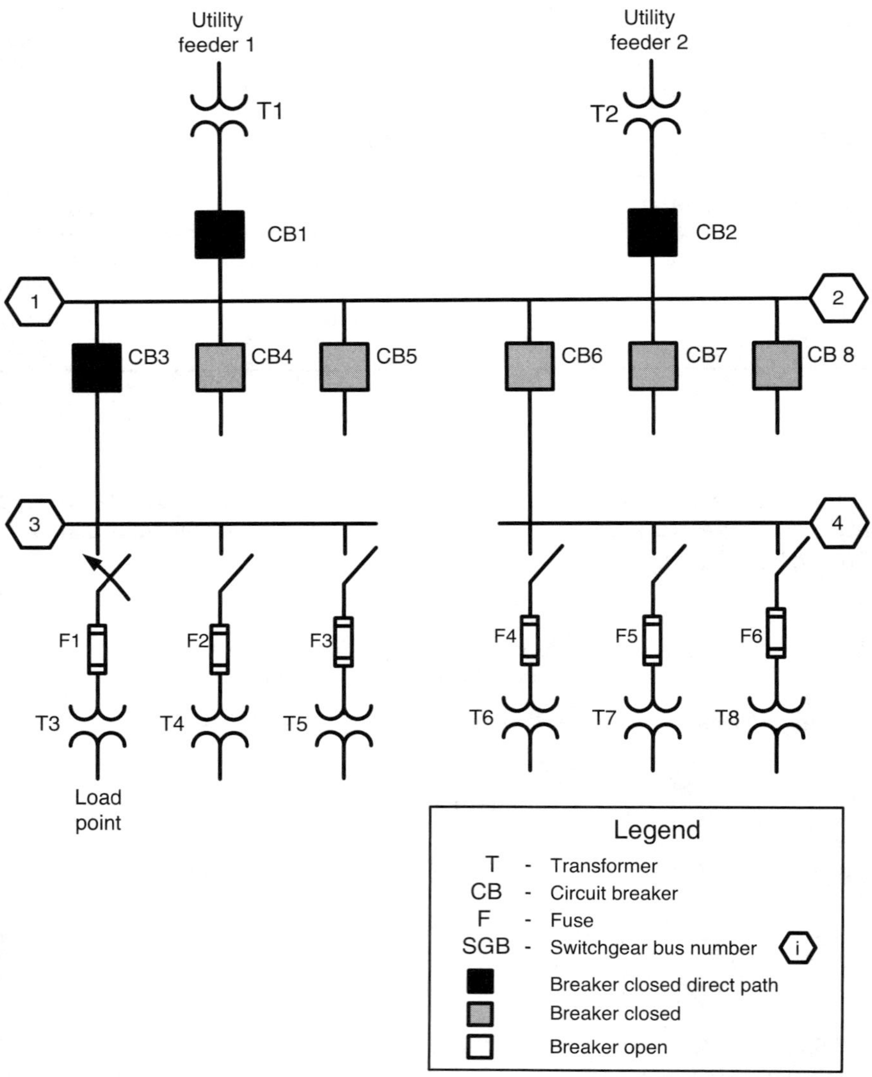


Figure 8.9. Single-line diagram of substation Design "B" (both supplies connected).

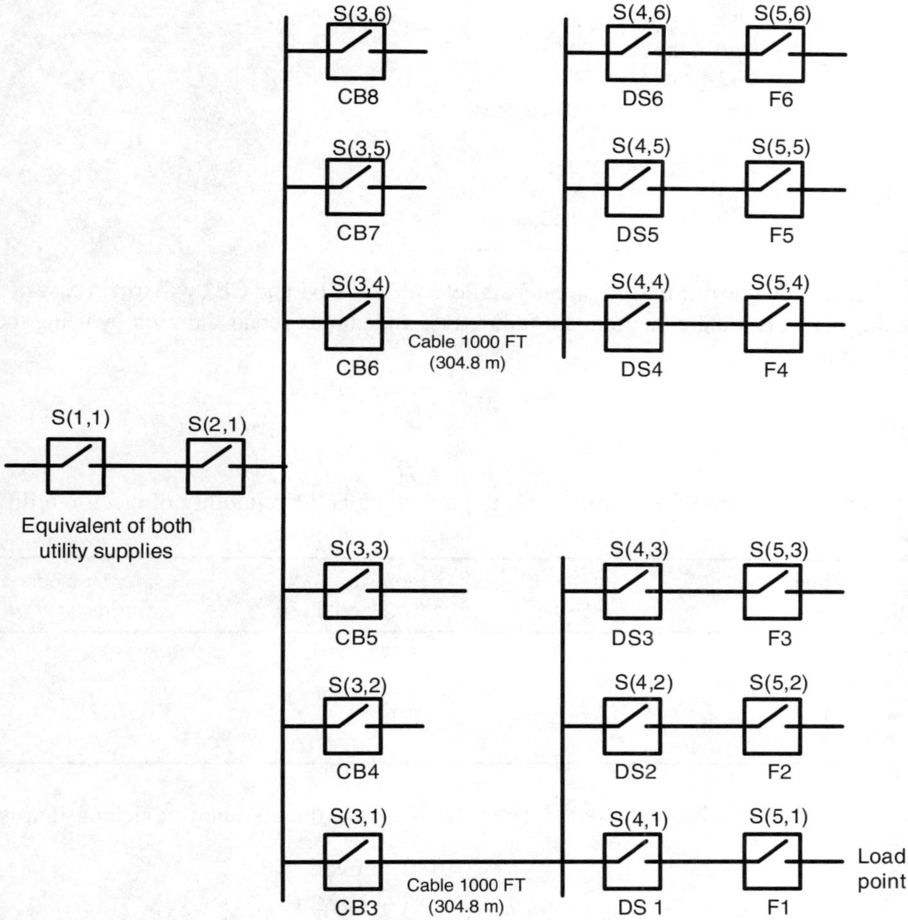


Figure 8.10. Zone branch diagram of substation Design "B" (both supplies connected).

#### *8.4.2.5 Zone Branch Calculations—Design "B"—Both Utility Supplies Connected.*

---

Zone 1—Branch 1

$$\begin{aligned}
\lambda_1 &= \lambda_s + \lambda_{T1} + 0.50\ \lambda_{CB1} \\
&= 1.956 + 0.0062 + 0.002 \\
&= 1.9642 \text{ outages/year} \\
\lambda_1 r_1 &= \lambda_s r_s + \lambda_{T1} r_{T1} + 0.50\ \lambda_{CB1} r_{CB1} \\
&= 5.050392 + 2.20782 + 0.2 \\
&= 7.458212 \text{ h/year} \\
r_1 &= \lambda r_1 / \lambda_1 \\
&= 3.797073618 \text{ h/interruption}
\end{aligned}$$

$$\begin{aligned}
\lambda_2 &= \lambda_s + \lambda_{T2} + 0.50 \times \lambda_{CB2}\\
&= 1.956 + 0.0062 + 0.002\\
&= 1.9642 \text{ outages/year}\\
\lambda_2 r_2 &= \lambda_s r_s + \lambda_{T2} r_{T2} + 0.50 \times \lambda_{CB2} r_{CB2}\\
&= 5.050392 + 2.20782 + 0.2\\
&= 7.458212 \text{ h/year}\\
r_2 &= \lambda r_2/\lambda_2\\
&= 3.797073618 \text{ h/interruption}
\end{aligned}$$

---

Since both the utility feeds are in parallel and both CB1 and CB2 will trip in case of a fault, we can calculate their annual failure rate and annual repair duration by using the following formulas:

$$\lambda_s = \lambda_1 \lambda_2 (r_1 + r_2)/8760,$$

and

$$r_s = r_1 r_2/(r_1 + r_2)$$

From Table 7.3 of the *Gold Book*, the following is the reliability of electric utility power supplies:

| No. of Circuits (all Voltages) | $\lambda$ (outages/year) | $r$ (hours of downtime/failure) | $\lambda r$ (forced hours of downtime/year) |
|---|---|---|---|
| Failure of single circuit | 1.956 | 1.32 | 2.582 |
| Failure of both circuits | 0.312 | 0.52 | 0.1622 |

Also from IEEE Standard 493-1997 (*Gold Book*), the common-mode constant is

$$K_{CM} = 0.07856986$$

$$\lambda_{2\text{ supplies}} = \lambda_1 \lambda_2 (r_1 + r_2)/8760 + K_{CM}(\lambda_1 + \lambda_2)$$

$$\lambda(1,1) = \lambda_{2\text{supplies}}$$

A summary of the zone branch parameters is shown in Table 8.12.

Assume the herringbone network for the load point zone branch. The herringbone network for a given zone branch has the zone branches in the direct path as the main trunk and has all the zone branches immediately connected to the direct path as laterals.

#### 8.4.2.6 Calculating the Load Point Failure Rate and Average Repair Duration of Interruptions—Design "B"—Both Utility Supplies Connected.

*Frequency of interruptions at the load point*

$$\begin{aligned}
\lambda_{\text{load point}} &= \lambda(1,1) + \lambda(2,1) + \lambda(3,1) + \lambda(4,1) + \lambda(5,1)\\
&= 0.356234 \text{ interruptions/year}
\end{aligned}$$

TABLE 8.12. Summary of Zone Branch Calculations for Failure Rate, Average Repair Duration, and Annual Outage Time of the Load for Fig. 8.8 System

| Zone Branch | Annual Outage Rate ($\lambda$) (outages/year) | Annual Outage Duration ($\lambda r$) (h/year) | Average Outage Duration ($r$) (h/outage) |
|---|---|---|---|
| (1, 1) | 0.003344616 | 0.006349877 | 1.898536809 |
| (2, 1) | 0.014802 | 1.8411 | 124.3818403 |
| (3, 1) | 0.018282 | 0.841485 | 46.02806039 |
| (3, 4) | 0.018282 | 0.841485 | 46.02806039 |
| (4, 1) | 0.004 | 0.016205 | 4.05125 |
| (4, 2) | 0.004 | 0.016205 | 4.05125 |
| (4, 3) | 0.004 | 0.016205 | 4.05125 |
| (4, 4) | 0.004 | 0.016205 | 4.05125 |
| (4, 5) | 0.004 | 0.016205 | 4.05125 |
| (4, 6) | 0.004 | 0.016205 | 4.05125 |
| (5, 1) | 0.00715 | 2.213045 | 309.5167832 |
| (5, 2) | 0.00715 | 2.213045 | 309.5167832 |
| (5, 3) | 0.00715 | 2.213045 | 309.5167832 |
| (5, 4) | 0.00715 | 2.213045 | 309.5167832 |
| (5, 5) | 0.00715 | 2.213045 | 309.5167832 |
| (5, 6) | 0.00715 | 2.213045 | 309.5167832 |

*Annual duration of interruptions to the load point*

$$\begin{aligned}\lambda r_{\text{load point}} &= \lambda r(1,1) + \lambda r(2,1) + \lambda r(3,1) + \lambda r(4,1) + \lambda r(5,1)\\ &= 6.810371809 \text{ h/year}\end{aligned}$$

*Average duration of interruptions per failure to load point*

$$\begin{aligned}r_{\text{load point}} &= \lambda r_{\text{load point}} / \lambda_{\text{load point}}\\ &= 19.11769177 \text{ h/interruption}\end{aligned}$$

### 8.4.3 Case 3: Design "C"—Dual Supply Radial with Tiebreaker

This configuration is similar to the previous one with the difference that the supply to the load is more reliable since there is a normally open tiebreaker (CB3), which can be closed to maintain supply to the load from utility feed no. 2 in case of a fault in utility no. 1. The single-line diagram of substation Design "B"—dual supply radial with tiebreaker is provided in Fig. 8.11 and its zone branch diagram is given in Fig. 8.12.

First, let us consider that normally open circuit breaker CB3 is ideal (i.e., has zero failure rate) and is not switched when there is a fault in utility feed no. 1.

***8.4.3.1 Zone Branch Calculations—Design "C"—When the Tiebreaker is Ideal and Switching Activity is not Included.*** A summary of the zone branch calculations for Design "C" is shown in Table 8.13.

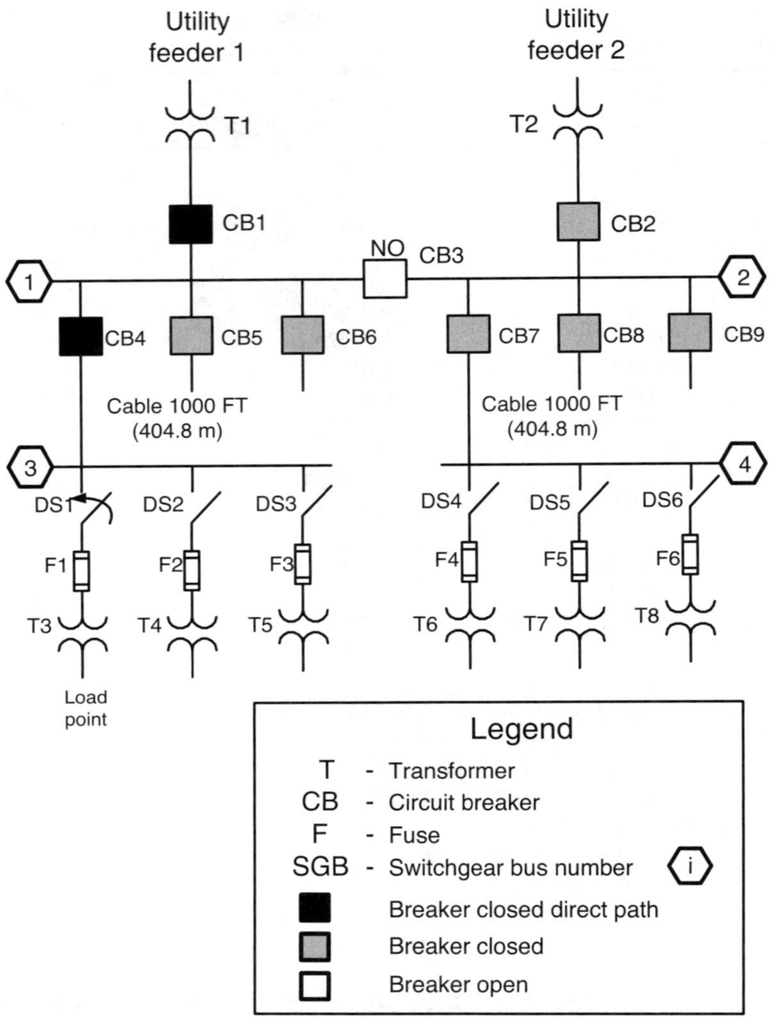


Figure 8.11. Single-line diagram of substation Design "C".

#### 8.4.3.2 Calculating the Load Point Failure Rate and Average Repair Duration—Design "C"—When Tiebreaker is Ideal and Switching Activity is not Included.

*Frequency of interruptions at the load point*

$$\lambda_{\text{load point}} = \lambda(1,1) + \lambda(2,1) + \lambda(3,1) + \lambda(4,1) + \lambda(5,1)$$
$$= 2.002434 \text{ interruptions/year}$$

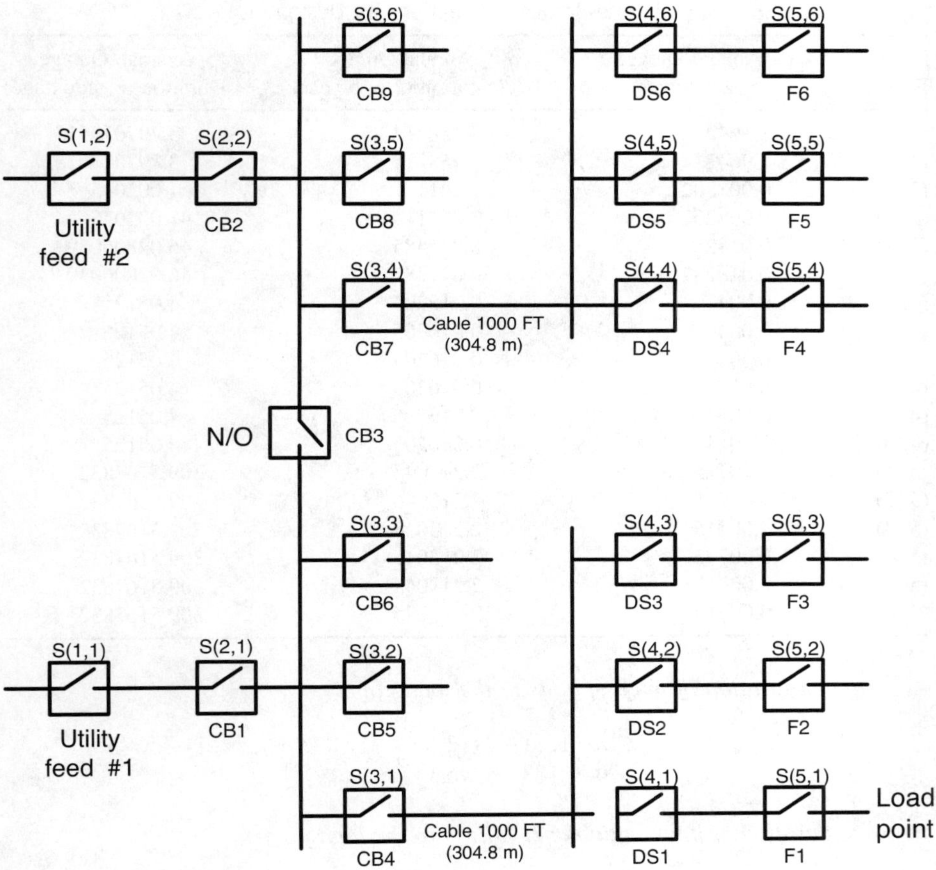


Figure 8.12. Zone branch diagram of substation Design "C".

*Annual duration of interruptions to the load point*

$$\begin{aligned}\lambda r_{\text{load point}} &= \lambda r(1,1) + \lambda r(2,1) + \lambda r(3,1) + \lambda r(4,1) + \lambda r(5,1) \\ &= 11.770047 \text{ h/year}\end{aligned}$$

*Average duration of interruptions per failure to the load point*

$$\begin{aligned}r_{\text{load point}} &= \lambda r_{\text{load point}} / \lambda_{\text{load point}} \\ &= 5.877870132 \text{ h/interruption}\end{aligned}$$

#### 8.4.3.3 Calculating the Load Point Failure Rate and Average Repair Duration—Design "C"—When Tiebreaker is Ideal and is Switched Within a Negligible Period of Time.

*Frequency of interruptions at the load point*

$$\begin{aligned}\lambda_{\text{load point}} &= \lambda(1,1) + \lambda(2,1) + \lambda(3,1) + \lambda(4,1) + \lambda(5,1) \\ &= 2.002434 \text{ interruptions/year}\end{aligned}$$

TABLE 8.13. Summary of Zone Branch Calculations for Design "C"

| Zone Branch | Annual Outage Rate ($\lambda$) (outages/year) | Annual Outage Duration ($\lambda r$) (h/year) | Average Outage Duration ($r$) (h/outage) |
|---|---|---|---|
| (1, 1) | 1.9642 | 7.458212 | 3.797073618 |
| (1, 2) | 1.9642 | 7.458212 | 3.797073618 |
| (2, 1) | 0.008802 | 1.2411 | 141.002045 |
| (2, 2) | 0.008802 | 1.2411 | 141.002045 |
| (3, 1) | 0.018282 | 0.841485 | 46.02806039 |
| (3, 4) | 0.018282 | 0.841485 | 46.02806039 |
| (4, 1) | 0.004 | 0.016205 | 4.05125 |
| (4, 2) | 0.004 | 0.016205 | 4.05125 |
| (4, 3) | 0.004 | 0.016205 | 4.05125 |
| (4, 4) | 0.004 | 0.016205 | 4.05125 |
| (4, 5) | 0.004 | 0.016205 | 4.05125 |
| (4, 6) | 0.004 | 0.016205 | 4.05125 |
| (5, 1) | 0.00715 | 2.213045 | 309.5167832 |
| (5, 2) | 0.00715 | 2.213045 | 309.5167832 |
| (5, 3) | 0.00715 | 2.213045 | 309.5167832 |
| (5, 4) | 0.00715 | 2.213045 | 309.5167832 |
| (5, 5) | 0.00715 | 2.213045 | 309.5167832 |
| (5, 6) | 0.00715 | 2.213045 | 309.5167832 |

*Annual duration of interruptions to the load point*

$$\begin{aligned}\lambda r_{\text{load point}} &= \lambda r(1,1) + \lambda r(2,1) + \lambda r(3,1) + \lambda r(4,1) + \lambda r(5,1)\\ &= 4.311835 \text{ h/year}\end{aligned}$$

*Average duration of interruptions per failure to the load point*

$$\begin{aligned}r_{\text{load point}} &= \lambda r_{\text{load point}} / \lambda_{\text{load point}}\\ &= 2.153296938 \text{ h/interruption}\end{aligned}$$

***8.4.3.4 Zone Branch Calculation—Design "C"—When the Tiebreaker has the Same Failure Rate as Other Breakers and When Switching Activity is Included.*** A summary of the zone branch calculations for Design "C" is shown in Table 8.14.

#### 8.4.3.5 Calculating the Load Point Failure Rate and Average Repair Duration—Design "C"—When the Tiebreaker has the Same Failure Rate as Other Breakers and When Switching Activity is Included.

*Frequency of interruptions at the load point*

$$\begin{aligned}\lambda_{\text{load point}} &= \lambda(1,1) + \lambda(2,1) + \lambda(3,1) + \lambda(4,1) + \lambda(5,1)\\ &= 2.002434 \text{ interruptions/year}\end{aligned}$$

TABLE 8.14. Summary of Zone Branch Calculations for Design "C"

| Zone Branch | Annual Outage Rate ($\lambda$) (outages/year) | Annual Outage Duration ($\lambda r$) (h/year) | Average Outage Duration ($r$) (h/outage) |
|---|---|---|---|
| (1, 1) | 1.9642 | 7.458212 | 3.797073618 |
| (1, 2) | 1.9642 | 7.458212 | 3.797073618 |
| (2, 1) | 0.010802 | 1.4411 | 141.002045 |
| (2, 2) | 0.010802 | 1.4411 | 141.002045 |
| (3, 1) | 0.018282 | 0.841485 | 46.02806039 |
| (3, 4) | 0.018282 | 0.841485 | 46.02806039 |
| (4, 1) | 0.004 | 0.016205 | 4.05125 |
| (4, 2) | 0.004 | 0.016205 | 4.05125 |
| (4, 3) | 0.004 | 0.016205 | 4.05125 |
| (4, 4) | 0.004 | 0.016205 | 4.05125 |
| (4, 5) | 0.004 | 0.016205 | 4.05125 |
| (4, 6) | 0.004 | 0.016205 | 4.05125 |
| (5, 1) | 0.00715 | 2.213045 | 309.5167832 |
| (5, 2) | 0.00715 | 2.213045 | 309.5167832 |
| (5, 3) | 0.00715 | 2.213045 | 309.5167832 |
| (5, 4) | 0.00715 | 2.213045 | 309.5167832 |
| (5, 5) | 0.00715 | 2.213045 | 309.5167832 |
| (5, 6) | 0.00715 | 2.213045 | 309.5167832 |

*Annual duration of interruptions to the load point*

$$\begin{aligned}\lambda r_{\text{load point}} &= \lambda r(1,1) \times \text{Rsw} + \lambda r(2,1) + \lambda r(3,1) + \lambda r(4,1) + \lambda r(5,1) \\ &= 4.511835 \text{ h/year}\end{aligned}$$

*Average duration of interruptions per failure to the load point*

$$\begin{aligned}r_{\text{load point}} &= \lambda r_{\text{load point}} / \lambda_{\text{load point}} \\ &= 2.250927194 \text{ h/interruption}\end{aligned}$$

### 8.4.4 Case 4: Design "D" —Dual Supply Loop with Tiebreaker

This design is more flexible than the previous design (design "C"). Fuses have been replaced with circuit breakers and two manual disconnect switches and a loop switch (LS) have been introduced. The single-line diagram of substation design "D" dual supply loop with tiebreaker is given in Fig. 8.13 and its zone branch diagram in Fig. 8.14.

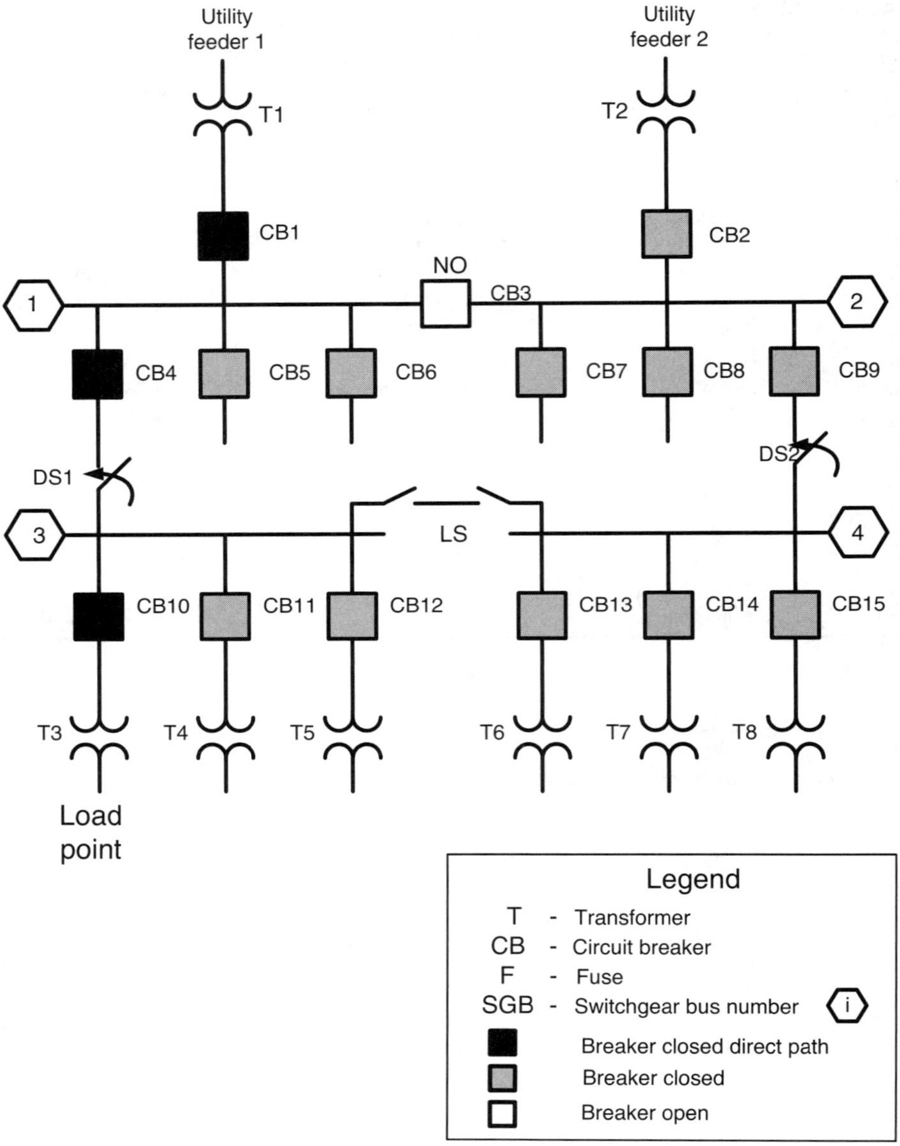


Figure 8.13. Single-line diagram of substation Design "D"—no switching activity.

This arrangement has many options for restoring the supply to the load in case of failure.

- In case of failure of utility feed no. 1, CB1 is opened and normally open circuit breaker CB3 is closed to supply the load.

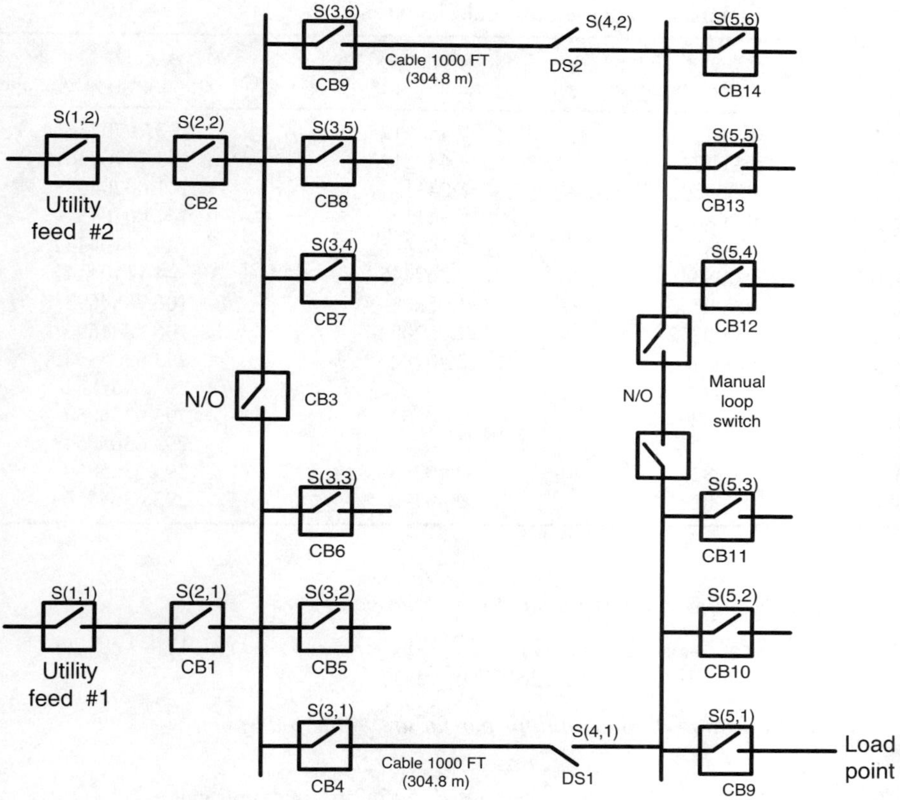


Figure 8.14. Zone branch diagram of substation Design "D".

- In case of failure of CB3 or fault above disconnect switch no. 1 (DS1), loop switch can be closed to supply the load.

***8.4.4.1 Zone Branch Calculations—Design "D"—Switching Activity is not Included.*** Figure 8.14 and Table 8.15 show the situation when no switching is done in case of any fault.

#### 8.4.4.2 Calculating the Load Point Failure Rate and Average Repair Duration—Design "D"—Switching Activity is not Included.

*Frequency of interruptions at the load point*

$$\begin{aligned}\lambda_{\text{load point}} &= \lambda(1,1)+\lambda(2,1)+\lambda(3,1)+\lambda(4,1)+\lambda(5,1)\\ &= 1.999384 \text{ interruptions/year}\end{aligned}$$

TABLE 8.15. Summary of Zone Branch Calculations

| Zone Branch | Annual Outage Rate ($\lambda$) (outages/year) | Annual Outage Duration ($\lambda r$) (h/year) | Average Outage Duration ($r$) (h/outage) |
|---|---|---|---|
| (1, 1) | 1.9642 | 7.458212 | 3.797073618 |
| (1, 2) | 1.9642 | 7.458212 | 3.797073618 |
| (2, 1) | 0.008802 | 1.2411 | 141.002045 |
| (2, 2) | 0.008802 | 1.2411 | 141.002045 |
| (3, 1) | 0.00833 | 0.367445 | 44.11104442 |
| (3, 6) | 0.00833 | 0.367445 | 44.11104442 |
| (4, 1) | 0.009852 | 1.05208 | 106.7884693 |
| (4, 2) | 0.009852 | 1.05208 | 106.7884693 |
| (5, 1) | 0.0082 | 2.40782 | 293.6365854 |
| (5, 2) | 0.0082 | 2.40782 | 293.6365854 |
| (5, 3) | 0.0082 | 2.40782 | 293.6365854 |
| (5, 4) | 0.0082 | 2.40782 | 293.6365854 |
| (5, 5) | 0.0082 | 2.40782 | 293.6365854 |
| (5, 6) | 0.0082 | 2.40782 | 293.6365854 |

*Annual duration of interruptions to the load point*

$$\begin{aligned}\lambda r_{\text{load point}} &= \lambda r(1,1) + \lambda r(2,1) + \lambda r(3,1) + \lambda r(4,1) + \lambda r(5,1) \\ &= 12.526657 \text{ h/year}\end{aligned}$$

*Average duration of interruptions per failure to the load point*

$$\begin{aligned}r_{\text{load point}} &= \lambda r_{\text{load point}}/\lambda_{\text{load point}} \\ &= 6.2652582 \text{ h/interruption}\end{aligned}$$

**8.4.4.3 Calculating the Load Point Failure Rate and Average Repair Duration—Design "D"—When the Tiebreaker is Switched Within a Negligible Period of Time(<5 s).** Here, it is assumed that CB3 has a zero failure rate and that it can be switched within zero time (Fig. 8.15).

*Frequency of interruptions at the load point*

$$\begin{aligned}\lambda_{\text{load point}} &= \lambda(1,1) + \lambda(2,1) + \lambda(3,1) + \lambda(4,1) + \lambda(5,1) \\ &= 1.999384 \text{ interruptions/year}\end{aligned}$$

*Annual duration of interruptions to the load point*

$$\begin{aligned}\lambda r_{\text{load point}} &= \lambda r(1,1) \times \text{Rsw} + (2,1) + \lambda r(3,1) + \lambda r(4,1) + \lambda r(5,1) \\ &= 4.511835 \text{ h/year}\end{aligned}$$

*Average duration of interruptions per failure to the load point*

$$\begin{aligned}r_{\text{load point}} &= \lambda r_{\text{load point}}/\lambda_{\text{load point}} \\ &= 2.535003281 \text{ h/interruption}\end{aligned}$$

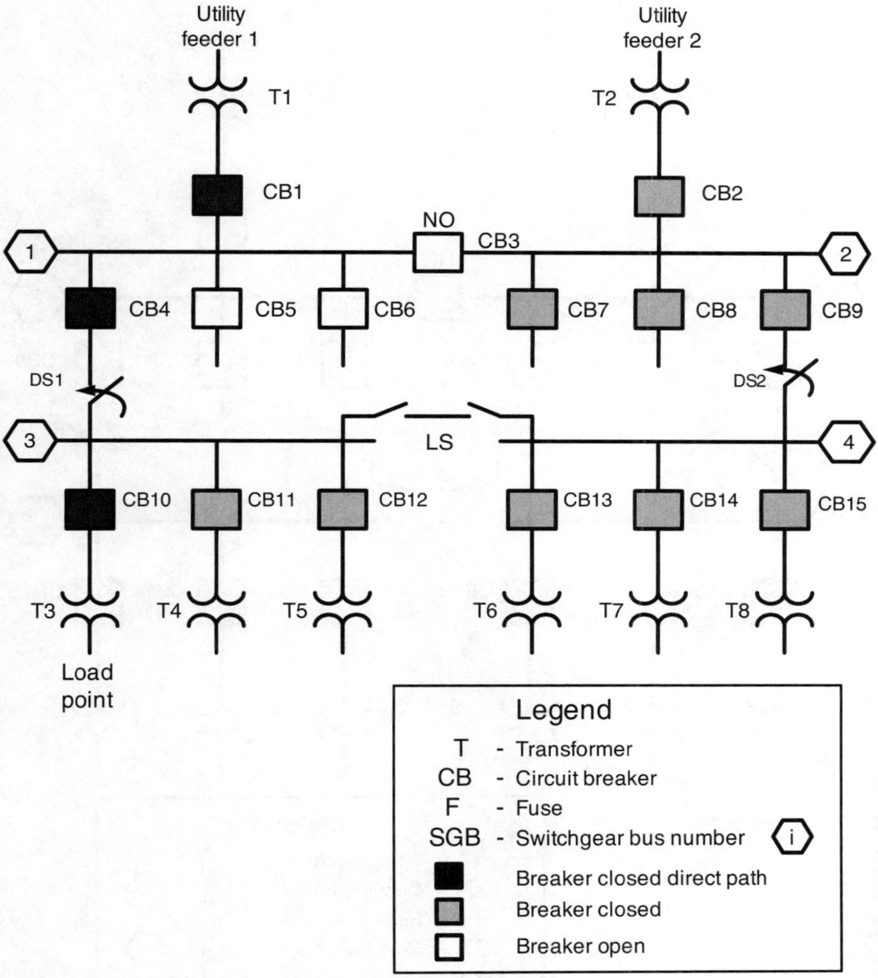


Figure 8.15. Single-line diagram of substation Design "D"—CB3 switching.

***8.4.4.4 Calculating the Load Point Failure Rate and Average Repair Duration—When the Loop Switch is Operated Within 15 min.*** Here, it is assumed that the fault is above disconnect switch DS1 and that the loop switch can be operated within 15 min of time (Fig. 8.16).

*Frequency of interruptions at the load point*

$$\begin{aligned}\lambda_{\text{load point}} &= \lambda(1,1) + \lambda(2,1) + \lambda(3,1) + \lambda(4,1) + \lambda(5,1) \\ &= 1.999384 \text{ interruptions per year}\end{aligned}$$

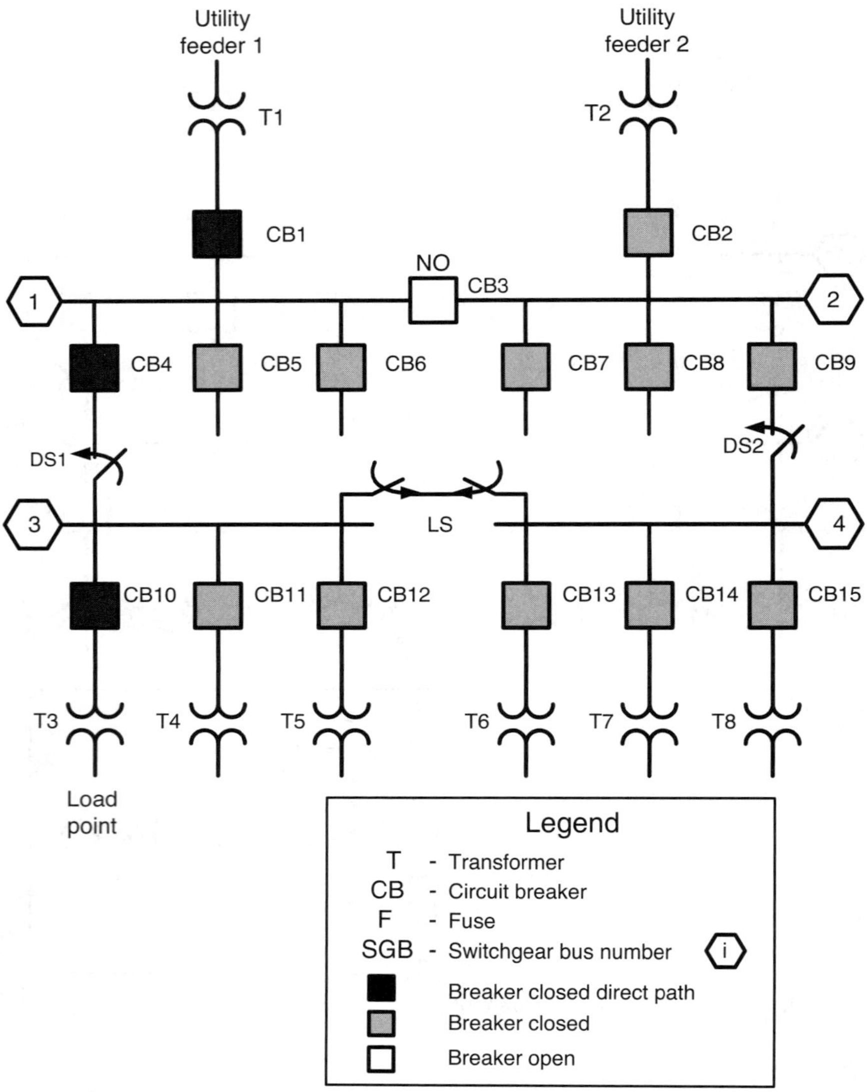


Figure 8.16. Single-line diagram of substation Design "D"—with loop switching activity.

*Annual duration of interruptions to the load point*

$$\begin{aligned}\lambda r_{\text{load point}} &= \lambda(1,1)\times \text{Rsw}+\lambda(2,1)\times \text{Rsw}+\lambda(3,1)\times \text{Rsw}+\lambda r(4,1)+\lambda r(5,1)\\ &= 3.4606625\ \text{h/year}\end{aligned}$$

*Average duration of interruptions per failure to the load point*

$$\begin{aligned}r_{\text{load point}} &= \lambda r_{\text{load point}}/\lambda_{\text{load point}}\\ &= 1.730864356\ \text{h/interruption}\end{aligned}$$

TABLE 8.16. Summary of Zone Branch Calculations

| Zone Branch | Annual Outage Rate ($\lambda$) (outages/year) | Annual Outage Duration ($\lambda r$) (h/year) | Average Outage Duration ($r$) (h/outage) |
|---|---|---|---|
| (1, 1) | 1.9642 | 7.458212 | 3.797073618 |
| (1, 2) | 1.9642 | 7.458212 | 3.797073618 |
| (2, 1) | 0.010802 | 1.4411 | 133.4104795 |
| (2, 2) | 0.010802 | 1.4411 | 133.4104795 |
| (3, 1) | 0.00833 | 0.367445 | 44.11104442 |
| (3, 6) | 0.00833 | 0.367445 | 44.11104442 |
| (4, 1) | 0.015952 | 1.07404 | 67.32948847 |
| (4, 2) | 0.015952 | 1.07404 | 67.32948847 |
| (5, 1) | 0.0082 | 2.40782 | 293.6365854 |
| (5, 2) | 0.0082 | 2.40782 | 293.6365854 |
| (5, 3) | 0.0082 | 2.40782 | 293.6365854 |
| (5, 4) | 0.0082 | 2.40782 | 293.6365854 |
| (5, 5) | 0.0082 | 2.40782 | 293.6365854 |
| (5, 6) | 0.0082 | 2.40782 | 293.6365854 |

**8.4.4.5 Calculating the Load Point Failure Rate and Average Repair Duration—When the Tiebreaker has the Same Failure Rate as Other Breakers and the Failure Rate of the Loop Switch is Included.** Here, it is assumed that tiebreaker CB3 is not ideal and has the same failure rate as other breakers and the failure rate of loop switch is also included. It is assumed that CB3 has negligible switching time and that the loop switch has a switching time of 15 min (Table 8.16 and Fig. 8.17).

*Frequency of interruptions at the load point*

$$\begin{aligned}\lambda_{\text{load point}} &= \lambda(1,1)+\lambda(2,1)+\lambda(3,1)+\lambda(4,1)+\lambda(5,1)\\ &= 2.007484 \text{ interruptions/year}\end{aligned}$$

*Annual duration of interruptions to the load point*

$$\begin{aligned}\lambda r_{\text{load point}} &= \lambda(1,1)\text{Rsw}+\lambda(2,1)\text{Rsw}+\lambda(3,1)\text{Rsw}+\lambda r(4,1)+\lambda r(5,1)\\ &= 3.485323 \text{ h/year}\end{aligned}$$

*Average duration of interruptions per failure to the load point*

$$\begin{aligned}r_{\text{load point}} &= \lambda r_{\text{load point}}/\lambda_{\text{load point}}\\ &= 1.736164771 \text{ h/interruption}\end{aligned}$$

## 8.4.5 Case 5: Design "E"—Dual Supply Primary Selective

This arrangement has two utility feeds and each feeder is assumed to have the capability to supply the whole network. The single-line diagram of substation design "E"—dual supply primary selective is given in Fig. 8.18 and its zone branch diagram in Fig. 8.19.

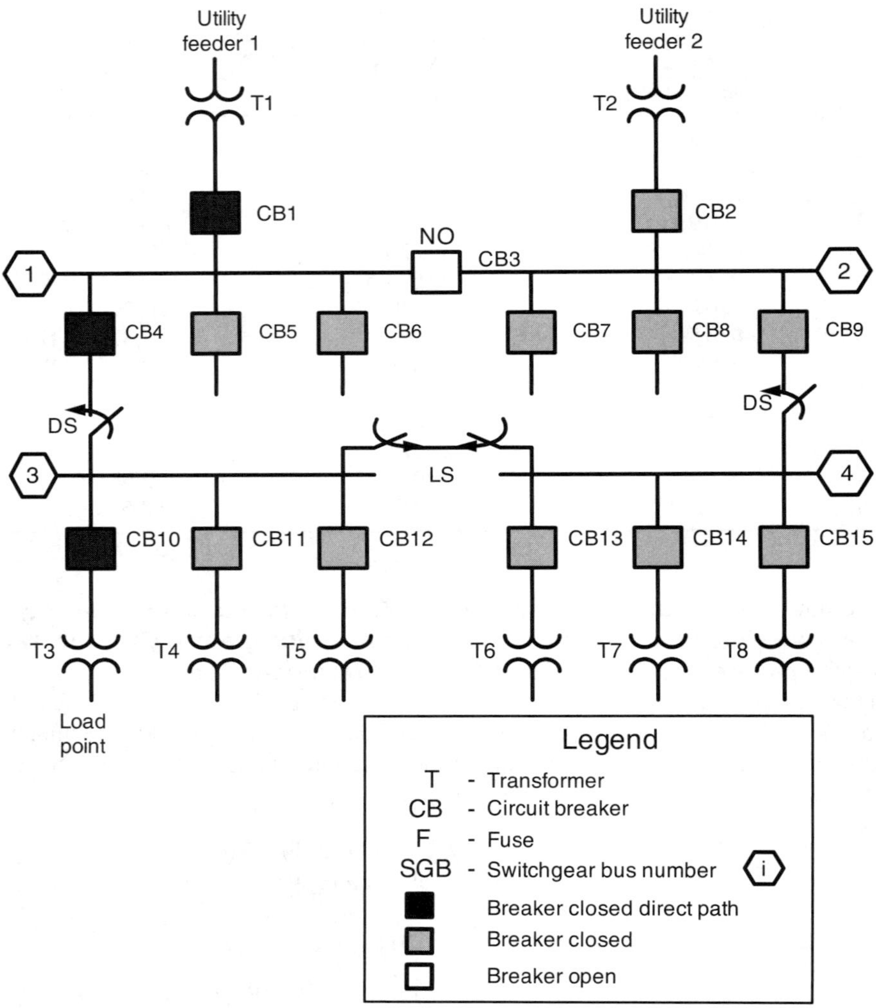


Figure 8.17. Single-line diagram of substation Design "D"—CB3 is not ideal.

There are many options for a reliable supply to the load point.

1. The load is primarily fed from utility feeder no. 1.
2. Normally open circuit breaker, CB3, can be switched on if there is a fault in the utility feeder no. 1 supply system.
3. In case of a faulty CB3, the load can be supplied through CB7–SGB6–DS4 from utility feeder no. 2.

***8.4.5.1 Zone Branch Calculations—Design "E"—Switching Activity is not Included.*** In this case, it has been assumed that normally open circuit breaker CB3 is not operated in case of a fault above switch gear no. 1 (Fig. 8.18). Therefore, CB3 is not

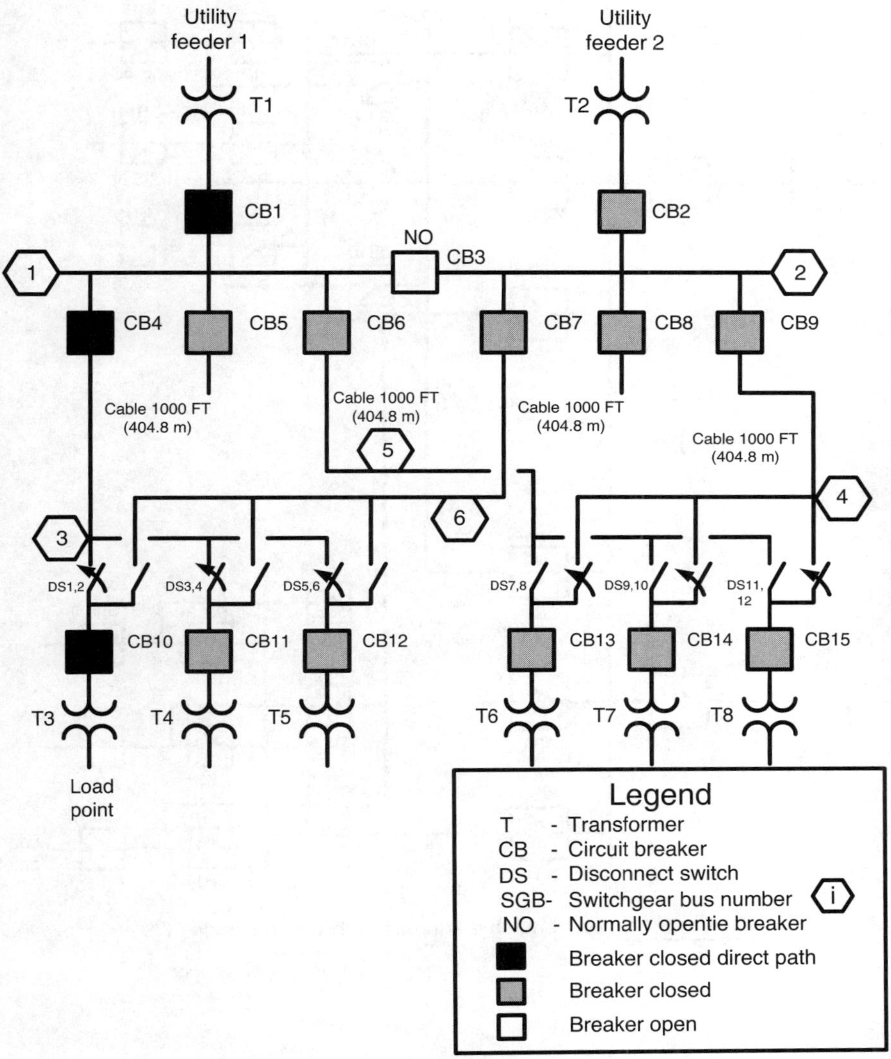


Figure 8.18. Single-line diagram of substation Design "E".

included in zone branch calculations. Moreover, load is not supplied though switchgear bus no. 6 in case of a fault above disconnect switches (Table 8.17).

#### 8.4.5.2 Calculating the Load Point Failure Rate and the Average Repair Duration—Switching Activity is not Included.

*Frequency of interruptions at the load point*

$$\begin{aligned}\lambda_{\text{load point}} &= \lambda(1,1)+\lambda(2,1)+\lambda(3,1)+\lambda(4,1)+\lambda(5,1)\\ &= 2.004534 \text{ interruptions/year}\end{aligned}$$

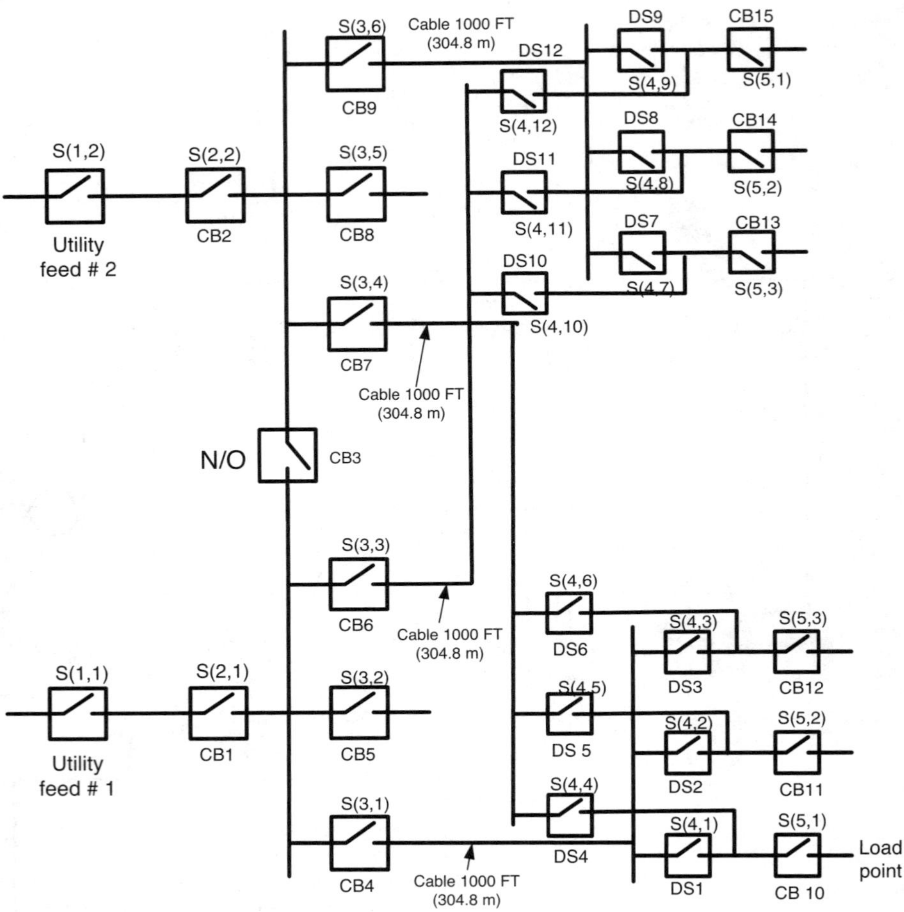


Figure 8.19. Zone branch diagram of substation Design "E".

*Annual duration of interruptions to the load point*

$$\begin{aligned}\lambda r_{\text{load point}} &= \lambda r(1,1) + \lambda r(2,1) + \lambda r(3,1) + \lambda r(4,1) + \lambda r(5,1)\\ &= 12.157097 \text{ h/year}\end{aligned}$$

*Average duration of interruptions per failure to the load point*

$$\begin{aligned}r_{\text{load point}} &= \lambda r_{\text{load point}}/\lambda_{\text{load point}}\\ &= 6.064799599 \text{ h/interruption}\end{aligned}$$

TABLE 8.17. Summary of Zone Branch Calculations

| Zone Branch | Annual Outage Rate ($\lambda$) (outages/year) | Annual Outage Duration ($\lambda r$) (h/year) | Average Outage Duration ($r$) (h/outage) |
|---|---|---|---|
| (1, 1) | 1.9642 | 7.458212 | 3.797073618 |
| (1, 2) | 1.9642 | 7.458212 | 3.797073618 |
| (2, 1) | 0.008802 | 1.2411 | 141.002045 |
| (2, 2) | 0.008802 | 1.2411 | 141.002045 |
| (3, 1) | 0.018282 | 0.838985 | 45.89131386 |
| (3, 3) | 0.018282 | 0.838985 | 45.89131386 |
| (3, 4) | 0.018282 | 0.838985 | 45.89131386 |
| (3, 6) | 0.018282 | 0.838985 | 45.89131386 |
| (4, 1) | 0.00505 | 0.21098 | 41.77821782 |
| (4, 2) | 0.00505 | 0.21098 | 41.77821782 |
| (4, 3) | 0.00505 | 0.21098 | 41.77821782 |
| (4, 4) | 0.00505 | 0.21098 | 41.77821782 |
| (4, 5) | 0.00505 | 0.21098 | 41.77821782 |
| (4, 6) | 0.00505 | 0.21098 | 41.77821782 |
| (4, 7) | 0.00505 | 0.21098 | 41.77821782 |
| (4, 8) | 0.00505 | 0.21098 | 41.77821782 |
| (4, 9) | 0.00505 | 0.21098 | 41.77821782 |
| (4, 10) | 0.00505 | 0.21098 | 41.77821782 |
| (4, 11) | 0.00505 | 0.21098 | 41.77821782 |
| (4, 12) | 0.00505 | 0.21098 | 41.77821782 |
| (5, 1) | 0.0082 | 2.40782 | 293.6365854 |
| (5, 2) | 0.0082 | 2.40782 | 293.6365854 |
| (5, 3) | 0.0082 | 2.40782 | 293.6365854 |
| (5, 4) | 0.0082 | 2.40782 | 293.6365854 |

***8.4.5.3 Calculating the Load Point Failure Rate and the Average Repair Duration—When the Tiebreaker has the Same Failure Rate as Other Breakers and When Switching Activity is Included.*** To feed the load point, the following are the options:

1. The load point is fed from utility feed no. 1 through CB10, DS1, CB4, and CB1.
2. In case of a fault in utility feed no. 1, the load point is fed through utility feed no. 2 by instantly switching in CB3.
3. In case of failure of CB3, the load point is fed from utility feed no. 2 through CB7, switchgear bus no. 6, and manual switch DS1.
4. In case of failure of manual switch DS1, supply can be restored within 15 min (Rsw = 0.25 h) using DS4.

Here, it is assumed that CB3 is closed instantly (Rsw = 0 h) when there is a fault in or before CB1 or CB2 (any one of the supplies fails) and the affected side is fed by closing the CB3. It is also assumed that CB3 has the same failure rate as other circuit breakers (Table 8.18 and Fig. 8.20).

TABLE 8.18. Summary of Zone Branch Calculations

| Zone Branch | Annual Outage Rate ($\lambda$) (outages/year) | Annual Outage Duration ($\lambda r$) (h/year) | Average Outage Duration ($r$) (h/outage) |
|---|---|---|---|
| (1, 1) | 1.9642 | 7.458212 | 3.797073618 |
| (1, 2) | 1.9642 | 7.458212 | 3.797073618 |
| (2, 1) | 0.010802 | 1.4411 | 133.4104795 |
| (2, 2) | 0.010802 | 1.4411 | 133.4104795 |
| (2, 3) | 0.008802 | 1.2411 | 141.002045 |
| (3, 1) | 0.018282 | 0.838985 | 45.89131386 |
| (3, 3) | 0.018282 | 0.838985 | 45.89131386 |
| (3, 4) | 0.018282 | 0.838985 | 45.89131386 |
| (3, 6) | 0.018282 | 0.838985 | 45.89131386 |
| (4, 1) | 0.00505 | 0.21098 | 41.77821782 |
| (4, 2) | 0.00505 | 0.21098 | 41.77821782 |
| (4, 3) | 0.00505 | 0.21098 | 41.77821782 |
| (4, 4) | 0.00505 | 0.21098 | 41.77821782 |
| (4, 5) | 0.00505 | 0.21098 | 41.77821782 |
| (4, 6) | 0.00505 | 0.21098 | 41.77821782 |
| (4, 7) | 0.00505 | 0.21098 | 41.77821782 |
| (4, 8) | 0.00505 | 0.21098 | 41.77821782 |
| (4, 9) | 0.00505 | 0.21098 | 41.77821782 |
| (4, 10) | 0.00505 | 0.21098 | 41.77821782 |
| (4, 11) | 0.00505 | 0.21098 | 41.77821782 |
| (4, 12) | 0.00505 | 0.21098 | 41.77821782 |
| (5, 1) | 0.0082 | 2.40782 | 293.6365854 |
| (5, 2) | 0.0082 | 2.40782 | 293.6365854 |
| (5, 3) | 0.0082 | 2.40782 | 293.6365854 |
| (5, 4) | 0.0082 | 2.40782 | 293.6365854 |
| (5, 5) | 0.0082 | 2.40782 | 293.6365854 |
| (5, 6) | 0.0082 | 2.40782 | 293.6365854 |

#### 8.4.5.4 Calculating the Load Point Failure Rate and Average Repair Duration—CB3 is Closed Instantly (Rsw = 0 h).

*Frequency of interruptions at the load point*

$$\begin{aligned}\lambda_{\text{load point}} &= \lambda(1,1)+\lambda(2,1)+\lambda(3,1)+\lambda(4,1)+\lambda(5,1)\\ &= 2.006534 \text{ interruptions/year}\end{aligned}$$

*Annual duration of interruptions to the load point*

$$\begin{aligned}\lambda r_{\text{load point}} &= \lambda(1,1)\times \text{Rsw}+\lambda(2,1)\times \text{Rsw}+\lambda(3,1)\times \text{Rsw}+\lambda r(4,1)+\lambda r(5,1)\\ &= 1.9642\times 0+0.010802\times 0+0.018282\times 0.25+0.21098+2.40782\\ &= 2.6233705 \text{ h/year}\end{aligned}$$

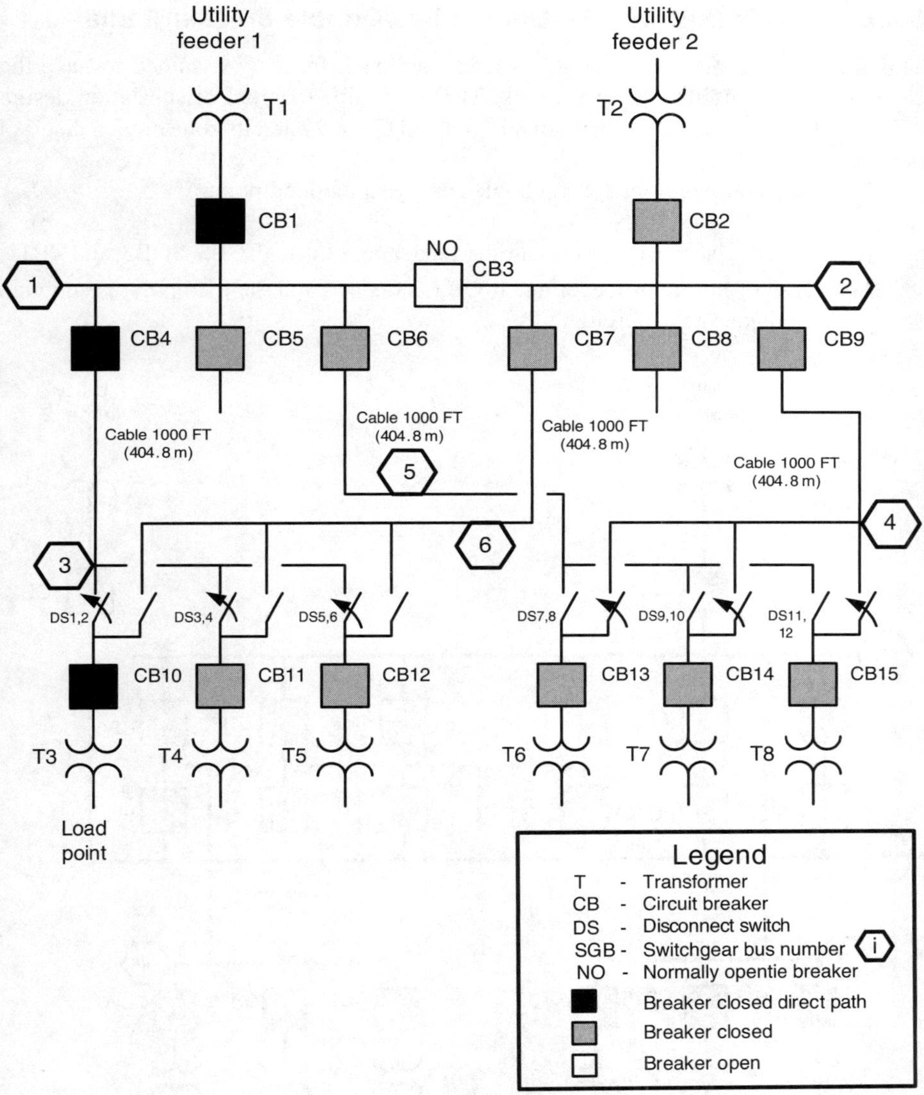


Figure 8.20. Single-line diagram of substation Design "E"—switching activity is included.

*Average duration of interruptions per failure to the load point*

$$
\begin{aligned}
r_{\text{load point}} &= \lambda r_{\text{load point}} / \lambda_{\text{load point}} \\
&= 1.307413929 \text{ h/interruption}
\end{aligned}
$$

### 8.4.6 Case 6: Design "F"—Double Bus/Double Breaker Radial

This arrangement also has two utility feeds and each feeder is assumed to have the capability to supply the whole network. The single-line diagram of substation design "F"—double bus/double breaker radial is given in Fig. 8.21 and its zone branch diagram is shown in Fig. 8.22.

There are many options for a reliable supply to the load point.

1. The load is primarily fed from utility feeder no. 1 through CB3, SGB3, and CB15.
2. In case of failure of feeder no. 1, CB3 is opened and the load point is supplied from feeder no. 2 through CB9.

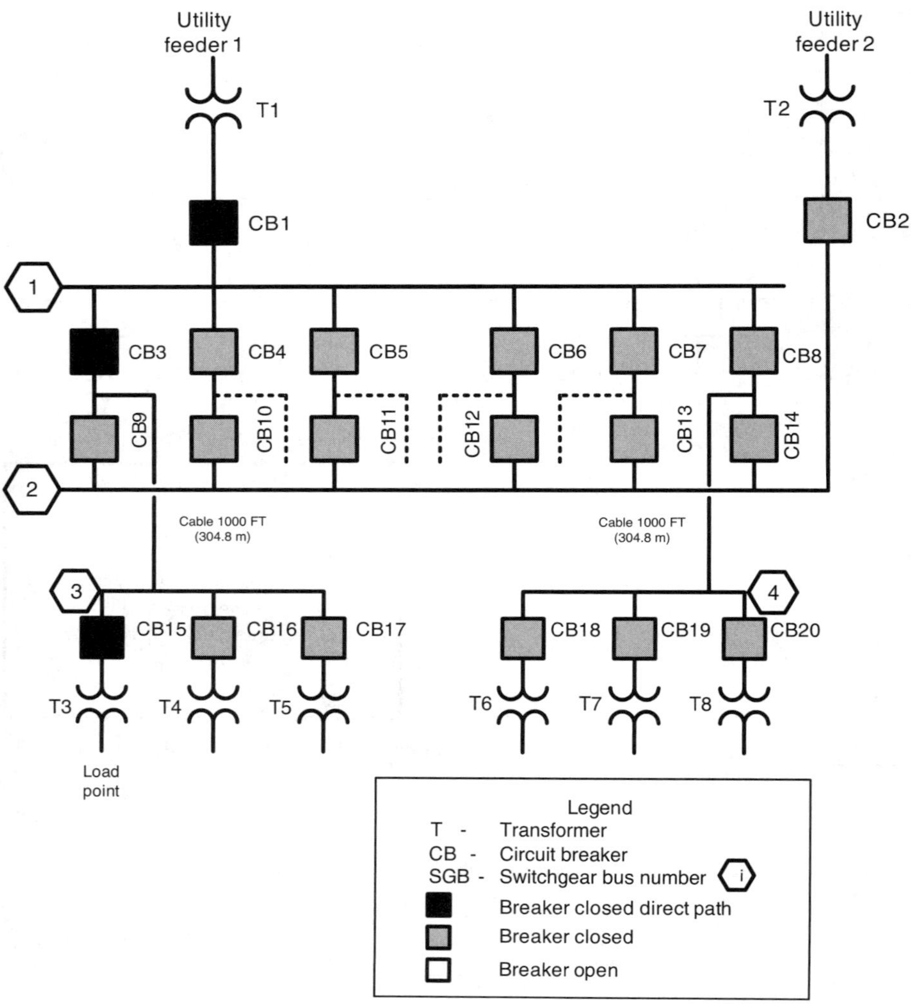


Figure 8.21. Single-line diagram of substation Design "F".

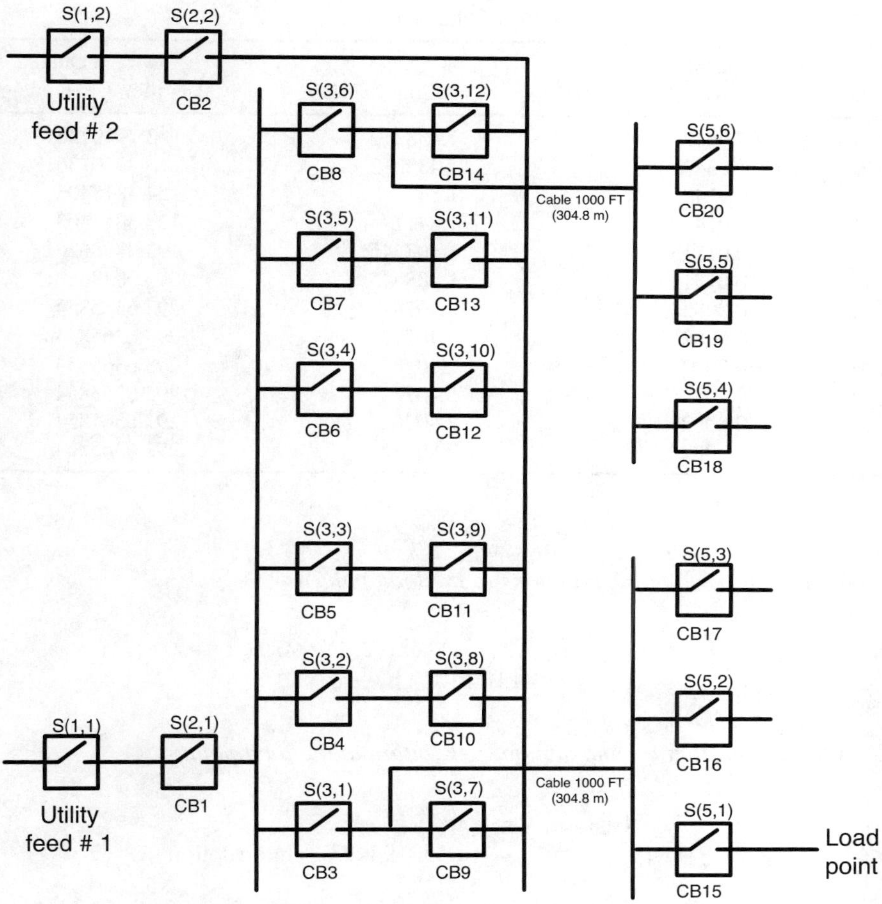


Figure 8.22. Zone branch diagram of substation Design "F".

3. CB4 to CB7 and CB10 to CB13 are normally open and are for future use, but they can be used in the situation when utility feed no. 1, CB8 and CB9 fail. Then any set of circuit breakers (e.g., CB4 and CB10) can be closed to supply power to the load.

***8.4.6.1 Zone Branch Calculations—Switching Activity is not Included.*** In this case, it has been assumed that circuit breakers CB8 and CB9 are not operated in case of a fault. Following are the reliability calculations for load point (Table 8.19).

### *8.4.6.2 Calculating the Load Point Failure Rate and Average Repair Duration—Design "F"—Switching Activity is not Included.*

*Frequency of interruptions at the load point*

$$\begin{aligned}\lambda_{\text{load point}} &= \lambda(1,1)+\lambda(2,1)+\lambda(3,1)+\lambda(4,1)\\ &= 2.002334 \text{ interruptions/year}\end{aligned}$$

TABLE 8.19. Summary of Zone Branch Calculations

| Zone Branch | Annual Outage Rate ($\lambda$) (outages/year) | Annual Outage Duration ($\lambda r$) (h/year) | Average Outage Duration ($r$) (h/outage) |
|---|---|---|---|
| (1, 1) | 1.9642 | 7.458212 | 3.797073618 |
| (1, 2) | 1.9642 | 7.458212 | 3.797073618 |
| (2, 1) | 0.014802 | 1.8411 | 124.3818403 |
| (2, 2) | 0.014802 | 1.8411 | 124.3818403 |
| (3, 1) | 0.015132 | 1.408545 | 93.08386201 |
| (3, 12) | 0.015132 | 1.408545 | 93.08386201 |
| (4, 1) | 0.0082 | 2.40782 | 293.6365854 |
| (4, 2) | 0.0082 | 2.40782 | 293.6365854 |
| (4, 3) | 0.0082 | 2.40782 | 293.6365854 |
| (4, 4) | 0.0082 | 2.40782 | 293.6365854 |
| (4, 5) | 0.0082 | 2.40782 | 293.6365854 |
| (4, 6) | 0.0082 | 2.40782 | 293.6365854 |

*Annual duration of interruptions to the load point*

$$\begin{aligned}\lambda r_{\text{load point}} &= \lambda r(1,1) + \lambda r(2,1) + \lambda r(3,1) + \lambda r(4,1) \\ &= 13.115677 \text{ h/year}\end{aligned}$$

*Average duration of interruptions per failure to the load point*

$$\begin{aligned}r_{\text{load point}} &= \lambda r_{\text{load point}} / \lambda_{\text{load point}} \\ &= 6.550194423 \text{ h/interruption}\end{aligned}$$

**8.4.6.3 Zone Branch Calculations—Design "F"—Switching Activity is Included.** In this case, it has been assumed that circuit breakers CB8 and CB9 are operated in case of a fault in or above CB1, CB1 is opened and CB8 is closed to supply power from utility feed no. 2 to the load. In case of a fault above CB3, CB3 is opened and CB9 is closed to supply power from utility feed no. 2 through SGB no. 2. The reliability calculations are presented subsequently (Table 8.20 and Fig 8.23).

**8.4.6.4 Calculating the Load Point Failure Rate and Average Repair Duration—Design "F"—Switching Activity is Included.**

*Frequency of interruptions at the load point*

$$\begin{aligned}\lambda_{\text{load point}} &= \lambda(1,1) + \lambda(2,1) + \lambda(3,1) + \lambda(4,1) + \lambda(5,1) \\ &= 2.004334 \text{ interruptions/year}\end{aligned}$$

TABLE 8.20. Summary of Zone Branch Calculations

| Zone Branch | Annual Outage Rate ($\lambda$) (outages/year) | Annual Outage Duration ($\lambda r$) (h/year) | Average Outage Duration ($r$) (h/outage) |
|---|---|---|---|
| (1, 1) | 1.9642 | 7.458212 | 3.797073618 |
| (1, 2) | 1.9642 | 7.458212 | 3.797073618 |
| (2, 1) | 0.014802 | 1.8411 | 124.3818403 |
| (2, 2) | 0.014802 | 1.8411 | 124.3818403 |
| (3, 1) | 0.017132 | 1.608545 | 93.89125613 |
| (3, 6) | 0.017132 | 1.608545 | 93.89125613 |
| (3, 7) | 0.017132 | 1.608545 | 93.89125613 |
| (3, 12) | 0.017132 | 1.608545 | 93.89125613 |
| (4, 1) | 0.0082 | 2.40782 | 293.6365854 |
| (4, 2) | 0.0082 | 2.40782 | 293.6365854 |
| (4, 3) | 0.0082 | 2.40782 | 293.6365854 |
| (4, 4) | 0.0082 | 2.40782 | 293.6365854 |
| (4, 5) | 0.0082 | 2.40782 | 293.6365854 |
| (4, 6) | 0.0082 | 2.40782 | 293.6365854 |

*Annual duration of interruptions to the load point*

$$\begin{aligned}\lambda r_{\text{load point}} &= \lambda(1,1) \times \text{Rsw} + \lambda r(2,1) + \lambda r(3,1) + \lambda r(4,1) \\ &= 4.016365 \text{ h/year}\end{aligned}$$

*Average duration of interruptions per failure to the load point*

$$\begin{aligned}r_{\text{load point}} &= \lambda r_{\text{load point}} / \lambda_{\text{load point}} \\ &= 2.003840178 \text{ h/interruption}\end{aligned}$$

### 8.4.7 Case 7: Design "G"—Double Bus/Double Breaker Loop

This arrangement is quite similar to case no. 6, but with the addition of disconnect switches and a loop switch for switching purposes. It has two utility feeds and each feeder is assumed to have capability to supply the whole network. The single-line diagram of substation design "G"—double bus/double breaker loop is shown in Fig. 8.24 and its zone branch diagram is shown in Fig. 8.25.

There are many options for reliable supply to the load point.

1. The load is primarily fed from utility feeder no. 1 through CB3, SGB3, DS1, and CB15.
2. In case of failure of feeder no. 1, CB3 is opened and the load point is supplied from feeder no. 2 through CB9.

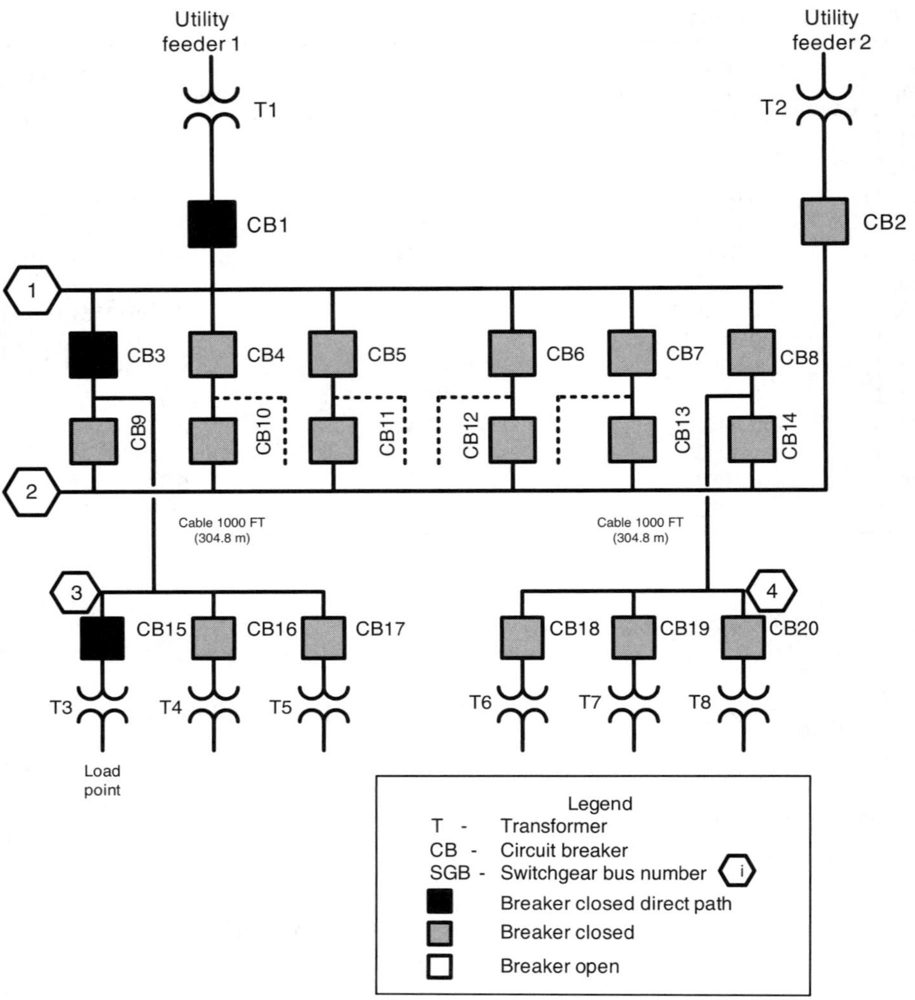


Figure 8.23. Single-line diagram of substation Design "F"—switching activity included.

3. In case of a fault in the cable or failure of DS1, the load is fed from feeder no. 1 through loop switch.
4. CB4 to CB7 and CB10 to CB13 are for future extensions.

***8.4.7.1 Zone Branch Calculation—Switching Activity is not Included.*** In this case, it has been assumed that normally open-circuit tiebreakers CB8 and CB9 and loop switch are not operated in case of any fault (Table 8.21).

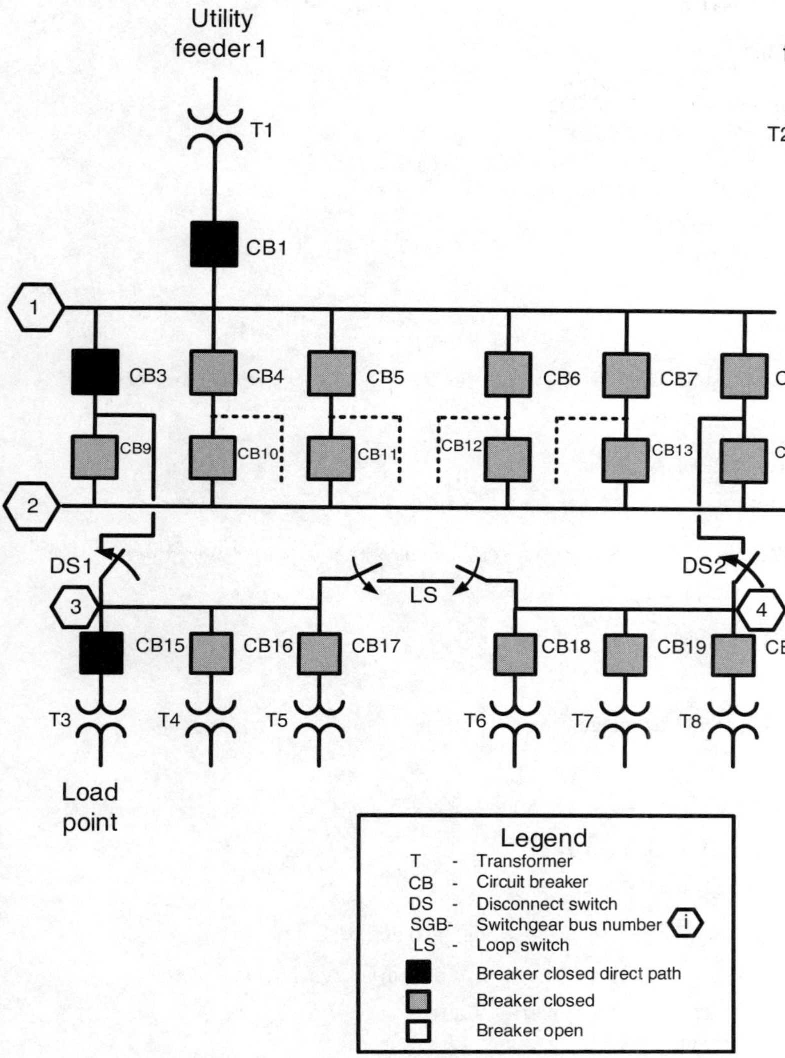


Figure 8.24. Single-line diagram of substation Design "G".

---

Zone 1—Branch 1

$$\begin{aligned}
\lambda(1,1) &= \lambda s + \lambda_{T1} + 0.50 \times \lambda_{CB1} \\
&= 1.956 + 0.0062 + 0.002 \\
&= 1.9642 \text{ outages/year} \\
\lambda r(1,1) &= \lambda_s r_s + \lambda_{T1} r_{T1} + 0.50 \times \lambda_{CB1} r_{CB1} \\
&= 5.050392 + 2.20782 + 0.2 \\
&= 7.458212 \text{ h/year} \\
r(1,1) &= \lambda r(1,1)/\lambda(1,1) \\
&= 3.797073618 \text{ h/outage}
\end{aligned}$$

Zone 1—Branch 2

$$\begin{aligned}
\lambda(1,2) &= \lambda_s + \lambda_{T2} + 0.50 \times \lambda_{CB2} \\
&= 1.956 + 0.0062 + 0.002 \\
&= 1.9642 \text{ outages/year} \\
\lambda r(1,2) &= \lambda_s r_s + \lambda_{T2} \times r_{T2} + 0.50 \times \lambda_{CB2} \times r_{CB2} \\
&= 5.050392 + 2.20782 + 0.2 \\
&= 7.458212 \text{ h/year} \\
r(1,2) &= \lambda r(1,2)/\lambda(1,2) \\
&= 3.797073618 \text{ h/outage}
\end{aligned}$$

Zone 2—Branch 1

$$\begin{aligned}
\lambda(2,1) &= 0.50\lambda_{CB1} + \lambda_{SGB1} + 0.50 \times \{\lambda_{CB3} + \lambda_{CB4} + \lambda_{CB5} + \lambda_{CB6} + \lambda_{CB7} + \lambda_{CB8}\} \\
&= 0.002 + 0.000802 + 0.012 \\
&= 0.014802 \text{ outages/year} \\
\lambda r(2,1) &= 0.50\lambda_{CB1} \times r_{CB1} + \lambda_{SGB1} \times r_{SGB1} + 0.50 \\
&\quad \times \{\lambda_{CB3} r_{CB3} + \lambda_{CB4} r_{CB4} + \lambda_{CB5} r_{CB5} + \lambda_{CB6} \times r_{CB6} + \lambda_{CB7} \times r_{CB7} + \lambda_{CB8} \times r_{CB8}\} \\
&= 0.2 + 0.4411 + 1.2 \\
&= 1.8411 \text{ h/year} \\
r(2,1) &= \lambda r(2,1)/\lambda(2,1) \\
&= 124.3818403 \text{ h/outage}
\end{aligned}$$

Zone 2—Branch 1

$$\begin{aligned}
\lambda(2,2) &= 0.50 \times \lambda_{CB2} + \lambda_{SGB2} + 0.50 \times \{\lambda_{CB9} + \lambda_{CB10} + \lambda_{CB11} + \lambda_{CB12} + \lambda_{CB13} + \lambda_{CB14}\} \\
&= 0.002 + 0.000802 + 0.012 \\
&= 0.014802 \text{ outages/year} \\
\lambda r(2,2) &= 0.50 \times \lambda_{CB2} \times r_{CB2} + \lambda_{SGB2} \times r_{SGB2} + 0.50 \\
&\quad \times \{\lambda_{CB9} \times r_{CB9} + \lambda_{CB10} \times r_{CB10} + \lambda_{CB11} \times r_{CB11} + \lambda_{CB12} \times r_{CB12} \\
&\quad + \lambda_{CB13} \times r_{CB13} + \lambda_{CB14} \times r_{CB14}\} \\
&= 0.2 + 0.4411 + 1.2 \\
&= 1.8411 \text{ h/year} \\
r(2,2) &= \lambda r(2,1)/\lambda(2,1) \\
&= 124.3818403 \text{ h/outage}
\end{aligned}$$

Zone 3—Branch 1

$$\begin{aligned}
\lambda(3,1) &= 0.50 \times \lambda_{CB3} + \lambda_C + 2\lambda_{CT} + 0.50 \times \lambda_{DS1} \\
&= 0.002 + 0.00613 + 0.0002 + 0.00305 \\
&\quad 0.01138 \text{ outages/year} \\
\lambda r(3,1) &= 0.50 \times \lambda_{CB3} \times r_{CB3} + \lambda_C r_C + 2\lambda_{CT} r_{CT} + 0.50 \times \lambda_{DS1} \times r_{DS1} \\
&= 0.2 + 0.162445 + 0.005 + 0.01098 \\
&= 0.378425 \text{ h/year} \\
r(3,1) &= \lambda r(3,1)/\lambda(3,1) \\
&= 33.25351494 \text{ h/outage}
\end{aligned}$$

*Note:* As circuit breakers CB4 through CB13 are assumed to be normally open so their reliability has not been calculated.

Zone 3—Branch 12

$$\begin{aligned}
\lambda(3,12) &= 0.50 \times \lambda_{CB14} + \lambda_C + 2\lambda_{CT} + 0.50 \times \lambda_{DS2} \\
&= 0.002 + 0.00613 + 0.0002 + 0.00305 \\
&= 0.01138 \text{ outages/year} \\
\lambda r(3,12) &= 0.50 \times \lambda_{CB14} r_{CB14} + \lambda_C \times r_C + 2\lambda_{CT} \times r_{CT} + \lambda_{DS2} \times r_{DS2} \\
&= 0.2 + 0.162445 + 0.005 + 0.01098 \\
&= 0.378425 \text{ h/year} \\
r(3,12) &= \lambda r(3,12)/\lambda(3,12) \\
&= 33.25351494 \text{ h/outage}
\end{aligned}$$

Zone 4—Branch 1

$$\begin{aligned}
\lambda(4,1) &= 0.50 \times \lambda_{DS1} + \lambda_{SGB3} + 0.50 \times \{\lambda_{CB15} + \lambda_{CB16} + \lambda_{CB17}\} \\
&= 0.00305 + 0.000802 + 0.006 \\
&= 0.009852 \text{ outages/year} \\
\lambda r(4,1) &= 0.50 \times \lambda_{DS1} \times r_{DS1} + \lambda_{SGB3} \times r_{SGB3} + 0.50 \\
&\quad \times \{\lambda_{CB15} \times r_{CB15} + \lambda_{CB16} \times r_{CB16} + \lambda_{CB17} \times r_{CB17}\} \\
&= 0.01098 + 0.4411 + 0.6 \\
&= 1.05208 \text{ h/year} \\
r(4,1) &= \lambda r(4,1)/\lambda(4,1) \\
&= 106.7884693 \text{ h/outage}
\end{aligned}$$

Zone 4—Branch 2

$$\begin{aligned}
\lambda(4,2) &= 0.50 \times \lambda_{DS2} + \lambda_{SGB4} + 0.50 \times \{\lambda_{CB18} + \lambda_{CB19} + \lambda_{CB20}\} \\
&= 0.00305 + 0.000802 + 0.006 \\
&= 0.009852 \text{ outages/year} \\
\lambda r(4,2) &= 0.50 \times \lambda_{DS2} \times r_{DS2} + \lambda_{SGB4} \times r_{SGB4} + 0.50 \\
&\quad \times \{\lambda_{CB18} \times r_{CB18} + \lambda_{CB19} \times r_{CB19} + \lambda_{CB20} \times r_{CB20}\} \\
&= 0.01098 + 0.4411 + 0.6 \\
&= 1.05208 \text{ h/year} \\
r(4,2) &= \lambda r(4,2)/\lambda(4,2) \\
&= 106.7884693 \text{ h/outage}
\end{aligned}$$

Zone 5—Branch 1

$$\begin{aligned}
\lambda(5,1) &= 0.50 \times \lambda_{CB15} + \lambda_{T3} \\
&= 0.002 + 0.0062 \\
&= 0.0082 \text{ outages/year} \\
\lambda r(5,1) &= 0.50 \times \lambda_{CB15} \times r_{CB15} + \lambda_{T3} \times r_{T3} \\
&= 0.2 + 2.20782 \\
&= 2.40782 \text{ h/year} \\
r(5,1) &= \lambda r(5,1)/\lambda(5,1) \\
&= 293.6365854 \text{ h/outage}
\end{aligned}$$

Zone 5—Branch 2

$$\begin{aligned}
\lambda(5,2) &= 0.50 \times \lambda_{CB16} + \lambda_{T4} \\
&= 0.002 + 0.0062 \\
&= 0.0082 \text{ outages per year} \\
\lambda r(5,2) &= 0.50 \times \lambda_{CB16} \times r_{CB16} + \lambda_{T4} \times r_{T4} \\
&= 0.2 + 2.20782 \\
&= 2.40782 \text{ hours per year} \\
r(5,2) &= \lambda r(5,2)/\lambda(5,2) \\
&= 293.6365854 \text{ hours per outage}
\end{aligned}$$

Zone 5—Branch 3

$$\begin{aligned}
\lambda(5,3) &= 0.50 \times \lambda_{CB17} + \lambda_{T5} \\
&= 0.002 + 0.0062 \\
&= 0.0082 \text{ outages/year} \\
\lambda r(5,3) &= 0.50 \times \lambda_{CB17} \times r_{CB17} + \lambda_{T5} \times r_{T5} \\
&= 0.2 + 2.20782 \\
&= 2.40782 \text{ h/year} \\
r(5,3) &= \lambda r(5,3)/\lambda(5,3) \\
&= 293.6365854 \text{ h/outage}
\end{aligned}$$

Zone 5—Branch 4

$$\begin{aligned}
\lambda(5,4) &= 0.50 \times \lambda_{CB18} + \lambda_{T6} \\
&= 0.002 + 0.0062 \\
&= 0.0082 \text{ outages/year} \\
\lambda r(5,4) &= 0.50 \times \lambda_{CB18} \times r_{CB18} + \lambda_{T6} \times r_{T6} \\
&= 0.2 + 2.20782 \\
&= 2.40782 \text{ h/year} \\
r(5,4) &= \lambda r(5,4)/\lambda(5,4) \\
&= 293.6365854 \text{ h/outage}
\end{aligned}$$

Zone 5—Branch 5

$$\begin{aligned}
\lambda(5,5) &= 0.50 \times \lambda_{CB19} + \lambda_{T7} \\
&= 0.002 + 0.0062 \\
&= 0.0082 \text{ outages/year} \\
\lambda r(5,5) &= 0.50 \times \lambda_{CB19} \times r_{CB19} + \lambda_{T7} \times r_{T7} \\
&= 0.2 + 2.20782 \\
&= 2.40782 \text{ h/year} \\
r(5,5) &= \lambda r(5,5)/\lambda(5,5) \\
&= 293.6365854 \text{ h/outage}
\end{aligned}$$

Zone 5—Branch 6

$$\begin{aligned}
\lambda(5,6) &= 0.50 \times \lambda_{CB20} + \lambda_{T8} \\
&= 0.002 + 0.0062 \\
&= 0.0082 \text{ outages/year} \\
\lambda r(5,6) &= 0.50 \times \lambda_{CB20} \times r_{CB20} + \lambda_{T8} \times r_{T8} \\
&= 0.2 + 2.20782 \\
&= 2.40782 \text{ h/year} \\
r(5,6) &= \lambda r(5,6)/\lambda(5,6) \\
&= 293.6365854 \text{ h/outage}
\end{aligned}$$

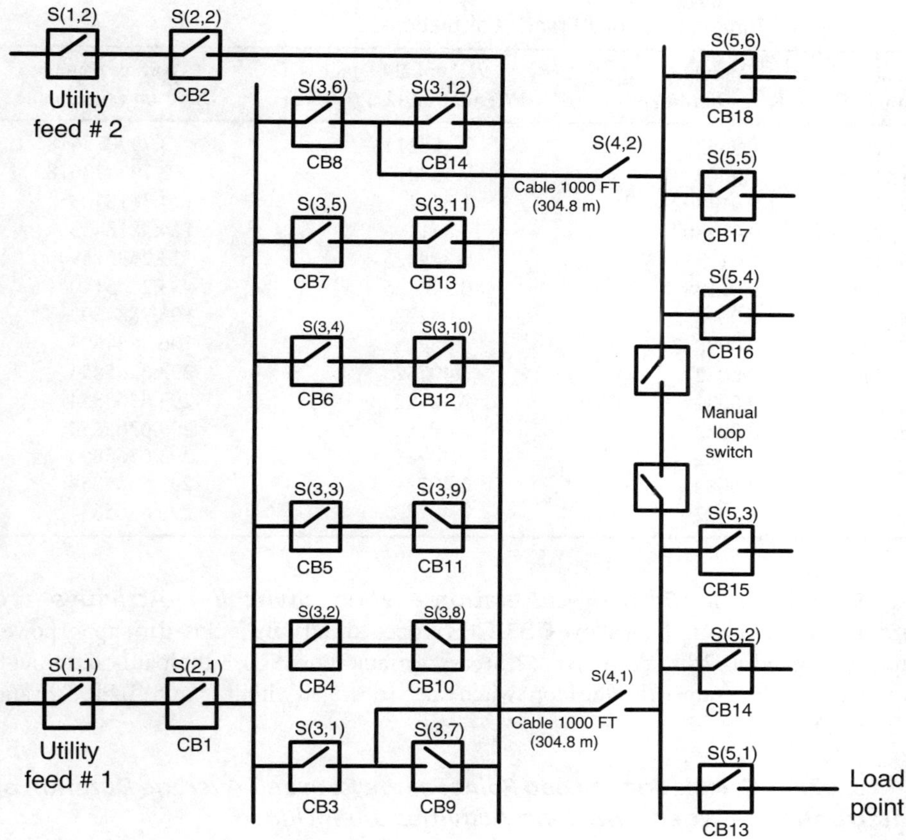


Figure 8.25. Zone branch diagram of substation Design "G".

### 8.4.7.2 Calculation of Load Point Failure Rate and Average Duration of Interruptions—Switching Activity is not Included.

*Frequency of interruptions at the load point*

$$\begin{aligned}\lambda_{\text{load point}} &= \lambda(1,1)+\lambda(2,1)+\lambda(3,1)+\lambda(4,1)+\lambda(5,1)\\ &= 1.998582 \text{ interruptions/year}\end{aligned}$$

*Annual duration of interruptions to the load point*

$$\begin{aligned}\lambda r_{\text{load point}} &= \lambda r(1,1)+\lambda r(2,1)+\lambda r(3,1)+\lambda r(4,1)+\lambda(5,1)\\ &= 12.085557 \text{ h/year}\end{aligned}$$

*Average duration of interruptions per failure to the load point*

$$\begin{aligned}r_{\text{load point}} &= \lambda r_{\text{load point}}/\lambda_{\text{load point}}\\ &= 6.04706587 \text{ h/interruption}\end{aligned}$$

TABLE 8.21. Summary of Zone Branch Calculations

| Zone Branch | Annual Outage Rate ($\lambda$) (outages/year) | Annual Outage Duration ($\lambda r$) (h/year) | Average Outage Duration ($r$) (h/outage) |
|---|---|---|---|
| (1, 1) | 1.9642 | 7.458212 | 3.797073618 |
| (1, 2) | 1.9642 | 7.458212 | 3.797073618 |
| (2, 1) | 0.014802 | 1.8411 | 124.3818403 |
| (2, 2) | 0.014802 | 1.8411 | 124.3818403 |
| (3, 1) | 0.01138 | 0.378425 | 33.25351494 |
| (3, 12) | 0.01138 | 0.378425 | 33.25351494 |
| (4, 1) | 0.009852 | 1.05208 | 106.7884693 |
| (4, 2) | 0.009852 | 1.05208 | 106.7884693 |
| (5, 1) | 0.0082 | 2.40782 | 293.6365854 |
| (5, 2) | 0.0082 | 2.40782 | 293.6365854 |
| (5, 3) | 0.0082 | 2.40782 | 293.6365854 |
| (5, 4) | 0.0082 | 2.40782 | 293.6365854 |
| (5, 5) | 0.0082 | 2.40782 | 293.6365854 |
| (5, 6) | 0.0082 | 2.40782 | 293.6365854 |

#### 8.4.7.3 Zone Branch Calculation—When Switching Activities are Included.

In case of a fault above CB3, CB3 is opened and CB9 is closed to supply power from utility feed no. 2 through SGB no. 2. In case of a fault above SGB3, the load is fed through loop switch. It is assumed that the loop switch can be operated within 15 min (Table 8.22 and Fig. 8.26).

#### 8.4.7.4 Calculation of Load Point Failure Rate and Average Duration of Interruptions—When Switching Activities are Included.

*Frequency of interruptions at the load point*

$$\begin{aligned}\lambda_{\text{load point}} &= \lambda(1,1)+\lambda(2,1)+\lambda(3,1)+\lambda(4,1)+\lambda(5,1)\\ &= 2.016534 \text{ interruptions/year}\end{aligned}$$

*Annual duration of interruptions to the load point*

$$\begin{aligned}\lambda r_{\text{load point}} &= \lambda(1,1)\times \text{Rsw}+\lambda(2,1)\text{Rsw}+\lambda(3,1)\text{Rsw}+\lambda r(4,1)\,\lambda r(5,1)\\ &= 3.485205 \text{ h/year}\end{aligned}$$

*Average duration of interruptions per failure to the load point*

$$\begin{aligned}r_{\text{load point}} &= \lambda r_{\text{load point}}/\lambda_{\text{load point}}\\ &= 1.728314524 \text{ h/interruption}\end{aligned}$$

### 8.4.8 Case 8: Design "H"—Double Bus/Breaker Primary Selective

Half of the portion of this configuration is similar to the case no. 7. It has two utility feeds and each feeder is assumed to have the capability to supply the whole network. The single-line diagram of substation design "H"—double bus/breaker primary selective is provided in Fig. 8.27 and its zone branch diagram in Fig. 8.28.

TABLE 8.22. Summary of Zone Branch Calculations

| Zone Branch | Annual Outage Rate ($\lambda$) (outages/year) | Annual Outage Duration ($\lambda r$) (h/year) | Average Outage Duration ($r$) (h/outage) |
|---|---|---|---|
| (1, 1) | 1.9642 | 7.458212 | 3.797073618 |
| (1, 2) | 1.9642 | 7.458212 | 3.797073618 |
| (2, 1) | 0.014802 | 1.8411 | 124.3818403 |
| (2, 2) | 0.014802 | 1.8411 | 124.3818403 |
| (3, 1) | 0.01338 | 0.578425 | 43.23056801 |
| (3, 6) | 0.01338 | 0.578425 | 43.23056801 |
| (3, 7) | 0.01338 | 0.578425 | 43.23056801 |
| (3, 12) | 0.01338 | 0.578425 | 43.23056801 |
| (4, 1) | 0.015952 | 1.07404 | 67.32948847 |
| (4, 2) | 0.015952 | 1.07404 | 67.32948847 |
| (4, 3) | 0.015952 | 1.07404 | 67.32948847 |
| (5, 1) | 0.0082 | 2.40782 | 293.6365854 |
| (5, 2) | 0.0082 | 2.40782 | 293.6365854 |
| (5, 3) | 0.0082 | 2.40782 | 293.6365854 |
| (5, 4) | 0.0082 | 2.40782 | 293.6365854 |
| (5, 5) | 0.0082 | 2.40782 | 293.6365854 |
| (5, 6) | 0.0082 | 2.40782 | 293.6365854 |

The following are the ways the load can be fed:

1. The load is primarily fed from utility feeder no. 1 through CB3, SGB3, and CB17.
2. In case of a failure of feeder no. 1, CB3 is opened and load point is supplied from feeder no. 2 through CB9.
3. In case of a fault in cable, SGB3 or failure of CB3 and CB9, the load is fed from feeder no. 2 through CB23.
4. CB4 to CB7, CB10 to CB13, CB15 to CB18, and CB23 to CB26 are for future use but these can be used to restore supply to the load point. For example, if CB1, CB8, and CB9 are out of order then by closing any set of circuit breakers (e.g., CB4 and CB10), the load can be fed through CB3.

***8.4.8.1 Zone Branch Calculations—Switching Activity is Not Included.*** In this case, it has been assumed that normally open circuit breakers CB8, CB9, and CB23 are not operated in case of a fault (Table 8.23).

### *8.4.8.2 Calculation of Load Point Failure Rate and Average Duration of Interruptions—Switching Activity is not Included.*

*Frequency of interruptions at the load point*

$$\begin{aligned}\lambda_{\text{load point}} &= \lambda(1,1) + \lambda(2,1) + \lambda(3,1) + \lambda(4,1) + \lambda(5,1) \\ &= 2.012334 \text{ interruptions/year}\end{aligned}$$

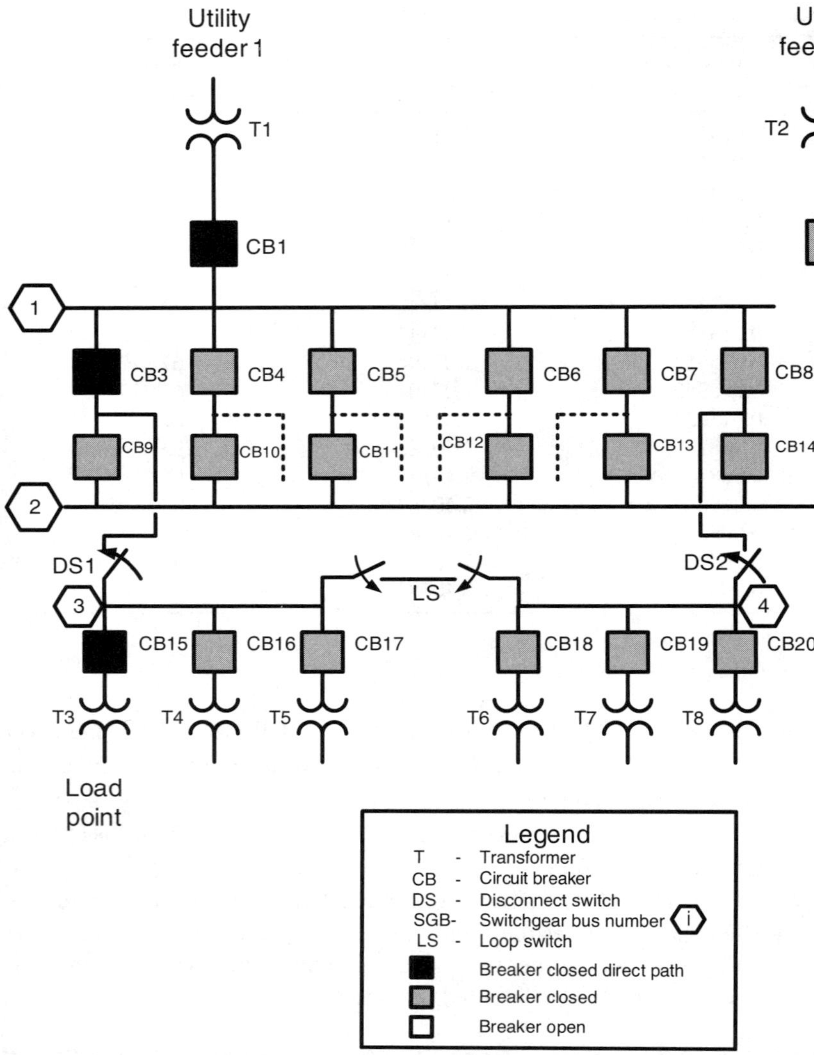


Figure 8.26. Single-line diagram of substation Design "G"—switching activity included.

*Annual duration of interruptions to the load point*

$$\begin{aligned}\lambda r_{\text{load point}} &= \lambda r(1,1) + \lambda r(2,1) + \lambda r(3,1) + \lambda r(4,1) + \lambda r(5,1) \\ &= 14.115677 \text{ h/year}\end{aligned}$$

*Average duration of interruptions per failure to the load point*

$$\begin{aligned}r_{\text{load point}} &= \lambda r_{\text{load point}} / \lambda_{\text{load point}} \\ &= 7.014579588 \text{ h/interruption}\end{aligned}$$

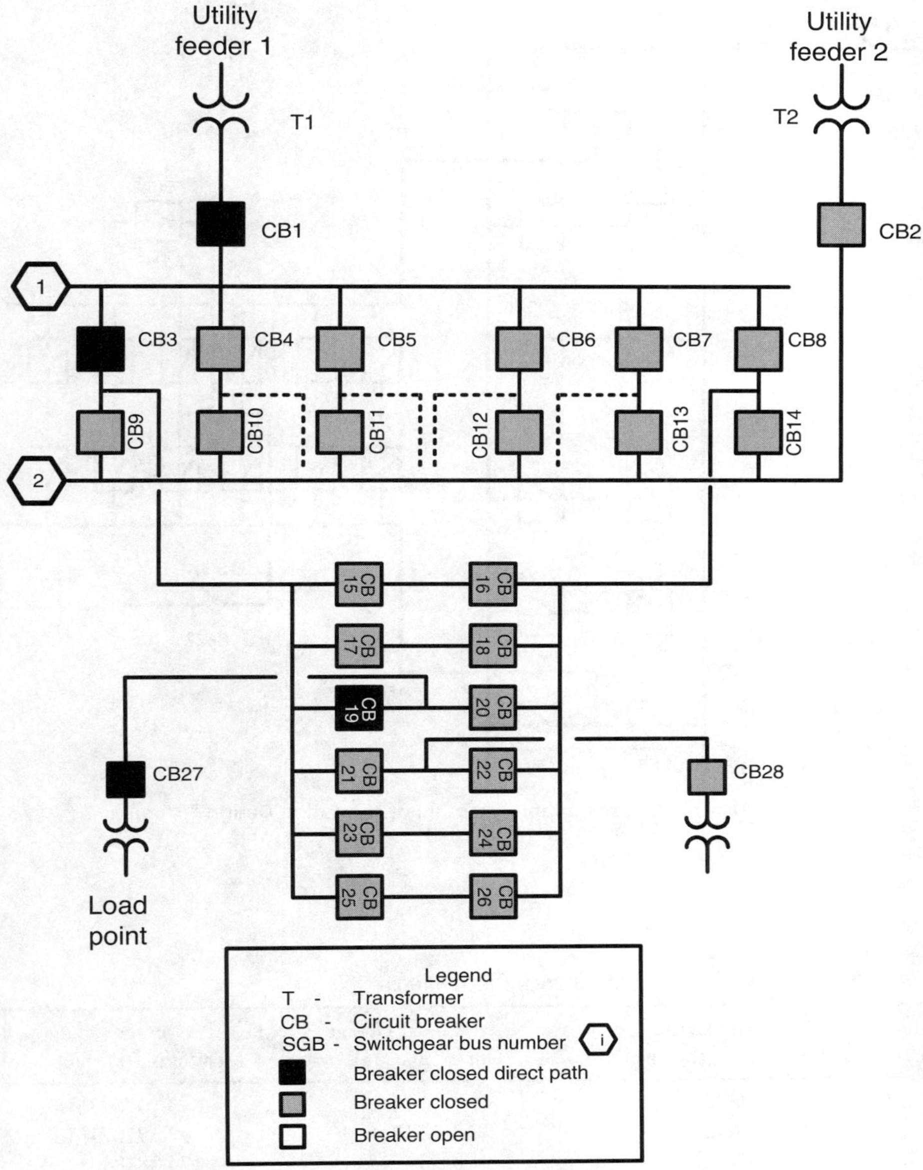


Figure 8.27. Single-line diagram of substation Design "H"

***8.4.8.3 Zone Branch Calculations—When Switching Activities are Included.*** In case of a fault above CB3, CB3 is opened and CB9 is closed to supply power from utility feed no. 2 through SGB no. 2. In case of a fault in or above SGB3, the load is fed from feed no. 2 by closing CB23. It is assumed that the circuit

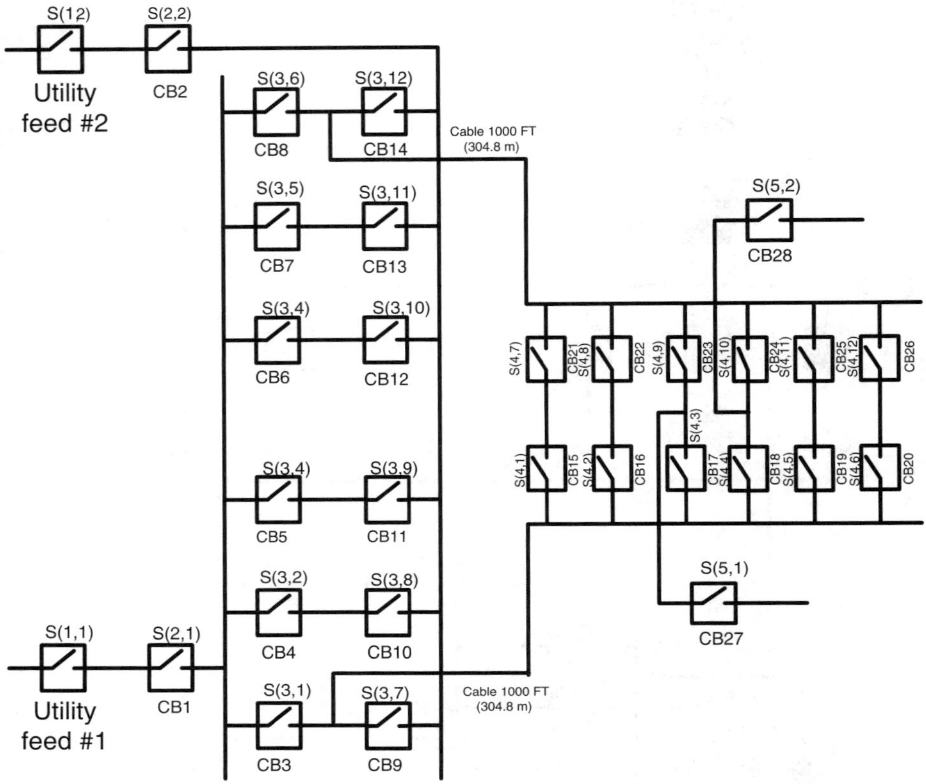


Figure 8.28. Zone branch diagram of substation Design "H"

TABLE 8.23. Summary of Zone Branch Calculations

| Zone Branch | Annual Outage Rate ($\lambda$) (outages/year) | Annual Outage Duration ($\lambda r$) (h/year) | Average Outage Duration ($r$) (h/outage) |
|---|---|---|---|
| (1, 1) | 1.9642 | 7.458212 | 3.797073618 |
| (1, 2) | 1.9642 | 7.458212 | 3.797073618 |
| (2, 1) | 0.014802 | 1.8411 | 124.3818403 |
| (2, 2) | 0.014802 | 1.8411 | 124.3818403 |
| (3, 1) | 0.021132 | 2.008545 | 95.04755821 |
| (3, 2) | 0.021132 | 2.008545 | 95.04755821 |
| (4, 1) | 0.004 | 0.4 | 100 |
| (4, 2) | 0.004 | 0.4 | 100 |
| (5, 1) | 0.0082 | 2.40782 | 293.6365854 |
| (5, 2) | 0.0082 | 2.40782 | 293.6365854 |

TABLE 8.24. Summary of Zone Branch Calculations

| Zone Branch | Annual Outage Rate ($\lambda$) (outages/year) | Annual Outage Duration ($\lambda r$) (h/year) | Average Outage Duration ($r$) (h/outage) |
|---|---|---|---|
| (1, 1) | 1.9642 | 7.458212 | 3.797073618 |
| (1, 2) | 1.9642 | 7.458212 | 3.797073618 |
| (2, 1) | 0.014802 | 1.8411 | 124.3818403 |
| (2, 2) | 0.014802 | 1.8411 | 124.3818403 |
| (3, 1) | 0.023132 | 2.208545 | 95.47574788 |
| (3, 6) | 0.023132 | 2.208545 | 95.47574788 |
| (3, 7) | 0.023132 | 2.208545 | 95.47574788 |
| (3, 12) | 0.023132 | 2.208545 | 95.47574788 |
| (4, 1) | 0.006 | 0.6 | 100 |
| (4, 2) | 0.006 | 0.6 | 100 |
| (5, 1) | 0.0082 | 2.40782 | 293.6365854 |
| (5, 2) | 0.0082 | 2.40782 | 293.6365854 |

breakers can be operated within a negligible period of time (Rsw = 0 for circuit breakers) (Table 8.24 and Fig. 8.29).

---

Zone 1—Branch 1

$$
\begin{aligned}
\lambda(1,1) &= \lambda_s + \lambda_{T1} + 0.50 \times \lambda_{CB1} \\
&= 1.956 + 0.0062 + 0.002 \\
&= 1.9642 \text{ outages/year} \\
\lambda r(1,1) &= \lambda_s \times r_s + \lambda_{T1} \times r_{T1} + 0.50 \times \lambda_{CB1} \times r_{CB1} \\
&= 5.050392 + 2.20782 + 0.2 \\
&= 7.458212 \text{ h/year} \\
r(1,1) &= \lambda r(1,1)/\lambda(1,1) \\
&= 3.797073618 \text{ h/outage}
\end{aligned}
$$

Zone 1—Branch 2

$$
\begin{aligned}
\lambda(1,2) &= \lambda_s + \lambda_{T2} + 0.50 \times \lambda_{CB2} \\
&= 1.956 + 0.0062 + 0.002 \\
&= 1.9642 \text{ outages/year} \\
\lambda r(1,2) &= \lambda_s \times r_s + \lambda_{T2} \times r_{T2} + 0.50 \times \lambda_{CB2} \times r_{CB2} \\
&= 5.050392 + 2.20782 + 0.2 \\
&= 7.458212 \text{ h/year} \\
\mathrm{r}(1,2) &= \lambda r(1,2)/\lambda(1,2) \\
&= 3.797073618 \text{ h/outage}
\end{aligned}
$$

Zone 2—Branch 1

$$\lambda(2,1) = 0.50 \times \lambda_{CB1} + \lambda_{SGB1} + 0.50 \times \{\lambda_{CB3} + \lambda_{CB4} + \lambda_{CB5} + \lambda_{CB6} + \lambda_{CB7} + \lambda_{CB8}\}$$

$$= 1.956 + 0.0062 + 0.002$$

$$= 1.9642 \text{ outages/year}$$

$$\begin{aligned}\lambda r(2,1) &= 0.50\times\lambda_{CB1}\times r_{CB1}+\lambda_{SGB1}\times r_{SGB1}+0.50\{\lambda_{CB3}\times r_{CB3}+\lambda_{CB4}\times r_{CB4}+\lambda_{CB5}\\ &\quad\times r_{CB5}+\lambda_{CB6}\times r_{CB6}+\lambda_{CB7}\times r_{CB7}+\lambda_{CB8}\times r_{CB8}\}\\ &= 5.050392+2.20782+0.2\\ &= 7.458212\ \text{h/year}\\ r(2,1) &= \lambda r(2,1)/\lambda(2,1)\\ &= 3.797073618\ \text{h/outage}\end{aligned}$$

Zone 2—Branch 2

$$\begin{aligned}\lambda(2,2) &= 0.50\times\lambda_{CB2}+\lambda_{SGB2}+0.50\times\{\lambda_{CB9}+\lambda_{CB10}+\lambda_{CB11}+\lambda_{CB12}+\lambda_{CB13}+\lambda_{CB14}\}\\ &= 1.956+0.0062+0.002\\ &= 1.9642\ \text{outages/year}\\ \lambda r(2,2) &= 0.50\times\lambda_{CB2}\times r_{CB2}+\lambda_{SGB2}\times r_{SGB2}+0.50\times\{\lambda_{CB9}\times r_{CB9}+\lambda_{CB10}\\ &\quad\times r_{CB10}+\lambda_{CB11}\times r_{CB11}+\lambda_{CB12}\times r_{CB12}+\lambda_{CB13}\times r_{CB13}+\lambda_{CB14}\times r_{CB14}\}\\ &= 5.050392+2.20782+0.2\\ &= 7.458212\ \text{h/year}\\ r(2,2) &= \lambda r(2,2)/\lambda(2,2)\\ &= 3.797073618\ \text{h/outage}\end{aligned}$$

Zone 3—Branch 1

$$\begin{aligned}\lambda(3,1) &= 0.50\times\{\lambda_{CB3}+\lambda_{CB9}\}+\lambda_C+2\lambda_{CT}+\lambda_{SGB3}+0.50\times\{\lambda_{CB15}\ \text{to}\ \lambda_{CB20}\}\\ &= 0.004+0.00613+0.0002+0.000802+0.012\\ &= 0.023132\ \text{outages/year}\\ \lambda r(3,1) &= 0.50\times\{\lambda_{CB3}\times r_{CB3}+\lambda_{CB9}\times r_{CB9}\}+\lambda_C\times r_C+2\lambda_{CT}\times r_{CT}+\lambda_{SGB3}\times r_{SGB3}\\ &\quad+0.50\times\{\lambda_{CB17}\times r_{CB17}\ \text{to}\ \lambda_{CB17}\times r_{CB17}\}\\ &= 0.4+0.162445+0.005+0.4411+1.2\\ &= 2.208545\ \text{h/year}\\ r(3,1) &= \lambda r(3,1)/\lambda(3,1)\\ &= 95.47574788\ \text{h/outage}\end{aligned}$$

*Note*: Since circuit breakers CB4 through CB7 and CB10 through CB13 are assumed to be normally open, there zone branch calculations have not been done.

Zone 3—Branch 6

$$\begin{aligned}\lambda(3,6) &= 0.50\times\{\lambda_{CB8}+\lambda_{CB14}\}+\lambda_C+2\lambda_{CT}+\lambda_{SGB4}+0.50\times\{\lambda_{CB21}\ \text{to}\ \lambda_{CB26}\}\\ &= 0.004+0.00613+0.0002+0.000802+0.012\\ &= 0.023132\ \text{outages/year}\end{aligned}$$

$$\lambda r(3,6) = 0.50 \times \{\lambda_{CB3} \times r_{CB3} + \lambda_{CB14} \times r_{CB14}\} + \lambda_C \times r_C$$
$$+ 2\lambda_{CT} \times r_{CT} + \lambda_{SGB4} \times r_{SGB4} + 0.50 \times \{\lambda_{CB21} \text{ to } \lambda_{CB26}\}$$
$$= 0.4 + 0.162445 + 0.005 + 0.4411 + 1.2$$
$$= 2.208545 \text{ h/year}$$
$$r(3,6) = \lambda r(3,6)/\lambda(3,6)$$
$$= 95.47574788 \text{ h/outage}$$

Zone 3—Branch 7

$$\lambda(3,7) = 0.50 \times \{\lambda_{CB3} + \lambda_{CB9}\} + \lambda_C + 2\lambda_{CT} + \lambda_{SGB3} + 0.50 \times \{\lambda_{CB15} \text{ to } \lambda_{CB20}\}$$
$$= 0.004 + 0.00613 + 0.0002 + 0.000802 + 0.012$$
$$= 0.023132 \text{ outages/year}$$
$$\lambda r(3,7) = 0.50 \times \{\lambda_{CB3} \times r_{CB3} + \lambda_{CB9} \times r_{CB9}\} + \lambda_C \times r_C + 2\lambda_{CT} \times r_{CT} + \lambda_{SGB3}$$
$$\times r_{SGB3} + 0.50 \times \{\lambda_{CB17} \times r_{CB17} \text{ to } \lambda_{CB17} \times r_{CB17}\}$$
$$= 0.4 + 0.162445 + 0.005 + 0.4411 + 1.2$$
$$= 2.208545 \text{ h/year}$$
$$r(3,7) = \lambda r(3,7)/\lambda(3,7)$$
$$= 95.47574788 \text{ h/outage}$$

Zone 3—Branch 12

$$\lambda(3,12) = 0.50\{\lambda_{CB8} + \lambda_{CB14}\} + \lambda_C + 2\lambda_{CT} + \lambda_{SGB4} + 0.50\{\lambda_{CB21} \text{ to } \lambda_{CB26}\}$$
$$= 0.004 + 0.00613 + 0.0002 + 0.000802 + 0.012$$
$$= 0.023132 \text{ outages/year}$$
$$\lambda r(3,12) = 0.50 \times \{\lambda_{CB3} \times r_{CB3} + \lambda_{CB14} \times r_{CB14}\} + \lambda_C \times r_C + 2\lambda_{CT} \times r_{CT} + \lambda_{SGB4} \times r_{SGB4}$$
$$+ 0.50 \times \{\lambda_{CB21} \times r_{CB21} \text{ to } \lambda_{CB26} \times r_{CB26}\}$$
$$= 0.4 + 0.162445 + 0.005 + 0.4411 + 1.2$$
$$= 2.208545 \text{ h/year}$$
$$r(3,12) = \lambda r(3,12)/\lambda(3,12)$$
$$= 95.47574788 \text{ h/outage}$$

Zone 4—Branch 1

$$\lambda(4,1) = 0.50 \times \{\lambda_{CB17} + \lambda_{CB23} + \lambda_{CB27}\}$$
$$= 0.006 \text{ outages/year}$$
$$\lambda r(4,1) = 0.50 \times \{\lambda_{CB17} \times r_{CB17} + \lambda_{CB23} \times r_{CB23} + \lambda_{CB27} \times r_{CB27}\}$$
$$= 0.6 \text{ h/year}$$
$$r(4,1) = \lambda r(4,1)/\lambda(4,1)$$
$$= 100 \text{ h/outage}$$

Zone 4—Branch 2

$$\begin{aligned}
\lambda(4,2) &= 0.50 \times \{\lambda_{CB24} + \lambda_{CB18} + \lambda_{CB28}\} \\
&= 0.006 \text{ outages/year} \\
\lambda r(4,2) &= 0.50 \times \{\lambda_{CB24} \times r_{CB24} + \lambda_{CB18} \times r_{CB18} + \lambda_{CB28} \times r_{CB28}\} \\
&= 0.6 \text{ h/year} \\
r(4,2) &= \lambda r(4,2)/\lambda(4,2) \\
&= 100 \text{ h/outage}
\end{aligned}$$

Zone 5—Branch 1

$$\begin{aligned}
\lambda(5,1) &= 0.50 \times \lambda_{CB27} + \lambda_{T3} \\
&= 0.002 + 0.0062 \\
&= 0.0082 \text{ outages/year} \\
\lambda r(5,1) &= 0.50 \times \lambda_{CB27} \times r_{CB27} + \lambda_{T3} \times r_{T3} \\
&= 0.2 + 2.20782 \\
&= 2.40782 \text{ h/year} \\
r(5,1) &= \lambda r(5,1)/\lambda(5,1) \\
&= 293.6365854 \text{ h/outage}
\end{aligned}$$

Zone 5—Branch 2

$$\begin{aligned}
\lambda(5,2) &= 0.50 \times \lambda_{CB28} + \lambda_{T4} \\
&= 0.002 + 0.0062 \\
&= 0.0082 \text{ outages/year} \\
\lambda r(5,2) &= 0.50 \times \lambda_{CB28} \times r_{CB28} + \lambda_{T4} \times r_{T4} \\
&= 0.2 + 2.20782 \\
&= 2.40782 \text{ h/year} \\
r(5,2) &= \lambda r(5,2)/\lambda(5,2) \\
&= 293.6365854 \text{ h/outage}
\end{aligned}$$

---

#### *8.4.8.4 Calculation of Load Point Failure Rate and Average Duration of Interruptions—When Switching Activities are Included.*

*Frequency of interruptions at the load point*

$$\begin{aligned}
\lambda_{\text{load point}} &= \lambda(1,1) + \lambda(2,1) + \lambda(3,1) + \lambda(4,1) + \lambda(5,1) \\
&= 2.016334 \text{ interruptions/year}
\end{aligned}$$

*Annual duration of interruptions to the load point*

$$\begin{aligned}
\lambda r_{\text{load point}} &= \lambda r(1,1) \times \text{Rsw} + \lambda(2,1) \times \text{Rsw} + \lambda(3,1) \times \text{Rsw} + \lambda r(4,1) + \lambda r(5,1) \\
&= 3.00782 \text{ h/year}
\end{aligned}$$

*Average duration of interruptions per failure to the load point*

$$\begin{aligned}
r_{\text{load point}} &= \lambda r_{\text{load point}}/\lambda_{\text{load point}} \\
&= 1.491727065 \text{ h/interruption}
\end{aligned}$$

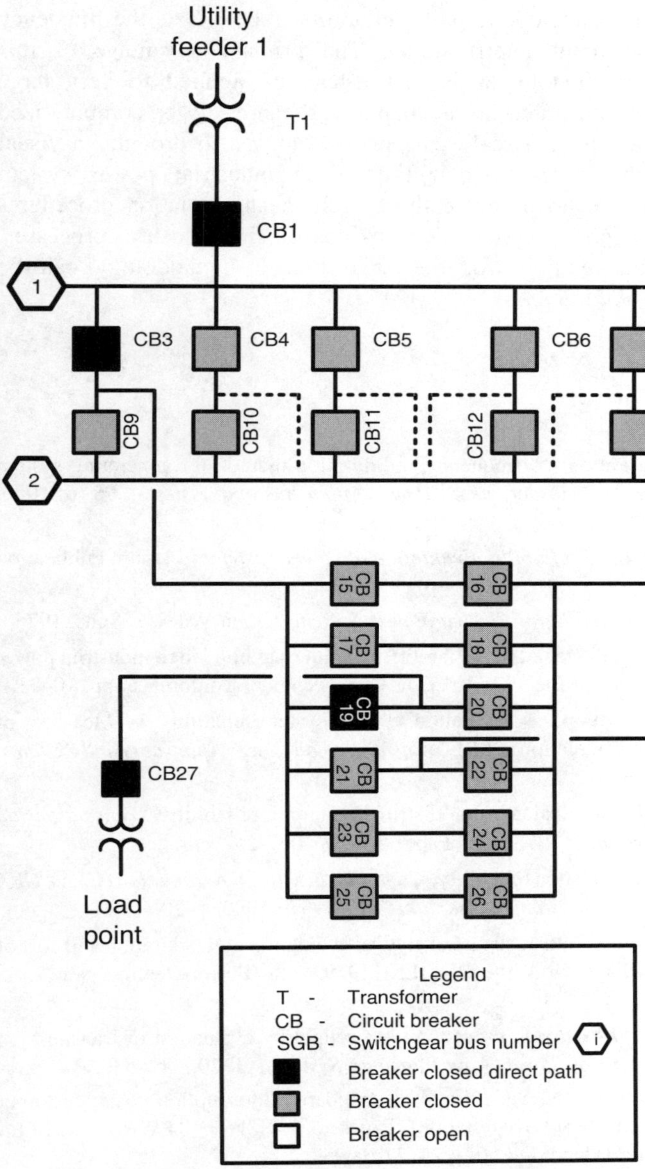


Figure 8.29. Single-line diagram of substation Design "H"—switching activity included.

## 8.5 CONCLUSIONS

This chapter has presented a zone branch methodology that can readily identify potential faulty protection schemes involving all the components of an industrial power system and can also evaluate load point reliability indices (i.e., the frequency and the duration of load point interruptions). The primary advantage of this methodology over other methodologies is that it can be applied to very large industrial power systems as illustrated in the chapter and can easily be computerized on any spreadsheet program (e.g., Excel). The methodology also provides a visual picture (i.e., zone branch single-line diagram) of an industrial power system configuration and can be used to optimize the switching and isolation procedures to minimize the impact of power system interruptions on the industrial processes. Equipment reliability data required for the zone branch methodology can be obtained from the *IEEE Gold Book*.

## REFERENCES


1. A. K. Ayoub and A. D. Patton, A frequency and duration method for generating system reliability evaluation, *IEEE Transactions on Power Apparatus and Systems*, **95** (6) 1976, 1929–1933.
2. R. Billinton and R. N. Allan, *Reliability Evaluation of Power Systems,* Plenum Publication Corp., 1983.
3. J. Endrenyi, *Reliability Modeling in Electric Power Systems*, John Wiley & Sons, 1978.
4. W. H. Dickinson, et al., Fundamentals of reliability techniques as applied to industrial power systems, Conference Record of the 1971 IEEE I&CPS Technical Conference, pp. 10–31.
5. D. P. Gaver, F. E. Montmeat and A. D. Patton, Power system reliability, I—Measures of reliability and methods of calculation, *IEEE Transactions on Power Apparatus and Systems*, 1964, 727–737.
6. D. O. Koval and R. Billinton, Evaluation of distribution circuit reliability, *Transactions on Power Apparatus and Systems*, 1978, 1–9, Paper No. F77-067-2,
7. D. O. Koval, Distribution substation reliability, *Canadian Electrical Association (CEA), CEA Engineering and Operating Division Transactions*, **19**, 1979/1980, 1–20.
8. D. O. Koval and R. Billinton, Statistical and analytical evaluation of the duration and cost of consumer's interruptions, Paper NoA79-057-1, IEEE 1979 Winter Power Meeting, New York, NY.
9. R. J. Ringlee and S. D. Goode, On procedures for reliability evaluation of transmission systems, *IEEE Transactions on Power Apparatus and Systems*, 1970, 527–537.
10. M. J. Sullivan, T. Vardell, B. N. Sudeth and A. Vojdani, Interruption costs, customer satisfaction and expectations for service reliability, Paper no. 95 SM 572-8 PWRS, IEEE-PES Summer Power Meeting, Portland, OR, July 24–27, 1995.
11. W. H. Dickinson, Report of reliability of electrical equipment in industrial plants, *AIEE Transactions*, Part II, 1962, 132–151.
12. D. W. McWilliams, A. D. Patton and C. R. Heising, Reliability of electrical equipment in industrial plants—comparison of results from 1959 and 1971 survey, IEEE-I&CPS Technical Conference Record, 74CH0855-71A, Denver, CO, June 2–6, 1974, pp. 105–115.
13. P. O'Donnell, Report of large motor reliability survey of industrial plants and commercial buildings, *IEEE Transactions on Industry Applications*, Parts 1 and 2, July/August, 1985,

853–872; Part 3, January/February 1987, pp. 153–158 (See Appendix H, IEEE Standard 493-1997).

14. D. J. Love, Reliability of utility supply configurations for industrial power systems, *IEEE Transactions on Industry Applications*, **30** (5), 1994, 1303–1308.
15. L. M. Shooman, *Probabilistic Reliability: An Engineering Approach,* McGraw-Hill, 1968.
16. C. Singh, On the behavior of failure frequency bounds, *IEEE Transactions on Reliability*, **R-26**, 1977, 63–66.
17. C. Singh and R. Billinton, *System Reliability Modeling and Evaluation*, Hutchinson Educational, 1977.
18. IEEE Standard 493-1997 (IEEE Gold Book), IEEE Recommended Practice for Design of Reliable Industrial and Commercial Power Systems, The Institute of Electrical and Electronics Engineers, Inc. New York, NY, USA.

# 9

# EQUIPMENT OUTAGE STATISTICS

## 9.1 INTRODUCTION

The distribution system is an important part of the total electric supply system, as it provides the final link between a utility's bulk transmission system and its ultimate customers. It has been reported in many technical publications that 80% of all customer interruptions occur due to failures in the distribution systems. Historical assessment generally analyzes discrete interruption events occurring at specific locations over specific time periods. Predictive assessment determines the long-term behavior of systems by combining component failure rates and the duration of repair, restoration, switching, and isolation activities of the electric utility's distribution system for given system configurations to calculate average reliability performance indices. Accurate component outage data are therefore the key to distribution system predictive performance analysis. In addition to the physical configuration of the distribution network, the reliability characteristics of system components, the operation of protection equipment, and the availability of alternative supplies with adequate capacity also have a significant impact on service reliability.

All quantitative reliability assessments require numerical data. This chapter deals with the types of data needed for predictive reliability assessments of distribution

*Power Distribution System Reliability.* By Ali A. Chowdhury and Don O. Koval

systems. Distribution system reliability assessment can be divided into two basic segments: measuring past performance and predicting future performance. The historical data are very useful when analyzed to ascertain what went wrong in the past and correct it, and they are also useful as input to predict future service reliability. Data are valuable for assessing past performance of components and systems because they identify the chronological changes in performance and therefore help determine weak areas needing reinforcements or network modifications. In addition, past performance yields reliability indices that serve as a guide for acceptable values in future system reliability assessments and enables previous predictions to be compared with actual operating experience.

The historical assessment procedure, therefore, looks back at the past behavior of the distribution system. The predictive procedure looks forward at future system behavior. To perform predictive assessment, it is necessary to transform past experience into the required future prediction. Collection and reporting of distribution equipment data is therefore invaluable as it forms the input to relevant reliability models, techniques, and equations that estimate the future performance of the system, the benefits of alternative system designs, reinforcements and expansion plans, the effects of alternative operational and maintenance policies, and the related reliability cost/benefit of the alternatives being investigated. This chapter recommends outage statistics for different distribution system equipment for use in predictive reliability planning.

## 9.2 INTERRUPTION DATA COLLECTION SCHEME

To perform reliability and many other kinds of analysis on equipment, a lot of data are required. The quality of the analysis depends, among other things, on the amount of data available. The more the data, the more accurate and representative the findings and conclusions; for example, failure rate determined from 10 years' operating experience on 5 units is not as accurate as that from 50 years' operating experience on 100 units.

The amount of data collected by a single utility or manufacturer is limited. If data from other sources can be pooled together, the amount of data will increase manifold. The result is beneficial to all participants.

This can be done by linking the databases of various sources together. There are technical problems to be solved, of course, such as security, protocol, compatibility, and so on; however, there is none that cannot be solved by data processing specialists. Such activities have started in some areas on some kinds of data and will gain pace and support in future. The ultimate goal, of course, is to have a nationwide or continent-wide data bank.

Among the equipment data that receive priority on data pooling are those on failures of generators, transformers, power lines, cables, and so on, as a result of some of the research activities in organizations such as Canadian Electricity Association (CEA) and Edison Electric Institute. The first step, of course, is to establish a corporate data bank. This has been in progress for some time in many North American utilities.

While system disturbances originating in generation and transmission segments, such as the Blackout of August 14, 2003, draw national attention and scrutiny, localized service interruptions at the distribution level are the primary concern to the end-use customers. Benchmarking of distribution reliability performance has become a common practice in the electric utility industry over the past several years, even though it is often difficult to make a meaningful comparison between reliability performances of disparate utilities. This is due to the differences in data collection methods employed, differences in system design and operation, and differences in the environments associated with each individual utility. It has been recognized that to arrive at meaningful conclusions, consistent interruption event data and categorization of that data are desirable. IEEE Standard 1366-2003 has defined a methodology that, if used, will provide a common way to segment data, thereby eliminating many hurdles to benchmarking.

This chapter presents recommendations for a uniform approach to outage data collection and analysis for comparison of utility reliability performance based on a high-level categorization of interruption related data. The basic objective is to define a minimum set of data collection categories required for benchmarking purposes and to give consistency to those categories. The approach presented in this chapter is based on approaches to data categorization contained in CEA's Annual Service Continuity Report. However, it is worth noting that when performing a comparison of reliability statistics between utilities, the differences between the collection methods, the locations, the differences in system design, and operating and maintenance policies can make a huge impact on the calculated statistics.

Among different utilities, the outage data collection methods can differ in terms of the differences in the interruption data collection systems (ranging from manually entered paper systems to completely automated computer-based systems), the ability to collect interruption data from the system (ranging from the substation level down to the customer service drop), the use or nonuse of step restoration when collecting interruption data, the determination of the start time, the definition of sustained interruption (ranging from >1 to >5 min), the definition of a customer (account, meter, premise, etc.), and interruption delineations (forced interruptions, scheduled interruptions, major events, etc.). Localization of systems by location such as system characterization (rural, suburban, and urban) and climatic information (hot, cold, wet, dry, lightning, etc.) is important. In addition, classification of distribution systems by system design characteristics such as system layout (radial, loop, two-transformer station, etc.) and system placement (underground, overhead, etc.) makes a significant impact on system reliability performance.

This chapter presents a minimal set of data and a consistent categorization structure necessary for a meaningful comparison of distribution system performance. Included are categories for system characterization, interruption causes, responsible systems, conditions, voltages, devices, device initiation, and restorations.

A utility system is usually characterized into three categories, namely, *rural*, *suburban*, and *urban* systems, based on some arbitrary target value of customer density per mile. Seven general interruptions cause categories that are suggested for data

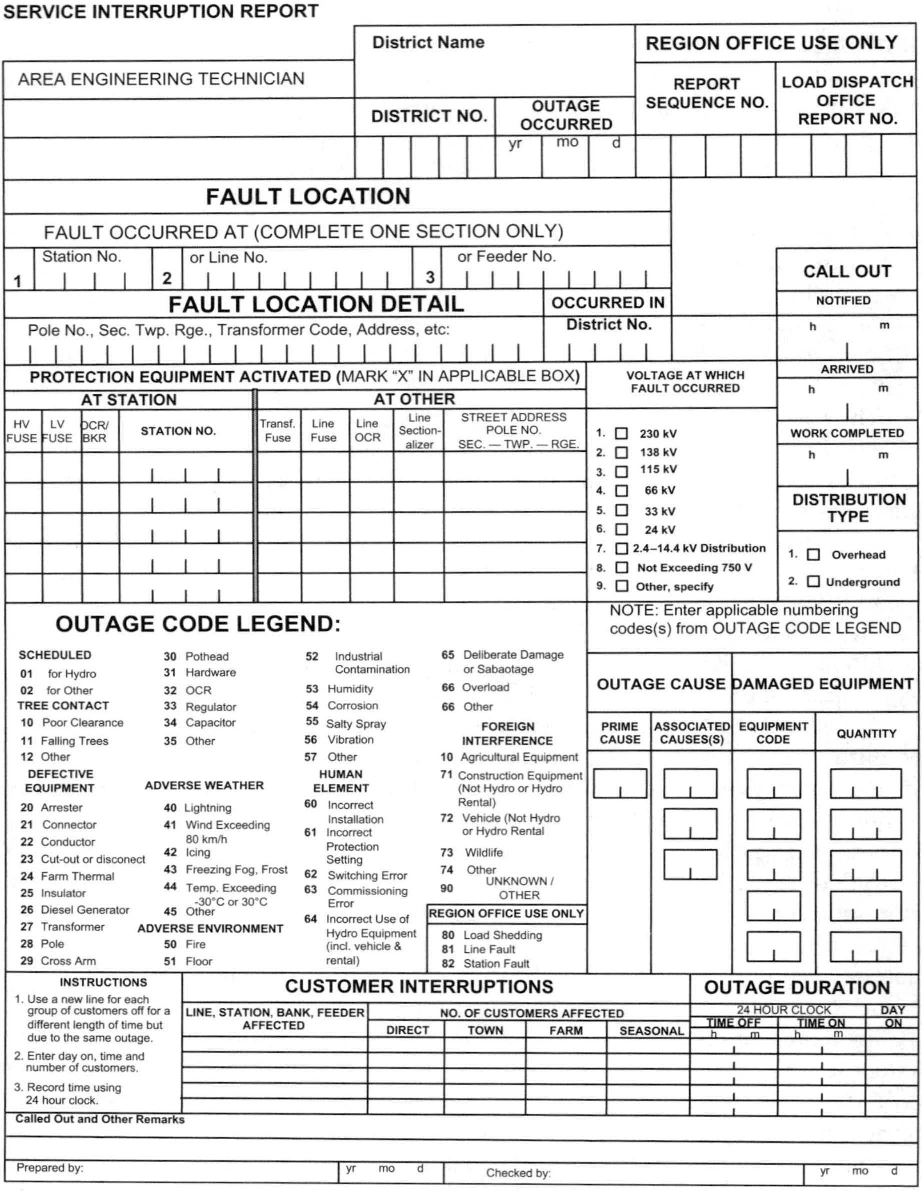

SERVICE INTERRUPTION REPORT

District Name

REGION OFFICE USE ONLY

AREA ENGINEERING TECHNICIAN

REPORT SEQUENCE NO. | LOAD DISPATCH OFFICE REPORT NO.

DISTRICT NO. | OUTAGE OCCURRED: yr mo d

FAULT LOCATION

FAULT OCCURRED AT (COMPLETE ONE SECTION ONLY)

1 Station No. | 2 or Line No. | 3 or Feeder No.

CALL OUT

FAULT LOCATION DETAIL | OCCURRED IN District No.

Pole No., Sec. Twp. Rge., Transformer Code, Address, etc:

NOTIFIED h m

ARRIVED h m

WORK COMPLETED h m

PROTECTION EQUIPMENT ACTIVATED (MARK "X" IN APPLICABLE BOX)

AT STATION: HV FUSE | LV FUSE | OCR/BKR | STATION NO.

AT OTHER: Transf. Fuse | Line Fuse | Line OCR | Line Sectionalizer | STREET ADDRESS POLE NO. SEC. — TWP. — RGE.

VOLTAGE AT WHICH FAULT OCCURRED

1. ☐ 230 kV
2. ☐ 138 kV
3. ☐ 115 kV
4. ☐ 66 kV
5. ☐ 33 kV
6. ☐ 24 kV
7. ☐ 2.4–14.4 kV Distribution
8. ☐ Not Exceeding 750 V
9. ☐ Other, specify

DISTRIBUTION TYPE

1. ☐ Overhead
2. ☐ Underground

OUTAGE CODE LEGEND:

NOTE: Enter applicable numbering codes(s) from OUTAGE CODE LEGEND

SCHEDULED
01 for Hydro
02 for Other

TREE CONTACT
10 Poor Clearance
11 Falling Trees
12 Other

DEFECTIVE EQUIPMENT
20 Arrester
21 Connector
22 Conductor
23 Cut-out or disconect
24 Farm Thermal
25 Insulator
26 Diesel Generator
27 Transformer
28 Pole
29 Cross Arm
30 Pothead
31 Hardware
32 OCR
33 Regulator
34 Capacitor
35 Other

ADVERSE WEATHER
40 Lightning
41 Wind Exceeding 80 km/h
42 Icing
43 Freezing Fog, Frost
44 Temp. Exceeding -30°C or 30°C
45 Other

ADVERSE ENVIRONMENT
50 Fire
51 Floor
52 Industrial Contamination
53 Humidity
54 Corrosion
55 Salty Spray
56 Vibration
57 Other

HUMAN ELEMENT
60 Incorrect Installation
61 Incorrect Protection Setting
62 Switching Error
63 Commissioning Error
64 Incorrect Use of Hydro Equipment (incl. vehicle & rental)
65 Deliberate Damage or Sabaotage
66 Overload
66 Other

FOREIGN INTERFERENCE
10 Agricultural Equipment
71 Construction Equipment (Not Hydro or Hydro Rental)
72 Vehicle (Not Hydro or Hydro Rental
73 Wildlife
74 Other
90 UNKNOWN / OTHER

REGION OFFICE USE ONLY
80 Load Shedding
81 Line Fault
82 Station Fault

OUTAGE CAUSE: PRIME CAUSE | ASSOCIATED CAUSES(S)

DAMAGED EQUIPMENT: EQUIPMENT CODE | QUANTITY

INSTRUCTIONS
1. Use a new line for each group of customers off for a different length of time but due to the same outage.
2. Enter day on, time and number of customers.
3. Record time using 24 hour clock.

CUSTOMER INTERRUPTIONS

LINE, STATION, BANK, FEEDER AFFECTED | NO. OF CUSTOMERS AFFECTED: DIRECT | TOWN | FARM | SEASONAL

OUTAGE DURATION: 24 HOUR CLOCK: TIME OFF h m | TIME ON h m | DAY ON

Called Out and Other Remarks

Prepared by: yr mo d | Checked by: yr mo d

Figure 9.1. A typical service interruption event collection scheme.

collection purposes. These are intentionally broad categories that will make possible more precise benchmark comparisons between different distribution utilities. There are numerous categories that could be chosen, but with the goal of uniformity for comparison purposes, the following seven broad categories can be used for data collection:

1. Scheduled
2. Tree contact
3. Defective equipment
4. Adverse weather
5. Adverse environment
6. Human element
7. Foreign interference

The Service Interruption Report (SIR) identifies many subcategories as shown in Fig. 9.1. The SIR is primarily based on the CEA outage system. The suggested categories do not prevent a utility from collecting more detailed data. However, the data collected should be able to be classified into one of the seven categories recommended. The SIR shown in Fig. 9.1 describes the types of interruptions that should be put into each category.

With the advent of high-speed digital computing systems, outage data collection and analysis became easier. The purpose of the SIR system is twofold: to record the interruption data as an indication of customer service performance and to trace the causes and record other selected events such as equipment damage so that a utility can improve its operation. For the first purpose, the basic data required are time of outage, place, number of customers affected, and cause. For the second purpose, a utility should record the voltage, equipment damaged, protective device activated, and related activities such as pole fires and icing conditions. A lot of effort is spent on filling out the SIR form, so a utility might as well get the most use out of it by getting all the pertinent information. A representative copy of SIR is shown in Fig. 9.1.

There will be errors in filling out the service interruption reports in the field. This kind of error can be minimized if the district operators who fill out the forms are given sufficient instructions and training. With a self-explanatory form and the accompanying computer system, a lot of information can be collected that is useful not only in producing an annual reliability index but also in designing, planning, and other decision making. The whole effort will then be worthwhile.

## 9.3 TYPICAL DISTRIBUTION EQUIPMENT OUTAGE STATISTICS

Future performance assessment of distribution systems requires realistic distribution equipment data that include relevant failure rates and restoration times. In this section, industry operating experience-based outage statistics for different distribution equipment are presented.

The urban areas with high-density commercial, industrial, residential, government, and institutional loads are serviced by a number of meshed distribution supply systems such as primary selective systems, primary loop systems, and secondary grid networks. However, the sparsely populated rural service areas with a mix of commercial and residential customers are normally serviced by overhead radial distribution systems. There are many technical publications that detail the different urban area distribution systems and provide reliability comparisons of the schemes. To familiarize the reader with salient design features of urban distribution supply schemes, each design is briefly described in the following.

*The primary selective system* is frequently used by many utilities to supply concentrated commercial and industrial loads. In this primary selective system, each transformer is supplied from two independent primary feeders through a normally open and a normally closed switch to enable primary feeder selection. The transformer is normally supplied from the preferred feeder through the normally closed switch. Upon interruption of voltage on the preferred feeder, the normally closed switch is opened and the normally open switch is closed, thereby transferring the power supply to the standby feeder. The feeder switchover could be made through either manual or automatic operation of the switches. For some large transformer locations, application of primary circuit breakers instead of the switches may be desirable. The automatic primary selective system results in only a momentary supply cessation during system faults on the primary feeders. The manually operated primary selective system, however, results in a permanent outage during faults on the primary feeders, where the time of the outage is equal to the time required to perform the switching operation.

*The primary loop system* is generally used for underground residential distribution; however, it is also used by many utilities for serving commercial and industrial loads. In the primary loop system, each transformer is looped into the primary circuit through two normally closed fuses. One of the transformer primary cables located approximately in the middle of the loop is left in normally open position, which allows for each side of the open loop to be supplied separately. The two sides of the loop are typically supplied from the same circuit.

For urban areas, where higher capacity feeders are used, branch loops known as subloops are often used. Two ends of the subloop can be serviced from two separate nodes on the main loop through fused disconnects. The substation breaker clears a fault on either side of the main loop, while its fuse clears a fault on the subloop. The faulty section of the loop or subloop is then located and isolated by opening switches, cutouts, underground cable elbows, jumpers, and so on at both ends. The normally open point in the loop is then closed and the isolated faulty section of the loop becomes the open point, until the faulted section is repaired. The primary loop system results in an outage during each system fault, the duration of the outage being equal to the time required to locate the faulty section and perform necessary switching operations to isolate the faulted section.

*The secondary grid network system* provides the highest level of reliability and operating flexibility. This type of network system is usually used in downtown commercial districts of most North American major metropolitan areas. In the secondary grid network system, several primary feeders simultaneously supply the network grid through a number of transformers connected in parallel. The number, sizes, locations, and supply feeders for the grid transformers are selected so that power supply continuity to all customers can be maintained without equipment overloading during a fault on any one of the feeders or transformers. Each transformer is connected to the secondary grid through a network protector, a low-voltage circuit breaker equipped with automatic controls and reverse power protection. Each section of the secondary grid comprises two or more underground cable circuits in parallel, each cable circuit being protected at both ends by cable limiters.

Due to the multiplicity of equipment, the customers supplied from the network grid do not experience an interruption during a planned or forced outage on a primary feeder, network transformer, or secondary grid cables. Faults on primary feeders are cleared by opening of the primary feeder breaker and network protectors on all of the transformers supplied from that feeder. Secondary network conductors use an electrical connector with a reduced and fusible midsection called a limiter. This time–current fusing characteristic is predetermined so that fusing will coordinate with the time–current insulation damage characteristics of the individual cable. These limiters normally clear secondary network faults. There have been some faults reported as burn free without limiter interruption; however, these occurrences are exceptional. In this way, the power supply to the customers is maintained under most fault conditions.

Because of design differences, the restoration times for failures are much shorter in urban networks compared to those in rural systems. In this chapter, different equipment outage statistics for rural and urban supply systems are proposed owing to the inherent significant design differences between urban and rural areas and the long distances with higher exposure problems in rural and sparsely populated areas.

Tables 9.1–9.10 present a summary of average failure and repair statistics for different distribution system equipment, which have been synthesized and derived from

TABLE 9.1. Distribution Feeder Average Failure Rate Statistics—Urban Locations

| Equipment | Construction Type | Failure Rate (failures/mile year) | | |
|---|---|---|---|---|
| | | 1-Φ | 2-Φ | 3-Φ |
| O/H line | Crossarm | 0.03 | 0.06 | 0.09 |
| O/H line | Armless | 0.03 | 0.06 | 0.09 |
| O/H line | Spacer cable | 0.03 | 0.06 | 0.09 |
| Direct buried cable | XLPE—lateral | 0.02 | 0.03 | 0.035 |
| | XLPE—three-phase feeder | 0.02 | 0.03 | 0.035 |
| Direct buried cable | TRXLPE—lateral | 0.013 | 0.023 | 0.028 |
| | TRXLPE—three-phase feeder | 0.013 | 0.023 | 0.028 |
| Cable in duct | XLPE—lateral | 0.02 | 0.03 | 0.035 |
| | XLPE—three-phase feeder | 0.02 | 0.03 | 0.035 |
| Cable in duct | TRXLPE—lateral | 0.013 | 0.023 | 0.028 |
| | TRXLPE—three-phase feeder | 0.013 | 0.023 | 0.028 |

TABLE 9.2. Distribution Feeder Average Repair Duration Statistics—Urban Locations

| | | Repair Activity Times (h) | | | | | Total Repair Activity Time (h) | | |
|---|---|---|---|---|---|---|---|---|---|
| | | | | Repair | | | | | |
| Equipment | Construction Type | Callout | Isolation | 1-Φ | 2-Φ | 3-Φ | 1-Φ | 2-Φ | 3-Φ |
| O/H line | Crossarm | 1.0 | 0.5 | 1.5 | 2.0 | 2.5 | 3.0 | 3.5 | 4.0 |
| O/H line | Armless | 1.0 | 0.5 | 1.5 | 2.0 | 2.5 | 3.0 | 3.5 | 4.0 |
| O/H line | Spacer cable | 1.0 | 1.0 | 2.5 | 3.0 | 3.5 | 4.5 | 5.0 | 5.5 |
| Direct buried cable | XLPE—lateral | 1.0 | 2.0 | 7.0 | 7.5 | 8.0 | 10.0 | 10.5 | 11.0 |
| | XLPE—three-phase feeder | 1.0 | 2.0 | 10.0 | 12.0 | 15.0 | 13.0 | 15.0 | 18.0 |
| Direct buried cable | TRXLPE—lateral | 1.0 | 2.0 | 7.0 | 7.5 | 8.0 | 10.0 | 10.5 | 11.0 |
| | TRXLPE—three-phase feeder | 1.0 | 2.0 | 10.0 | 12.0 | 15.0 | 13.0 | 15.0 | 18.0 |
| Cable in duct | XLPE—lateral | 1.0 | 2.0 | 8.0 | 10.0 | 12.0 | 11.0 | 13.0 | 15.0 |
| | XLPE—three-phase feeder | 1.0 | 2.0 | 21.0 | 21.0 | 21.0 | 24.0 | 24.0 | 24.0 |
| Cable in duct | TRXLPE—lateral | 1.0 | 2.0 | 8.0 | 10.0 | 12.0 | 11.0 | 13.0 | 15.0 |
| | TRXLPE—three-phase feeder | 1.0 | 2.0 | 21.0 | 21.0 | 21.0 | 24.0 | 24.0 | 24.0 |

TABLE 9.3. Distribution Feeder Average Failure Rate Statistics—Rural Locations

| | | Failure Rate (failures/mile year) | | |
|---|---|---|---|---|
| Equipment | Construction Type | 1-Φ | 2-Φ | 3-Φ |
| O/H line | Crossarm | 0.05 | 0.10 | 0.13 |
| O/H line | Armless | 0.05 | 0.10 | 0.13 |
| Direct buried cable | XLPE—lateral | 0.025 | 0.035 | 0.040 |
| | XLPE—three-phase feeder | 0.025 | 0.035 | 0.040 |
| Direct buried cable | TRXLPE—lateral | 0.018 | 0.027 | 0.035 |
| | TRXLPE—three-phase feeder | 0.018 | 0.027 | 0.035 |
| Cable in duct | XLPE—lateral | 0.025 | 0.035 | 0.040 |
| | XLPE—three-phase feeder | 0.025 | 0.035 | 0.040 |
| Cable in duct | TRXLPE—lateral | 0.018 | 0.028 | 0.035 |
| | TRXLPE—three-phase feeder | 0.018 | 0.028 | 0.035 |

TABLE 9.4. Distribution Feeder Average Repair Duration Statistics—Rural Locations

| | | Repair Activity Times (h) | | | | | Total Repair Activity Time (h) | | |
|---|---|---|---|---|---|---|---|---|---|
| | | | | Repair | | | | | |
| Equipment | Construction Type | Callout | Isolation | 1-Φ | 2-Φ | 3-Φ | 1-Φ | 2-Φ | 3-Φ |
| O/H line | Crossarm | 1.5 | 1.0 | 1.5 | 2.0 | 2.5 | 4.0 | 4.5 | 5.0 |
| O/H line | Armless | 1.5 | 1.0 | 1.5 | 2.0 | 2.5 | 4.0 | 4.5 | 5.0 |
| Direct buried cable | XLPE—lateral | 1.5 | 2.0 | 7.0 | 7.5 | 8.0 | 10.5 | 11.0 | 11.5 |
| | XLPE—three-phase feeder | 1.5 | 2.0 | 10.0 | 12.0 | 15.0 | 13.5 | 15.5 | 18.5 |
| Direct buried cable | TRXLPE—lateral | 1.5 | 2.0 | 7.0 | 7.5 | 8.0 | 10.5 | 11.0 | 11.5 |
| | TRXLPE—three-phase feeder | 1.5 | 2.0 | 10.0 | 12.0 | 15.0 | 13.5 | 15.5 | 18.5 |
| Cable in duct | XLPE—lateral | 1.5 | 2.0 | 8.0 | 10.0 | 12.0 | 11.5 | 13.5 | 15.5 |
| | XLPE—three-phase feeder | 1.5 | 2.0 | 21.0 | 21.0 | 21.0 | 24.5 | 24.5 | 24.5 |
| Cable in duct | TRXLPE—lateral | 1.5 | 2.0 | 8.0 | 10.0 | 12.0 | 11.5 | 13.5 | 15.5 |
| | TRXLPE—three-phase feeder | 1.5 | 2.0 | 21.0 | 21.0 | 21.0 | 24.5 | 24.5 | 24.5 |

TABLE 9.5. Distribution Transformer Average Failure Rate Statistics—Urban Locations

| Phase | Construction Type | Failure Rate (failures/unit per year) |
|---|---|---|
| Single | Polemount | 0.010 |
| Single | Padmount | 0.007 |
| Three | Padmount | 0.012 |

TABLE 9.6. Distribution Transformer Average Repair Duration Statistics—Urban Locations

| | | Repair Activity Times (h) | | | |
|---|---|---|---|---|---|
| Phase | Construction Type | Callout | Isolation | Repair/ Replace | Total Repair Activity Time (h) |
| Single | Polemount | 1.0 | 0.0 | 2.5 | 3.5 |
| Single | Padmount | 1.0 | 2.0 | 2.5 | 5.5 |
| Three | Padmount | 1.0 | 2.0 | 3.5 | 6.5 |

TABLE 9.7. Distribution Transformer Average Failure Rate Statistics—Rural Locations

| Phase | Construction Type | Failure Rate (failures/unit per year) |
|---|---|---|
| Single | Polemount | 0.015 |
| Single | Padmount | 0.007 |
| Three | Padmount | 0.010 |

TABLE 9.8. Distribution Transformer Average Repair Duration Statistics—Rural Locations

| | | Repair Activity Times (h) | | | |
|---|---|---|---|---|---|
| Phase | Construction Type | Callout | Isolation | Repair/ Replace | Total Repair Activity Time (h) |
| Single | Polemount | 1.5 | 0.0 | 2.5 | 4.0 |
| Single | Padmount | 1.5 | 2.0 | 2.5 | 6.0 |
| Three | Padmount | 1.5 | 2.0 | 3.5 | 7.0 |

TABLE 9.9. Distribution System Miscellaneous Equipment Average Failure Rate Outage Statistics

| Equipment | Construction Type | Failure Rate (failures/unit per year) |
|---|---|---|
| Circuit breaker | Substation | 0.0010 |
| Recloser (one- or three-phase) | Feeder | 0.0150 |
| Fuse (replace) | Polemount | 0.0030 |
| Cutout (100 or 200 A) | Polemount | 0.0030 |
| Switch (600 A single pole) | Polemount | 0.0010 |
| Cable elbow (750 Al "T" body connector, radial) | Vault | 0.0010 |
| Load break elbow (200 A distribution elbow, radial) | These would not typically be in vault | 0.0020 |
| Splice (750 Al, radial) | Vault | 0.0010 |
| Lightning arrester | Pole-top | 0.0005 |

TABLE 9.10. Distribution System Miscellaneous Equipment Average Repair Duration Statistics

| | | Repair Activity Times (h) | | | |
|---|---|---|---|---|---|
| Equipment | Construction Type | Callout | Isolation | Repair/ Replace | Total Repair Activity Time (h) |
| Circuit breaker | Substation | 1.5 | 2.0 | 28.5 | 32.0 |
| Recloser (one- or three-phase) | Feeder | 1.5 | 1.0 | 5.0 | 7.5 |
| Fuse (replace) | Polemount | 1.5 | 0.0 | 0.5 | 2.0 |
| Cutout (100 or 200 A) | Polemount | 1.5 | 0.5 | 1.0 | 3.0 |
| Switch (600 A single pole) | Polemount | 1.0 | 1.5 | 3.0 | 5.5 |
| Cable elbow (750 Al "T" body connector, radial) | Vault | 1.0 | 3.0 | 3.0 | 7.0 |
| Load break elbow (200 A distribution elbow, radial) | These would not typically be in vault | 1.0 | 2.0 | 2.0 | 5.0 |
| Splice (750 Al, radial) | Vault | 1.0 | 2.0 | 3.0 | 6.0 |
| Lightning arrester | Pole-top | 1.0 | 0.5 | 1.0 | 2.5 |

published literature reflecting different utilities' operating experience. These outage statistics can be used in comparative predictive reliability analyses of distribution systems.

## 9.4 CONCLUSIONS

Two approaches to reliability evaluation of distribution systems are frequently used, namely, historical assessment and predictive assessment. Historical reliability assessment involves the collection and analysis of a distribution system's outage and interruption data. Historical assessment generally analyzes discrete interruption events occurring at specific locations over specific time periods. Predictive assessment determines the long-term behavior of distribution systems by combining component failure rates and the duration of repair, restoration, switching, and isolation activities for given network configurations of a distribution system.

Accurate component outage data are the key to distribution system predictive performance analysis. To use the interruption data effectively in the planning and designing of new distribution circuits, a utility-specific equipment outage statistics will be required. In the absence of utility-specific outage statistics, the industry average outage statistics can be used in comparative reliability planning and design analysis of distribution improvement projects. It is, however, strongly suggested that every effort should be made to derive accurate utility-specific average failure and repair statistics for different distribution equipment for use in predictive system analyses.

## REFERENCES

1. Canadian Electricity Association, *Guide to Value-Based Distribution Reliability Planning*, CEA Report 273 D 887, Vol. I, January 1996.
2. IEEE Standard 493-1990, *Design of Reliable Industrial and Commercial Power Systems*, IEEE Gold Book, 1990.
3. IEEE Standard 1366™-2003, *IEEE Guide for Electric Power Distribution Reliability Indices*, 2003.
4. W. F. Horton, S. Goldberg, and V. Backman, The failure rates of underground distribution system components, NAPS Proceedings, Auburn, AL, October 15–16, 1990, pp. 306–316.
5. L. H. Stember, M. M. Epstein, G. B. Gaines, G. C. Derringers, and R. E. Thomas, Analysis of field failure data on HMWPE and XLPE-insulated high voltage distribution cable, *IEEE Transactions on Power Apparatus and Systems*, **104** (8), 1985, 1979–1985.
6. S. Goldberg and W.F. Horton, Analysis of feeder service reliability using component failure rates, *IEEE Transactions on Power Delivery*, **2** (4), 1987.

# 10

# HISTORICAL ASSESSMENT

## 10.1 INTRODUCTION

The methodology used to assess the historical reliability performance of a practical utility's electric distribution system is outlined in this chapter. Included is an overview of the process used to collect and organize the required interruption data as well as a description of the performance indices calculated for use in the causal assessment. The various parts of the reliability performance assessment are described. This includes a description of reliability indices, comparisons between years of operation, comparisons to the average at different voltage levels of the system, and comparisons by outage cause and component failure. Finally, results from the 2004 reliability assessment of the utility's electric distribution system are summarized, and the application of the calculated performance statistics in planning, operating, and maintaining distribution systems is described.

Two approaches to reliability evaluation of distribution systems are normally used, namely, historical assessment and predictive assessment. Historical assessment involves the collection and analysis of distribution system outage and customer interruption data. It is essential for electric utilities to measure actual distribution system reliability performance levels and define performance indicators to assess the basic function of providing a cost-effective and reliable power supply to all customer types. The

*Power Distribution System Reliability.* By Ali A. Chowdhury and Don O. Koval

distribution system is an important part of the total electrical supply system. This is due to the fact that the distribution system provides the final link between a utility's transmission system and its customers. It has been reported elsewhere that more than 80% of all customer interruptions occur as a result of failures in the distribution system.

Historical assessment is generally described as measuring the past performance of a system by consistently logging the frequency, duration, and causes of system component failures and customer interruptions. Predictive reliability assessment, on the contrary, combines historical component outage data and mathematical models to estimate the performance of designated configurations. Predictive techniques therefore rely on two basic types of data to compute service reliability: component reliability parameters and network physical configurations.

Historical data are very useful when analyzed to ascertain what went wrong in the past and thereby correcting it and are also used as input to predict future service reliability. Both historical and predictive assessments, therefore, involve the collection of system outage data. Historical models summarize the actual performance of a distribution system during some time period, for example, quarterly, semiannually, or annually. The basic data item in this case is a system failure, which is a component outage or a customer interruption. Each failure event is taken into consideration and analyzed according to the causes of failure, duration of outage, and area of the system affected.

A variety of customer- and load-oriented system performance indices can be derived by manipulating the recorded data. These indices are very useful for assessing the severity of interruption events. Assessment of past performance is useful in the sense that it helps identify weak areas of the system and the need for reinforcement. It enables previous predictions to be compared with actual field experience. It can also serve as a guide for acceptable values in future reliability assessments. A variety of performance indices that express interruption statistics in terms of system customers can be computed using the service continuity data.

This chapter is concerned with aspects of historical reliability assessment. It briefly describes the different characteristics of an automatic outage management system (AOMS) used for collecting and analyzing the distribution supply interruptions and also presents a summary of service continuity statistics for the practical utility distribution system for the 5-year period from 2000 to 2004. In addition to the analysis by primary causes, analysis of failure data by subcomponents that fail on the distribution system and what contributions they make to the total unreliability is also presented.

## 10.2 AUTOMATIC OUTAGE MANAGEMENT SYSTEM

The practical utility of concern in this chapter uses its automatic outage management system to collect outage data for its distribution system. The current system was installed in the late 1990s. The practical utility is an integrated utility consisting of generation, transmission, and distribution facilities with urban, fringe, and rural networks. Analysis was done using AOMS data for the period 2000–2004. For the purpose of analyzing this data, the service area has been divided into three regions, namely, Region 1, Region 2, and Region 3. The general practical utility distribution system characteristics are summarized in Table 10.1.

TABLE 10.1. Utility Distribution System Characteristics—2004

| Load Density ($kW/km^2$) | Load Density (kW/km circuit) | Load Density (kW/customer served) |
|---|---|---|
| 229.20 | 113.62 | 5.72 |

## 10.2.1 Definitions of Terms and Performance Indices

The definitions of terms and indices presented in the following are used in the utility AOMS and are compatible with those presented in the latest IEEE Standard 1366–2003.

*Distribution System*: A distribution system is that portion of an electric power system that links the bulk power source or sources to the customer's facilities. Subtransmission lines, distribution substations, primary feeders, distribution transformers, secondaries, and customer's services all form different parts of what can generally be termed a distribution system.

*Customers*: This means the number of customer meters fed at secondary and primary voltages.

*Interruption*: An interruption is the loss of service to one or more customers and is the result of one or more component outages.

*Interruption Duration*: This is the period from the recorded initiation of an interruption to a customer until service has been restored to that customer.

*Customer-Minutes of Interruptions*: This is the product of the customer services interrupted by the period of interruption.

*Customer Interruption*: This is the sum of products of the customer services interrupted and the number of interruptions that affect those customer services

## 10.2.2 Customer-Oriented Indices

A variety of performance indices that express interruption statistics in terms of system customers are defined in the following.

The System Average Interruption Frequency Index (SAIFI) is the average number of times that a system customer is interrupted during a time period. In this chapter, the time period considered in computing performance indices is 1 year. SAIFI is therefore determined by dividing the total number of customer interruptions in a year by the total number of customers served at the end of the year. A customer interruption is considered to be one interruption to one customer.

$$\text{SAIFI} = (\text{total customer interruptions})/(\text{total customers served})$$

The System Average Interruption Duration Index (SAIDI) is the average interruption duration per customer served. It is determined by dividing the sum of all customer interruption durations during a year by the number of customers served.

$$\text{SAIDI} = (\text{total customer hours of interruptions})/(\text{total customers served})$$

The Customer Average Interruption Duration Index (CAIDI) is the average interruption duration for those customers interrupted during a year. It is determined by dividing the sum of all customer interruption durations by the total number of customer interruptions over a 1-year period.

$$\text{CAIDI} = (\text{total customer hours of interruption})/(\text{total customer interruptions})$$

The Average Service Availability Index (ASAI) is the ratio of the total number of customer hours that the service was available during a year to the total customer hours demanded. Customer hours demanded are determined as the year-end number of customer served times 8760 h. This is sometimes known as the Index of Reliability (IOR). The complementary value to this index, that is, the Average Service Unavailability Index may also be used. This is the ratio of the total number of customer hours that service was unavailable during a year to the total customer hours demanded.

$$\text{ASAI} = (\text{customer hours available for service})/(\text{customer hours demanded})$$

The results obtained from two surveys dealing with the United States and Canadian utility activities in regard to service continuity data collection and utilization show that a large number of utilities calculate the customer-based indices of SAIFI, SAIDI, and CAIDI for their systems. ASAI is widely reported by many US utilities but appears to have relatively less appeal in Canada as a basic utility index. This index has, however, been calculated on a national basis for over 20 years and is designated as the IOR in the Canadian Electricity Association (CEA) reporting system. It provides a very general indication of Canadian performance.

### 10.2.3 Classification of Interruption as to Causes

A customer interruption has been defined in terms of the following primary causes of the interruption:

*Transmission/Substation*: Customer interruption resulting from problems in the bulk electricity supply system such as under-frequency load shedding, transmission system transients, or system frequency excursions. During a rotating load-shedding cycle, the duration is the total outage time until normal operating conditions resume. The number of customers affected is the average number of customers interrupted per rotating cycle.

*Tree Contacts*: Customer interruptions caused by faults due to trees or tree limbs contacting energized circuits.

*Equipment Overhead (OH)/Underground (UG)*: Customer interruptions resulting from equipment failures due to deterioration from age, incorrect maintenance, or imminent failures detected by maintenance.

*Weather*: Customer interruptions resulting from lightning, rain, ice storms, snow, winds, extreme ambient temperatures, freezing fog, or frost and other extreme conditions.

*Personnel Error*: Customer interruptions due to the interface of the utility staff with the system such as incorrect records, incorrect use of equipment, incorrect construction or installation, incorrect protection settings, switching errors, and commissioning errors.

*Public*: Customer interruptions beyond the control of the utility such as vehicles, dig-ins, vandalism, sabotage, and foreign objects.

*Animal*: Customer interruptions due to animals such as birds and squirrels.

*Unknown/Others*: Customer interruptions with no apparent or defined cause or reason, which could have contributed to the outage.

In addition to the analysis by primary causes, analysis of failure data by subcomponents that fail on the distribution system and what contribution they make to the unreliability is also performed.

## 10.3 HISTORICAL ASSESSMENT

A variety of customer- and load-oriented system performance indices can be derived by manipulating the recorded data. In this chapter, however, only the most commonly used customer-oriented indices are presented.

This section discusses a statistical summary of the practical utility distribution system performance for the years 2000 through 2004 at the utility corporate level, region level, crew center level, and circuit level. Figure 10.1 depicts the hierarchical deeper level performance analysis process.

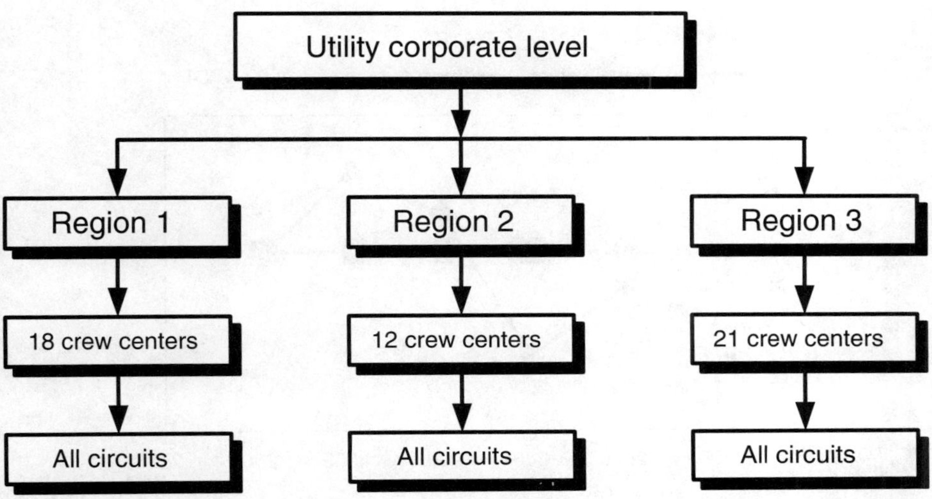


Figure 10.1. Hierarchical level in-depth causal analyses of distribution supply interruptions.

TABLE 10.2. Utility Distribution System Performance Indices for 2000–2004

| Year | SAIFI (occurrence/year) | SAIDI (h/year) | CAIDI (h/year) | IOR | IOU (upm) |
|---|---|---|---|---|---|
| 2000 | 1.168 | 1.93 | 1.65 | 0.999780 | 220 |
| 2001 | 0.991 | 1.58 | 1.59 | 0.999819 | 181 |
| 2002 | 1.352 | 1.65 | 1.22 | 0.999812 | 188 |
| 2003 | 1.455 | 1.81 | 1.25 | 0.999793 | 207 |
| 2004 | 1.227 | 1.54 | 1.26 | 0.999824 | 176 |
| 2000–2004 Average | 1.256 | 1.77 | 1.41 | 0.999797 | 203 |

## 10.3.1 A Utility Corporate Level Analysis

Table 10.2 shows the historical values for SAIFI, SAIDI, CAIDI, IOR, and IOU for the years 2000–2004. The overall IOR provides a very useful general indication of utility performance. This index is the per unit value of annual customer hours that service was available. In addition to IOR, Table 10.2 shows the complement to IOR. This index has been known as the Index of Unreliability (IOU). When multiplied by a million, this IOU index is expressed in units per million and labeled as "upm."

Figures 10.2–10.4 show the overall trends in utilities SAIFI, SAIDI, and CAIDI over the 2000–2004 operating period.

Figures 10.2–10.4 illustrate that utilities SAIFI and SAIDI have downward trends in 2004 compared to those of 2003. As shown in Table 10.1, the index of reliability, or the per unit annual customer hours that service is available for 2003 and 2004 were 0.999793 and 0.999824, respectively. The complementary index, IOU, for 2003 and 2004 were 207 and 176 upm respectively. Table 10.1 shows slightly increased overall distribution

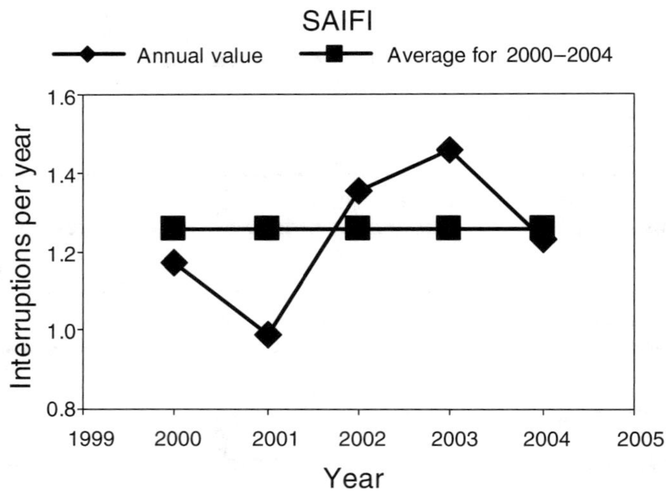


Figure 10.2. Overall trends for utility SAIFI values for the operating period 2000–2004.

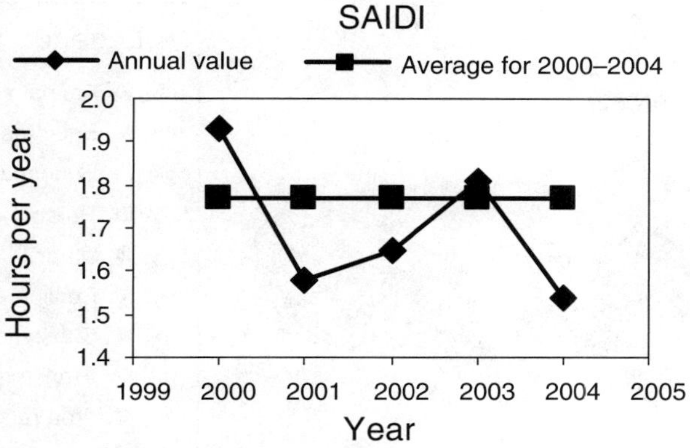


Figure 10.3. Overall trends for utility SAIDI values for the operating period 2000–2004.

reliability for the utility system in 2004 compared to that of 2003. This section presents a statistical summary of the utility distribution system performance for the year 2004 and compares it with 2003 and then with the 2000–2004 year average.

*SAIFI*: The average number of interruptions per year for 2004 was 1.227, which represents a 15.67% decrease over the 2003 figure of 1.455 interruptions per customer per year. The 2000–2004 average value for SAIFI is 1.256 interruptions per customer per year.

*SAIDI*: The system average interruption duration for customers served per year for 2004 was 1.54 h/year, which represents a decrease of 14.81% over the 2003 figure of 1.81 h. The 2000–2004 average SAIDI figure is 1.77 h.

*CAIDI*: The average customer interruption duration per interruption for 2004 was 1.26 h, which is an increase of 0.95% from the 2003 figure of 1.25 h. The 2000–2004 average CAIDI figure is 1.41 h.

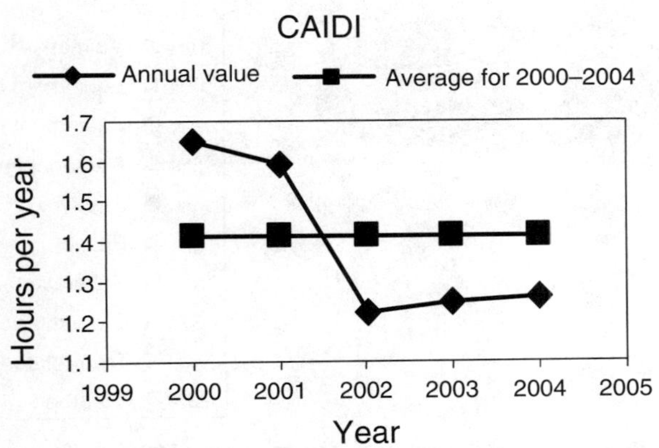


Figure 10.4. Overall trends for utility CAIDI values for the operating period 2000–2004.

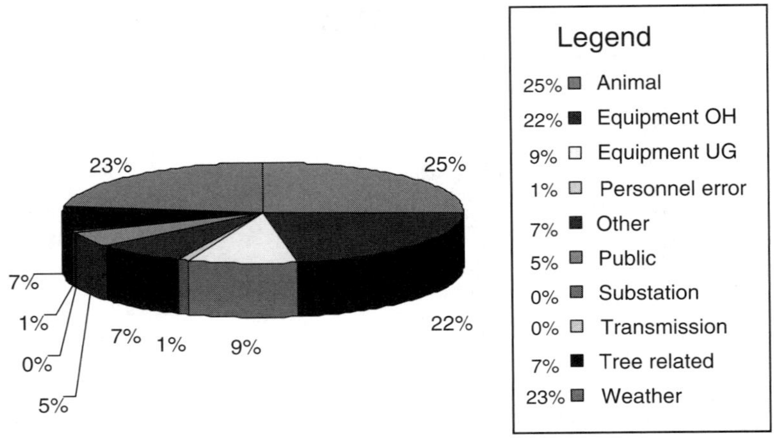


Figure 10.5. Major causes of interruptions for 2004 at utility corporate level.

The five major causes of service interruption for 2004 at the corporate level are equipment overhead (21.75%), animal (25.51%), weather (22.63%), tree related (7.48%), and underground equipment (8.91%). Figure 10.5 illustrates the causes of interruptions for 2004 at the corporate level.

The five major causes of customer interruption for 2004 at the corporate level are weather (23.85%), equipment overhead (17.32%), animal (11.70%), other (15.80%), and public (7.76%). Figure 10.6 illustrates causes of customer interruptions at the corporate level for 2004.

As shown in Fig. 10.7, the five major causes of customer minutes of interruption at the corporate level for 2004 are weather (23.98%), equipment overhead (17.99%), other (12.62%), tree related (10.80%), and animal (9.49%).

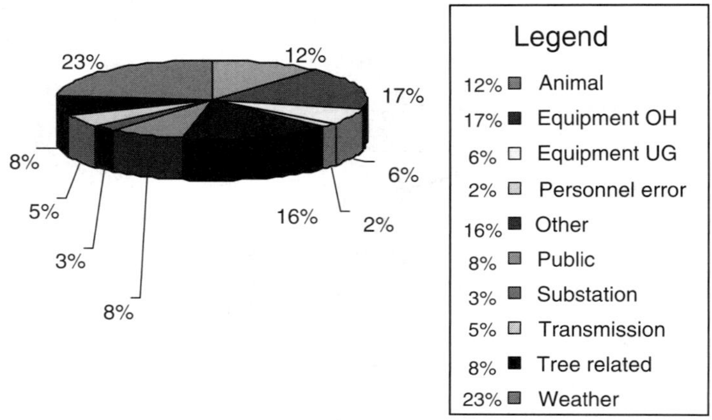


Figure 10.6. Major causes of customer interruptions for 2004 at utility corporate level.

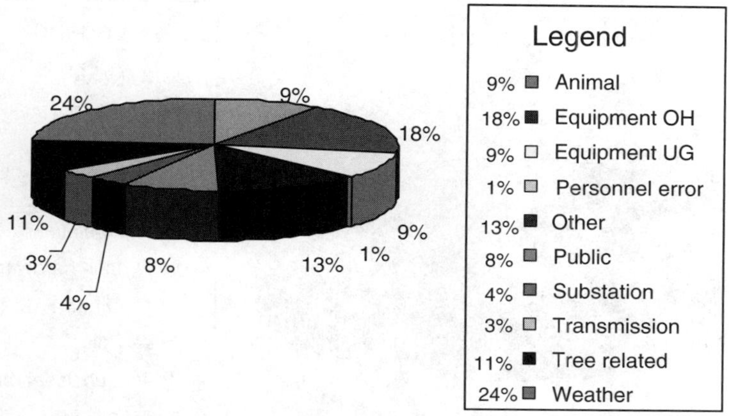


Figure 10.7. Major causes of customer minutes of interruption for 2004 at the utility corporate level.

As shown in Fig. 10.8, the five major contributing components of service interruption at utility corporate level for 2004 are none (69.73%), line hardware (8.25%), underground cable (6.82%), transformer (4.93%), and conductor (3.03%).

The five major contributing components for customer interruption at the corporate level for 2004 are none (74.24%), underground cable (4.92%), line hardware (3.26%), insulator (3.56%), and conductor (3.50%). Figure 10.9 illustrates major component contribution to customer interruption at the corporate level for 2004. The six major contributing components for customer minutes of interruption for 2004 at the corporate level are none (69.61%), underground cable (7.59%), line hardware (3.46%), insulator

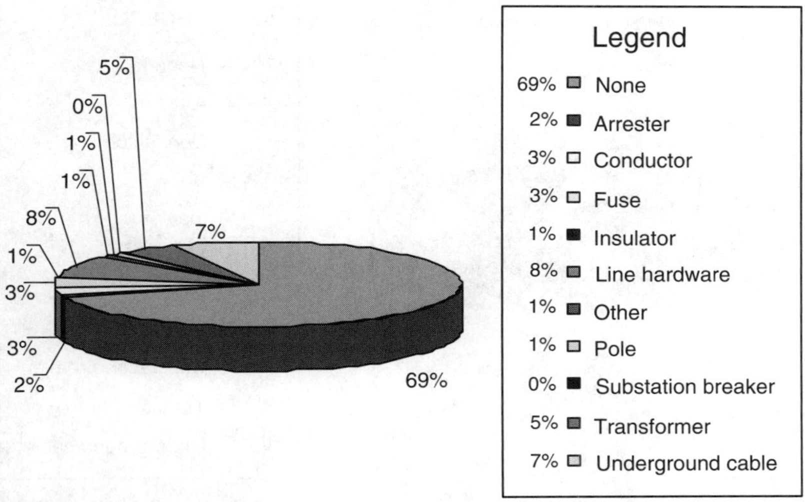


Figure 10.8. Major contributing components of interruptions at utility corporate level for 2004.

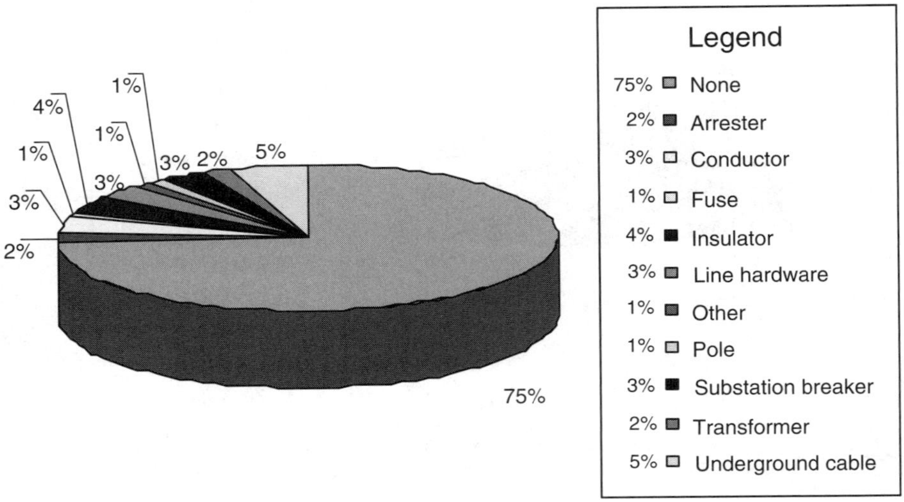


Figure 10.9. Major contributing components of customer interruptions for 2004 at utility corporate level.

(3.46%), conductor (4.02%), and substation breaker (4.59%). These component contributions are shown in Fig. 10.10.

Figures 10.11–10.14 show the values of SAIFI and SAIDI for the main causes of outage and components that failed in the distribution system and for the contributions they represent in the total value of the indices for years 2000 through 2004.

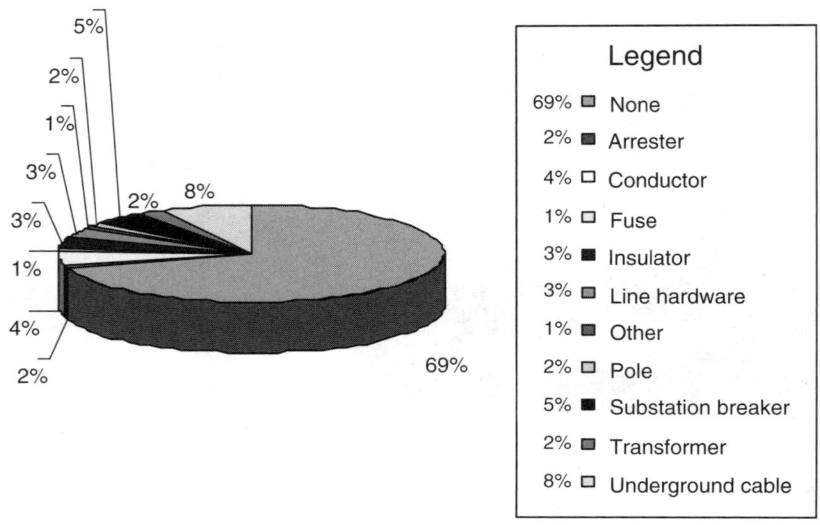


Figure 10.10. Major component contributions to customer minutes of interruptions at the utility corporate level for 2004.

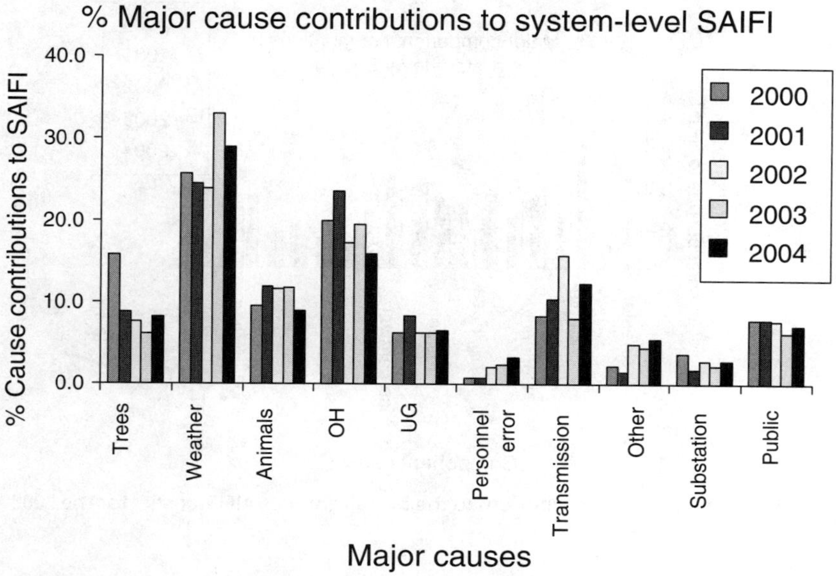


Figure 10.11. Cause contributions to the overall annual SAIFI statistics for the 2000–2004 period.

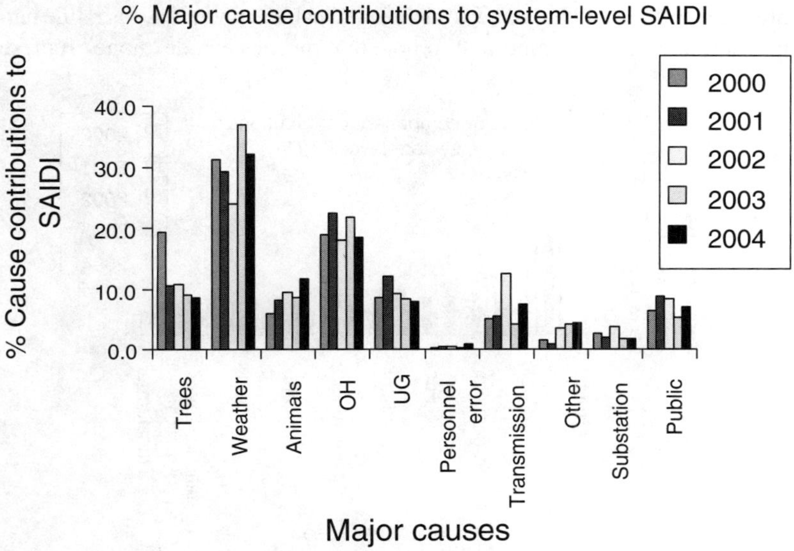


Figure 10.12. Cause contributions to the overall annual SAIDI statistics for the 2000–2004 period.

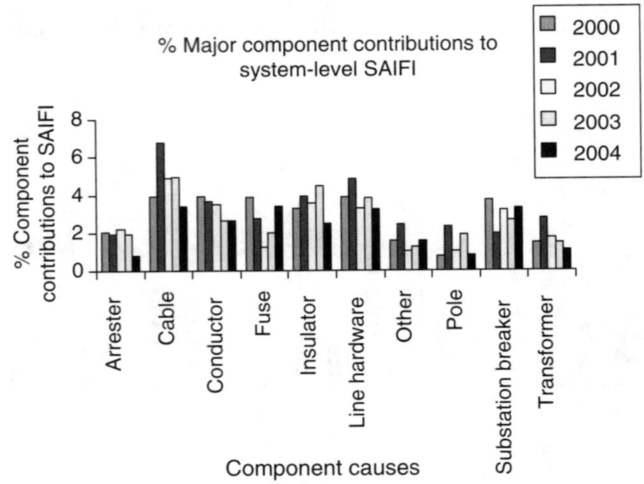


Figure 10.13. Component contributions to the overall annual SAIFI statistics for the 2000–2004 period.

As shown in Fig. 10.11, the dominant cause contributions to utility corporate-level 2004 SAIFI are weather, equipment overhead, animal, other, and trees. Figure 10.12 depicts the cause contribution to the overall annual SAIDI statistics for the 2000–2004 period. The dominant cause contribution to the utility corporate level 2004 SAIDI are weather, equipment overhead, animal, tree related, underground equipment, and other.

As depicted in Fig. 10.13, the dominant component contributions to the utility corporate level 2004 SAIFI are none, underground cable, conductor, fuse, line hardware, and substation breaker. It is important to note that the cause code "none" represents all

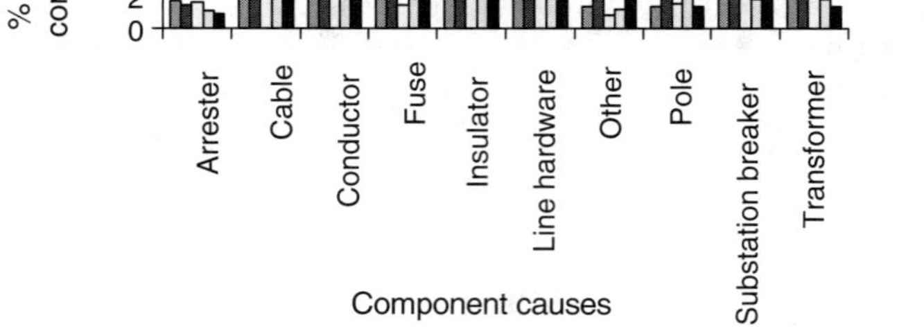


Figure 10.14. Component contributions to the overall annual SAIDI statistics for the 2000–2004 period.

other causes presented in Figs. 10.11 and 10.12, that is, noncomponent-related causes of interruptions.

Figure 10.14 indicates that the dominant component contributions to the utility corporate level 2004 SAIDI are none, underground cable, line hardware, insulator, fuse, and conductor. As shown in Figs. 10.11 and 10.13, in terms of SAIFI, majority of the causes have a downward trend in 2004 compared to previous years. The category "none" is not depicted in Figs. 10.13 and 10.14 because of its significantly higher value compared to other components.

### 10.3.2 Utility Region-Level Analysis

Detailed analyses of the causal and component contributions to regional SAIFI and SAIDI indices have been performed in a similar manner as was done for the utility corporate-level analysis. For the purpose of this chapter, the utility service area has been divided into three regions, namely, Region 1, Region 2, and Region 3. The calculated performance indices are summarized for each of the regions.

***10.3.2.1 Region 1 Statistics for 2004.*** Table 10.3 summarizes the major indices at the Region 1 level for the years 2000 through 2004.

Table 10.3 shows improved distribution reliability for the Region 1 system in 2004 compared to that of 2003 in terms of SAIFI, SAIDI, and CAIDI. A statistical summary of the Region 1 distribution system performance for the year 2004 is presented in the following and compares it to 2003 and then to the 2000–2004 year average.

*SAIFI*: The average number of interruptions per year for 2004 for Region 1 was 1.390, which represents an 11.30% decrease over the 2003 figure of 1.567 interruptions per customer per year. The 2000–2004 average value for SAIFI is 1.401 interruptions per customer per year.

*SAIDI*: The system average interruption duration for customers served per year for the 2004 for Region 1 was 1.53 h/year, which represents a 16.52% decrease over the 2003 figure of 1.83 h. The 2000–2004 average SAIDI figure is 1.92 h.

*CAIDI*: The average customer interruption duration per interruption for 2004 for Region 1 was 1.10 h, which is a 5.9% decrease from the 2003 figure of 1.17 h. The 2000–2004 average CAIDI figure is 1.37 h.

TABLE 10.3. Region 1 Distribution System Performance Indices for 2000–2004

| Year | SAIFI (occurrence/year) | SAIDI (h/year) | CAIDI (h/year) |
|---|---|---|---|
| 2000 | 1.369 | 2.07 | 1.51 |
| 2001 | 1.187 | 1.91 | 1.61 |
| 2002 | 1.428 | 1.66 | 1.16 |
| 2003 | 1.567 | 1.83 | 1.70 |
| 2004 | 1.390 | 1.53 | 1.10 |
| 2000–2004 Average | 1.401 | 1.92 | 1.37 |

TABLE 10.4. Region 2 Distribution System Performance Indices for 2000–2004

| Year | SAIFI (occurrence/year) | SAIDI (h/year) | CAIDI (h/year) |
|---|---|---|---|
| 2000 | 1.218 | 2.15 | 1.77 |
| 2001 | 0.868 | 1.44 | 1.66 |
| 2002 | 1.265 | 1.53 | 1.21 |
| 2003 | 1.418 | 1.75 | 1.23 |
| 2004 | 1.186 | 1.66 | 1.40 |
| 2000–2004 Average | 1.208 | 1.78 | 1.47 |

The dominant causes of Region 1 SAIFI contribution for 2004 are equipment overhead, weather, animal, tree related, and other. The dominant causes of Region 1 SAIDI contribution for 2004 are weather, equipment overhead, tree related, animal, other, and underground equipment. The dominant component contributions to Region 1 SAIFI for 2004 are none, line hardware, conductor, underground cable, and insulator. The dominant component contributions to Region 1 SAIDI for 2004 are none, underground cable, insulator, fuse, conductor, and line hardware.

***10.3.2.2 Region 2 Statistics for 2004.*** Table 10.4 summarizes the major indices at the Region 2 level for years 2000–2004.

Table 10.4 shows increased distribution reliability for the Region 2 system in 2004 compared to that of 2003 in terms of SAIFI and SAIDI but not CAIDI.

A statistical summary of the Region 2 distribution system performance for the year 2004 is presented in the following and compares it with 2003 and then with the 2000–2004 year average.

*SAIFI*: The average number of interruptions per year for 2004 Region 2 was 1.186, which represents a 16.36% decrease over the 2003 figure of 1.418 interruptions per customer per year. The 2000–2004 average value for SAIFI is 1.208 interruptions per customer per year.

*SAIDI*: The system average interruption duration for customers served per year for the 2004 Region 2 was 1.66 h/year, which represents a decrease of 4.62% over the 2003 figure of 1.75 h. The 2000–2004 average SAIDI figure is 1.78 h.

*CAIDI*: The average customer interruption duration per interruption for 2004 Region 2 was 1.40 h, which is an increase of 14.04% from the 2003 figure of 1.23 h. The 2000–2004 average CAIDI figure is 1.47 h.

The dominant causes of Region 2 SAIFI for 2004 contribution are equipment overhead, weather, tree related, underground equipment, public, and other. The dominant causes of Region 2 SAIDI for 2004 are weather, equipment underground, equipment overhead, public, animal, and tree related. The dominant component contribution to Region 2 SAIFI for 2004 are none, underground cable, fuse, line hardware, and substation breaker. The dominant component contribution to Region 2 SAIDI for 2004 are none, underground cable, fuse, line hardware, conductor, and other.

TABLE 10.5. Region 3 Distribution System Performance Indices for 2000–2004

| Year | SAIFI (occurrence/year) | SAIDI (h/year) | CAIDI (h/year) |
|---|---|---|---|
| 2000 | 0.750 | 1.35 | 1.79 |
| 2001 | 0.884 | 1.32 | 1.49 |
| 2002 | 1.370 | 1.84 | 1.35 |
| 2003 | 1.343 | 1.89 | 1.41 |
| 2004 | 1.049 | 1.37 | 1.30 |
| 2000–2004 Average | 1.097 | 1.54 | 1.41 |

***10.3.2.3 Region 3 Statistics for 2004.*** Table 10.5 summarizes the major indices at the Region 3 level for years 2000–2004.

Table 10.5 shows increased distribution reliability for the Region 3 system in 2004 compared to that of 2003 in terms of SAIFI, SAIDI, and CAIDI. A statistical summary of the Region 3 distribution system performance for the year 2004 is presented in the following and compares it with 2003 and then with the 2000–2004 year average.

*SAIFI*: The average number of interruptions per year for 2004 for Region 3 was 1.049, which represents a 21.89% decrease over the 2003 figure of 1.343 interruptions per customer per year. The 2000–2004 average value for SAIFI is 1.097 interruptions per customer per year.

*SAIDI*: The system average interruption duration for customers served per year for the 2004 for Region 3 was 1.37 h/year, which represents a 27.88% decrease over the 2003 figure of 1.89 h. The 2000–2004 average SAIDI figure is 1.54 h.

*CAIDI*: The average customer interruption duration per interruption for 2004 for Region 3 was 1.30 h, which is a 7.67% decrease from the 2003 figure of 1.41 h. The 2000–2004 average CAIDI figure is 1.40 h.

The dominant cause contributions to Region 3 SAIFI for 2004 are weather, equipment overhead, animal, substation, public, and other. The dominant causes of Region 3 SAIDI for 2004 are weather, equipment overhead, substation, animal, and other. The dominant component contributions to Region 3 SAIFI for 2003 are none, insulator, fuse, substation breaker, line hardware, and conductor. The dominant component contributions to Region 3 SAIDI for 2004 are none, insulator, pole, conductor, fuse, and substation breaker.

Although outages in a distribution system have a localized effect, analysis of the customer failure statistics of most utilities shows that the distribution system makes the greatest individual contribution to the unavailability of supply to a customer. As in the case of this Canadian utility, it can be seen from Figs. 10.11 and 10.12 that contributions from the "transmission" and "substation" categories, that is, outages associated with generation and transmission systems, to the overall average SAIFI and SAIDI values in the 2000–2004 period range from 3.33 to 8.39% and 2.74 to 7.17%, respectively. This supports the general statement that over 80% of all customer interruptions occur due to failures in the distribution system.

A major observation is that since the "other" cause category consists primarily of unknown or otherwise undefined outage causes, it is not possible to draw any definite conclusions on system improvements that might be undertaken. The personnel responsible for entering the outage data should make every effort to identify, where possible, the cause of the outage. In addition, new cause categories should be developed and implemented, as warranted, to better identify the outages causes. This will aid in determining possible system improvements to increased reliability.

## 10.4 CREW CENTER-LEVEL ANALYSIS

The ability to generate reliability indices at the regional and crew center levels permits these entities to be flagged for critical review and for localized mitigating actions before severe outages and customer complaints occur. This deeper level of system analysis allows to proactively improve the reliability of poorly performing service areas by more efficiently assigning capital and O&M budgets and schedules for maintenance, tree trimming, aging infrastructure replacements, and backup supply considerations, while tracking work crew response and performance levels.

A detailed causal analysis of interruptions, customer interruptions, customer minutes of interruptions, as well as cause contribution to SAIFI and SAIDI has been performed for 51 crew centers. Also, performed are detailed component contributions to previously mentioned performance indices. The 2004 SAIFI for crew centers ranged from a low of 0.080 interruptions per year to a high of 2.588 interruptions per year. The 2004 SAIDI value for crew centers ranged from a low of 0.17 h/year to a high of 6.59 h/year. The 2004 CAIDI value for crew centers ranged from a low of 0.71 h/year to a high of 2.82 h/year.

## 10.5 DEVELOPMENT OF A COMPOSITE INDEX FOR RELIABILITY PERFORMANCE ANALYSIS AT THE CIRCUIT LEVEL

There are a number of factors to be considered when determining circuit reliability performance. Momentary outages can become more important on a circuit than some other index, depending on the customer mix the circuit serves. In these circumstances, a composite index that includes multiple reliability indices is useful. In reliability assessments, it is prudent to optimize circuit ranking and hence improvements using a model that minimizes a composite reliability index that is based on popularly used indices such as SAIFI, SAIDI, and Momentary Average Interruption Frequency Index (MAIFI). A useful common composite statistical model is defined as follows:

$$\begin{aligned}\text{Composite Index(CI)} = {} & \text{SAIFI}_{\text{weight}} \times (\text{SAIFI} - \text{SAIFI}_{\text{target}})/\text{SAIFI}_{\text{target}} \\ & + \text{SAIDI}_{\text{weight}}(\text{SAIDI} - \text{SAIDI}_{\text{target}})/\text{SAIDI}_{\text{target}} \\ & + \text{MAIFI}_{\text{weight}}(\text{MAIFI} - \text{MAIFI}_{\text{target}})/\text{MAIFI}_{\text{target}}\end{aligned} \tag{10.1}$$

SAIFI, SAIDI, and MAIFI are average historical values for a given circuit configuration. $SAIFI_{target}$, $SAIDI_{target}$, and $MAIFI_{target}$ are the guideline values for a distribution utility system. $SAIFI_{weight}$, $SAIDI_{weight}$, and $MAIFI_{weight}$ are the weightings for each of the three indices.

A combined ranking for each circuit can be calculated by adding up the individual index ranking values. The fixed number of worst-ranked circuits was identified based on the above ranking, and recommended upgrades for each of the identified worst circuits in a fiscal year proposed.

## 10.6 CONCLUSIONS

An example calculation of service continuity data for the year 2004 for a practical utility distribution system has been analyzed and results have been presented in this chapter. A mathematical model to rank individual circuit reliability performance based on historical records has also been proposed. The computed statistics and causal analyses from the utility corporate level to the individual circuit level will aid in judiciously allocating reliability dollars to deserving feeders and areas.

Almost all historical performance analyses reported in the literature are carried out at the system level. This chapter reveals that system-level reliability indices mask the unusually good or poor reliability performances at different system levels. It can therefore be concluded that the system reliability improvement plans based solely on system level reliability indices would not result in expected customer service enhancements.

It is important to note that the utilization of historical reliability indices as to their causes at the system, region, crew center, and individual circuit levels would permit a utility to more accurately pinpoint trouble areas and take accurate mitigating actions to enhance service reliability. The multilevel interruption reporting and analysis will enable utilities and state regulators to set reliability performance standards at different system levels to minimize outages and reduce their impact on customer services. In addition, the historical reliability performance indices can be used by state regulators in designing a performance-based rate (PBR) mechanism in a deregulated environment.

## REFERENCES

1. R. Billinton and R. N. Allan, *Reliability Evaluation of Power Systems*, Plenum Press, New York, 1996.
2. A. A. Chowdhury and D. O. Koval, Value-based distribution system reliability planning, *IEEE/IAS Transactions*, **4**, 1996, 2359–2365.
3. U.S. Department of Energy The National Electric Reliability Study: Executive Summary, DOE/EP—0003 Dist. Category UC-97C, April 1981.
4. Westinghouse Electric Corporation Development of Distribution System Reliability and Risk Analysis Models, EPRI, EL—2018, August 1981.
5. IEEE Standard 1366–2003 Guide for Electric Power Distribution Reliability Indices, November 2003.

6. M. H. J. Bollen, *Understanding Power Quality Problems: Voltage Sags and Interruptions*, IEEE Press, New York, 2000.
7. Canadian Electricity Association Distribution System Reliability Statistics, Research and Development Report number 148 D255, October 1984.
8. A. A. Chowdhury and D. O. Koval *Distribution system reliability: historical assessment*, G. E. Lasker (ed.), *Advances in Power System Research*, Vol. 2, International Institute for Advancement Studies in Systems Research and Cybernetics, Baden, Germany, 1997.
9. R. Billinton and J. E. Billinton, Distribution system reliability indices, *IEEE Transactions on Power Delivery*, **4** (1), 1989, 561–568.
10. R. E. Brown, *Electric Power Distribution Reliability*, Marcel Dekker, New York, 2004.
11. Edison Electric Institute, 2004 Reliability Report, May 2005.
12. Canadian Electricity Association, 2005 Annual Service Continuity Report on Distribution System Performance in Canadian Electrical Utilities, May 2005.
13. G. F. L. Dixon and H. Hammersely, Reliability and its costs on distribution systems, International Conference on Reliability of Power Supply Systems, IEE Conference Publication No. 148, 1977.

# 11

# DETERMINISTIC CRITERIA

## 11.1 INTRODUCTION

Among the major issues facing utilities in today's competitive electricity market is the pressure to hold the line on rates and provide electricity with adequate quality and reliability. Utilities are increasingly recognizing that the level of supply reliability planned and designed into a system has to evolve away from levels determined basically on a technical framework using deterministic criteria toward a balance between minimizing costs and achieving a sustainable level of customer complaints. Assessment of the cost of maintaining a certain level of supply reliability, or making incremental changes therein, must include not only the utility's cost of providing such reliability and the potential revenue losses during outages but also the interruption costs incurred by the affected customers during utility power outages. Such a cost–benefit analysis constitutes the focal point of value-based reliability planning. Value-based reliability planning provides a rational and consistent framework for answering the fundamental economic question of how much reliability is adequate from the customers' perspective and where a utility should spend its reliability dollars to optimize efficiency and satisfy customers' electricity requirements at the lowest cost. Costs to customers associated with varying levels of service reliability are significant factors that cannot be ignored.

*Power Distribution System Reliability.* By Ali A. Chowdhury and Don O. Koval

Explicit considerations of these customer interruption costs in developing supply reliability targets and in evaluating alternative proposals for network upgrade, maintenance, and system design must, therefore, be included in the system planning and designing process. This chapter provides a brief overview of current deterministic planning practices in utility distribution system planning and introduces a probabilistic, customer value-based approach to the planning of alternative feeder requirements for overhead distribution networks.

The determination of acceptable levels of supply reliability is presently achieved by comparing actual interruption frequency and duration indices with arbitrary targets. Inherent in this approach is the perception of customer satisfaction level for supply cessation. This type of implicit reliability criteria are inadequate for rationally evaluating the validity of suggested capital investment to materialize improvements that optimize utility efficiency. It is therefore a foregone conclusion that rules of thumb and vague criteria cannot be applied in a consistent manner to the very large amount of capital investment and operating and maintenance decisions that electric utilities routinely make. The ultimate impact is a likely misallocation of resources within distribution systems.

To provide a rational and consistent means of prudent decision making on the necessity of changing supply reliability levels experienced by utility customers in a given service area, quantifiable factors other than utility revenue losses are required to be modeled. In particular, explicit modeling of customer damage costs in establishing supply reliability criteria has to be incorporated in the regular planning practices.

A value-based reliability planning methodology attempts to ascertain the minimum cost solution, where costs are identified as the sum of investment cost plus operating and maintenance cost plus customer outage costs. This minimum cost point is normally defined by the marginality condition where the marginal cost of reliability enhancement is equal to the marginal benefit, which is the expected reduction in customer damage costs due to the marginal investment. This chapter presents a brief overview of current utility practices in distribution planning. In particular, deterministic planning criteria used in alternative feed requirements planning and designing for overhead networks are detailed. A probabilistic, customer value-added, alternative feeder planning approach that could complement the current deterministic criteria is illustrated with a numerical example system.

## 11.2 CURRENT DISTRIBUTION PLANNING AND DESIGN CRITERIA

The primary criteria for distribution system planning and design used by most utilities are loading and voltage. It is important to note that utilities have traditionally considered reliability in some qualitative fashion in their system planning and design. Losses and environment also are considered by some utilities as a secondary criterion. The planning criteria were established by utilities, based on a host of considerations including industry practices, electrical and construction codes, manufacturers' equipment ratings, and rules of thumb developed from long-term operating experience. It is worth noting that at present, although limited, some utilities have started using customer value-based distribution reliability methodologies in regular planning activities for project justification.

### 11.2.1 Outage Data Collection and Reporting

Almost all North American utilities have some form of a computerized database and interruption reporting system to log key data elements on component and feeder failures and customer outages. Customer outages are identified by time of occurrence, duration, weather condition, and causes that can be generally categorized as planned, equipment failure, trees, foreign objects, human error, lightning, supply, and so on. Operating experience records from the field are normally entered into a database and a number of reports are generated.

### 11.2.2 Reliability Indices

The most commonly used indices of distribution system reliability are one or more of the following system and customer indices:

$$\begin{aligned}\text{SAIFI}(&\text{System Average Interruption Frequency Index})\\ &= (\text{total number of customer interruptions})\\ &\quad /(\text{total number of customers served})\end{aligned}$$

SAIFI is the average number of interruptions per customer served.

$$\begin{aligned}\text{SAIDI}(&\text{System Average Interruption Duration Index})\\ &= (\text{sum of customer interruption durations})\\ &\quad /(\text{total number of customers served})\end{aligned}$$

SAIDI is the average duration of a customer outage per customer served.

$$\begin{aligned}\text{CAIDI}(&\text{Customer Average Interruption Duration Index})\\ &= (\text{sum of customer interruption durations})\\ &\quad /(\text{total number of interrupted customers})\end{aligned}$$

CAIDI measures the average duration of a customer outage within the class of customers that experienced at least one sustained interruption.

$$\begin{aligned}\text{CAIFI}(&\text{Customer Average Interruption Frequency Index})\\ &= (\text{total number of customer interruptions})\\ &\quad /(\text{total number of interrupted customers})\end{aligned}$$

CAIFI defines the conditional average number of interruptions among the class of customers who experience at least one interruption.

ASAI (Average System Availability Index) is the ratio of total customer hours that service was available divided by the total customer hours in the period for which the index is calculated. On an annual basis, it can be shown that

$$\text{ASAI} = \left[1 - \frac{\text{SAIDI}}{8760\ \text{h per year}}\right] \times 100$$

To illustrate, an ASAI of 99.9543 indicates that the average customer had service available for 8756.0 h out of a total 8760 h in the year, that is, SAIDI = 4 h/year. These indices are routinely reported on an annual basis and are typically computed for each feeder. Aggregate reports present comparable statistics on a district or regional basis and on a system-wide basis. These indices can be readily computed down to the distribution transformer level.

### 11.2.3 Targets for Customer Service Reliability

The historical reliability indices are used to establish guidelines by many utilities in planning and design. The reliability of service is measured by SAIFI, CAIDI, SAIDI, and ASAI in most cases. Utilities use different target levels for their overall system, district, and regional service reliability performance. It is, however, worth noting that none of the utilities has minimum standards for customer service reliability. Most utilities prefer to have objectives or targets they strive toward. The basic distinction is that a standard is perceived to be a "hard constraint," whereas "guidelines," "objectives," and "targets" refer to performance levels that are considered to be desirable and that a utility strives toward. Typical numerical values of the reliability indices used as desired targets by different utilities are summarized in Table 11.1.

As shown in Table 11.1, some utilities formally use quantitative targets for SAIFI, SAIDI, CAIDI, or ASAI. The rationale for the target values is that these figures are consistent with actual performance during a certain historical period when service reliability to customers was considered to be adequate by the utility and the associated level of customer complaints was viewed not excessive. Two examples of mandated distribution reliability standards by regulatory bodies in a deregulated environment are presented in Section 11.2.4.

### 11.2.4 Examples of Distribution Reliability Standards in a Deregulated Market

Customers are increasingly looking for improved service reliability from electricity suppliers. This has been recognized by regulators who incorporate customer reliability measures such as "how often" and "how long" the supply to the customers may be interrupted in a year. For example, as part of the deregulation of electric utilities in California, the state established a new rate mechanism in 1996, called "performance-based rate making" (PBR). To guarantee reliable service to the customers, PBR includes performance incentives with a specific system of rewards or penalties for each utility. With respect to the Southern California Edison Company, the PBR rewards or penalties are based on the 2-year rolling averages of the annual customer minutes of interruptions (CMI or SAIDI) and the annual number of distribution circuit interruptions (DCI). Catastrophic events are excluded in both CMI and DCI computations, but events of

TABLE 11.1. Typical Distribution Reliability Targets Used by Utilities

| Different Desired Target Types | | SAIDI | SAIFI |
|---|---|---|---|
| 1 | Urban | 2 h | 1.5 outages/year |
| | Rural | 4 h | 3.0 outages/year |
| 2 | SAIDI: Varies between 4.4 and 8 h/year depending on region, with a system-wide average of about 5.5 h/year | | |
| 3 | SAIDI: Less than or equal to 3.75 h/year, which is the rural distribution average | | |
| 4 | SAIFI: Less than or equal to two outages per year and SAIDI less than or equal to 4 h/year | | |
| 5 | Strive to ensure that individual feeder reliability moves toward company averages. These are approximately as follows depending on region: SAIFI 2.5–3.5 outages/year and CAIDI 1.5–4 h/year | | |
| 6 | Implicit goals: (1) improve reliability on rural radial systems; (2) improve reliability if there are too many customer complaints | | |
| 7 | Identify poorly performing feeders for mitigating actions based on performance indicators such as (SAIFI × SAIDI) and customer density per kilometer on feeder. For example, if the product (SAIFI × SAIDI) for a district or region is greater than four times the average value, then take mitigating action | | |
| 8 | ASAI greater than or equal to 0.9994, that is, SAIDI is less than or equal to 5.3 h/year | | |
| 9 | As mentioned earlier, some utilities use different target levels for the overall system and for the districts or regions. In addition to the total company targets, each district and service point has target indices. The typical district and the overall company targets are as follows: | | |
| | District | SAIFI | CAIDI |
| | A | 2.5 | 1.2 |
| | B | 2.8 | 1.5 |
| | C | 2.2 | 1.5 |
| | D | 2.9 | 1.3 |
| | E | 4.4 | 1.5 |
| | F | 2.5 | 1.9 |
| | G | 3.1 | 1.0 |
| | Total company | 2.8 | 1.4 |

storms are included. The PBR guidelines include three ranges, upper, middle, and lower for the CMI and DCI. The CMI and DCI figures for Southern California Edison are as follows:

| Customer Minutes of Interruptions | |
|---|---|
| Upper range: penalty | \$1 million per CMI above 65 up to \$18 millions at 83 and above |
| Dead band: no penalties or rewards | From 53 to 65 |
| Lower range: rewards | \$1 million per CMI below 53 up to \$18 millions at 35 and below |

Number of Distribution Circuit Interruptions

| | |
|---|---|
| Upper range: penalty | $1 million per 183 interruptions above 12,400 up to $18 millions at 15,700 and above |
| Dead band: no penalties or rewards | From 10,200 to 12,400 |
| Lower range: rewards | $1 million per 183 interruptions below 10,200 up to $18 millions at 6900 and below |

These limits became effective in 1997 and are being reduced by two CMI for the next 5 years.

In the Australian State of Victoria, the regulatory bodies have written the following performance requirements into the Distribution Code:

1. A local distribution company must use reasonable endeavors to ensure that the duration of interruption of the supply of electricity to a customer's installation does not exceed on average 500 min (8.3 h) per annum in rural areas and 250 min (4.15 h) per annum in other areas.
2. On request a local distribution company must make individual customer targets and actual performance information available to the customer.

It is important to note that these deregulated criteria as well as conventional targets presented in Table 11.1, in no instance, are linked to or derived from estimates of the value of service reliability of customers. The value of service information has not been used explicitly to rank order the reliability performance of different feeders and identify poor performance for upgrading on the basis of total interruption costs incurred by customers on the feeder.

## 11.3 RELIABILITY COST VERSUS RELIABILITY BENEFIT TRADE-OFFS IN DISTRIBUTION SYSTEM PLANNING

Utilities routinely make many reliability-related investment and operating decisions. This section presents typical cost–benefit trade-off situations in distribution system reliability planning, where using customer interruption cost data can enhance decision making to a more rational, consistent, and economic framework. The distribution system consists of a very large number of individual components (e.g., subtransmission circuits, distribution stations, primary feeders, distribution transformers, secondary circuits, and customers' connections). As a result, a large number of standards such as design, construction, operation, maintenance, and so on have been developed over the years that entirely define supply reliability provided to utility customers. It is important to note that the current different distribution planning practices can be expanded to a value-based planning framework. A host of distribution planning and operating practices can benefit from a value-based planning approach. For example, there are many distribution system reliability problems, where customer interruption cost information could be explicitly used to support rational investment decision making, such as facilities design and configuration, feeder upgrade, distribution system upgrading, tree trimming, pole maintenance, protection design, service restoration, underground cables design, cable replacement, distribution system automation, general system-wide utilization, and so on.

As a result of load growth, completely new distribution facilities are required in certain instances. This may be the case in surrounding areas of large cities. In these situations, decisions have to be made regarding substation configuration, size and number of transformers, capacity planning, selecting a route for a new line, and so on. Regarding route selection, if two routes are available for an overhead line extension, and one route is more costly but has less exposure to traffic, a decision has to be made whether the more expensive route is justified from a customers' benefit point of view.

Utilities normally identify poorly performing feeders as potential candidates for feeder upgrading using one or more implicit criteria. These may include the number of complaints received or other deterministic decision rules based on deviations from calculated SAIFI, SAIDI, and so on. Normally, the number of feeders identified for upgrading and the investment made for this purpose is further constrained by capital, operation and maintenance budget, and manpower resources.

Feeder upgrade decisions made in the context of poorly performing feeders could be made on a more rational and consistent manner even within severe budget-constrained situations by ranking the feeders on the framework of the total costs of improvements relative to the benefits. The benefit would be assessed by quantifying expected reductions in interruption costs to all customers as a consequence of the upgrade being considered. Using the cost–benefit strategy, the available resources could be more objectively and consistently used across the feeders since such a value-based method facilitates establishing an economically rational prioritization of the feeders requiring upgrades. It is important to note that in value-based approach, there are no absolute standards of SAIFI, SAIDI, and CAIDI that could be used uniformly across the feeders. In a value-based planning framework, rather customers' preferences and mix unique to each feeder and/or geographical service area determine an inadequate or poor performance. In a competitive electricity market, a value-based approach, therefore, makes perfect sense.

With load growth and customer mix change, the distribution system requires upgrading. One aspect of such changes is the current trend toward a higher distribution voltage level, where possible. Conductor upgrading, installation of line regulators to boost voltages when required, transformer upgrading to accommodate load increases, replacement of underloaded transformers with smaller transformers, and considering substation loading criteria (e.g., high loadings can result in extremely poor reliability) are a few of the typical decisions related to system upgrading.

Vegetation control normally affects the service reliability level of a distribution system. The amount of work done on each occasion and the frequency of maintenance are the two major decisions that have to be made in regard to tree trimming. Reliability--centered tree trimming could result in optimum tree trimming schedules that balance the marginal cost and the marginal benefit from the customers' perspective.

Normally, most lines in a distribution system are overhead wires on wooden poles. The poles require continuous maintenance. The level of maintenance for poles and their replacement programs affect the level of distribution system reliability.

An increase in protection, such as the number of devices, the type of devices, and the location of devices, will inevitably result in changes to the reliability of the distribution system supply. Increases in protection spending will protect the distribution system from unreliability resulting from sabotage, lightning, and equipment failure by limiting the area impacted by these events and by limiting their frequency and duration. In turn, this will result in lower total outage costs to customers. Examples of investment decisions related to protection include fuse coordination, installation of primary fuses, installation of switches to facilitate sectionalizing, dual feed substations versus single feed, and so on.

From the customers' point of view, supply reliability means how quickly voltage is restored following an interruption. It could be different for different customer types. Strategies related to sizing, location, and scheduling of service restoration crew as well as service restoration sequencing by feeder can have an important impact on the outage costs incurred by customers.

Most distribution system lines are overhead lines requiring regular tree trimming and pole maintenance. The reliability trade-offs in the overhead versus underground installation involve a lower frequency of customer interruptions in the underground case. Customer outages due to underground equipment failures, however, can be longer in duration due to the time required to locate and rectify the problem.

Virtually all utilities use rules of thumb in their cable replacement strategies. Typical considerations for cable replacement are replacements after a certain number of failures, number of customers and cost to replace, moisture content, number of failures, number of customers served, length, age, commercial or industrial customers, and so on.

Electricity could be more cost-effectively and reliably delivered to customers through distribution feeders by automation of the feeders. Generally, most feeder interrupting devices are static and require line crew to open and close switches manually to locate and isolate faults and restore voltage to customers. In addition, as the distribution network becomes more and more complex due to expansion, conventional feeder maintenance practice becomes inadequate and customers have to endure longer outages. Because of the time required in locating faults manually, revenue losses to the utility could also be significantly high. Extended duration outages have serious monetary impacts from the customer's perspective as well. Distribution automation can provide a cost-effective means for reducing the frequency and duration of potential service interruptions.

A number of typical decisions where customer interruption costs and value-based methodology could be used are as follows:

1. For assessing the appropriate level of reliability for feeders in each district/region. This should be based on the expectations of the customer mix on the feeder/region and consistent with their willingness to pay.
2. For assessing the level of distribution reliability that customers are willing to pay for. This means, analyze the distribution system as a whole and compute the total dollars required to raise or lower the reliability of the distribution system to the level customers are willing to pay for.
3. For determining the amount of corporate reliability funds to be spent on distribution versus other segments of the power system.

Under current planning and design practices, investment decisions in the context of any of the number of reliability-related problems identified earlier are based on implicit rules-of-thumb criteria normally employed by utilities. Explicit cost–benefit analysis is not always undertaken. Although, for the purpose of ranking problem feeders, avoided lost revenues to the utility provide a measure of the benefits of upgrading, present planning and operating practices, as they relate to distribution service reliability, are not based on a direct and objectively specified linkage between the level of reliability that is planned and delivered and the level of reliability that customers want. This section presents a typical utility's current practices in the planning of alternative feed requirements for overhead distribution systems. An illustrative customer cost–customer benefit assessment for justification of an alternative feed to a major load center is performed to demonstrate the underlying concepts of reliability cost–reliability worth analysis that could complement the current deterministic planning practices in distribution investment decision making.

## 11.4 ALTERNATIVE FEED REQUIREMENTS FOR OVERHEAD DISTRIBUTION SYSTEMS

The standard service provided to all overhead customers is the radial feed. Full or partial alternative service is normally considered when the additional cost is paid by the customer. A rudimentary evaluation of the advantages to the customers and the utility is looked at and a comparison of the costs of providing the alternative feed is made. Evaluation of costs normally includes the additional costs of the transmission, the substation, and the breakers required. A partial alternative supply having an 80% capacity on peak is deemed as fulfilling the customer demands. Utility personnel look at acceptable restoration times for different types of customers under various outage conditions. Situations where outage times exceed the acceptable restoration times are examined and a contingency plan developed to deal with the specific situation. For communities with a certain load demand (e.g., 3000 kVA and higher), this could mean an alternative feed. Customers fed from underground residential subdivisions, by nature of the design and development of the system, are usually provided with an alternative feed.

The existing criteria for determining the provision of alternative feed are primarily based on a deterministic approach. As stated earlier, deterministic approaches are not based on a formal framework but rather on the planner's experience and intuition that do not and cannot account for the probabilistic nature of the distribution system behavior, of customer demands, or of system component failures. Since the basic objective of distribution planning criteria is to provide a consistent approach to obtaining a balance between the distribution system performance and the total cost to satisfy both customer and a utility need, alternative feed requirements identified solely based on deterministic criteria have to be verified through probabilistic value-based analyses that include customer interruption costs.

This involves the recognition of reliability cost–reliability benefit. A cost–benefit analysis to determine when an alternative distribution feed should be planned is needed to complement the deterministic guidelines for distribution planning. The alternative source would likely be provided if the analysis indicates that the improvement in supply reliability would be cost-effective. Ideally, outage probability and outage cost data specific to a utility distribution system are required for cost–benefit analysis. In this chapter, an example of using customer interruption cost data in evaluating the economics of an alternative distribution feed is described. It is important to note that this analysis includes only forced outages. Maintenance outages are not modeled in the analysis because the impacts of maintenance outages can be controlled and minimized by communicating in advance maintenance schedules to customers, scheduling maintenance outages during light load periods, performing hotline maintenance, bringing in mobile substations, scheduling maintenance during customer downtimes, and so on.

## 11.5 EXAMPLES OF DETERMINISTIC PLANNING GUIDELINES FOR ALTERNATIVE FEED REQUIREMENTS

### 11.5.1 Reliability of Supply to 25 kV Buses

***11.5.1.1 Peak Loads up to 10 MW.*** The distribution supply system is planned with backup capacity to a minimum of 30% of coincident peak demand by means of a standby transformer or through the ability to import standby capacity from adjacent substations or by a combination of both. It is anticipated that a failed transformer should normally be replaced within 12 h. This guideline is based on the assumption that 20% of the feeder peak demand would cover all essential services for different customer types. A typical 25 kV distribution network is depicted in Fig. 11.1. According to this guideline, in the event of a fault on 25 kV line "A" or upstream from it, while operating at its peak demand, the 25 kV line "B" would be capable of providing supply to 20% of the peak demand of 25 kV line "A". During a contingency, voltage will be allowed to drop the extreme values.

Whenever possible the distribution system is arranged to accommodate rotational supply to those customers who are interrupted.

***11.5.1.2 Peak Loads 10–15 MW.*** As in Section 11.5.1.1 but up to 50% of coincident peak demand.

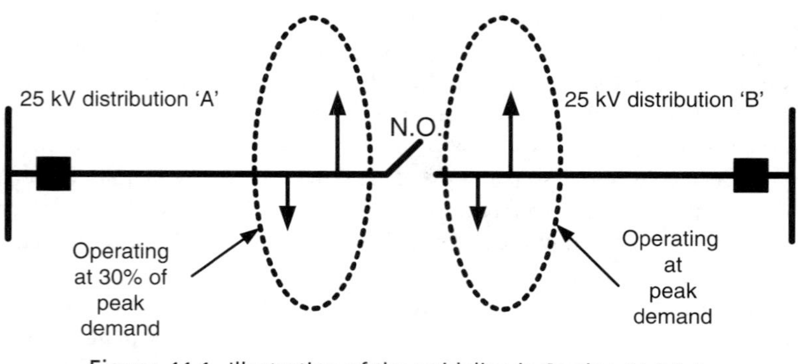


Figure 11.1. Illustration of the guideline in Section 11.5.1.1.

***11.5.1.3 Peak Loads 15–25 MW.*** As in Section 11.5.1.1 but up to 75% of coincident peak demand.

***11.5.1.4 Peak Loads in Excess of 25 MW.*** As in Section 11.5.1.1 but up to 90% of coincident peak demand.

### 11.5.2 Reliability of Supply to Towns/Cities

The level of backup supply to a town or a city suggested in the following reflects the size of the town or the city. The underlying assumption is that as a population center gets larger, a greater number of customers get impacted due to cessation of their power supply, and the cost per customer to alternative supply generally is lower.

***11.5.2.1 Towns with a Population of up to 1000.*** No backup supply needs to be planned. The distribution system should be arranged to accommodate rotational supply to interrupted customers.

***11.5.2.2 Towns with a Population Between 1000 and 5000.*** Twenty percent of the town load should have a backup supply.

***11.5.2.3 Towns with a Population Between 5000 and 10,000.*** Forty percent of the town load should have a backup supply.

***11.5.2.4 Cities with a Population Between 10,000 and 25,000.*** Sixty percent of the city load should have a backup supply.

***11.5.2.5 Cities with a Population over 25,000.*** Eighty percent of the city load should have a backup supply.

### 11.5.3 Reliability of Supply to Large Users and Industrial Customers

The normal supply to large users and industrial customers is the radial feed. An alternative feed may be provided to these customer types at full cost to the customer.

## 11.6 VALUE-BASED ALTERNATIVE FEEDER REQUIREMENTS PLANNING

Deterministic criteria given in Sections 11.2, 11.4 and 11.5 alone are insufficient for rationally assessing the validity of suggested alternative feed requirements. There is increasing recognition in the electric utility industry that investments related to the provision of electric service reliability should be more explicitly evaluated considering their cost and benefit implications. The underlying intention is to relate the benefit of uninterrupted power supply as a means to rationalize the cost of alternative feed additions. Such a cost–benefit assessment is the focal point of the probabilistic value-based

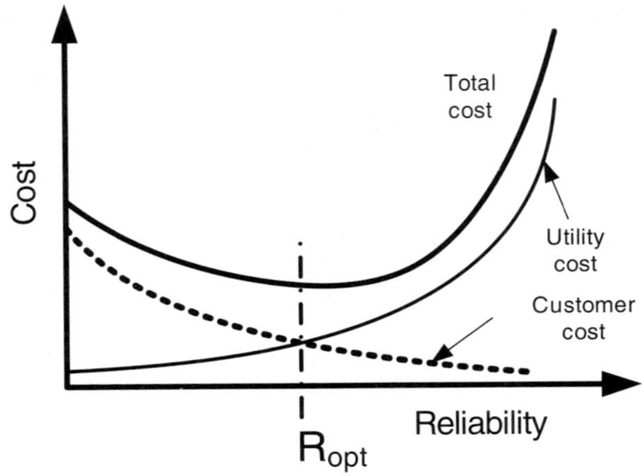


Figure 11.2. Reliability cost–reliability worth concept.

reliability planning. As mentioned earlier, a value-based reliability planning approach attempts to locate the minimum cost solution where the total cost includes the utility investment cost plus the operating and maintenance cost plus the customer interruption costs. The underlying principle of value-based reliability planning is illustrated in Fig. 11.2.

Figure 11.2 illustrates how utility costs reflected in customer rates and customer interruption costs are combined to give total customer cost. The utility cost curve shows how customer rates go up as more money is spent to increase distribution system reliability levels. The customer interruption cost curve shows how customer cost of interruptions decreases as the distribution system reliability increases. It is also important to note that for low levels of distribution system reliability levels, the customer interruption costs are significant. However, the utility cost can also increase significantly via the additional costs of restoring the system to a normal operating state and the loss of revenue (i.e., the utility cost curve shown in Fig. 11.2 is based on the belief that increased costs will achieve higher levels of distribution system reliability). When the combined utility and customer interruption costs are minimized, the utility customers will receive the least cost service. Therefore, using the concept of value-based distribution system reliability planning, a given level of service reliability can be examined in terms of the costs and the worth to the customer of providing the electric service from various proposed distribution operating configurations.

The basic assumption in the value-based assessment is that planning practices, as they relate to distribution service reliability, should be based on a direct and objectively specified relationship between the level of reliability that is planned and delivered and the level of reliability that customers expect. Any investment decision related to distribution reliability, however, could easily be addressed on a more rational and quantitative basis within a cost-benefit framework, explicitly using data on the value of changes in service reliability to customers. The ultimate determination of poor or adequate distribution reliability performance in value-based planning framework is based on customer preferences and customer mix that are unique to each feeder. In a competitive market environment, such an approach makes perfect sense.

### 11.6.1 Customer Interruption Cost Data

The value of service, that is, the worth of reliability expressed in terms of customer interruption costs can be established on the basis of actual surveys of customer perception regarding the level of service reliability they are willing to pay for. By establishing a method of giving a dollar value to various levels of service reliability, it is possible to ascertain the balance where distribution system reliability is best matched. The data compiled from customer surveys lead to the creation of sector damage functions. The cost of interruptions at a single customer load point depends entirely on the cost characteristics of that customer. The sector damage function presents the sector interruption costs as a function of the duration of service interruptions. The customer costs associated with an interruption at any load point in the distribution system involve the combination of costs associated with all customer types affected by the distribution system outage. This combination leads to the generation of a composite customer damage function. The customer damage functions after escalation to 2000 Canadian dollars for each sector are shown in Fig. 11.3. The cost of interruptions is expressed in dollars per kilowatt.

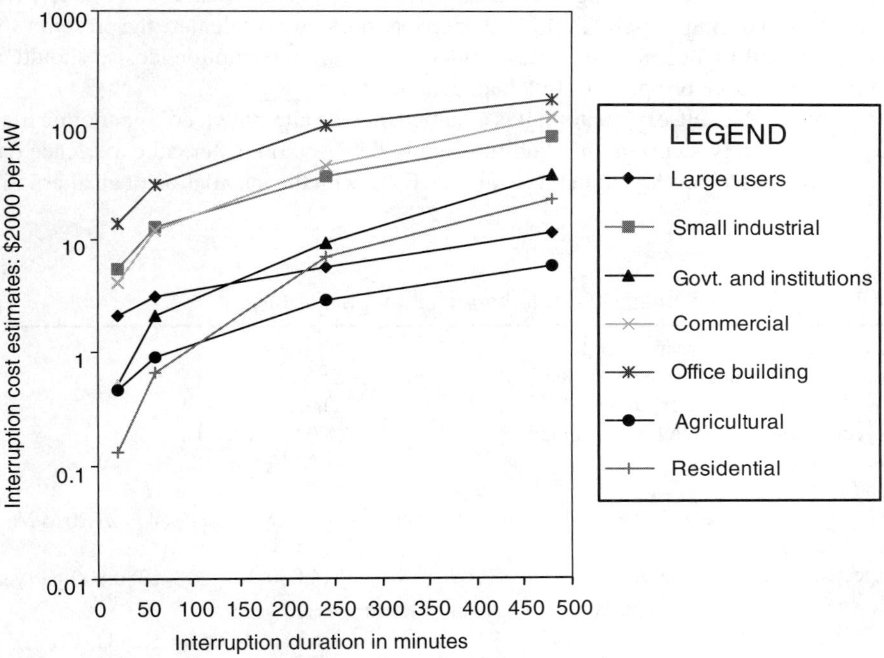


Figure 11.3. Utility sector customer damage functions.

In this chapter, one simple illustrative example showing the practical applications of the reliability cost–reliability benefit methodology is described. A reliability cost–reliability benefit assessment to determine when an alternative feed to a major city should be added should be performed to complement the deterministic distribution reliability planning criteria. The alternative feed should be provided if the analysis indicates that the improvement in service reliability would be cost-effective. In practice, outage probability and interruption cost data specific to the distribution network of study, or the best representative data available at the time of analysis, should be used for the reliability cost–reliability worth analysis. An actual application should consider the amount of load growth and should show numbers on an escalated and present worth basis for years to come.

### 11.6.2 An Illustrative Example for Justification of an Alternate Feed to a Major City

A numerical example illustrating the application of the reliability cost–reliability benefit approach in evaluating the cost-effectiveness of an alternative feed to a major city is discussed here. A very simplified representation of the investment decision is adopted in this example. The example uses generic outage and cost information data. The basic objective of this example is to focus on the framework for cost–benefit analysis in similar situations.

For this example, assume that there is a major city with a population of over 25,000 people. Deterministic planning criteria from Section 11.4 state that 80% of this city load should have a backup supply. Table 11.2 describes the steps to calculate the present value of the reliability benefits, that is, avoided customer interruption costs, should an alternative feed be provided to this load center.

Table 11.2 is self-explanatory. It is assumed that the alternate feed's economic life is 10 years. The city's current unreliability should be based on historical experience data that is assumed to be 5 h/year in this example. Even with the installation of an alternative

TABLE 11.2. An Alternative Feed to be Added to a Major City

| | |
|---|---|
| Economic life of alternative feed | 10 years |
| Current annual supply unavailability | 5 h/year |
| Switching time for alternative feed | 1 h |
| Expected annual outage hours avoided | $(5.0 - 1.0) = 4$ h |
| Unserved load | 10 MW |
| Load factor | 85% |
| Expected unserved energy avoided | $10 \times 0.85 \times 0.80 \times 4 = 27.20$ MWh |
| Customer interruption cost | \$14/kWh |
| Expected reliability benefit | \$14,000 × 27.20 = \$380,800 per year |
| Expected cumulative present value of reliability benefit over 10 years | |
| Load growth = 0% | \$380,800 × 7.57 = \$2,882,656 |
| Load growth = 2% | \$380,800 × 8.20 = \$3,122,560 |

feed, it will take 1 h to switch the power supply from the alternative source. Therefore, the expected number of outage hours avoided is (5.0 − 1.0) = 4 h/year.

If the unserved load is 10 MW with a load factor of 85%, then the avoided expected unserved energy is 27.20 MWh considering only 80% of the city load would have a backup supply. Assuming that the customer interruption cost for the mix of customers served by the city to be $14/kWh yields a present value of the reliability benefit of $2,882,656 at 0% load growth and $3,122,560 at a 2% load growth. The present value figures were calculated assuming a 10% discount rate and a 3% inflation rate. These benefits, together with loss reduction benefits if any, can be compared to the present value revenue requirements to install the alternative feed in question.

## 11.7 CONCLUSIONS

Current utility practices in the distribution system reliability planning are presented in this chapter. It is important to note that neither the traditional planning criteria nor the current criteria mandated by regulators in a deregulated market are linked to customer preferences.

Typical utility distribution planning practices in determining alternative feed requirements for overhead distribution systems are discussed. Deterministic planning practices are complemented with a probabilistic value-based planning analysis in the assessment of cost-effectiveness of an alternative feed to a major load center.

In a competitive energy market in which reliability of service does influence customer purchasing decisions, a utility cannot afford to ignore customer preferences. Today's energy market is characterized by intense price competition that puts utilities under continuous pressure to hold the line on rate increases. Value-based planning renders a rational solution to these emerging pressures and will permit service reliability to evolve to a level that customers would perceive to be fair value.

## REFERENCES


1. M. Munasinghe and M. Gellerson, Economic criteria for optimizing power system reliability levels, *Bell Journal of Economics*, Spring 1979, 353–365.
2. Electric Power Research Institute, Cost–Benefit Analysis of Power System Reliability: Determi-nation of Interruption Costs. Vol. 1: Measurement Methods and Potential Applications in Reliability Cost–Benefit Analysis, EPRI IL-6791, Palo Alto, CA, April 1990.
3. M. J. Sullivan, B. N. Suddeth, T. Vardell, and A. Vojdani, Interruption costs, customer satisfaction and expectations for service reliability, *IEEE Transactions on Power Systems*, **11** (2), 1996, 989–995.
4. Canadian Electrical Association, Economic Penalties Including Customer Costs for Loss of Service Continuity, Research and Development Report No. SD-273, July 1991.
5. R. Billinton and J. E. Billinton, Distribution system reliability indices, *IEEE Transactions of Power Delivery*, **4** (1), 1989, 561–586.

6. Canadian Electrical Association, 1993 Annual Service Continuity Report on Distribution System Performance in Canadian Electric Utilities, May 1992.
7. P. Save and T. Velarde, Incorporation of performance based rating (PBR) incentives in probabilistic transmission planning criteria, in Proceedings of the 5th International Conference on the Probabilistic Methods Applied to Power Systems (PMAPS '97), Vancouver, BC, Canada, September 21–25, 1997, pp. 17–24.
8. Victorian Distribution Code, Section 19, Interruption to Supply, 19.1(a) and (b), Victoria, Australia.
9. P. Lyons and G. Nourbakhs, Applied subtransmission and distribution reliability assessment of SEQEB network in Queensland, Proceedings of 1997 AUPEC Conference, Sydney, Australia, 1997.
10. L. Goel and R. Billinton, A procedure for evaluating interrupted energy assessment rates in an overall electric power system, *IEEE Transactions on Power Systems*, **8** (3), 1993, 929–936.
11. G. Wacker, R. K. Subramaniam, and R. Billinton, Using cost of electric service interruption surveys in the determination of a composite customer damage function, in Presented at the IASTED Energy, Power and Environmental Systems Conference, San Francisco, CA, June 4–6, 1984.
12. G. Tollefson, G. Wacker, R. Billinton, E. Chan and J. Awaya, NSERC customer cost of interruption study preliminary results, *CEA Transactions*, **31**, 1992.
13. R. Billinton, G. Wacker, G. Tollefson, J. Awaya, and E. Chan, Customer requirements for electric service reliability, *CEA Transactions*, **31**, 1992.

# 12

# IMPORTANT FACTORS RELATED TO DISTRIBUTION STANDARDS

## 12.1 INTRODUCTION

The development of standard distribution reliability metric values, for example, System Average Interruption Frequency Index (SAIFI), System Average Interruption Duration Index (SAIDI), and Customer Average Interruption Duration Index (CAIDI), against which all utilities can compare their performances, can be problematic without strict adherence to a national or international standard (e.g., IEEE Standard 1366-2003). At present, there are many differences between data collection processes and characteristics of utility systems making comparisons between utilities against such standard metric values impossible for many utilities. Rather, the development of uniform standard metric values, which utilities compare to their own historical reliability performance indices, is more practical. If cross-comparisons between utilities are desirable, a number of issues and factors associated with individual utilities must be taken into consideration when establishing distribution reliability standards. This chapter deals with a number of such pertinent factors and issues related to establishing distribution reliability standards and illustrates the issues and factors of using historical reliability performance data for a sample of Canadian utilities.

*Power Distribution System Reliability.* By Ali A. Chowdhury and Don O. Koval

Since the late 1980s, electricity supply has been deregulated in many countries in an attempt to create a competitive market for power generation and transmission services with the expectation that deregulation would ultimately benefit the consumers in the lower electricity rates. In the competitive market process, the distribution system remains regulated to ensure that the customer supply system is operated reliably and cost-effectively. This is due to the regulatory position in different countries that an introduction of competition in the generation and transmission segments of an integrated power system could have a negative impact on system reliability and service received by ultimate customers.

Customers are connected to a regulated distribution system that determines system reliability experienced by the customers. As customers of regulated systems, customers cannot switch distribution systems at their will if reliability becomes unacceptable. For this reason, regulatory agencies are looking for ways to define and establish distribution reliability standards. Normally, these standards include reliability performance indices at the corporate levels, region levels, and crew center (CC) levels and a list of worst performing feeders. Some state regulatory agencies financially penalize and/or reward utilities based on preset reliability standards. A map of U.S. state reporting requirements as of 2001 is shown in Fig. 12.1. As shown in Fig. 12.1, 11 states reward or penalize utilities based on reliability performance. Sixteen states require annual reliability reporting and five are considering some form of reporting requirements.

Regulators should continue to ensure customer service reliability performance, at a reasonable cost. Setting standards without a complete understanding of performance drivers can have significant ramifications, including increased costs to customers and/or utilities shareholders. The basic objective for establishing reliability standards should be to provide electric utility customers with some assurance that reliability levels would not

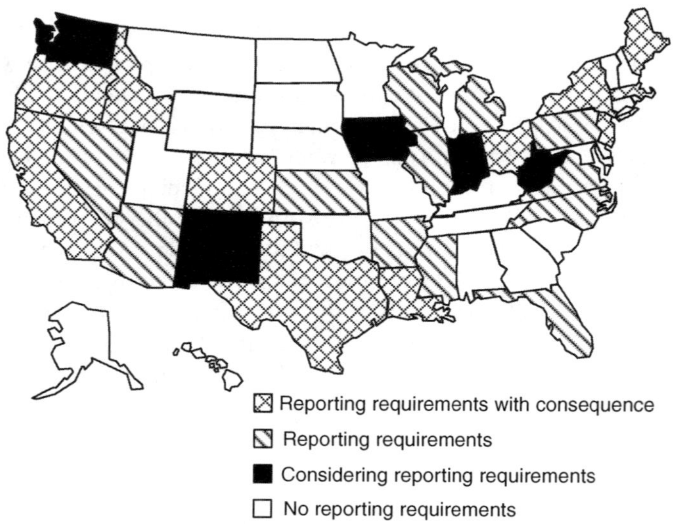


Figure 12.1. Reliability reporting requirements in the United States as of 2001.

drop below a predetermined level. The most widely used reliability indices are averages that treat every customer equally. Customer-oriented reliability indices such as SAIFI, SAIDI, and CAIDI are widely used by utilities and regulatory agencies. Most regulatory agencies choose to use SAIFI in combination with either SAIDI or CAIDI, since using two of the three indices provides meaningful information about all three measures. The index that reflects multiple interruptions experienced by customers is Customers Experiencing Multiple Interruptions (CEMI) [2]. The CEMI index indicates the ratio of individual customers experiencing more than $n$ sustained interruptions to the total number of customers served. This index is sometimes used in a series of computations with $n$ incremented from a value of 1 to the highest value of interest. SAIFI, SAIDI, and CAIDI indices have limitations. These indices are normally considered good aggregate measures of reliability performance when used as reliability performance indicators for system reliability improvements.

Benchmarking or comparing distribution reliability indices across utilities is difficult due to a host of reasons, including system design and operating characteristics, maintenance policies, definition of terms, geography, and major event exclusion or inclusion. Edison Electric Institute (EEI) and the Canadian Electricity Association (CEA) report distribution reliability indices for North American utilities annually. It is inevitable that comparisons between individual utility reliability performances will be made.

Virtually, all utilities report SAIFI and SAIDI indices. Each index is shown to have wide variation among utilities due to differences in system design, operations, and maintenance practices. Such differences do not account for all the variations among utilities and therefore the indices cannot indicate that utilities with better reliability indices have higher reliability of service.

From a utility perspective, there are a host of important issues and factors that should be addressed when developing distribution reliability standards for application across utilities operating under the jurisdiction of a state/provincial regulatory agency. In general, standards should be uniformly applied so that if one entity is subject to a particular set of standards, all entities will be subject to those standards regardless of the size of the utility and customer and physical characteristics of systems, such as urban, rural, or mixed systems. By applying standards in this way, all customers, whether served by a small municipal utility or by a large investor-owned utility, would have a reasonable assurance of comparable levels of service reliability. It is therefore required that all entities use the same terms and definitions (e.g., IEEE Standard 1366-2003) in reliability performance reporting procedures.

There is an ongoing debate in electric utilities as to whether or not the reliability performance indicators are to be defined as performance "standards," performance "guidelines," or "target values." The general perception among the utility personnel is that rather than standards, it is desirable that minimum reliability performance guidelines or target levels should be used, since in the utility environment, standards are normally perceived as rigid requirements cut in stone and noncompliance with standards implies financial risks to the utilities. However, performance guidelines, or targets, are considered to be continuously evolving performance indices. Violations of performance guidelines require a rigorous review of a system's weak areas to identify areas needing reliability enhancements. Therefore, standards become risks in an evolving market while guidelines provide solution opportunities. In this chapter, the word "standards" is used for reliability performance indicators unless otherwise noted.

Distinctions in system characteristics based on rural, urban, and suburban, as well as overhead versus underground, systems need to be recognized and incorporated in establishing distribution reliability standards. Minimum performance standards should be utility specific using historical outage data from that utility. This chapter identifies different pertinent issues and factors that need to be considered in developing consistent distribution reliability standards. The relevant issues and factors that impact reliability performance indices are illustrated using historical reliability data from a sample of Canadian utilities.

## 12.2 RELEVANT ISSUES AND FACTORS IN ESTABLISHING DISTRIBUTION RELIABILITY STANDARDS

Many regulatory agencies have indicated that they would use historic utility performance to establish specified service reliability standards and would require a utility to have at least 2–3 years worth of outage data. Each utility should, at the minimum, remain within the range of its average historic performance level. In cases where a utility has not monitored service reliability in the past, the utility would be required to initiate monitoring and reporting of the service reliability performance indices to regulatory agencies.

As noted earlier, SAIFI, SAIDI, and/or CAIDI indices are widely used by regulatory agencies. SAIFI is a measure of how many sustained interruptions an average customer will experience over a predefined period of time. For a specified number of customers served, the only way to improve the SAIFI index is to reduce the number of sustained interruptions experienced by customers served.

SAIDI is a measure of how many interruption hours an average customer will experience over a predefined period of time. For a specified number of customers served, SAIDI can be improved by reducing the number of interruptions, by reducing the duration of the interruptions, or by reducing the average number of customers affected by each interruption.

CAIDI represents the average time required to restore electric service. A reduction in the SAIFI, SAIDI, or CAIDI indices indicates an improvement in system reliability.

The increasing sensitivity of customer loads to brief disturbances has created a need for indices measuring momentary interruptions. The IEEE Standard 1366 introduces two momentary indices. The Momentary Average Interruption Frequency Index (MAIFI) indicates the average frequency of momentary interruptions. The other index is the Momentary Average Interruption Event Frequency Index ($MAIFI_E$), which measures the average frequency of momentary interruption events. Events immediately preceding a lockout are not included.

At present, the MAIFI index is used in the electric utility industry to a lesser extent. The MAIFI index does not provide the information most customers seek. Also, most utilities are able to collect only circuit breaker operation counts at the substation level. In some cases, utilities do not have automatic collection schemes. In addition, momentary interruptions are more pertinent to industrial customers who normally make up less than 1% of all customers on a utility system. Measuring at the system level therefore is not useful.

$MAIFI_E$ is a better measure of customer satisfaction with distribution system reliability. This is due to the fact that multiple closely spaced momentary interruptions have much less impact on customer systems than the same number of momentary interruptions spaced days or weeks apart.

Among the issues and factors that should be standardized to establish distribution reliability standards for cross-comparison purposes among similar utilities are data pool, definitions of terms, system characteristics, and outage data collection systems. Each of these four major items is briefly discussed.

### 12.2.1 Data Pool

An examination of the contributions to the service continuity indices from various system factors provides considerable insight into how the system reliability performance can be improved. There are a number of well-defined customer outage causes that are integrated into the data pooling process. For example, the CEA reporting system, which is used by all Canadian utilities and some U.S. utility companies operating in Canada, divides the customer outages into the following broad codes: scheduled outages, loss of supply, tree contact, lightning, defective equipment, adverse weather, adverse environment, human element, foreign interference, and unknown. These cause codes are defined in Canadian Electricity Association's Service Continuity Report. Some individual utilities have an extensive list of outage causes defined in their outage management schemes including dig-ins, animals, customer caused outages, foreign utility or alternative third-party energy supplier caused outages, and many others. For the computations of the historic standard metric values and the current year standard metric values on a uniform basis with respect to outage causes included, a well-defined set of outage causes and rules for their applications (e.g., as presented in IEEE Standard 1366) must be strictly implemented during the outage data collection process by all utilities. In addition, any outage causes that will be excluded during the calculation of the standard metric values must be well defined and understood by all utilities. This is critical if any cross-comparisons between similar utilities are to be performed.

For the 2- to 3-year historic standard metric values and the current year standard metric values to be used to correctly measure a utility's ability to impact the reliability of its system, outages beyond the utility's control must be identified and included or excluded from the calculation of the metric values. It is critical that a precise, uniform definition be developed that is consistent with the outage cause codes to identify controllable and uncontrollable outages such that all utilities are uniformly including or excluding the same types of uncontrollable outages. Examples of uncontrollable outages include acts of God exceeding system design strength, such as hurricane, earthquake, or airplanes hitting transmission towers.

Also, for the 2- to 3-year historic standard metric values and the current year standard metric values to be used to correctly measure a utility's ability to impact the reliability of its system, outages due to major events that severely skew the metric calculation must be identified and excluded. This will require development of a precise, uniform definition that accurately identifies major events such that all utilities can uniformly apply the definitions to exclude major events. The IEEE Standard 1366 presents the 2.5 Beta method that defines a major event for use in standard reliability metric calculations.

At present, however, in the United States, comparisons of reliability indices between similar utilities are difficult because of incongruity in both reporting and definitions. More than 80% of U.S. utilities exclude major events and the definition of a major event varies widely. However, Canadian utilities include major events in their reporting of reliability performance indices whether the major event is within a utility's control or not.

There are differing views about the appropriateness of excluding major events from reliability index calculations. From a customer perspective, it does not matter whether interruptions occur during normal or mild/severe inclement weather. Reliability standards should be set to maximize societal welfare. Though major event outages due to natural phenomena are normally understood as being the events beyond a utility's control, the major event such as the August 14, 2003 Northeast Blackout occurring in normal weather conditions is not perceived to be a noncontrollable major event by the general public. For any cross-comparison of reliability performance between similar utilities, a precise, simple, easy-to-use method that uniformly classifies major events such that all utilities can apply a model to include or exclude major events is required. IEEE Standard 1366-2003 includes a simple model, namely the 2.5 Beta method, that defines major events adequately. At present, many utilities normalize a severe weather outage event by substituting the severe outage event with an average outage event in computing system performance indicators.

Another very important factor that needs attention in developing reliability standards is data sufficiency. Sufficient data must be collected on a system (e.g., 5- to 10-year data) such that the calculation of the historical standard metric value is statistically valid. If sufficient data are not available, comparison of current year standard metric values against historical standard metric values will not be valid. As a part of the deregulation of electric utilities in California, the State of California in 1996 established a new rate mechanism, called "performance-based rate making (PBR)." To ensure reliable service to customers, PBR includes performance incentives, with a specific system of rewards or penalties for each utility. For example, for Southern California Edison (SCE), the PBR rewards or penalties are based on the 2-year rolling averages of the annual SAIDI value as defined in the IEEE Standard 1366. There are instances of using 1-year, 5-year, and 10-year rolling average SAIFI and SAIDI indices in different jurisdictions. For example, Ontario Energy Board in its first-generation PBR uses 3-year rolling average values for reliability indices. IEEE Standard 1366-2003 suggests the 5-year data that are statistically significant. Therefore, a definite need exists for a uniform data sufficiency period to be set by regulatory agencies in establishing reliability standards. The authors suggest at least 5-year historical data, which would be statistically significant and sufficient in computing historical performance indices.

### 12.2.2 Definitions of Terms

Nonconformity of reliability index definitions can make it difficult to compare reliability performance between utilities. A good example is the definition of the time frame for a sustained interruption. If a utility defines a sustained interruption based on 5 min intervals, automatic switching will be effective in reducing SAIFI as most switching could be performed within this time frame. However, if a utility defines a sustained interruption based on a 1 min interval, automation cannot reduce SAIFI, as most automated switching would require more than 1 min to be successful. IEEE Standard 1366-2003 defines a sustained interruption as an outage of 5 min or longer duration, whereas some individual utilities in the United States and all Canadian utilities define a sustained interruption as being an outage of 1 min or longer duration. A uniformly applied definition of "sustained outage" must be developed so that the calculation of the long-term historic standard metric values and the current year metric values is done on a uniform basis by all utilities with respect to exclusion of outages that do not meet the definition of sustained outage. These two different definitions of sustained interruption result in significantly different SAIFI values.

Another important term that requires explicit clarification is what a "customer" in the index calculation means. IEEE Standard 1366-2003 defines a customer as being a metered electrical service point for which an active bill account is established at a specific location, that is, premise. The Canadian Service Continuity report defines a customer as the number of customer services fed at secondary, primary, and subtransmission voltages. Individual utilities may use different definitions for a customer. It is important to note that a uniformly applied definition of "customer" must be developed so that during the data collection process, each utility is identifying and capturing customers involved in an outage in the same manner. Again, differences in calculated performance indicators would be significant due to differences in definitions of a customer.

Perhaps, the greatest difficulty in comparing reliability indices between similar utilities is the exclusion of major events. Some utilities include all interruptions when computing reliability indices and others have widely varying exclusion criteria. Some utilities also exclude scheduled and bulk power events, which constitute 10–15% of total customer outages. Interruptions from bulk power system operations are due to problems in the bulk electricity supply system such as underfrequency load shedding, transmission system transients, or system frequency excursions. These interruptions are generally beyond the control of the distribution utilities. However, from the customer's perspective, the interruption origination point is the utility's concern to resolve. It is therefore important to note that in most circumstances, major events contribute most often to customer interruption duration, and the exclusion of major events will completely change the reliability index characteristics for a distribution system. The 1998 ice storm in Quebec and Ontario left a large number of customers without electric power for weeks. Canadian Electricity Association's Annual Service Continuity Report presented the reliability performance indices for the Canadian utilities including and excluding the major outage event of 1998 ice storm. The difference in calculated indices is profound when the 1998 ice storm is included.

### 12.2.3 System Characteristics

The identification of the system or part of the system as rural, urban, suburban, or a mixture of these demographics is important. The system reliability characteristics of individual utilities differ due to diversities in service areas, load densities, circuit ratios, system topologies, weather environments, and service standards. Urban systems usually have short supply feeders, underground circuits, and alternative power supplies, while rural systems typically have long supply feeders, overhead circuits, and dedicated power supplies.

Distribution systems across the country cover widely varying terrain and are exposed to widely varying weather patterns. Some distribution systems are routed through dense vegetation while others are routed through open fields. Some areas of the country have high lightning and windstorm activity while other areas experience little. Some distribution systems serve dense populations and are primarily underground while others serving rural populations are primarily overhead. Dense areas can have many feeder interconnections while sparse areas are typically radial. All of these factors can have a major impact on reliability index values, and these types of differences should always be taken into consideration when comparing reliability indices among different distribution systems.

It is important that the identification of the system or part of the system as rural, urban, suburban, or a mixture of demographics is crucial in setting reliability standards. Within a utility system, different parts of the system will have different performance levels due to demographics that will impact the calculated long-term historic and current year standard metric values. In addition, any cross-comparison between utilities would need to take these differences into account to make such comparisons valid.

The identification of the system, or part of the system, as overhead, underground, or a mixture will also have significant impact on distribution system reliability indices. Within a utility system, these different construction types result in different performance levels that impact the calculated long-term historic and current year standard metric values. As stated earlier, any cross-comparison between utilities would need to take these system design differences into account to make such comparisons valid.

### 12.2.4 Outage Data Collection Systems

In addition to geography, reliability indices can vary substantially based on a utility's data gathering practices. At present, many utilities compute reliability indices based on outage reports that are manually filled by field crew. Manual data collection tends to omit a significant percentage of interruptions and inaccurately account for customer restoration due to system reconfiguration procedures. To overcome data collection problems, many utilities are moving toward the installation of outage management systems that automatically track customer interruptions as they occur. After the installation of the electric outage management systems, some utilities have found that their SAIDI values increased. This does not imply that reliability worsens for those utilities that installed outage management systems. Rather, reliability indices are now more accurate. Ideally, the outage data collection system should be fully automated if cross-comparisons of reliability performances between similar utilities are to be performed.

System coverage in the automatic outage management system can have a profound impact on the calculated reliability performance indices. The range of system coverage can be from the entire distribution system (i.e., entire system from substation breaker down to customer meter), the substation breaker down to the distribution transformer, or just the substation breaker only. Some utility outage data collection systems collect outage data only at the feeder breaker level. Collecting only at the feeder breaker level distorts a utility's reliability performance since those outages that occur beyond the feeder breaker level are not included. This important aspect of data collection must also be taken into account when making cross-comparisons of reliability performance between utilities. Ideally, the system coverage in the electric outage management system should be extended to the individual customer level.

## 12.3 PERFORMANCE INDICES AT DIFFERENT SYSTEM LEVELS OF A UTILITY

System-wide performance indices provide a good indication of the long-term average system performance; however, these system level indices tend to mask unusually good or poor performance in regions, crew service areas, or individual feeders of the same utility. To illustrate performance indices variations at different system levels of a utility, Table 12.1 presents the average SAIFI and SAIDI indices for a Canadian integrated utility (IU) at the system and region levels for the operating period 1997–2001. An integrated utility denotes a utility with urban, suburban, rural, and a mixture of systems for the purpose of this chapter. The identity of the utility is unknown as per Canadian Electricity Association's confidentiality rules. In Tables 12.1–12.4, SAIFI and SAIDI are given as interruptions per year and hours per year, respectively.

TABLE 12.1. Reliability Indices at System and Regional Levels of an Integrated Utility

| | | |
|---|---|---|
| System level | | |
| SAIFI = 1.17, SAIDI = 1.93 | | |
| Regional level | | |
| Region 1 | Region 2 | Region 3 |
| SAIFI = 1.37 | SAIFI = 0.75 | SAIFI = 1.22 |
| SAIDI = 2.07 | SAIDI = 1.35 | SAIDI = 2.15 |

TABLE 12.2. Reliability Indices at Crew Center Level of an Integrated Utility

| | | |
|---|---|---|
| Crew center (CC) level | | |
| Region 1 | Region 2 | Region 3 |
| CC1-Urban | CC1-Urban | CC1-Urban |
| SAIFI = 1.68 | SAIFI = 0.97 | SAIFI = 1.16 |
| SAIDI = 2.32 | SAIDI = 1.07 | SAIDI = 1.91 |
| CC2-Rural | CC2-Rural | CC2-Rural |
| SAIFI = 2.60 | SAIFI = 1.16 | SAIFI = 1.06 |
| SAIDI = 4.93 | SAIDI = 4.44 | SAIDI = 1.63 |

TABLE 12.3. Reliability Indices at Feeder Level of an Integrated Utility

| | | |
|---|---|---|
| Feeder level indices | | |
| Feeder 1: rural | Feeder 2: urban | Feeder 3: urban |
| SAIFI = 3.39 | SAIFI = 3.41 | SAIFI = 3.59 |
| SAIDI = 13.23 | SAIDI = 12.98 | SAIDI = 9.97 |
| Feeder 4: urban | Feeder 5: urban | Feeder 6: urban |
| SAIFI = 3.04 | SAIFI = 1.90 | SAIFI = 3.14 |
| SAIDI = 7.42 | SAIDI = 7.47 | SAIDI = 6.85 |

Table 12.2 provides SAIFI and SAIDI indices for the same integrated utility at the crew center level. Table 12.3 presents the reliability performance indices for the same integrated utility at the feeder level. An examination of Tables 12.1–12.3 indicates that the reliability indices for the same integrated utility at different system levels are different.

At the customer level, the SAIFI and SAIDI values are at a magnitude higher than that of the system level. Also, different regions and crew centers have different reliability

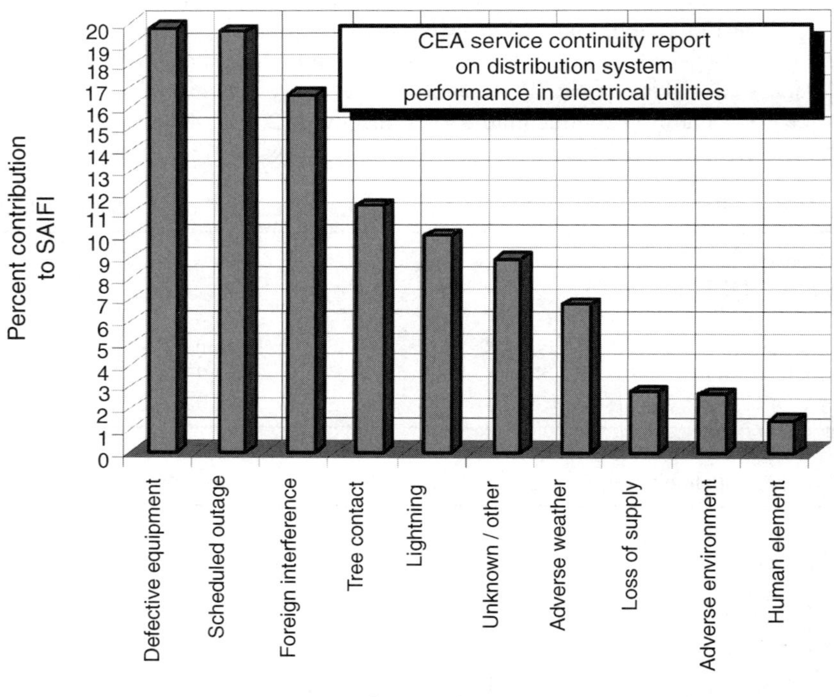


Figure 12.2. Major cause contribution to overall Canadian SAIFI.

levels because of variations in system operating configurations, operating and maintenance policies, age of the infrastructure, crew center proximity, weather conditions, and whether the system is urban or rural. It is extremely important to consider these variations in reliability performances at different levels of system hierarchy in setting reliability standards by regulatory agencies. A region in Tables 12.1 and 12.2 consists of a number of crew centers. There are crew centers with higher and lower reliability levels than the region level performance within the same region depending on the age of the equipment, weather pattern, system design, or maintenance practices. In Table 12.2, only indices for one urban crew center and one rural crew center in a region are presented to illustrate differences in reliability at different system levels. Although the figures shown in Table 12.2 are higher than the region level figures, there are crew centers with higher reliability than the region level reliability in the same region.

Figures 12.2-12.4 present major cause contributions to SAIFI, SAIDI, and CAIDI indices at the overall Canadian national level. Figures 12.5-12.7 present the trends in an integrated Western Canada utility versus overall Canadian SAIFI, SAIDI, and CAIDI indices for the period 1989–2002. For example, the 2001 CEA results are based on 23 U. S. companies along with 39 Canadian utilities.

An integrated utility includes rural, urban, and a mixture of urban/rural systems. The particular utility has long stretched transmission lines from south to north and is a voltage-constrained system. Its crew centers are scattered all over the service territories

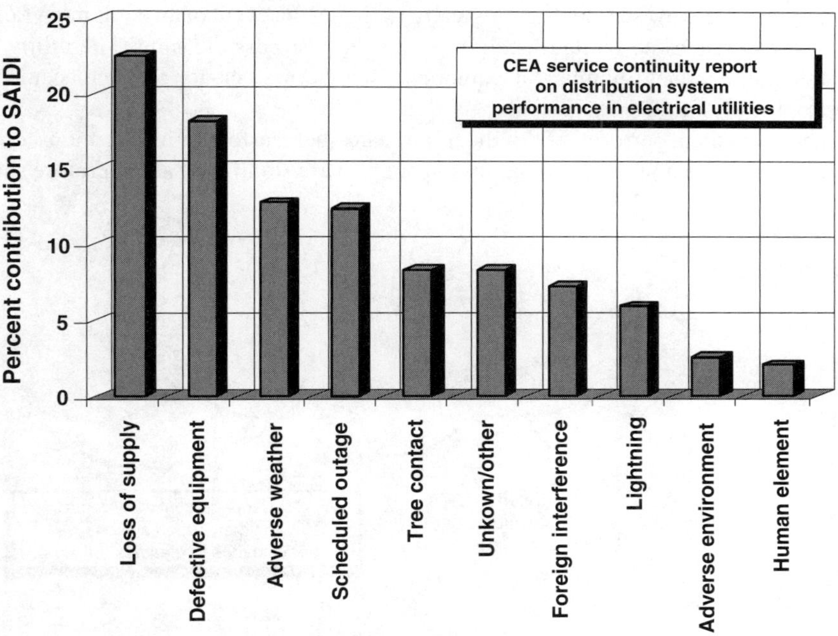


Figure 12.3. Major cause contribution to overall Canadian SAIDI.

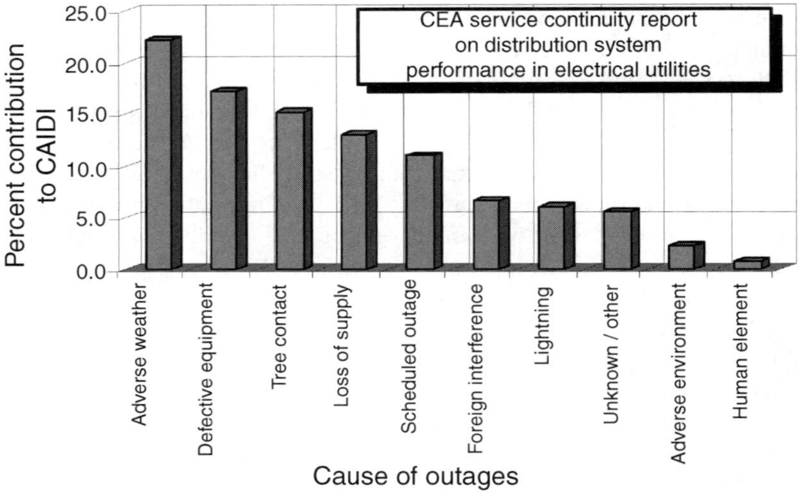


Figure 12.4. Major cause contribution to overall Canadian CAIDI.

unlike urban utilities such as City of Winnipeg, York Hydro, and Edmonton Power. The integrated Western Canada utility is a low load density system. As mentioned earlier, it has maintenance crew stationed strategically so that in the event of an equipment failure, the crew can reach the faulted location in the shortest possible time. The utility also maintains an adequate number of equipment spares in stock for fast replacement of catastrophically faulted equipment.

The Canadian performance indices indicate that customer interruptions due to transmission system problems (e.g., loss of supply) are significant and therefore should

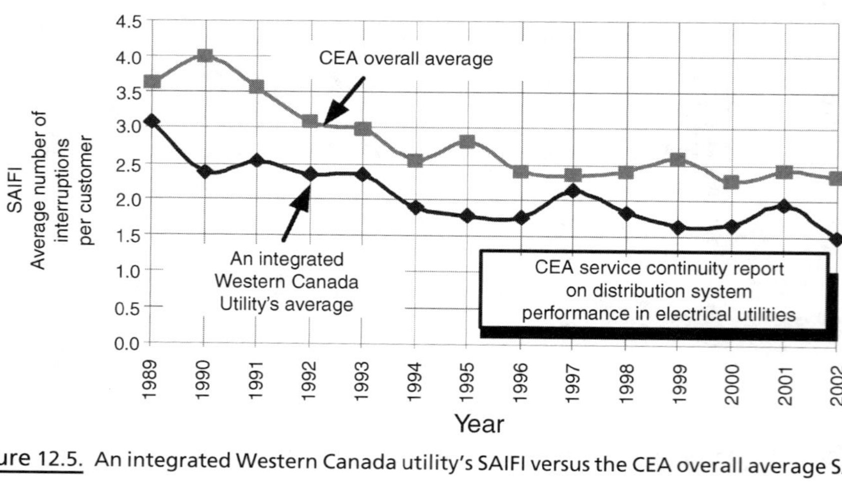


Figure 12.5. An integrated Western Canada utility's SAIFI versus the CEA overall average SAIFI.

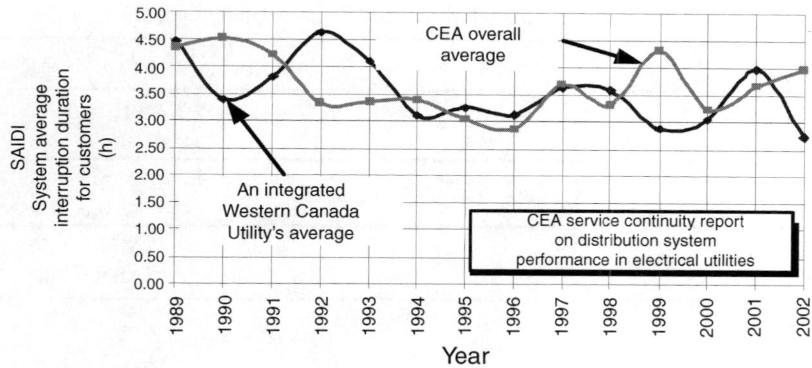


Figure 12.6. An integrated Western Canada utility's SAIDI versus the CEA overall average SAIDI.

be taken into consideration in setting distribution system reliability standards. In addition, Figs. 12.5-12.7 indicate that an integrated Western Canada utility's reliability performance is significantly different from that of the Canadian Electricity Association overall average. The computed indices depicted in Figs. 12.5-12.7 do not include the 1998 major ice storm event. As shown, the reliability performance for the integrated Western Canada utility is better than that of the Canadian national averages. However, there are utilities with different system characteristics with higher than national average SAIFI, SAIDI, or CAIDI values. Figures 12.5-12.7 do not represent a norm, but rather illustrate those utilities with different system characteristics and operations and maintenance policies that have unique levels of system performance.

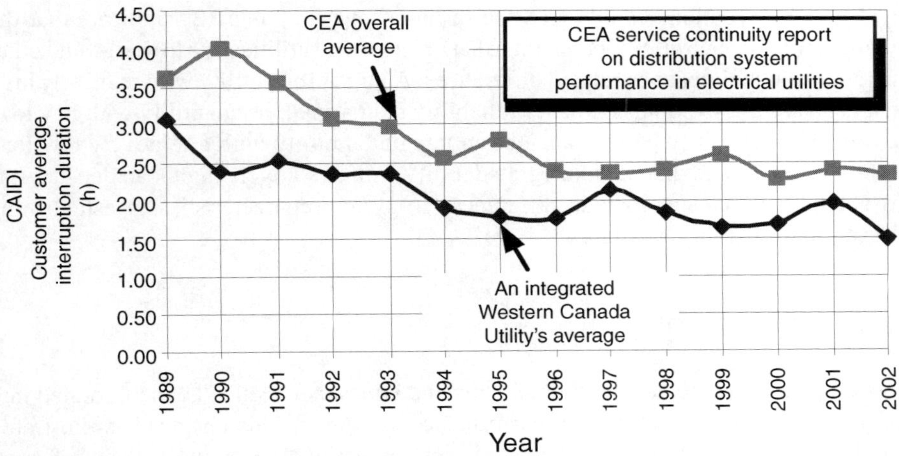


Figure 12.7. An integrated Western Canada utility's CAIDI versus the CEA overall average CAIDI.

TABLE 12.4. Reliability Performance Indices for Different Utility Types

| Small Urban Utility (UU-1) | Large Urban Utility (UU-2) |
|---|---|
| SAIFI = 1.87<br>SAIDI = 1.29 | SAIFI = 0.94<br>SAIDI = 0.72 |
| **High Load Density Integrated Utility (IU-1)** | **Low Load Density Integrated Utility (IU-2)** |
| SAIFI = 3.47<br>SAIDI = 4.31 | SAIFI = 1.50<br>SAIDI = 3.68 |

## 12.4 PERFORMANCE INDICES FOR DIFFERENT UTILITY TYPES

The Canadian Electricity Association maintains a comprehensive service continuity outage database on behalf of the Canadian reporting utilities and U.S. companies. The service continuity report on distribution system performance in Canadian electrical utilities is published annually. The CEA report presents annual reliability indices such as SAIFI and SAIDI, including the interruption cause contribution to the overall reliability indices for the participating utilities.

Section 12.3 presented reliability performance indices at different levels of a utility. This section presents average SAIFI and SAIDI indices for four Canadian utilities to illustrate the differences in performance indices for different utility types. UU-1 is a small urban utility with a comparatively high circuit ratio and low load density. UU-2 is a large urban utility with a relatively low circuit ratio and relatively high load density. IU-1 is an integrated utility with a relatively high load density, and IU-2 is an integrated utility with a relatively low load density. The identity of these four utilities is unknown as per CEA confidentiality rules. Table 12.4 presents average SAIFI and SAIDI indices for the four utilities based on the operating period 1992–2001.

Though exceptions could exist, the results shown in Table 12.4 indicate that the reliability of urban utilities is better than that of integrated utilities due to design and other operation and maintenance related differences. A large urban utility with relatively high load density have experienced better reliability than a small urban utility with low load density. The reliability performance of an integrated utility with a relatively low load density is better than that of an integrated utility with a relatively high load density. It is extremely important to include differences in system characteristics in establishing distribution reliability standards.

## 12.5 CONCLUSIONS

This chapter has presented pertinent factors and issues that need to be incorporated into the distribution system reliability performance standards. This chapter has illustrated that there are wide variations between utilities and within a given utility. Major factors and issues that need special considerations in setting reliability standards are geography, system design, operating and maintenance practices, weather conditions, physical age of electrical equipment, restoration practices, rural versus urban systems, high load density versus low load density systems, overhead versus underground systems, manual outage data collection versus automated outage data collection, major event exclusion or inclusion, definition of terms for different indices and their calculations, and variation in major event definitions. Exclusion of bulk power events, which constitute about 15% of outages experienced by customers, would have significant impact on the standard metrics. Since no two utilities are alike under the same regulatory jurisdictions, benchmarking is not possible between utilities. Therefore, performance standards should be set utility specific.

There are variations in weather normalization practices by different utilities. Some utilities substitute an extreme weather event with a normal outage event in computing the reliability indices, whereas some exclude the severe weather outage events altogether. IEEE Standard 1366-2003 suggests segmenting the severe weather outage events using the 2.5 Beta method. The CEA database containing 10 major causes of distribution system outages can be stratified and organized in many ways (e.g., with or without weather normalization, defective equipment reflecting aging of the distribution system, etc.).

Standards are perceived by utility personnel to be rigid requirements attached with penalties. On the contrary, performance guidelines or performance targets set a cutoff point, violation of which requires a rigorous review and causal analysis of poorly performing circuits or areas needing reinforcements. Utilities, in general, favor targets or guidelines over standards. Under the current uncertain time of deregulation, regulators and utilities should work together to continue to ensure customer service reliability performance at a reasonable cost. Setting reliability standards or guidelines without a full understanding of underlying performance issues and possible consequences can have significant societal cost consequences. Understanding the pertinent factor and issues is critical to achieving this ultimate goal.

## REFERENCES

1. C. A. Warren, The impact of regulatory policy on reliability, Transmission & Distribution World, December 2003, 16–23.
2. IEEE Standard 1366, Guide for Electric Power Distribution Reliability Indices, 2003.
3. Canadian Electricity Association, 2001 Annual Service Continuity Report on Distribution System Performance in Electrical Utilities, May 2002.
4. C. A. Warren and M. J. Adams, Reliability on the regulatory horizon, in IEEE T&D Conference and Exposition, Atlanta, GA, October 2001 (CD-ROM).
5. R. E. Brown and J. J. Burke, Managing the risk of performance based rates, *IEEE Transactions on Power Systems*, **15** (2), 2000, 893–898.
6. P. Save and T. Velarde, Incorporation of performance based ratemaking (PBR) incentives in probabilistic transmission planning criteria, in Proceedings of the 5th International Conference on Probabilistic Methods Applied to Power Systems, Vancouver, Canada, September 21–25, 1997, pp. 17–23.
7. Ontario Energy Board Electricity Distribution Rate Handbook, March 2000.

# 13

# STANDARDS FOR REREGULATED DISTRIBUTION UTILITY

## 13.1 INTRODUCTION

In an attempt to reregulate the distribution segment of an electric power system, public utility commissions (PUCs) are increasingly adopting a reward/penalty framework to guarantee acceptable electric supply reliability. This reward/penalty framework is commonly known as performance-based rate making (PBR). A PBR framework is introduced to provide distribution utilities with incentives for economic efficiency gains in the competitive generation and transmission markets. A distribution utility's historical reliability performance records could be used to create practical PBR mechanisms. This chapter presents actual reliability performance history from two different Canadian utilities used to develop PBR frameworks for use in a reregulated environment. An analysis of financial risk related to historic reliability data is presented by including reliability index probability distributions in a PBR plan. In addition, this chapter identifies a number of factors and issues that should be considered in generating a PBR plan for a distribution utility. A brief analysis of cause contributions to reliability indices is also performed and presented in this chapter. The historic reliability-based PBR framework developed in this chapter will find practical applications in the emerging deregulated electricity market.

*Power Distribution System Reliability.* By Ali A. Chowdhury and Don O. Koval

Electric utility industry has been deregulated in many jurisdictions since the late 1980s in an attempt to develop a competitive electricity market for power generation and transmission services. In deregulated markets, the distribution segment of the power supply system has been reregulated to guarantee that the electric service received by customers is reliable and the distribution system is planned, operated, and maintained adequately and efficiently. Public utility commissions are increasingly turning to distribution system reliability PBR to guarantee electric service reliability in competitive markets. The basic objective of a PBR framework is to provide distribution companies incentives for economic efficiency gains and discourage distribution utilities from compromising supply reliability while pursuing economic profits.

The distribution system historic reliability performance information is extremely useful in the sense that it renders an invaluable reference in initiating a performance-based mechanism. This approach has been used by a number of public utility commissions in the United States, Canada, Australia, Great Britain, and many other countries. Most of the PUCs use distribution system historic performance to establish specified electric supply reliability standards, and they also require that electric utilities maintain at least 2–5 years of reliability performance data to remain within the range of their historic system reliability performance levels.

Almost all PUCs who adopted performance-based regulation introduced a reward/penalty structure (RPS) to encourage distribution utilities to maintain acceptable reliability levels in the new competitive deregulated environment. In the new market environment, the performance-based regulation presents local distribution utilities with incentives to operate efficiently and to be innovated in system planning, design, operation, and maintenance. At the same time, a PBR also introduces a potential financial risk to distribution companies due to the uncertainty with future system reliability performance.

Canadian electric utilities have a long history of collecting and reporting information on the levels of electric service reliability to their customers. This chapter presents some selected historical data for two disparate Canadian utilities that have been maintaining the system reliability information for over two decades. Reliability index probability distributions are developed from the actual system reliability performance data and are used to demonstrate the potential financial risks related to prescribed reward/penalty frameworks. The selected reliability performance data are also categorized into the different cause codes and presented to display the historic contributions from these causes to service reliability. This chapter will prove very useful for distribution companies that are subject to performance-based regulation in the reregulated environment and will find practical applications in designing a utility-specific PBR plan.

## 13.2 COST OF SERVICE REGULATION VERSUS PERFORMANCE-BASED REGULATION

Traditionally, rates that electric utilities charge are based on the cost of generating, transmitting, and delivering electricity to its customers' point of utilization. For fulfilling their obligation to serve customers in a particular service territory, utilities were guaranteed by PUCs a reasonable return on their investments in the utility infrastructures.

Utilities normally designed their systems to very conservative and expensive design standards in the cost of service regulation framework. Under the traditional cost of service regulation plan, utilities aggressively handled reliability problems knowing that the costs could be recovered. Deregulation of the electricity market changed everything.

To be competitive, utilities are reducing costs by deferring or canceling capital projects and by increasing maintenance intervals. As a result, the reliability of these utility systems is starting to deteriorate. Regulatory agencies are well aware that competition might have a negative impact on system reliability. Competition in the electric power industry provides incentives for enhanced performance, but it is not the complete solution for a number of reasons. First, of the three segments of an electric power system, only generation is being opened up to competition. In certain jurisdictions, few for profit transmission companies have been established. The majority of transmission and all distribution companies are still being regulated.

Customers are connected to a regulated distribution system that determines system reliability experienced by them. As customers of regulated systems, customers cannot switch distribution systems at their will if their reliability becomes unacceptable. For this reason, regulatory agencies are looking for ways to define and establish distribution reliability standards, and more and more utilities are finding themselves subject to performance-based regulation. The basic steps associated with the traditional cost of service regulation are as follows: (1) utility report costs, (2) regulators audit costs, (3) regulators set rates to enable utility to recover costs plus a fair rate of return on the used and/or useful capital invested, and (4) rates are periodically adjusted to reflect market and cost conditions. The basic steps related to the performance-based regulation are as follows: (1) performance requirements such as price and reliability standards are set more or less independent of costs, (2) utilities invest in profitable cost reduction programs and improve efficiency, and (3) utilities keep all or part of the increased profits.

The main effects of the cost of service regulation are as follows: (1) rates held at the market rate, (2) profits are proportional to rate base, (3) if prudent, cost increases result in increased rates, (4) cost reductions result in rate reductions, and (5) costs drive prices. On the contrary, the main effects of performance-based regulation are as follows: (1) efficiency improvements rewarded with higher profitability, (2) inefficiency penalized with lower profitability, and (3) as opposed to the cost of service regulation, price drives costs in the PBR plan. Efficiency incentives in the cost of service regulation regime are in general weak if utility management is motivated by profits. In the PBR plan, efficiency incentives are strong if utility management is motivated by profits.

## 13.3 A REWARD/PENALTY STRUCTURE IN THE PERFORMANCE-BASED RATES

A PBR is a contract between a PUC and a utility that rewards a utility for providing good reliability and/or penalizes a utility for providing poor reliability. Performance is normally based on average customer interruption information at the system level or at the customer level. This usually takes the form of system level reliability indices such as SAIFI (System Average Interruption Frequency Index) and SAIDI (System Average

Interruption Duration Index). These indices are computed using the following equations:

$$\text{SAIFI} = \frac{\text{number of customer sustained interruptions}}{\text{number of customers served}} \text{ per year} \qquad (13.1)$$

$$\text{SAIDI} = \frac{\text{sum of customer interruption durations}}{\text{number of customers served}} \text{ hours per year} \qquad (13.2)$$

A normal approach to implementing a performance-based rate is to have a "dead zone" where neither a penalty nor a reward is assessed. If reliability is worse than the dead zone boundary, a penalty is assessed. Penalties increase as reliability worsens and are capped when a maximum penalty is reached. Similarly, if reliability is better than the dead zone boundary, a reward is assessed, and the reward grows as reliability increases and is capped at a maximum value.

A real reward/penalty structure integrated into a PBR plan is illustrated for a Californian utility data discussed here. This particular PBR performance incentive framework includes three ranges, upper, middle, and lower, for the annual SAIDI reliability index. This particular RPS is structured in the following manner:

*Upper Range* : Penalty—$1 million per 1 min SAIDI above 65 min up to $18 million at 83 min and above.

*Dead Zone* : No reward or penalty—from 53 min SAIDI to 65 min SAIDI.

*Lower Range* : Reward—$1 million per 1 min SAIDI below 53 min and up to $18 million at 35 min and below.

A common method of implementing an RPS in a PBR plan is depicted in Fig. 13.1 using the data given above.

As shown in Fig. 13.1, this performance-based rate structure has a "dead zone" from the SAIDI value of 53–65 min, where neither a penalty nor a reward is assessed.

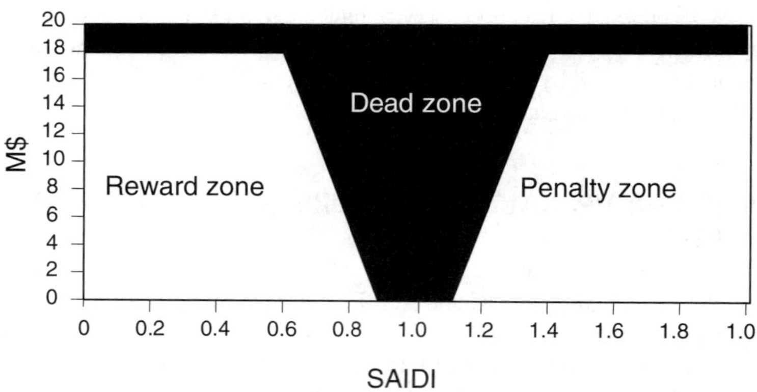


Figure 13.1. A general reward/penalty rate structure.

The RPS depicted in Fig. 13.1 can be expressed by a mathematical model, as shown in Equation (13.3). The financial penalty due to poor reliability associated with a reward/penalty structure can be computed by combining this RPS with related service reliability index expressed in the form of a probability distribution. The expected system reward/penalty payments could include both SAIFI and SAIDI contributions and are given by Equations (13.4) and (13.5).

The reward and penalty payments are computed as

$$\text{RP or PP} = f(\text{reliability index}) \tag{13.3}$$

$$\text{ERP} = \sum \text{RP}_i \times P_i \tag{13.4}$$

$$\text{EPP} = \sum \text{PP}_i \times P_i \tag{13.5}$$

where RP is reward payment, PP is the penalty payment, $\text{RP}_i$ is the reward payment at $\text{SAIFI}_i$ or $\text{SAIDI}_i$, $\text{PP}_i$ is the penalty payment at $\text{SAIFI}_i$ or $\text{SAIDI}_i$, and $P_i$ is the system probability of $\text{SAIFI}_i$ or $\text{SAIDI}_i$. ERP and EPP denote expected total reward and penalty payments, respectively.

Equations (13.4) and (13.5) indicate that the reward/penalty structure dictates the utility expected reward/penalty payments. It is therefore important that the reward/penalty policies should be designed with extreme care to encourage distribution utilities to maintain reliability levels in the dead zone. For example, if the RPS is designed using a single-point long-term average estimate for SAIFI or SAIDI without a dead zone, the utilities will be subject to frequent penalty payments due to the variability of annual system reliability performance. This situation is depicted in Fig. 13.2.

Figure 13.2 dictates that to design a PBR framework, the historical average reliability indices such as SAIFI and SAIDI should reside in the dead zone of the proposed reward/penalty structure and ideally in the middle of the dead zone. The dead zone spread should be related to the standard deviation of the SAIFI and SAIDI indices. The dead zone was set

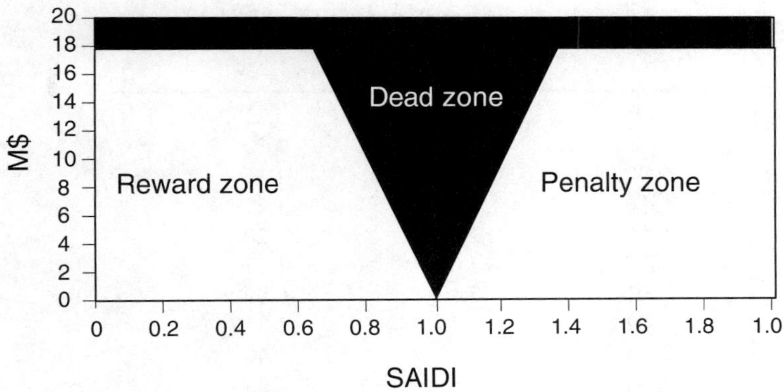


Figure 13.2. A reward/penalty structure without a dead zone.

at ±1 standard deviation in the studies discussed in this chapter. The impact of dead zone width on the reward/penalty structure was investigated using ±2 standard deviations. The utilities and the public utility commissions can negotiate the proper bandwidth for the dead zone of the RPS using utility-specific reliability performance and system characteristics. The financial parameters for reward and penalty in the PBR plan should be related to the incentive strategy established by the public utility commissions.

## 13.4 HISTORICAL SAIFI AND SAIDI DATA AND THEIR DISTRIBUTIONS

The Canadian Electricity Association (CEA) maintains a comprehensive service continuity outage database on behalf of the Canadian reporting utilities. The service continuity report on distribution system performance in Canadian electric utilities is published annually. The CEA report presents annual reliability indices such as SAIFI and SAIDI, including the interruption cause contributions to the overall reliability indices for the participating utilities.

Tables 13.1 and 13.2 show the annual SAIFI and SAIDI indices for two disparate integrated and urban Canadian utilities for the 10-year period from 1995 to 2004. For the purpose of this chapter, an integrated investor owned utility (IIOU) includes rural, urban, and mixture of urban/rural systems. The particular utility has long stretched transmission lines from south to north and is a voltage-constrained system. Its crew centers are scattered all over the service territories unlike a large urban utility (LUU) such as City of Winnipeg, York Hydro, Edmonton Power, and so on. The integrated investor owned utility is a low load density system. The large urban utility is a relatively big urban system with a normally low circuit ratio and high load density. The large urban system has short supply feeders, underground circuits, and alternative power supplies, while the integrated investor owned utility containing urban and rural systems have a mixture of short and long supply feeders, overhead circuits, and dedicated power supplies. The identity of these utilities is unknown as per company confidentiality rules.

Table 13.3 shows the average values of SAIFI and SAIDI and their standard deviations for each utility system based on the 10-year historical data. Figures 13.3 and 13.4 depict the utility reliability performance presented in Tables 13.1–13.3.

TABLE 13.1. System Performance—SAIFI

| Year | IIOU | LUU |
|---|---|---|
| 1995 | 3.08 | 1.21 |
| 1996 | 3.15 | 1.32 |
| 1997 | 3.52 | 1.16 |
| 1998 | 4.17 | 1.26 |
| 1999 | 2.68 | 1.20 |
| 2000 | 3.02 | 1.17 |
| 2001 | 2.40 | 0.99 |
| 2002 | 2.53 | 1.35 |
| 2003 | 2.35 | 1.46 |
| 2004 | 2.35 | 1.25 |
| Average | 2.925 | 1.237 |

TABLE 13.2. System Performance—SAIDI

| Year | IIOU | LUU |
|---|---|---|
| 1995 | 4.62 | 2.03 |
| 1996 | 3.78 | 2.21 |
| 1997 | 4.58 | 1.88 |
| 1998 | 6.67 | 2.05 |
| 1999 | 3.73 | 1.69 |
| 2000 | 4.42 | 1.93 |
| 2001 | 3.43 | 1.58 |
| 2002 | 3.85 | 1.65 |
| 2003 | 4.62 | 1.81 |
| 2004 | 4.09 | 1.84 |
| Average | 4.379 | 1.867 |

As shown in Tables 13.1 and 13.2, there are only 10 years' SAIFI and SAIDI values in this analysis. Histograms of these data have been developed and combined with the reward/penalty structure to create a PBR plan. The fundamental assumption made in this investigation is that each system remains virtually constant over the period of study in regard to design, maintenance, and operational changes. Obviously, this is a gross assumption and therefore the histograms for these indices provide approximate probability distributions of the indices. The historical SAIFI and SAIDI data, however, contain important information on the variation in the annual SAIFI and SAIDI indices and provide an insight into the variation that can be expected in later years.

## 13.5 COMPUTATION OF SYSTEM RISKS BASED ON HISTORICAL RELIABILITY INDICES

In the RPS of the PBR plan, the average historic values of the SAIFI and SAIDI should ideally be located in the middle of dead zone. The dead zone width would have impacts on the reward or penalty payments for a particular year reliability performance of a utility. The sensitivity analysis of the dead zone width was performed using the $\pm 1$ and $\pm 2$ standard deviations of the SAIFI and SAIDI indices in the following studies. This method was used to create the dead zones shown in Tables 13.4 and 13.5 for $\pm 1$ and $\pm 2$ standard deviations, respectively, using the historical data for the 1995–2004 period for the two disparate Canadian representative utility systems.

TABLE 13.3. Average Values with Standard Deviations

| | SAIFI | | SAIDI | |
|---|---|---|---|---|
| System Type | Average | SD | Average | SD |
| IIOU | 2.925 | 0.560 | 4.379 | 0.863 |
| LUU | 1.237 | 0.114 | 1.867 | 0.186 |

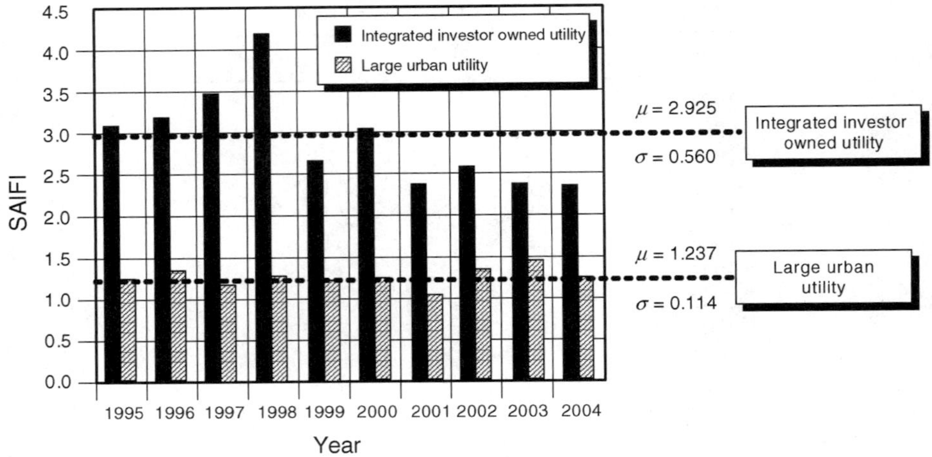


Figure 13.3. Utility performance—SAIFI levels.

Figures 13.5–13.12 show the combination of the historical SAIFI and SAIDI data for the two Canadian systems and four hypothetical reward/penalty structures. An infinite number of possible RPS could be designed using the historical information. The main focus in this analysis is on the determination of the appropriate location of the dead zone rather than on the computation of the expected reward or penalty payments.

In Fig. 13.5, for the integrated investor owned utility with $\pm 1$ standard deviation of historical average value of SAIFI, there is a 20% probability that the system SAIFI will lie in the penalty zone. The system SAIDI has 10% probability of residing in the penalty zone in Fig. 13.6. Both structures presented in Figs. 13.5 and 13.6 show that the utility's performance does qualify for reward payments, 20% for SAIFI and 10% for SAIDI. The utility should expect some future penalty payments according to the historic performance. It can be seen from Figs. 13.5 and 13.6 that 40% outcomes are close to the penalty

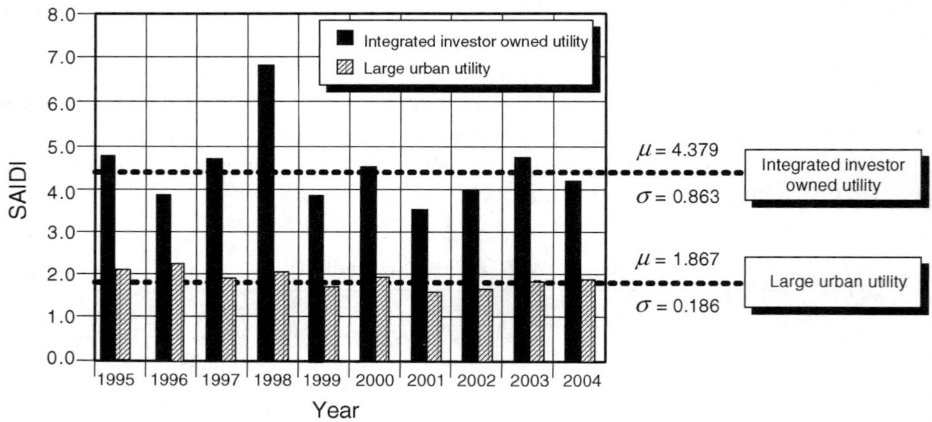


Figure 13.4. Utility performance—SAIDI levels.

TABLE 13.4. Dead Zones Using ±1 Standard Deviation for the Two Canadian Representative Utilities

| | Dead Zones | | | |
|---|---|---|---|---|
| System Type | SAIFI | | SAIDI | |
| IIOU | 2.365 | 3.485 | 3.316 | 5.242 |
| LUU | 1.123 | 1.351 | 1.681 | 2.053 |

TABLE 13.5. Dead Zones Using ±2 Standard Deviations for the Two Canadian Representative Utilities

| | Dead Zones | | | |
|---|---|---|---|---|
| System Type | SAIFI | | SAIDI | |
| IIOU | 1.805 | 4.045 | 2.653 | 6.105 |
| LUU | 1.009 | 1.465 | 1.495 | 2.239 |

boundaries. The utility should make improvements that would move its performance toward the center of the dead zone and avoid financial penalties from the regulator.

Figure 13.7 indicates that for the large urban utility with ±1 standard deviation of historical average value of SAIFI, there is 10% probability of reward payments and 20% probability of penalty payments. Figure 13.7 also reveals that 30% outcomes are close to the penalty boundaries. In Fig. 13.8, for the large urban utility with ±1 standard deviation of historical average value of SAIDI, there is a 20% probability that the system SAIDI will lie in the reward zone and 50% outcome will lie in the penalty zone. The variability in individual year's performance subjects the utility to some financial penalties in the future

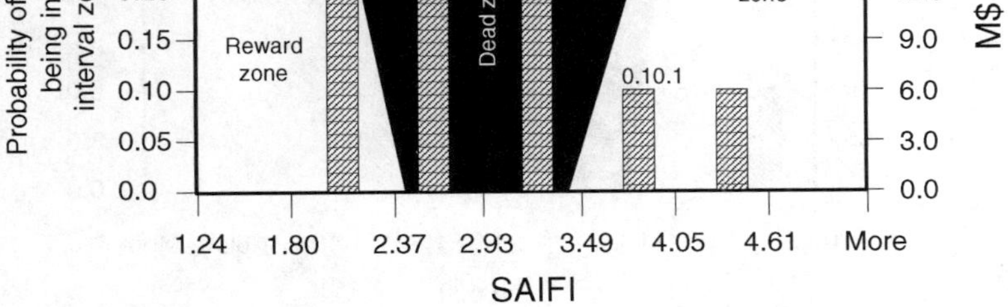


Figure 13.5. Combination of the SAIFI histogram with ±1 standard deviation and a hypothetical reward/penalty framework for the integrated investor owned utility.

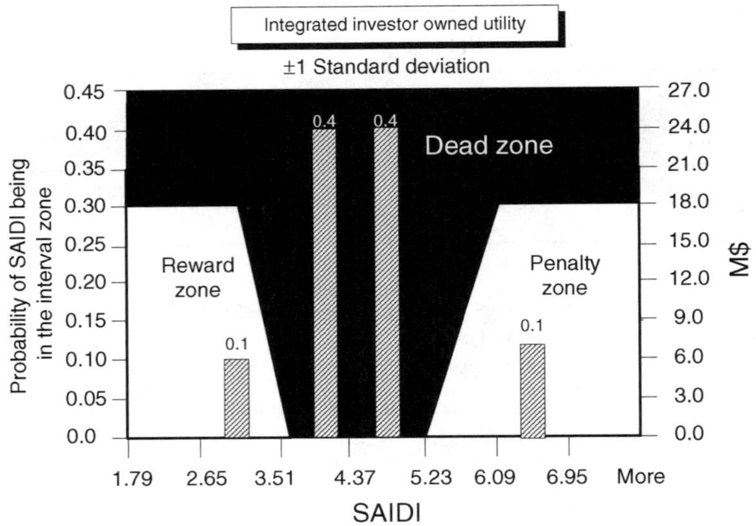


Figure 13.6. Combination of the SAIDI histogram with ±1 standard deviation and a hypothetical reward/penalty framework for the integrated investor owned utility.

unless improvements are made. The utility faces financial risks in the new PBR regime due to the considerable variation associated with its past performance. The utility could possibly earn rewards by making improvements.

It can be seen from Figs. 13.7 and 13.8 that the 40% of SAIFI outcomes and 30% of SAIDI outcomes lie in the center of the dead zones. The large urban utility should make

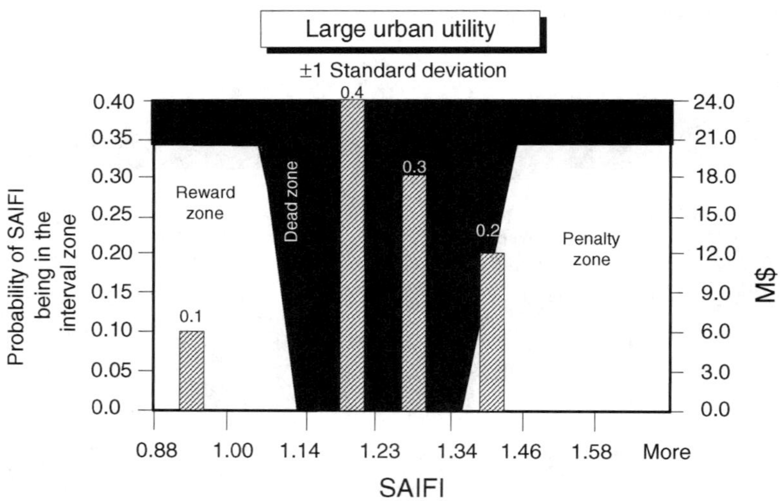


Figure 13.7. Combination of the SAIFI histogram with ±1 standard deviation and a hypothetical reward/penalty framework for the large urban utility.

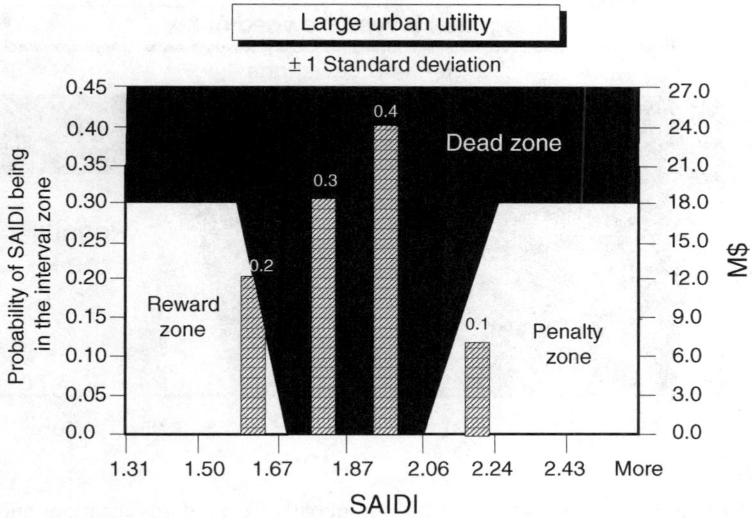


Figure 13.8. Combination of the SAIDI histogram with ±1 standard deviation and a hypothetical reward/penalty framework for the large urban utility.

significant system improvements that would move its performance toward the center of the dead zone and avoid financial penalties from the regulatory commissions.

The impact of dead zone width on the reward penalty structures was investigated by setting the dead zone at ±2 standard deviations for both SAIFI and SAIDI indices. Figures 13.9 and 13.10 show the combination of the SAIFI and SAIDI distributions with ±2 standard deviations and the hypothetical RPS for the integrated investor owned utility, respectively.

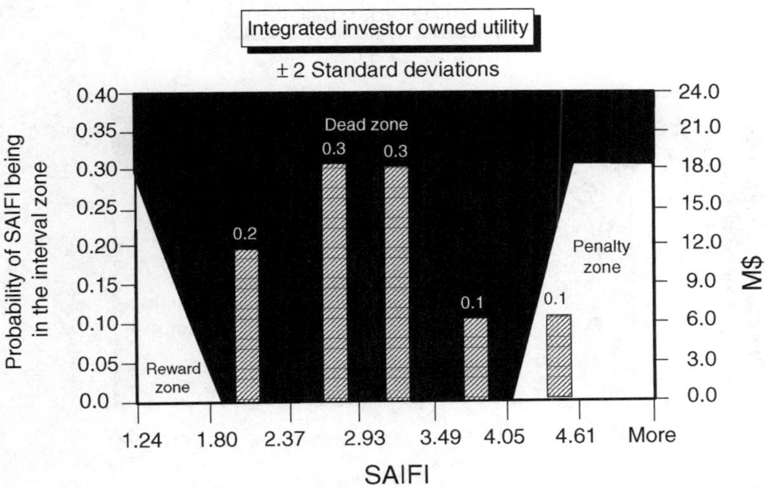


Figure 13.9. Combination of the SAIFI histogram with ±2 standard deviations and a hypothetical reward/penalty framework for the integrated investor owned utility.

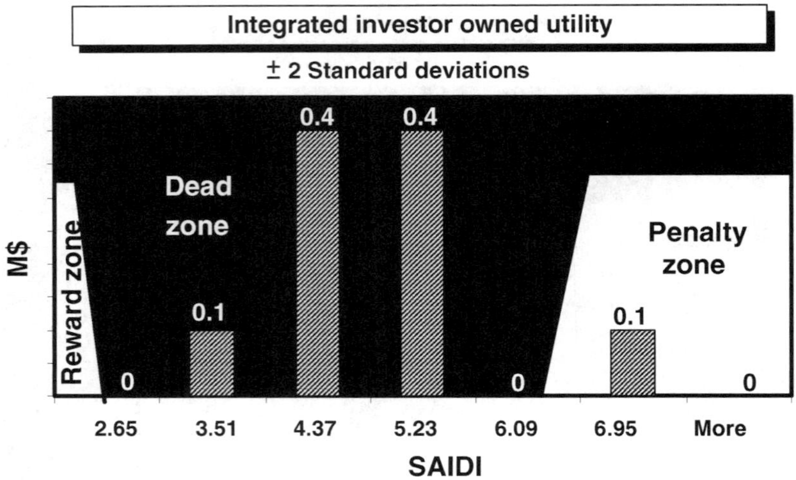


Figure 13.10. Combination of the SAIDI histogram with ±2 standard deviations and a hypothetical reward/penalty framework for the integrated investor owned utility.

Figures 13.11 and 13.12 show the combination of the SAIFI and SAIDI distributions with ±2 standard deviations and the hypothetical RPS for the large urban utility, respectively.

Figures 13.9 and 13.10 show that 10% of SAIFI and SAIDI outcomes for the integrated investor owned utility lie in the penalty zone for the RPS with ±2 standard deviations. The 60% of SAIFI outcomes and 80% of SAIDI outcomes lie in the center of the dead zones. Figures 13.11 and 13.12 show the RPS for the large urban utility with ±2 standard deviations. For both SAIFI and SAIDI indices, the RPS with ±2 standard deviations

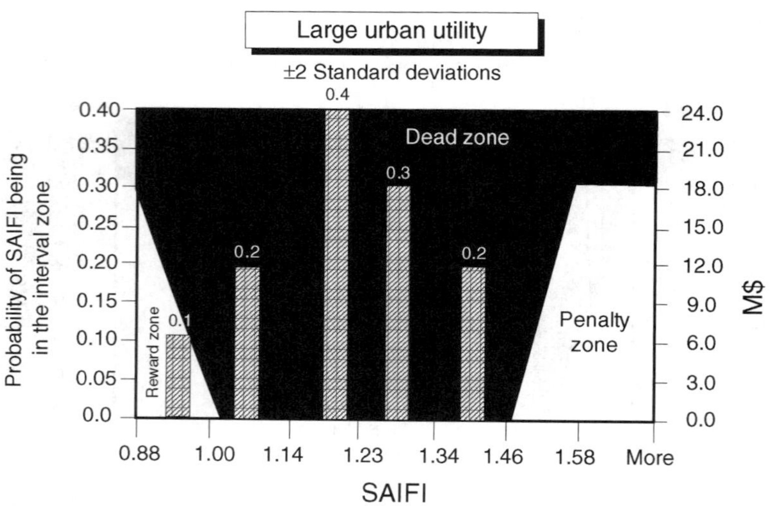


Figure 13.11. Combination of the SAIFI histogram with ±2 standard deviations and a hypothetical reward/penalty framework for the large urban utility.

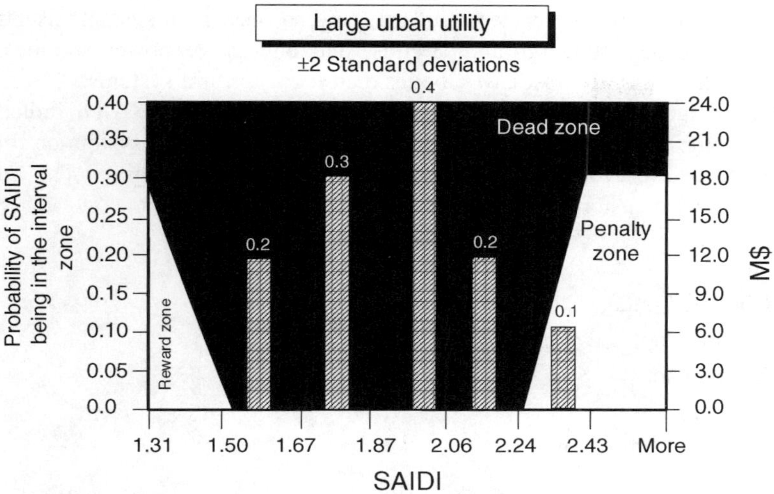


Figure 13.12. Combination of the SAIDI histogram with ±2 standard deviations and a hypothetical reward/penalty framework for the large urban utility.

indicates that the large urban utility reliability performance will lie in the dead zone areas. A significant outcome for the reliability performance for the large urban utility lies close to the penalty zone and none lies in the reward zone. It is, however, important to note that the width of the dead zone has significant impact on the reward/penalty structures, and utilities and regulators should pay close attention to this aspect of the PBR plan.

The methodology used to develop the dead zone values shown in Tables 13.4 and 13.5 provides a consistent framework to create the upper and lower bounds based on the utility's past performance. The decision to use ±1 or ±2 standard deviations is arbitrary and should be studied by both the utility and the regulator. As shown in Fig. 13.2, a single-point RPS for a system with no operating history or short operating history would result in relatively high financial risks due to the fact that there is no prescribed dead zone. It is obvious that in these cases, statistically significant historical data are required to create a reasonable dead zone.

It can be seen from the results presented in Figs. 13.5–13.12 that extreme care is required to develop appropriate dead zone width for both SAIFI and SAIDI. The bandwidth should not unduly penalize a utility and should provide appropriate incentives to encourage a utility to improve its reliability performance. As illustrated in this chapter, a reward/penalty structure based on the distributions associated with historical utility SAIFI and SAIDI indices could enable the investigation of the potential financial risks to a utility and provide a consistent framework to performance-based regulation.

## 13.6 CAUSE CONTRIBUTIONS TO SAIFI AND SAIDI INDICES

The system reliability characteristics of individual utilities differ due to the differences in service areas, load densities, system topologies, weather environments, company management philosophies, service standards, and so on. Urban systems usually have short supply feeders, underground circuits, and alternative power supplies. Their reliability indices are, in most cases, better than those in rural systems.

An investigation of the causal contributions to SAIFI and SAIDI indices from various system factors provides considerable insight into how the system performance can be improved to avoid financial penalties in the new PBR regime. The Canadian utilities divide the customer outages into the following codes:

- Unknown
- Scheduled outage
- Loss of supply
- Tree contact
- Lightning
- Defective equipment
- Adverse weather
- Adverse environment
- Human element
- Foreign interference

The major contributions to the service annual SAIFI and SAIDI indices can come from quite different causes in urban and integrated systems. This section presents the major interruption contributions for the two utility systems over the 1995–2004 period.

Figure 13.13 presents causal contributions to the annual SAIFI index for the integrated investor owned utility. Figure 13.14 presents causal contributions to annual SAIDI index for the integrated investor owned utility. Figure 13.15 presents the causal contributions to the annual SAIFI index for the large urban utility, and Fig. 13.16 presents the causal contributions to the annual SAIDI index for the large urban utility. The curves in Figs. 13.13–13.15 designated as AI, AI-Schd, AI-Tree, AI-Lightn, AI-Def. Eq., AI-AW, and AI-Forgn represent the annual index (AI), the annual index excluding the contribution from scheduled interruptions, the annual index excluding the contribution from tree-related interruptions, the annual index excluding the contribution from lightning-related interruptions, the annual index excluding the contribution from defective equipment, the annual index excluding the contribution from adverse weather-related interruptions, and the annual index excluding the contribution from foreign interference, respectively. Contributions from interruption causes such as human error, adverse environment, and loss of supply and unknown causes are insignificant compared to those from earlier noted causes, and therefore, contributions to SAIFI and SAIDI indices from these causes are not illustrated in Figs. 13.11–13.14.

Figures 13.17–13.20 present the percentage individual cause contributions to SAIF and SAIDI indices of the integrated investor owned and large urban utilities. In these figures, "all other causes" include interruption causes such as human error, adverse environment, and loss of supply and unknown.

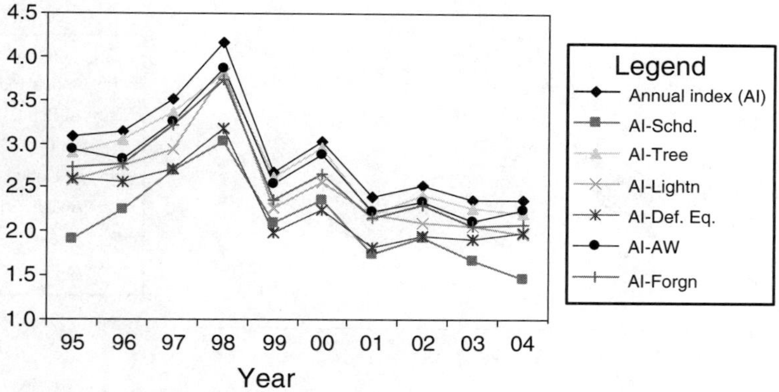


Figure 13.13. Causal contributions to SAIFI for the integrated investor owned utility.

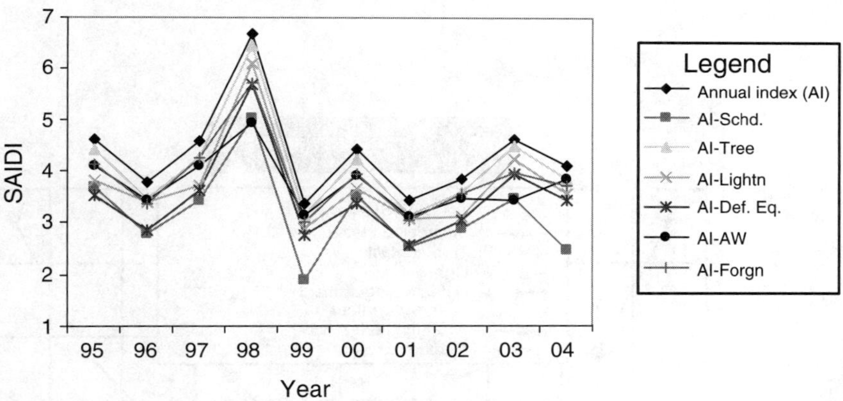


Figure 13.14. Causal contributions to SAIDI for the integrated investor owned utility.

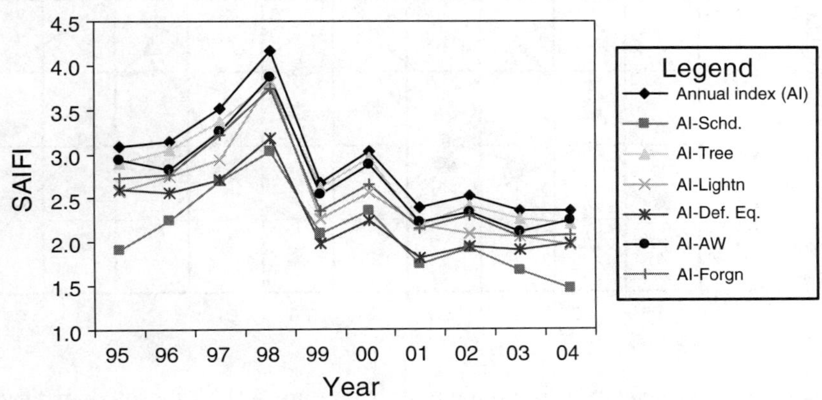


Figure 13.15. Causal contributions to SAIFI for the large urban utility.

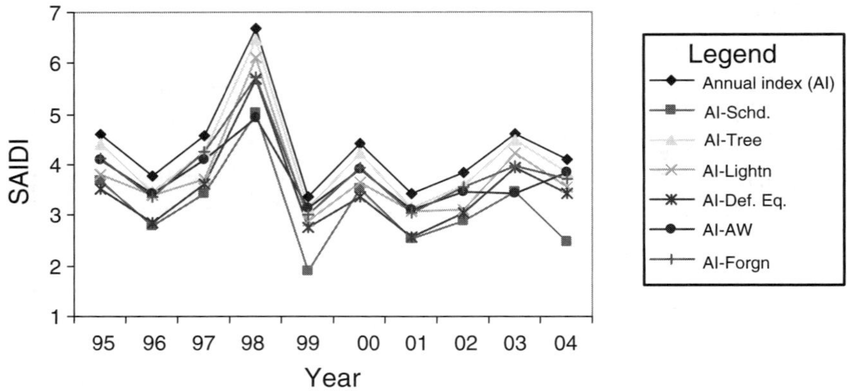


Figure 13.16. Causal contributions to SAIDI for the large urban utility.

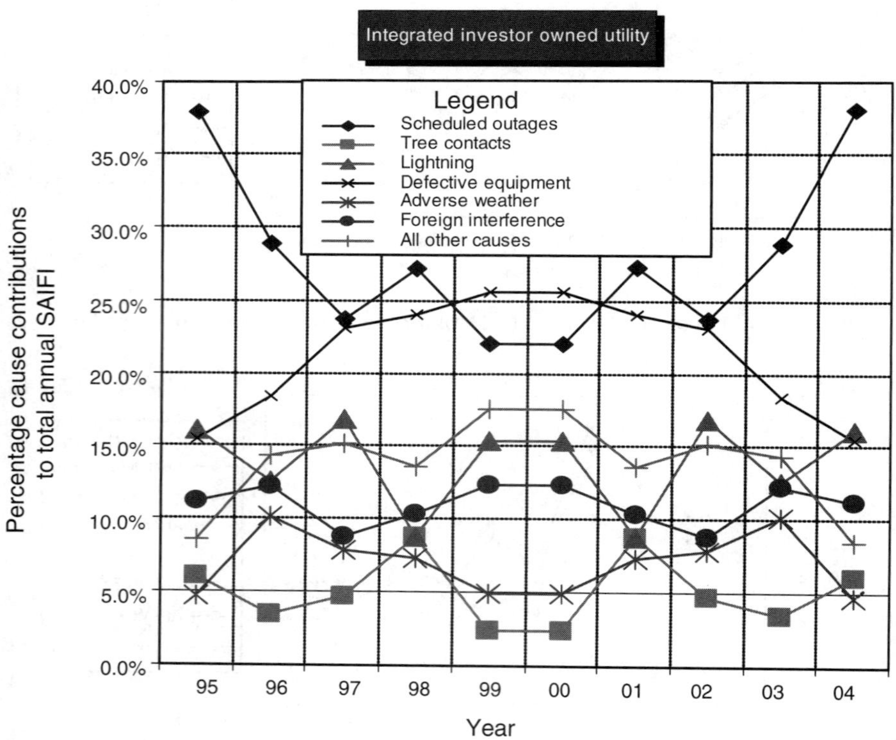


Figure 13.17. Percentage individual cause contributions to SAIFI for the integrated investor owned utility.

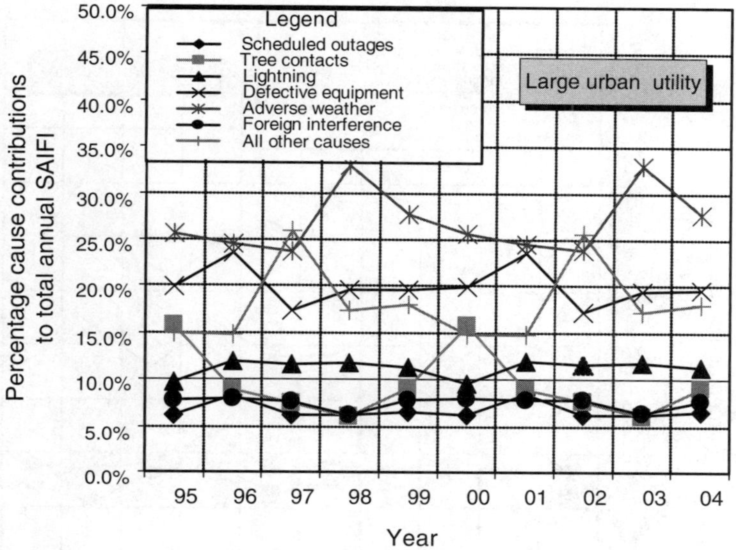


Figure 13.18. Percentage individual cause contributions to SAIFI for the large urban utility.

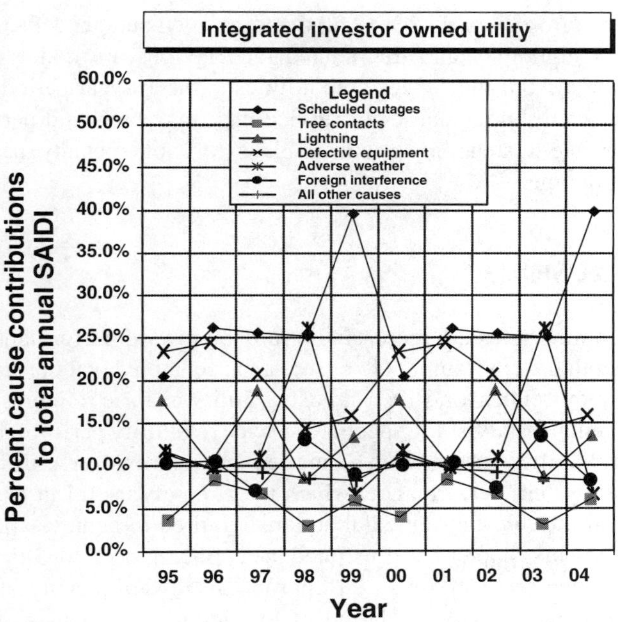


Figure 13.19. Percentage individual cause contributions to SAIDI for the integrated investor owned utility.

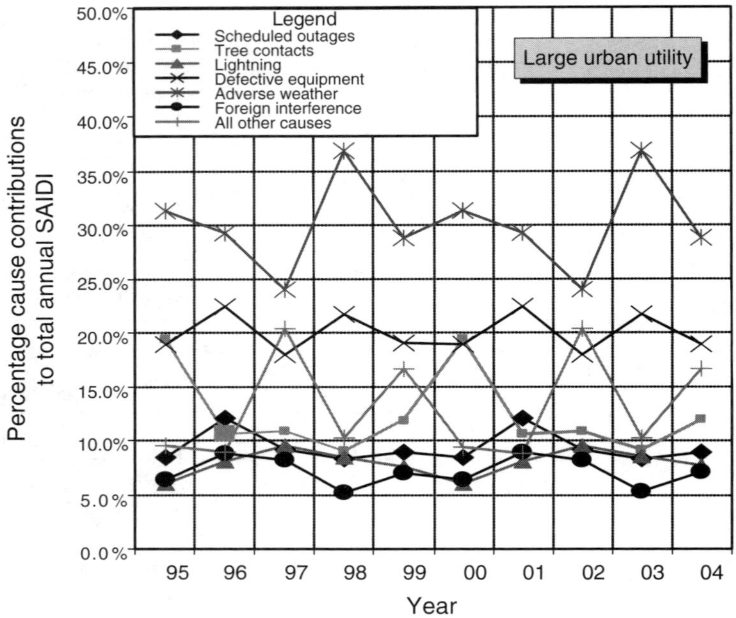


Figure 13.20. Percentage individual cause contributions to SAIDI for the large urban utility.

It can be seen from Figs. 13.13–13.20 that scheduled outage, defective equipment, adverse weather, lightning, and tree-related interruptions are major contributors to annual SAIFI and SAIDI indices for both utilities in the 10-year period. A knowledge base of primary contributing causes of service interruptions would permit a utility to identify appropriate system improvement plans to avoid penalty payments in the emerging PBR regime.

## 13.7 CONCLUSIONS

Public utility commissions are increasingly moving toward performance-based regulation in a deregulated environment to ensure an acceptable level of service reliability to customers. In this endeavor, PUCs are using utility historic reliability performance as a major element in establishing specified service reliability performance guidelines. The historic reliability information compiled by distribution utilities provides a measure of past system performance, which is extremely useful in predicting future system risks and the relevant remedial actions required to achieve specified service reliability levels. This chapter has illustrated the applications of historic utility service reliability performance to establish an appropriate reward/penalty structure in the emerging performance-based regulation of distribution companies. This RPS also includes incentives determined by the public utility commissions in regard to future desired performance.

The system reliability characteristics of individual utilities differ due to diversities in service areas, load densities, circuit ratios, system topologies, weather environments, and service standards. Urban systems usually have short supply feeders, underground circuits, and alternative power supplies, while rural systems typically have long supply feeders, overhead circuits, and dedicated power supplies and are subject to varied weather conditions. The historic reliability performance data for two different utility systems presented in Tables 13.1 and 13.2 demonstrate the effects of system diversities. It is therefore extremely important for regulatory commissions to include individual utility system characteristics in setting a reward/penalty structure. Finally, the approach of using reliability index probability distributions together with the average annual index values is an important tool in eliminating the impact of annual index variations and establishing appropriate reward/penalty structures in a PBR plan.

## REFERENCES


1. C. A. Warren, The impact of regulatory policy on reliability, Transmission and Distribution World, December 2003, 16–23.
2. C. A. Warren and M. J. Adams, Reliability on the regulatory horizon, in IEEE T&D Conference and Exposition, Atlanta, GA, October 2001 (CD-ROM).
3. R. E. Brown and J. J. Burke, Managing the risk of performance based rates, *IEEE Transactions on Power Systems*, **15** (2) 2000, 893–898.
4. Canadian Electricity Association, 2001 Annual Service Continuity Report on Distribution System Performance in Electrical Utilities, May 2002.
5. IEEE Standard 1366-2003, Guide for Electric Power Distribution Reliability Indices, 2003.
6. P. Save and T. Velarde, Incorporation of performance based ratemaking (PBR) incentives in probabilistic transmission planning criteria, in Proceedings of the 5th International Conference on Probabilistic Methods Applied to Power Systems, Vancouver, Canada, September 21–25, 1997, pp. 17–23.
7. R. Billinton and Z. Pan, Historic performance based distribution system risk assessment, *IEEE Transactions on Power Delivery*, **19** (4) 2004, 1759–1765.

# 14

# CUSTOMER INTERRUPTION COST MODELS FOR LOAD POINT RELIABILITY ASSESSMENT

## 14.1 INTRODUCTION

In a competitive energy market in which power supply reliability can influence customer purchasing decisions, electric utilities throughout the world are rapidly recognizing that they cannot ignore customer preferences. Today's energy market is characterized by intense price competition and electric utilities are faced with new challenges of large debts, budget constraints, safety, environment and economic issues, lower load growth than in the past, need for more involvement of different stakeholders in the planning and designing process, and more competitive nonconventional suppliers of electricity. In addition, in a deregulated competitive energy market, electric utilities are under conflicting pressures of providing even higher standards of service reliability and holding the line on rates. The reliability cost–reliability worth system facility planning approach offers a rational response to these conflicting customer and regulatory demands. This chapter will present the basic concepts and their applications to computing load point customer reliability indices and interruption costs. Case studies showing the applications of load point reliability index calculations including customer interruption costs in distribution system planning are described in detail in this chapter.

*Power Distribution System Reliability.* By Ali A. Chowdhury and Don O. Koval

At present, the electric utilities throughout the world have confronted many challenges in an increasingly competitive market. The move to competition and deregulation in the electric power sector is a global phenomenon. Utilities are already immersed in the battle for customers and are reorganizing to become more efficient. The electric utility industry is moving surprisingly rapidly in this direction, and utilities worldwide are adopting different innovative strategies to position themselves to operate effectively in a free market environment. Utilities are increasingly realizing that there is a need to embrace excellence in all areas of customer service, financing, and technology.

Deregulation is forcing electric utilities into uncharted waters. For the first time, the customer is looking for value-added services from their utilities, or they will start shopping around. Failure to recognize customer needs has caused a great number of business failures in numerous industries. The electric industries' movement toward a competitive market forces all related businesses to assess their focus, strengths, weaknesses, and strategies. One of the major challenges to electric utilities is to increase the market value of the services they provide with the right amount of reliability and to lower its costs for operation, maintenance, and construction to provide customers power at lower rates. For any power system supplying a specific mix of customers, there is an optimum value of reliability that would result in lowest combined costs. Value-based reliability planning is an attempt to achieve this optimum reliability in power systems.

## 14.2 CUSTOMER INTERRUPTION COST

The concept behind the cost–benefit reliability planning method and the application of customer interruption cost figures in system planning and designing are briefly discussed in Chapter 11. Reliability of electric service should be based on balancing the costs to a utility and the value of the benefits received by its customers. Therefore, to render a rational means of decision making in current planning and operating practices for changing service continuity levels experienced by customers, it is necessary to incorporate the utility costs and the costs incurred by customers associated with interruptions of service in the analysis.

The annual cost values shown in Fig. 11.2 pictorially depict the economic consequences of power interruptions as a function of reliability. A reliability cost/reliability worth assessment is performed to determine when additional system facilities should be planned. The additional source should be provided if the analysis indicates that the improvement in service reliability would be cost-effective.

The design of reliable *utility* distribution feeder configurations to supply power to industrial and commercial facilities is important because of the high costs associated with power outages. There is a need to be able to consider the cost of power outages when making *design decisions and defining operating procedures* for new and existing utility power distribution feeder systems and configurations and to be able to conduct quantitative cost versus reliability trade-off studies.

Sufficient information will be provided in this section so that reliability analyses can be performed on various distribution feeder configurations without referring to other texts. The discussion includes

1. Basic model equations for reliability analysis
2. Fundamentals of basic utility feeder configurations
3. Economic evaluation of reliability
4. Cost of power outage data
5. Equipment reliability data
6. Examples of reliability analysis of various distribution feeder configurations.

## 14.3 SERIES AND PARALLEL SYSTEM MODEL EQUATIONS

The basic equations used in IEEE Standard 493-2007 (IEEE Gold Book) for a two-component system whose component failure rates are $\lambda_1$ and $\lambda_2$ and whose individual component average repair times are $r_1$ and $r_2$, respectively, are summarized in Tables 14.1 and 14.2.

The key load point reliability indices used to define the reliability of the distribution power delivery system to a single load point are

1. $\lambda$, the frequency of load point interruptions per year
2. $r$, the average interruption duration expressed in hours per interruptions
3. $U$, the total annual interruption duration (i.e., hours per year).

*Two* reliability parameters are used to define distribution system electrical equipment and each distribution feeder section $i$:

1. The failure rate of the section ($\lambda_i$) or the failure rate of a specific electrical component.
2. The average time to repair a faulty section ($r_i$) or the average time to repair a specific electrical component.

TABLE 14.1. Failure Rate Equations for Series and Parallel Systems

| Failure Rate of a Series System | Failure Rate of a Parallel System |
|---|---|
| $\lambda_s = \lambda_1 + \lambda_2$ | $\lambda_p = \dfrac{\lambda_1\lambda_2[r_1 + r_2]}{1 + \lambda_1 r_1 + \lambda_2 r_2}$ |
| | $\lambda_p \sim \lambda_1\lambda_2[r_2 + r_2]$ |

TABLE 14.2. Average Repair Time Equations for Series and Parallel Systems

| Average Repair Time for a Series System | Average Repair Time for a Parallel System |
|---|---|
| $r_s = \dfrac{\lambda_1 r_1 + \lambda_2 r_2 + \lambda_1\lambda_2 r_1 r_2}{\lambda_1 + \lambda_2}$ | $r_p = \dfrac{r_1 r_2}{r_1 + r_2}$ |

TABLE 14.3. Component Forced Outage Data

| Component | $\lambda$ (failures/year) | Average Repair Time (h) |
|---|---|---|
| Utility substation (assumed ideal) | 0.0 | 0.0 |
| Utility feeder breaker (assumed ideal) | 0.0 | 0.0 |
| Distribution feeder section | 2.0 failures/100 km/year | 5.0 |
| Feeder isolating device (assumed ideal) | 0.0 | 0.0 |
| (Manual and automatic) | | |

*Note*: Manual isolating switching time = 1.0 h.

The cost of power interruptions to an individual load point depends on

1. $\lambda$, the frequency of load point interruptions per year
2. $L$, the magnitude of the average load expressed in kilowatts
3. CIPKW, cost of load point interruptions expressed in \$/kW of load interrupted

The cost of power interruptions to an individual load point (COLPI) is defined as

$$\text{COLPI} = \lambda \times L \times \text{CIPKW} \quad \$/\text{year for an individual load point}$$

The cost of distribution feeder outages to the customers (i.e., load points) being serviced by the feeder is the sum of the cost of interruptions to each individual load point and is defined as

$$\text{Cost of distribution feeder line outages to all load points} = \sum_{i=1}^{n} \text{COLPI}_i$$

where $n$ is the number of load points being serviced by the distribution feeder, $i$ is the $i$th load point, and $\text{COLPI}_i$ is the individual load point interruption cost.

The component forced outage data presented in Table 14.3 will be used in distribution feeder reliability analysis of all distribution feeder operating configurations.

## 14.4 DEDICATED DISTRIBUTION RADIAL FEEDER CONFIGURATION

The dedicated distribution radial feeder arrangement usually consists of three-phase sections, either overhead or underground feeder sections, connecting the utility substation directly to the industrial or commercial customer as shown in Fig. 14.1 for a two-section feeder.

Detailed calculations of the reliability indices as seen by industrial load point A (i.e., 1000 kW average load) are provided in Table 14.4. With reference to Fig. 14.1, the feeder line sections are connected in series; therefore, the frequency of interruptions ($\lambda_A$) seen by industrial load point A is the sum of the failure rates of the two line sections.

TABLE 14.4. Load Point "A" Reliability Indices and Annual Cost of Outages

| Section Number | Length (km) | $\lambda$ (failures /year) | $r$ (h) | $U$ (h/year) | Load (kW) | Cost ($/kW) | Cost ($/year) |
|---|---|---|---|---|---|---|---|
| 1 | 10 | 0.2 | 5 | 1.0 | 1000.0 | 20.00 | 4000.00 |
| 2 | 20 | 0.4 | 5 | 2.0 | 1000.0 | 20.00 | 8000.00 |
| Total | | 0.6 | $r_{av} = U/\lambda = 5.0$ | 3.0 | | | 12,000.00 |

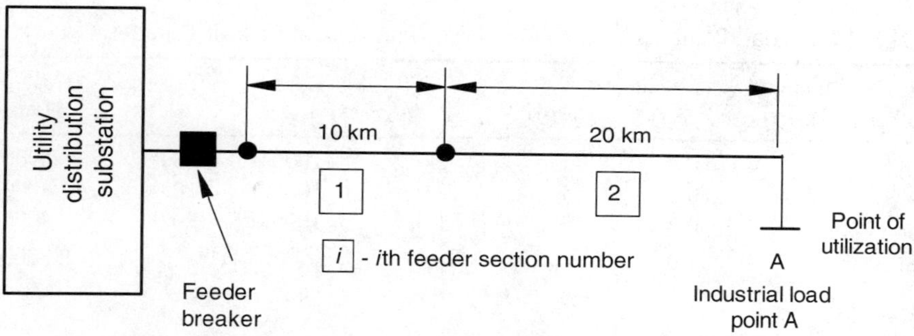


Figure 14.1. Dedicated distribution radial feeder configuration.

$$\lambda_A = \lambda_1 + \lambda_2 = 0.2 + 0.4 = 0.6 \text{ load point interruptions per year}$$

The annual outage duration of the distribution electric supply to customer A is

$$U_A = \lambda_1 r_l + \lambda_2 \times r_2 = 0.2(5.0) + 0.4(5.0)$$
$$= 3.0 \text{ h of load point interruptions per year}$$

The average interruption duration per outage at load point A is

$$r_A = U_A/\lambda_A = 3.0/0.6 = 5 \text{ h per load point interruption}$$

## 14.5 DISTRIBUTION RADIAL FEEDER CONFIGURATION SERVING MULTIPLE CUSTOMERS

A distribution radial feeder consisting of three feeder sections and serving two customer load points A and B is shown in Fig. 14.2. The average load of customers A and B is assumed to be 1000 kW each.

Detailed calculations of the reliability indices as seen by industrial load points A and B are provided in Tables 14.5 and 14.6, respectively.

TABLE 14.5. Load Point "A" Reliability Indices and Annual Cost of Outages

| Section Number | Length (km) | $\lambda$ (failures/ year) | $r$ (h) | $U$ (h/year) | Load (kW) | Cost ($/kW) | Cost ($/year) |
|---|---|---|---|---|---|---|---|
| 1 | 20 | 0.4 | 5 | 2.0 | 1000.0 | 20.00 | 8000.00 |
| 2 | 20 | 0.4 | 5 | 2.0 | 1000.0 | 20.00 | 8000.00 |
| 3 | 10 | 0.2 | 5 | 1.0 | 1000.0 | 20.00 | 4000.00 |
| Total | | 1.0 | $r_{av} = U/\lambda = 5.0$ | 5.0 | | | 20,000.00 |

TABLE 14.6. Load Point "B" Reliability Indices and Annual Cost of Outages

| Section Number | Length (km) | $\lambda$ (failures/ year) | $r$ (h) | $U$ (h/year) | Load (kW) | Cost ($/kW) | Cost ($/year) |
|---|---|---|---|---|---|---|---|
| 1 | 20 | 0.4 | 5 | 2.0 | 1000.0 | 20.00 | 8000.00 |
| 2 | 20 | 0.4 | 5 | 2.0 | 1000.0 | 20.00 | 8000.00 |
| 3 | 10 | 0.2 | 5 | 1.0 | 1000.0 | 20.00 | 4000.00 |
| Total | | 1.0 | $r_{av} = U/\lambda = 5.0$ | 5.0 | | | 20,000.00 |

*Note*: The reliability indices and the cost of interruptions for load points A and B are identical.

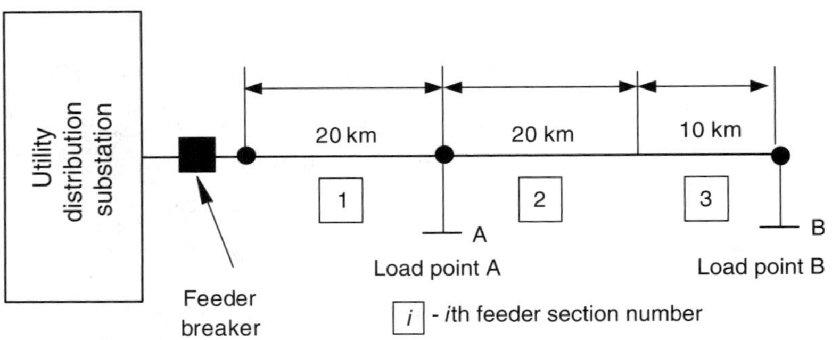


Figure 14.2. Distribution radial feeder configuration serving multiple customers.

## 14.6 DISTRIBUTION RADIAL FEEDER CONFIGURATION SERVING MULTIPLE CUSTOMERS WITH MANUAL SECTIONALIZING

To reduce the duration of customer interruptions caused by feeder section outages, the feeder sections can be interconnected by switching and isolation devices that can be either manual or automatic. A distribution radial feeder consisting of three feeder sections separated by normally closed, manually operated isolating devices and serving

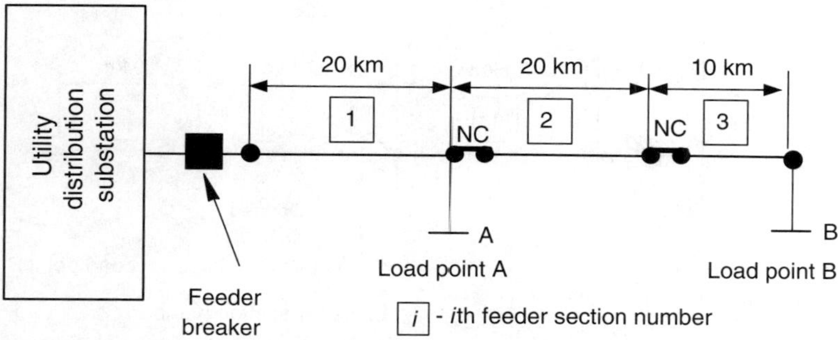


Figure 14.3. Distribution radial manual sectionalized feeder configuration serving multiple customers.

two customer load points A and B is shown in Fig. 14.3. The average load of customers A and B is assumed to be 1000 kW each.

Detailed calculations of the reliability indices as seen by industrial load point A are provided in Table 14.7.

*Note*: The manual switching and isolation devices are assumed to be ideals (i.e., $\lambda_{sw} = 0.0$); the time to open the manual switching and isolation devices will be assumed to be 1.0 h; and the cost of a 5 h interruption is assumed to be \$20.00/kW while the cost of a 1 h interruption will be assumed to be \$10.00/kW.

With reference to Fig. 14.3, a failure in feeder section 1 requires 5 h to repair and affects both customers. When a feeder section 2 outage occurs, it can be manually isolated (Fig. 14.4) in 1 h and the feeder reenergized to provide power to load point A. When a feeder section 3 outage occurs, it can be manually isolated (Fig. 14.5) in 1 h and the feeder reenergized to provide power to load point A.

*Note*: When the feeder sections are separated by manual sectionalizing devices, load point A in the first section has a significantly lower average annual interruption rate and a lower interruption duration (i.e., 2.6 h compared to 5 h/year with no feeder sectionalizing).

When a feeder section 2 outage occurs, it can be manually isolated in 1 h and the feeder reenergized to provide power to load point A.

TABLE 14.7. Load Point "A" Reliability Indices and Annual Cost of Outages

| Section Number | Length (km) | $\lambda$ (failures/ year) | $r$ (h) | $U$ (h/year) | Load (kW) | Cost (\$/kW) | Cost (\$/year) |
|---|---|---|---|---|---|---|---|
| 1 | 20 | 0.4 | 5 | 2.0 | 1000.0 | 20.00 | 8000.00 |
| 2 | 20 | 0.4 | 1 | 0.4 | 1000.0 | 10.00 | 4000.00 |
| 3 | 10 | 0.2 | 1 | 0.2 | 1000.0 | 10.00 | 2000.00 |
| Total | | 1.0 | $r_{av} = U/\lambda = 2.6$ | 2.6 | | | 14,000.00 |

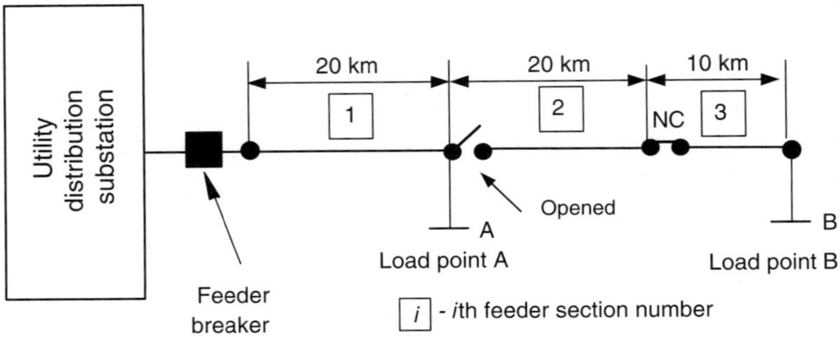


Figure 14.4. Feeder section 2 outage—manually isolated (i.e., switch opened).

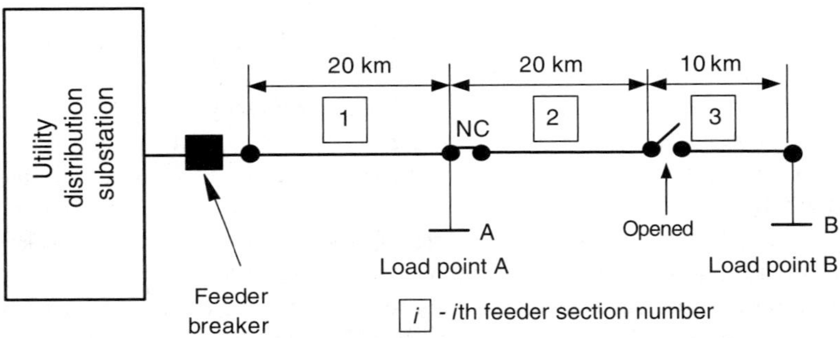


Figure 14.5. Feeder section 3 outage—manually isolated (i.e., switch opened).

When a feeder section 3 outage occurs, it can be manually isolated in 1 h and the feeder reenergized to provide power to load point A.

The detailed calculations of the reliability indices as seen by industrial load point B are provided in Table 14.8.

*Note*: The reliability indices and the cost of interruptions for load point B at the end of the feeder sections are the same as the distribution feeder delivery system with no sectionalizing devices.

TABLE 14.8. Load Point "B" Reliability Indices and Annual Cost of Outages

| Section Number | Length (km) | $\lambda$ (failures/ year) | $r$ (h) | $U$(h/year) | Load (kW) | Cost ($/kW) | Cost ($/year) |
|---|---|---|---|---|---|---|---|
| 1 | 20 | 0.4 | 5 | 2.0 | 1000.0 | 20.00 | 8000.00 |
| 2 | 20 | 0.4 | 5 | 2.0 | 1000.0 | 20.00 | 8000.00 |
| 3 | 10 | 0.2 | 5 | 1.0 | 1000.0 | 20.00 | 4000.00 |
| Total | | 1.0 | $r_{av} = U/\lambda = 5.0$ | 5.0 | | | 20,000.00 |

## 14.7 DISTRIBUTION RADIAL FEEDER CONFIGURATION SERVING MULTIPLE CUSTOMERS WITH AUTOMATIC SECTIONALIZING

To reduce both the frequency of customer interruptions and the duration of customer interruptions caused by feeder section outages, the feeder sections are interconnected with automatic isolating devices such as oil circuit recloser (OCR), circuit breaker, fuse, and so on that are coordinated with the feeder substation breaker and each other. An example of this type of distribution feeder is shown in Fig. 14.6. The average load of customers A, B, and C is assumed to be 1000 kW each.

The impact of individual feeder section failures on the load points that are interrupted is defined as follows:

1. When an outage occurs in distribution feeder section 1, the substation breaker trips and interrupts the power to load points A, B, and C.
2. When an outage occurs in distribution feeder section 2, the OCR 1 trips before the feeder breaker and interrupts the power to load points B and C. The power to load point A is not interrupted.
3. When an outage occurs in distribution feeder section 3, the OCR 2 trips before the feeder breaker and OCR 1 and interrupts the power to load point C. The power to load points A and B is not interrupted.

*Note*: In these examples, the relay and tripping time is assumed to be fast enough not to disrupt any of the feeder loads (i.e., particularly computer-controlled loads). If the equipment is susceptible to power anomalies of short duration, then the reliability calculations will be significantly different.

Detailed calculations of the reliability indices as seen by industrial load point A are provided in Table 14.9.

*Note*: The manual switching and isolation devices are assumed ideal (i.e., $\lambda_{OCR}$ 0.0). The time to open the automatic switching and isolation devices will be assumed to be approximately 0.0 h.

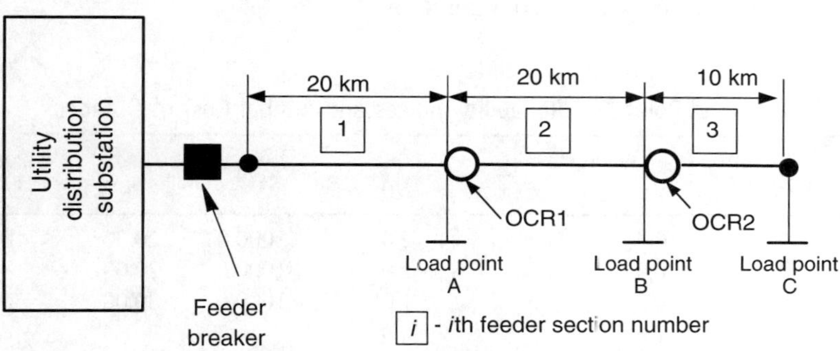


Figure 14.6. Distribution radial automatic sectionalized feeder configuration serving multiple customers.

TABLE 14.9. Load Point "A" Reliability Indices and Annual Cost of Outages

| Section Number | Length (km) | $\lambda$ (failures/ year) | $r$ (h) | $U$(h/year) | Load (kW) | Cost ($/kW) | Cost ($/year) |
|---|---|---|---|---|---|---|---|
| 1 | 20 | 0.4 | 5 | 2.0 | 1000.0 | 20.00 | 8000.00 |
| 2 | 20 | 0.0 | 0.0 | 0.0 | 1000.0 | 0.0 | 0.0 |
| 3 | 10 | 0.0 | 0.0 | 0.0 | 1000.0 | 0.0 | 0.0 |
| Total | | 0.4 | $r_{av} = U/\lambda = 5.0$ | 2.0 | | | 8000.00 |

TABLE 14.10. Load Point "B" Reliability Indices and Annual Cost of Outages

| Section Number | Length (km) | $\lambda$(failures/ year) | $r$ (h) | $U$(h/year) | Load (kW) | Cost ($/kW) | Cost ($/year) |
|---|---|---|---|---|---|---|---|
| 1 | 20 | 0.4 | 5 | 2.0 | 1000.0 | 20.00 | 8000.00 |
| 2 | 20 | 0.4 | 5.0 | 2.0 | 1000.0 | 20.00 | 8000.00 |
| 3 | 10 | 0.0 | 0.0 | 0.0 | 1000.0 | 0.0 | 0.0 |
| Total | | 0.8 | $r_{av} = U/\lambda = 5.0$ | 4.0 | | | 16,000.00 |

The cost of a 5 h interruption is assumed to be $20.00/kW, while the cost of a 1 h interruption will be assumed to be $10.00/kW.

Detailed calculations of the reliability indices as seen by industrial load points B and C are provided in Tables 14.10 and 14.11.

*Note*: In the distribution radial automatically sectionalized feeder, note the significant reduction in the frequency of load point interruptions and costs of interruptions seen by load points A and B. Load point C's reliability indices are unchanged compared to the other distribution feeder section configurations.

TABLE 14.11. Load Point "C" Reliability Indices and Annual Cost of Outages

| Section Number | Length (km) | $\lambda$ (failures/ year) | $r$ (h) | $U$(h/year) | Load (kW) | Cost ($/kW) | Cost ($/year) |
|---|---|---|---|---|---|---|---|
| 1 | 20 | 0.4 | 5 | 2.0 | 1000.0 | 20.00 | 8000.00 |
| 2 | 20 | 0.4 | 5 | 2.0 | 1000.0 | 20.00 | 8000.00 |
| 3 | 10 | 0.2 | 5 | 1.0 | 1000.0 | 20.00 | 4000.00 |
| Total | | 1.0 | $r_{av} = U/\lambda = 5.0$ | 5.0 | | | 20,000.00 |

## 14.8 DISTRIBUTION SYSTEM LOOPED RADIAL FEEDERS

The load of distribution system's industrial service area is supplied by two 25 kV distribution feeder circuits as shown in Fig. 14.7. The 25 kV feeder from substations A and B are operated as radial feeders although they can be interconnected by a normally open tie point. The disconnects, lateral distributors, step-down transformers, fuses, and the alternative supply are assumed to be 100% available in the analysis to simplify the reliability value-based planning methodology.

The loading conditions at each load point are provided in Table 14.12 for the operating year 2001. The service area is assumed to be entirely industrial. The interruption cost for a 1 h interruption to an industrial customer is \$9.62/average kW load interrupted and for a 4 h interruption \$18.544/average kW load interrupted.

### 14.8.1 Operating Procedures

If a fault occurs in any line section, the respective substation breaker is assumed to trip and deenergize the entire feeder. The faulty line section is first isolated, and whether some or the entire load is transferred to the adjacent substation depends on the following operating conditions.

The normally opened tie switch interconnecting substation A and B will be closed only if the sum of the peak loads of the energized substation plus the isolated peak loads from the deenergized substation does not exceed the rated capacity of the energized substation. If the sum exceeds the rated capacity of the substation, the normally opened tie switch will not be closed.

### 14.8.2 Feeder Characteristics: Looped Radial Feeders—Manual Sectionalizing

If both feeders are operated radially and are tied through a normally open tie switch, then any line section outage can be manually isolated and the remaining line sections can be

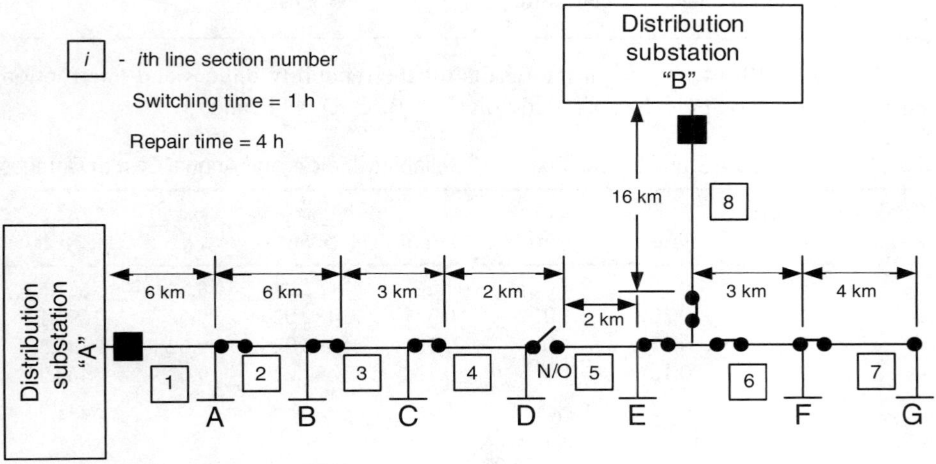


Figure 14.7. Distribution system looped radial feeders.

TABLE 14.12. Peak and Average Load Values for Fig. 14.7

| Load Point | A | B | C | D | E | F | G |
|---|---|---|---|---|---|---|---|
| Average load (kW) | 482.02 | 349.58 | 2107.49 | 3785.32 | 2468.62 | 2066.68 | 395.01 |
| Peak load (kW) | 626 | 545 | 2737 | 4916 | 3206 | 2684 | 513 |

energized from the alternative feeder or from the nearest continuous feeder. The load transfer is possible only when the feeder circuits and substations are unrestricted in capacity and the substations are not overloaded. If the substation is overloaded, then the load transfer will not occur. The reliability indices and the cost of interruptions for load point "C" are illustrated in Table 14.15 for an unrestricted substation capacity.

The reliability indices and the cost of interruptions for each load point for three case studies that will be evaluated are

1. $\lambda$, the number of feeder outages per year.
2. $U$, total duration of feeder outages in hours per year.
3. $r$, average duration of a feeder outage.
4. Total annual interruption cost per year at each load point and the total annual cost of interruptions for the distribution system.

*Note:* $\lambda$(25 kV feeder failure rate) = 5.0 failures/100 miles per year.

#### 14.8.2.1 *Case Study 1.* Refer to Fig. 14.7 for Case Study 1.

> The peak ratings for the 25 kV feeders from substations A and B are 10.0 and 7.5 MVA, respectively, at a power factor of 0.90 lagging.
> . . .manual feeder sectionalizing. . .

Tables 14.13–14.19 present the results for the reliability indices and interruption costs for the Case Study 1 at the load points A, B, C, D, E, F, and G.

TABLE 14.13. Case Study 1: Load Point "A" Reliability Indices and Annual Cost of Outages

| Section Number | Length (km) | $\lambda$(failures/year) | $r$ (h) | $U$(h/year) | Load (kW) | Cost ($/kW) | Cost ($/year) |
|---|---|---|---|---|---|---|---|
| 1 | 6 | 0.30 | 4.0 | 1.20 | 482.02 | 18.544 | 2681.57 |
| 2 | 6 | 0.30 | 1.0 | 0.30 | 482.02 | 9.62 | 1391.11 |
| 3 | 3 | 0.15 | 1.0 | 0.15 | 482.02 | 9.62 | 695.55 |
| 4 | 2 | 0.10 | 1.0 | 0.10 | 482.02 | 9.62 | 463.70 |
| Total | | 0.85 | $r_{av} = U/\lambda = 2.06$ | 1.75 | | | 5231.94 |

TABLE 14.14. Case Study 1: Load Point "B" Reliability Indices and Annual Cost of Outages

| Section Number | Length (km) | $\lambda$(failures/ year) | $r$ (h) | $U$(h/year) | Load (kW) | Cost ($/kW) | Cost ($/year) |
|---|---|---|---|---|---|---|---|
| 1 | 6 | 0.30 | 4.0 | 1.20 | 349.58 | 18.544 | 1944.78 |
| 2 | 6 | 0.30 | 4.0 | 1.20 | 349.58 | 18.544 | 1944.78 |
| 3 | 3 | 0.15 | 1.0 | 0.15 | 349.58 | 9.62 | 504.44 |
| 4 | 2 | 0.10 | 1.0 | 0.10 | 349.58 | 9.62 | 336.30 |
| Total | | 0.85 | $r_{av} = U/\lambda = 3.12$ | 2.65 | | | 4731.31 |

TABLE 14.15. Case Study 1: Load Point "C" Reliability Indices and Annual Cost of Outages

| Section Number | Length (km) | $\lambda$(failures/ year) | $r$ (h) | $U$(h/year) | Load (kW) | Cost ($/kW) | Cost ($/year) |
|---|---|---|---|---|---|---|---|
| 1 | 6 | 0.30 | 4.0 | 1.20 | 2107.49 | 18.544 | 11,724.39 |
| 2 | 6 | 0.30 | 4.0 | 1.20 | 2107.49 | 18.544 | 11,724.39 |
| 3 | 3 | 0.15 | 4.0 | 0.06 | 2107.49 | 18.544 | 5862.19 |
| 4 | 2 | 0.10 | 1.0 | 0.10 | 2107.49 | 9.62 | 2027.41 |
| Total | | 0.85 | $r_{av} = U/\lambda = 3.65$ | 3.10 | | | 31,338.38 |

TABLE 14.16. Case Study 1: Load Point "D" Reliability Indices and Annual Cost of Outages

| Section Number | Length (km) | $\lambda$ (failures/ year) | $r$ (h) | $U$ (h/year) | Load (kW) | Cost ($/kW) | Cost ($/year) |
|---|---|---|---|---|---|---|---|
| 1 | 6 | 0.30 | 4.0 | 1.20 | 3785.32 | 18.544 | 21,058.49 |
| 2 | 6 | 0.30 | 4.0 | 1.20 | 3785.32 | 18.544 | 21,058.49 |
| 3 | 3 | 0.15 | 4.0 | 0.60 | 3785.32 | 18.544 | 10,529.25 |
| 4 | 2 | 0.10 | 4.0 | 0.40 | 3785.32 | 18.544 | 7019.05 |
| Total | | 0.85 | $r_{av} = U/\lambda = 4.00$ | 3.10 | | | 59,665.73 |

TABLE 14.17. Case Study 1: Load Point "E" Reliability Indices and Annual Cost of Outages

| Section Number | Length (km) | $\lambda$ (failures/ year) | $r$ (h) | $U$ (h/year) | Load (kW) | Cost ($/kW) | Cost ($/year) |
|---|---|---|---|---|---|---|---|
| 1 | 6 | 0.10 | 4.0 | 0.40 | 2468.62 | 18.544 | 4577.81 |
| 2 | 3 | 0.15 | 1.0 | 0.15 | 2468.62 | 9.62 | 3562.22 |
| 3 | 4 | 0.20 | 1.0 | 0.20 | 2468.62 | 9.62 | 4749.62 |
| 4 | 16 | 0.80 | 4.0 | 3.20 | 2468.62 | 18.544 | 36,622.47 |
| Total | | 1.25 | $r_{av} = U/\lambda = 3.16$ | 3.10 | | | 49,512.12 |

TABLE 14.18. Case Study 1: Load Point "F" Reliability Indices and Annual Cost of Outages

| Section Number | Length (km) | $\lambda$ (failures/ year) | $r$ (h) | $U$ (h/year) | Load (kW) | Cost ($/kW) | Cost ($/year) |
|---|---|---|---|---|---|---|---|
| 1 | 6 | 0.10 | 1.0 | 0.10 | 2066.68 | 9.62 | 1988.15 |
| 2 | 3 | 0.15 | 4.0 | 0.60 | 2066.68 | 18.544 | 5748.68 |
| 3 | 4 | 0.20 | 1.0 | 0.20 | 2066.68 | 9.62 | 3976.29 |
| 4 | 16 | 0.80 | 4.0 | 3.20 | 2066.68 | 18.544 | 30,659.61 |
| Total | | 1.25 | $r_{av} = U/\lambda = 3.28$ | 4.10 | | | 42,372.73 |

TABLE 14.19. Case Study 1: Load Point "G" Reliability Indices and Annual Cost of Outages

| Section Number | Length (km) | $\lambda$ (failures/ year) | $r$ (h) | $U$ (h/year) | Load (kW) | Cost ($/kW) | Cost ($/year) |
|---|---|---|---|---|---|---|---|
| 1 | 6 | 0.10 | 1.0 | 0.10 | 395.01 | 9.62 | 380.00 |
| 2 | 3 | 0.15 | 4.0 | 0.60 | 395.01 | 18.544 | 1098.76 |
| 3 | 4 | 0.20 | 4.0 | 0.80 | 395.01 | 18.544 | 1465.01 |
| 4 | 16 | 0.80 | 4.0 | 3.20 | 395.01 | 18.544 | 5860.05 |
| Total | | 1.25 | $r_{av} = U/\lambda = 3.76$ | 4.70 | | | 8803.82 |

#### 14.8.2.2 *Case Study 2.* Refer to Fig. 14.7 for Case Study 2.

> The peak ratings for the 25 kV feeders from substations A and B are 20.0 and 17.5 MVA, respectively, at a power factor of 0.90 lagging.
> . . .manual feeder sectionalizing. . .

Tables 14.20–14.26 present the results for the reliability indices and interruption cost for the Case Study 2 at the load points A, B, C, D, E, F, and G.

TABLE 14.20. Case Study 2: Load Point "A" Reliability Indices and Annual Cost of Outages

| Section Number | Length (km) | $\lambda$ (failures/ year) | $r$ (h) | $U$ (h/year) | Load (kW) | Cost ($/kW) | Cost ($/year) |
|---|---|---|---|---|---|---|---|
| 1 | 6 | 0.30 | 4.0 | 1.20 | 482.02 | 18.544 | 2681.57 |
| 2 | 6 | 0.30 | 1.0 | 0.30 | 482.02 | 9.62 | 1391.11 |
| 3 | 3 | 0.15 | 1.0 | 0.15 | 482.02 | 9.62 | 695.55 |
| 4 | 2 | 0.10 | 1.0 | 0.10 | 482.02 | 9.62 | 463.70 |
| Total | | 0.85 | $r_{av} = U/\lambda = 2.06$ | 1.75 | | | 5231.94 |

TABLE 14.21. Case Study 2: Load Point "B" Reliability Indices and Annual Cost of Outages

| Section Number | Length (km) | $\lambda$ (failures/ year) | $r$ (h) | $U$ (h/year) | Load (kW) | Cost (\$/kW) | Cost (\$/year) |
|---|---|---|---|---|---|---|---|
| 1 | 6 | 0.30 | 1.0 | 0.30 | 349.58 | 9.62 | 1008.89 |
| 2 | 6 | 0.30 | 4.0 | 1.20 | 349.58 | 18.544 | 1944.78 |
| 3 | 3 | 0.15 | 1.0 | 0.15 | 349.58 | 9.62 | 504.44 |
| 4 | 2 | 0.10 | 1.0 | 0.10 | 349.58 | 9.62 | 336.30 |
| Total | | 0.85 | $r_{av} = U/\lambda = 2.06$ | 1.75 | | | 3794.41 |

TABLE 14.22. Case Study 2: Load Point "C" Reliability Indices and Annual Cost of Outages

| Section Number | Length (km) | $\lambda$ (failures/ year) | $r$ (h) | $U$ (h/year) | Load (kW) | Cost (\$/kW) | Cost (\$/year) |
|---|---|---|---|---|---|---|---|
| 1 | 6 | 0.30 | 1.0 | 0.30 | 2107.49 | 9.62 | 6082.22 |
| 2 | 6 | 0.30 | 1.0 | 0.30 | 2107.49 | 9.62 | 6082.22 |
| 3 | 3 | 0.15 | 4.0 | 0.06 | 2107.49 | 18.544 | 5862.19 |
| 4 | 2 | 0.10 | 1.0 | 0.10 | 2107.49 | 9.62 | 2027.41 |
| Total | | 0.85 | $r_{av} = U/\lambda = 1.53$ | 1.30 | | | 20,054.03 |

TABLE 14.23. Case Study 2: Load Point "D" Reliability Indices and Annual Cost of Outages

| Section Number | Length (km) | $\lambda$ (failures/ year) | $r$ (h) | $U$ (h/year) | Load (kW) | Cost (\$/kW) | Cost (\$/year) |
|---|---|---|---|---|---|---|---|
| 1 | 6 | 0.30 | 1.0 | 0.30 | 3785.32 | 9.62 | 10,924.43 |
| 2 | 6 | 0.30 | 1.0 | 0.30 | 3785.32 | 9.62 | 10,924.43 |
| 3 | 3 | 0.15 | 1.0 | 0.15 | 3785.32 | 9.62 | 5462.22 |
| 4 | 2 | 0.10 | 4.0 | 0.40 | 3785.32 | 18.544 | 7019.05 |
| Total | | 0.85 | $r_{av} = U/\lambda = 1.35$ | 1.15 | | | 34,330.58 |

TABLE 14.24. Case Study 2: Load Point "E" Reliability Indices and Annual Cost of Outages

| Section Number | Length (km) | $\lambda$ (failures/ year) | $r$ (h) | $U$ (h/year) | Load (kW) | Cost (\$/kW) | Cost (\$/year) |
|---|---|---|---|---|---|---|---|
| 1 | 6 | 0.10 | 4.0 | 0.40 | 2468.62 | 18.544 | 4577.81 |
| 2 | 3 | 0.15 | 1.0 | 0.15 | 2468.62 | 9.62 | 3562.22 |
| 3 | 4 | 0.20 | 1.0 | 0.20 | 2468.62 | 9.62 | 4749.62 |
| 4 | 16 | 0.80 | 1.0 | 0.80 | 2468.62 | 9.62 | 18,988.50 |
| Total | | 1.25 | $r_{av} = U/\lambda = 3.16$ | 1.55 | | | 31,888.15 |

TABLE 14.25. Case Study 2: Load Point "F" Reliability Indices and Annual Cost of Outages

| Section Number | Length (km) | $\lambda$ (failures/ year) | $r$ (h) | $U$ (h/year) | Load (kW) | Cost (\$/kW) | Cost (\$/year) |
|---|---|---|---|---|---|---|---|
| 1 | 6 | 0.10 | 1.0 | 0.10 | 2066.68 | 9.62 | 1988.15 |
| 2 | 3 | 0.15 | 4.0 | 0.60 | 2066.68 | 18.544 | 5748.68 |
| 3 | 4 | 0.20 | 1.0 | 0.20 | 2066.68 | 9.62 | 3976.29 |
| 4 | 16 | 0.80 | 1.0 | 0.80 | 2066.68 | 9.62 | 15,905.17 |
| Total | | 1.25 | $r_{av} = U/\lambda = 1.36$ | 1.70 | | | 27,618.28 |

TABLE 14.26. Case Study 2: Load Point "G" Reliability Indices and Annual Cost of Outages

| Section Number | Length (km) | $\lambda$ (failures/ year) | $r$ (h) | $U$ (h/year) | Load (kW) | Cost (\$/kW) | Cost (\$/year) |
|---|---|---|---|---|---|---|---|
| 1 | 6 | 0.10 | 1.0 | 0.10 | 395.01 | 9.62 | 380.00 |
| 2 | 3 | 0.15 | 4.0 | 0.60 | 395.01 | 18.544 | 1098.76 |
| 3 | 4 | 0.20 | 4.0 | 0.80 | 395.01 | 18.544 | 1465.01 |
| 4 | 16 | 0.80 | 1.0 | 0.80 | 395.01 | 9.62 | 3040.00 |
| Total | | 1.25 | $r_{av} = U/\lambda = 1.84$ | 2.30 | | | 5983.77 |

#### 14.8.2.3 *Case Study 3.* Refer to Fig. 14.7 for Case Study 3.

> The peak ratings for the 25 kV feeders from substations A and B are 20.0 and 17.5 MVA, respectively, at a power factor of 0.90 lagging.
>
> ...automatic feeder sectionalizing manual switches replaced with electronic reclosers...

Tables 14.27–14.33 present the results for the reliability indices and interruption costs for the Case Study 3 at the load points A, B, C, D, E, F, and G.

Tables 14.34 and 14.35 summarize the load point reliability and interruption cost indices for the three case studies.

TABLE 14.27. Case Study 3: Load Point "A" Reliability Indices and Annual Cost of Outages

| Section Number | Length (km) | $\lambda$ (failures/ year) | $r$ (h) | $U$ (h/year) | Load (kW) | Cost (\$/kW) | Cost (\$/year) |
|---|---|---|---|---|---|---|---|
| 1 | 6 | 0.30 | 4.0 | 1.20 | 482.02 | 18.544 | 2681.57 |
| 2 | 6 | 0.0 | 0.0 | 0.0 | 482.02 | 0.0 | 0.0 |
| 3 | 3 | 0.0 | 0.0 | 0.0 | 482.02 | 0.0 | 0.0 |
| 4 | 2 | 0.0 | 0.0 | 0.0 | 482.02 | 0.0 | 0.0 |
| Total | | 0.30 | $r_{av} = U/\lambda = 4.0$ | 1.20 | | | 2681.57 |

TABLE 14.28. Case Study 3: Load Point "B" Reliability Indices and Annual Cost of Outages

| Section Number | Length (km) | $\lambda$ (failures/ year) | $r$ (h) | $U$ (h/year) | Load (kW) | Cost ($/kW) | Cost ($/year) |
|---|---|---|---|---|---|---|---|
| 1 | 6 | 0.30 | 1.0 | 0.30 | 349.58 | 9.62 | 1008.89 |
| 2 | 6 | 0.30 | 4.0 | 1.20 | 349.58 | 18.544 | 1944.78 |
| 3 | 3 | 0.0 | 0.0 | 0.0 | 349.58 | 0.0 | 0.0 |
| 4 | 2 | 0.0 | 0.0 | 0.0 | 349.58 | 0.0 | 0.0 |
| Total | | 0.60 | $r_{av} = U/\lambda = 2.5$ | 1.50 | | | 2953.67 |

TABLE 14.29. Case Study 3: Load Point "C" Reliability Indices and Annual Cost of Outages

| Section Number | Length (km) | $\lambda$ (failures/ year) | $r$ (h) | $U$ (h/year) | Load (kW) | Cost ($/kW) | Cost ($/year) |
|---|---|---|---|---|---|---|---|
| 1 | 6 | 0.30 | 1.0 | 0.30 | 2107.49 | 9.62 | 6082.22 |
| 2 | 6 | 0.30 | 1.0 | 0.30 | 2107.49 | 9.62 | 6082.22 |
| 3 | 3 | 0.15 | 4.0 | 0.06 | 2107.49 | 18.544 | 5862.19 |
| 4 | 2 | 0.0 | 0.0 | 0.0 | 2107.49 | 0.0 | 0.0 |
| Total | | 0.75 | $r_{av} = U/\lambda = 1.60$ | 1.20 | | | 18,026.63 |

TABLE 14.30. Case Study 3: Load Point "D" Reliability Indices and Annual Cost of Outages

| Section Number | Length (km) | $\lambda$ (failures/ year) | $r$ (h) | $U$ (h/year) | Load (kW) | Cost ($/kW) | Cost ($/year) |
|---|---|---|---|---|---|---|---|
| 1 | 6 | 0.30 | 1.0 | 0.30 | 3785.32 | 9.62 | 10,924.43 |
| 2 | 6 | 0.30 | 1.0 | 0.30 | 3785.32 | 9.62 | 10,924.43 |
| 3 | 3 | 0.15 | 1.0 | 0.15 | 3785.32 | 9.62 | 5462.22 |
| 4 | 2 | 0.10 | 4.0 | 0.40 | 3785.32 | 18.544 | 7019.05 |
| Total | | 0.85 | $r_{av} = U/\lambda = 1.35$ | 1.15 | | | 34,330.58 |

TABLE 14.31. Case Study 3: Load Point "E" Reliability Indices and Annual Cost of Outages

| Section Number | Length (km) | $\lambda$ (failures/ year) | $r$ (h) | $U$ (h/year) | Load (kW) | Cost ($/kW) | Cost ($/year) |
|---|---|---|---|---|---|---|---|
| 1 | 6 | 0.10 | 4.0 | 0.40 | 2468.62 | 18.544 | 4577.81 |
| 2 | 3 | 0.0 | 0.0 | 0.0 | 2468.62 | 0.0 | 0.0 |
| 3 | 4 | 0.0 | 0.0 | 0.0 | 2468.62 | 0.0 | 0.0 |
| 4 | 16 | 0.80 | 1.0 | 0.80 | 2468.62 | 9.62 | 18,988.50 |
| Total | | 0.90 | $r_{av} = U/\lambda = 1.33$ | 1.20 | | | 23,576.31 |

TABLE 14.32. Case Study 3: Load Point "F" Reliability Indices and Annual Cost of Outages

| Section Number | Length (km) | $\lambda$ (failures/year) | $r$ (h) | $U$ (h/year) | Load (kW) | Cost ($/kW) | Cost ($/year) |
|---|---|---|---|---|---|---|---|
| 1 | 6 | 0.0 | 0.0 | 0.0 | 2066.68 | 0.0 | 0.0 |
| 2 | 3 | 0.15 | 4.0 | 0.60 | 2066.68 | 18.544 | 5748.68 |
| 3 | 4 | 0.0 | 0.0 | 0.0 | 2066.68 | 0.0 | 0.0 |
| 4 | 16 | 0.80 | 1.0 | 0.80 | 2066.68 | 9.62 | 15,905.17 |
| Total | | 1.15 | $r_{av} = U/\lambda = 1.47$ | 1.4 | | | 21,653.85 |

TABLE 14.33. Case Study 3: Load Point "G" Reliability Indices and Annual Cost of Outages

| Section Number | Length (km) | $\lambda$ (failures/year) | $r$ (h) | $U$ (h/year) | Load (kW) | Cost ($/kW) | Cost ($/year) |
|---|---|---|---|---|---|---|---|
| 1 | 6 | 0.0 | 0.0 | 0.0 | 395.01 | 0.0 | 0.0 |
| 2 | 3 | 0.15 | 4.0 | 0.60 | 395.01 | 18.544 | 1098.76 |
| 3 | 4 | 0.20 | 4.0 | 0.80 | 395.01 | 18.544 | 1465.01 |
| 4 | 16 | 0.80 | 1.0 | 0.80 | 395.01 | 9.62 | 3040.00 |
| Total | | 1.1 | $r_{av} = U/\lambda = 1.91$ | 2.20 | | | 5603.77 |

TABLE 14.34. Summary of Load Point Reliability Indices for the Three Case Studies

| Load Point | $\lambda$ (outages/year) | $U$(h/year) | $r$(h/outage) | Interrupt Cost ($/Year) | Substation A Capacity (MVA) | Substation B Capacity (MVA) | Type of Feeder Sectionalizing |
|---|---|---|---|---|---|---|---|
| A | 0.85 | 1.75 | 2.06 | 5231.94 | 10.0 | 7.5 | Manual |
| | 0.85 | 1.75 | 2.06 | 5231.94 | 20.0 | 17.5 | Manual |
| | 0.30 | 1.20 | 4.00 | 2681.57 | 20.0 | 17.5 | Automatic |
| B | 0.85 | 2.65 | 3.12 | 4730.31 | 10.0 | 7.5 | Manual |
| | 0.85 | 1.75 | 2.06 | 3792.41 | 20.0 | 17.5 | Manual |
| | 0.60 | 1.50 | 2.50 | 2953.67 | 20.0 | 17.5 | Automatic |
| C | 0.85 | 3.10 | 3.65 | 31,338.38 | 10.0 | 7.5 | Manual |
| | 0.85 | 1.30 | 1.53 | 20,054.03 | 20.0 | 17.5 | Manual |
| | 0.75 | 1.20 | 1.60 | 18,026.63 | 20.0 | 17.5 | Automatic |
| D | 0.85 | 3.40 | 4.00 | 59,665.73 | 10.0 | 7.5 | Manual |
| | 0.85 | 1.15 | 1.35 | 34,330.58 | 20.0 | 17.5 | Manual |
| | 0.85 | 1.15 | 1.35 | 34,330.58 | 20.0 | 17.5 | Automatic |

TABLE 14.35. Summary of Load Point Reliability Indices for the Three Case Studies

| Load Point | $\lambda$ (outages/ year) | $U$(h/ year) | $r$(h/ outage) | Interrupt Cost ($/Year) | Substation A Capacity (MVA) | Substation B Capacity (MVA) | Type of Feeder Sectionalizing |
|---|---|---|---|---|---|---|---|
| E | 1.25 | 3.95 | 3.16 | 49,512.12 | 10.0 | 7.5 | Manual |
| | 1.25 | 1.55 | 1.24 | 31,888.15 | 20.0 | 17.5 | Manual |
| | 0.90 | 1.20 | 1.33 | 23,576.31 | 20.0 | 17.5 | Automatic |
| F | 1.25 | 4.10 | 3.28 | 42,372.73 | 10.0 | 7.5 | Manual |
| | 1.25 | 1.70 | 1.36 | 27,618.28 | 20.0 | 17.5 | Manual |
| | 0.95 | 1.40 | 1.47 | 21,653.85 | 20.0 | 17.5 | Automatic |
| G | 1.25 | 4.70 | 3.76 | 8803.82 | 10.0 | 7.5 | Manual |
| | 1.25 | 2.30 | 1.84 | 5983.77 | 20.0 | 17.5 | Manual |
| | 1.15 | 2.20 | 1.91 | 5603.77 | 20.0 | 17.5 | Automatic |

## 14.9 CONCLUSIONS

This chapter has discussed the basic reliability concepts and models required for value-based distribution system planning. Three case studies clearly reveal how to evaluate the impact of manual and automatic switching in reliability modeling of distribution systems. The methodology presents the theory and models for assessing the cost of interruptions for individual customers load points serviced by a given distribution operating configuration. The total system interruption costs based on the aggregation of individual load point interruption costs provides the basic framework for value-based reliability planning in which the reliability cost to modify a system can be balanced against the cost of interruptions. The reliability methodology for determining the key consumer indices, that is, the frequency and duration of load point interruptions is discussed. Three case studies presented in this chapter are based on an actual distribution system.

## REFERENCES


1. J. E. Propst, Calculating electrical risk and reliability, *IEEE Transactions on Industry Applications*, **31** (5), 1995, PCIC Conference Record 94–2.
2. D. O. Koval, Zone-branch reliability methodology for analyzing industrial power systems, *IEEE Transactions on Industry Applications*, **36** (5), 2000, 1212–1219.
3. D. O. Koval, Evaluation of distribution system reliability, Ph.D. Thesis, University of Saskatchewan, Saskatoon, Saskatchewan, June 1978.
4. IEEE Standard 493-1997, IEEE Recommended Practice for Design of Reliable Industrial and Commercial Power Systems (IEEE Gold Book), The Institute of Electrical and Electronics Engineers, Inc., New York, 1997.

5. D. O. Koval, X. Zhang, J. E. Propst, T. Coyle, R. G. Arno and R. S. Hale, Jr, Reliability methodologies applied to The IEEE Gold Book Standard Network, *IEEE Industry Applications Magazine*, **9** (1), 2003, 32–41.
6. J. E. Propst and D. R. Doan, Improvements in modeling and evaluation of electrical power system reliability, *IEEE Transactions on Industry Applications*, **37** (5), 2001, 1413–1422.
7. J. E. Propst, PCIC 2000-2 Reliability Software and Documentation, Reliability Model Spreadsheet Software, Available from Equilon Enterprises, LLC and Web: www.ieee-pcic.org

# 15

# VALUE-BASED PREDICTIVE RELIABILITY ASSESSMENT

## 15.1 INTRODUCTION

Society is becoming increasingly dependent on a cost-effective reliable electric power supply. Unreliable electric power supplies can be extremely costly to electric utilities and their customers. Predictive reliability assessment combines historical outage data and mathematical models to estimate the performance of specific network and system configurations (e.g., IEEE Standard 493–2007). This chapter is concerned with the value-based assessment of proposed modifications to an existing industrial distribution system configuration to minimize the costs of interruptions to both utility and its industrial customers. It presents a series of case studies of an actual customer load area supplied by two feeder circuits originating from two alternative substations. Each case study reveals the impact on the cost of industrial load point interruptions and the frequency and duration of industrial load point interruptions when various system constraints (e.g., ideal and nonideal protection coordination schemes, substation capacity restrictions, etc.) are imposed on the distribution system. The chapter discusses in some detail the variance in reliability performance indices and its impact on the cost of load point interruptions. A basic conclusion of this chapter is that expansion plans of an

*Power Distribution System Reliability.* By Ali A. Chowdhury and Don O. Koval

industrial distribution system can be optimized in terms of reliability by using an economic criterion in which the sum of both the industrial facility interruptions and the utility system costs are minimized.

As stated earlier in Chapters 1 and 10, two approaches to reliability evaluation of power system distribution systems are frequently used, namely, historical assessment and predictive assessment. Historical reliability assessment involves the collection and analysis of an electric system's outage and interruption data. It is essential for electric utilities to measure actual distribution system reliability levels and define performance indicators to assess their basic functions of providing cost-effective reliable power supply to all sectors of society.

The distribution system is an important part of the total electrical supply system, as it provides the final link between a utility's bulk transmission system and its customers. It has been reported that 80% of all customer interruptions occur due to failures in distribution systems. Historical assessment generally analyzes discrete interruption events occurring at specific locations over specific time periods, whereas predictive assessment determines the long-term behavior of systems by combining component failure rates and the duration of repair, restoration, switching, and isolation activities that describe the central tendency of an entire utility's distribution system of the possible values for given network configurations. Accurate component outage data are therefore key to distribution system predictive performance analysis. In addition to the physical configuration of the distribution network, the reliability characteristics of system components, the operation of protection equipment, and the availability of alternative supplies with adequate capacity also have a significant impact on service reliability.

In practice, the determination of acceptable levels of service continuity is generally achieved by comparing the actual interruption frequency and duration indices with arbitrary targets. These targets are based on the perception of customer tolerance levels for service interruptions. It has, however, long been recognized that rules of the thumb and implicit criteria cannot be used in a consistent manner when a very large number of capital investments and operating decisions are routinely being made. As a result, there is a growing interest in economic optimization approaches to distribution planning and expansion. The basic concepts involved in utilizing customer interruption costs (CICs) in association with customer reliability indices in distribution system planning are illustrated in this chapter as well as in the subsequent chapters. The basic distribution system used in illustrating the value-based planning concept is the same distribution system that has been used in Chapter 14.

## 15.2 VALUE-BASED RELIABILITY PLANNING

Section 11.6 briefly discussed value-based reliability planning principle with a very simple illustrative example. This chapter provides an in-depth insight into the value-based reliability methodology using a practical distribution system that has been introduced in Chapter 14 for computing load point reliability index calculation purposes.

A value-based reliability planning approach attempts to locate the minimum cost solution where the total cost includes the utility investment costs plus the operating costs plus the customer interruption costs as shown in Fig. 11.2. Figure 11.2 illustrates how utility costs (i.e., reflected in customer rates) and customer interruption costs are combined to give "total cost."

The value of service, that is, the worth of reliability expressed in terms of costs of customer interruptions can be established on the basis of actual surveys of customer perception regarding the level of service reliability they are willing to pay for. By establishing a method of giving a dollar value to various levels of service reliability, it is possible to ascertain the balance where distribution system reliability is best matched. The data compiled from customer surveys lead to the creation of sector damage functions. The cost of interruptions at a single customer load point depends entirely on the cost characteristics of that customer. The sector damage function presents the sector interruption costs as a function of the duration of service interruptions. The customer costs associated with an interruption at any load point in the distribution system involve the combination of costs associated with all customer types affected by that distribution system outage. This combination leads to the generation of a composite customer damage function. Some references at the end of this chapter illustrate the approach involved in creating a composite customer damage function. The customer damage functions after escalation to 1995 Canadian dollars for each sector are shown in Fig. 15.1. The cost of interruptions is expressed in dollars per kilowatt.

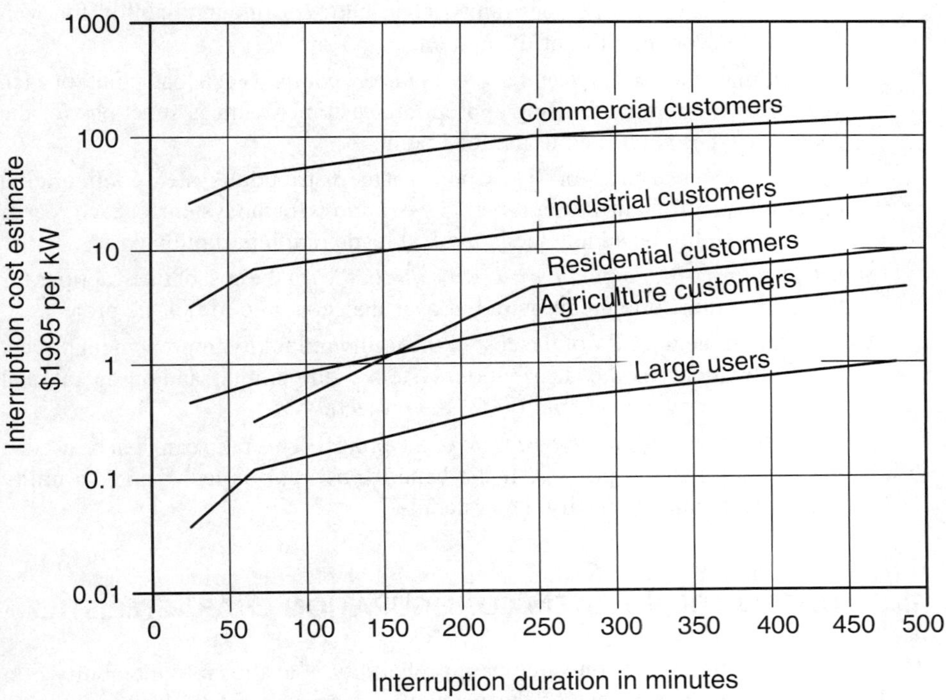


Figure 15.1. Utility customer cost damage functions.

Predicting distribution system reliability performance is normally concerned with the electric supply adequacy at the customer load point. The basic indices used in practice are load point average failure rate ($\lambda$), average outage duration ($r$), and the average annual outage time ($U$). For a radial system, the basic equations (IEEE Standard 493) for calculating the reliability indices at each load point "p" in a radial circuit are

$$\lambda_{\mathrm{p}} = \sum_{i=1}^{n} \lambda_i \quad \text{failures/year} \tag{15.1}$$

$$U_{\mathrm{p}} = \sum_{i=1}^{n} \lambda_i \times r_i \quad \text{h/year} \tag{15.2}$$

$$r_{\mathrm{p}} = \frac{U_{\mathrm{p}}}{\lambda_{\mathrm{p}}} \quad \text{h/failure} \tag{15.3}$$

where $n$ is the number of outage events affecting load point "p."

The steps associated with the value-based distribution system reliability planning approach are summarized as follows:

*Step 1*: Calculate the *reliability* of each load point being serviced by a given distribution system configuration considering all interruption events and system constraints (e.g., voltage constraints) contributing to its unreliability for each year of the economic life of the system.

*Step 2*: Estimate the *expected annual cost of interruptions* at each load point for each year of its economic life using appropriate customer damage functions for the customer types connected to the load point.

*Step 3*: Repeat Steps 1 and 2 for all load points of the distribution system configuration under study and obtain the *total cost of interruptions* for the system for each year of its economic life by adding the individual load point interruption costs.

*Step 4*: Determine the *cumulative present value* (CPV) of the cost of interruptions for the distribution system configuration over the economic life of the project.

*Step 5*: Determine the CPV of the cost of the utility reliability improvement project (e.g., alterations to the distribution system configuration and/or operational practices) over the economic life of the project.

*Step 6*: Determine the *benefit–cost ratio* of the project. This ratio can then be used in the decision-making process. If the benefit–cost ratio is less than 1, a utility cannot economically justify the system.

## 15.3 DISTRIBUTION SYSTEM CONFIGURATION CHARACTERISTICS

The objective of this chapter on value-based reliability is to illustrate the ability of a utility and its customers to assess the economic impact of modifying an existing distribution system configuration (e.g., adding an additional feeder) in an attempt to

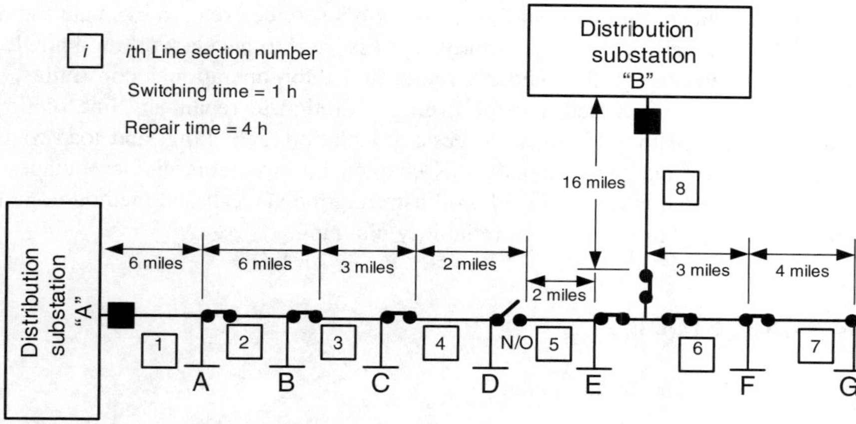


| | A | B | C | D | E | F | G |
|---|---|---|---|---|---|---|---|
| Average load (kW) | 482.02 | 349.58 | 2107.09 | 3785.32 | 2468.82 | 2066.68 | 395.01 |
| Peak load (kW) | 626 | 545 | 2737 | 4916 | 3206 | 2684 | 513 |

Figure 15.2. Distribution radial network configuration showing peak and average loads at each load point.

improve system reliability and to minimize the total cost of industrial customer interruptions and utility costs and to evaluate the benefit–cost ratio for the system modification. The load of a distribution system service area is supplied by two 25 kV distribution feeder circuits, as shown in Fig. 15.2. The 25 kV feeders from substations A and B are operated as radial feeders, although they can be interconnected by a normally open tie point. The disconnects, lateral distributors, step-down transformers, fuses, and the alternative supply are assumed to be 100% available in the analysis to illustrate the reliability value-based planning methodology.

The load factor for the industrial service area is assumed to be 77%. The loading conditions at each load point for 1995 are shown in Fig. 15.2. For the purposes of simplicity, the service area is assumed to be entirely industrial. The peak rating for the 25 kV feeders from substations A and B are 12.00 and 10.5 MVA at a power factor of 0.90 lagging. The 25 kV feeder failure rate is assumed to be 2.0 failures per 100 miles per year.

The CIC associated with an outage in section $j$ (i.e., SECT) outage is

$$\text{CIC} = \lambda_j \times L_{\text{p}} \times C(rj, p) \tag{15.4}$$

where $\lambda_j$ is the failure rate of feeder section $j$, $L_{\text{p}}$ is the average connected load at load point "p", and $C$ $(rj, p)$ is the cost of interruption in dollars per kilowatt for an outage duration of $rj$ associated with feeder section $j$.

The total cost of interruptions for load point "p" can be determined by adding the cost of all section outages (i.e., Steps 1 and 2). The total cost of customer interruptions for all customers (Step 3) can then be evaluated.

Four case studies involving the distribution system will be presented. Each case study will involve a set of different constraints imposed on the two feeder circuits serving the industrial loads within the distribution system's service area. To evaluate the load point reliability levels for each case study, it is essential to have a working knowledge base of the operation of the feeder circuits and their operational constraints. For simplicity, it will be assumed that the average duration to repair any line section is 4 h and the duration to perform the necessary isolation, switching, and load transfer activities to be an average of 1 h. In this chapter, multiple contingency outages are neglected and the emphasis is placed on illustrating the value-based methodology that can be used for distribution system reliability planning.

## 15.4 CASE STUDIES

*Case Study 1*: Feeder circuits radial

- Manual sectionalizing
- No feeder interties

If both feeders are operated radially with no interties, the calculation of the reliability indices and the cost of load point interruptions for load point "C" are illustrated in Table 15.1. With no feeder interties, the loads on unfaulted line sections cannot be transferred to the adjacent feeder circuit. The reliability indices and costs of individual load point interruptions are shown in Table 15.2 for Case Study 1.

Note that for a radial feeder, the failure rates for all load points on a feeder circuit are identical. The load points farther from the substation supply experience longer interruption durations. Also, note that, for this distribution system, the larger loads (e.g., "D," "E," and "F") are located close to the ends of the distribution feeder circuits.

*Case Study 2*: Looped radial feeders

- Manual sectionalizing
- Unrestricted feeder tie capacity

If both feeders are operated radially and are tied through a normally open tie switch, then any line section outage can be manually isolated and the remaining line sections can

TABLE 15.1. Load Point "C" Reliability Indices and Annual Cost of Outages

| Section | Length (miles) | $\lambda$ (failures/year) | $r$ (h) | Average Load (kW) | Cost ($/kW) | Interruption Cost ($/year) |
|---|---|---|---|---|---|---|
| 1 | 6 | 0.12 | 4 | 2107.49 | 18.544 | 4689.76 |
| 2 | 6 | 0.12 | 4 | 2107.49 | 18.544 | 4689.76 |
| 3 | 3 | 0.06 | 4 | 2107.49 | 18.544 | 2344.88 |
| 4 | 2 | 0.04 | 1 | 2107.49 | 9.620 | 810.96 |
| Total | | 0.34 | 3.12 | | | $12,535.35 |

TABLE 15.2. Distribution System Load Point Reliability Indices and Annual Cost of Outages: Case Study 1

| Load Point | $\lambda$ (outages/year) | $U$ (hours/year) | $r$ (h/outage) | Interruption Cost ($/year) |
|---|---|---|---|---|
| A | 0.34 | 0.70 | 2.06 | 2092.78 |
| B | 0.34 | 1.06 | 3.12 | 1892.12 |
| C | 0.34 | 1.24 | 3.65 | 12,535.35 |
| D | 0.34 | 1.36 | 4.00 | 23,866.29 |
| E | 0.50 | 1.58 | 3.16 | 19,804.85 |
| F | 0.50 | 1.64 | 3.28 | 16,949.09 |
| G | 0.50 | 1.88 | 3.76 | 3521.53 |
| Annual cost of interruptions | | | | $80,662.01 |

TABLE 15.3. Load Point "C" Reliability Indices and Annual Cost of Outages

| Section | Length (miles) | $\lambda$ (failures/year) | $r$ (h) | Average Load (kW) | Cost ($/kW) | Interruption Cost ($/year) |
|---|---|---|---|---|---|---|
| 1 | 6 | 0.12 | 1 | 2107.49 | 9.620 | 2432.89 |
| 2 | 6 | 0.12 | 1 | 2107.49 | 9.620 | 2432.89 |
| 3 | 3 | 0.06 | 4 | 2107.49 | 18.544 | 2344.88 |
| 4 | 2 | 0.04 | *1* | 2107.49 | 9.620 | *810.96* |
| Total | | 0.34 | 1.53 | | | $8021.61 |

be energized from the alternative feeder or from the nearest continuous feeder. The load transfer is possible because the feeder circuits and substation are assumed to be unrestricted in capacity and the substations will not be overloaded. The reliability indices and cost of interruptions for load point "C" are illustrated in Table 15.3. The reliability indices and costs of load point interruptions are shown in Table 15.4.

TABLE 15.4. Distribution System Load Point Reliability Indices and Annual Cost of Outages: Case Study 2

| Load Point | $\lambda$ (outages/year) | $U$ (hours/year) | $r$ (h/outage) | Interruption Cost ($/year) |
|---|---|---|---|---|
| A | 0.34 | 0.70 | 2.06 | 2092.78 |
| B | 0.34 | 0.70 | 2.06 | 1517.76 |
| C | 0.34 | 0.52 | 1.53 | 8021.61 |
| D | 0.34 | 0.46 | 1.35 | 13,732.23 |
| E | 0.50 | 0.62 | 1.24 | 12,755.26 |
| F | 0.50 | 0.74 | 1.48 | 11,416.17 |
| G | 0.50 | 0.92 | 1.84 | 2393.51 |
| Annual cost of interruptions | | | | $51,929.33 |

TABLE 15.5. Load Point "C" Reliability Indices and Annual Cost of Outages

| Section | Length (miles) | $\lambda$ (failures/year) | $r$ (h) | Average Load (kW) | Cost ($/kW) | Interruption Cost ($/year) |
|---|---|---|---|---|---|---|
| 1 | 6 | 0.12 | 1 | 2107.49 | 9.620 | 2432.89 |
| 2 | 6 | 0.12 | 1 | 2107.49 | 9.620 | 2432.89 |
| 3 | 3 | 0.06 | 4 | 2107.49 | 18.544 | 2344.88 |
| 4 | 2 | $\approx$0.0 | $\approx$0 | 2107.49 | $\approx$0.000 | 0.00 |
| Total | | 0.3 | 1.60 | | | $7210.65 |

Note that the significant difference in the total industrial cost of interruptions when the feeder circuits are operated strictly as radial feeders (i.e., Case Study 1) and unrestricted loop radial feeders (i.e., Case Study 2).

*Case Study 3*: Looped radial feeders

- Automatic sectionalizing
- Unrestricted feeder tie capacity

If both feeders are operated radially and are tied through a normally open tie switch, then any line section outage can be automatically sectionalized and isolated and the remaining line sections can be energized from the alternative feeder or from the nearest continuous feeder. For this case study, it is assumed that the sectionalizing activities do not disrupt any of the loads on the feeders (i.e., the feeder loads will ride through the sectionalizing activities). If this assumption is invalid, then the results of Case Study 2 are valid. The reliability indices and cost of interruptions for load point "C" are illustrated in Table 15.5. The reliability indices and costs of load point interruptions are listed in Table 15.6.

TABLE 15.6. Distribution System Load Point Reliability Indices and Annual Cost of Outages: Case Study 3

| Load Point | $\lambda$ (outages/year) | $U$ (h/year) | $r$ (h/outage) | Interruption Cost ($/year) |
|---|---|---|---|---|
| A | 0.12 | 0.48 | 4.00 | 1072.63 |
| B | 0.24 | 0.60 | 2.50 | 1181.47 |
| C | 0.30 | 0.48 | 1.60 | 7210.65 |
| D | 0.34 | 0.46 | 1.35 | 13,732.23 |
| E | 0.36 | 0.48 | 1.33 | 9430.52 |
| F | 0.38 | 0.38 | 1.00 | 7544.96 |
| G | 0.46 | 0.88 | 1.91 | 2241.51 |
| Annual cost of interruptions | | | | $42,423.97 |

TABLE 15.7. Load Point "D" Reliability Indices and Annual Cost of Outages "Transfer Load"

| Section | Length (miles) | $\lambda$ (failures/year) | $r$ (h) | Average Load (kW) | Cost ($/kW) | Interruption Cost ($/year) |
|---|---|---|---|---|---|---|
| 1 | 6 | 0.12 | 1 | 2316.49 | 9.620 | 2708.44 |
| 2 | 6 | 0.12 | 1 | 2316.49 | 9.620 | 2708.44 |
| 3 | 3 | 0.06 | 1 | 2316.49 | 9.620 | 1354.22 |
| 4 | 2 | *0.04* | *4* | 2316.49 | 18.544 | *1740.31* |
| Total | | 0.34 | 1.35 | | | $8511.61 |

Note that with automated sectionalizers placed in a looped radial feeder, there is a significant reduction in the cost of industrial point interruptions compared to when the feeder circuits are operated as pure radial circuits with no interties.

*Case Study 4*: Looped radial feeders

- Manual sectionalizing
- Restricted feeder tie capacity

If both feeders are operated radially and are tied through a normally open tie switch, then any line section outage can be manually isolated and the load on the remaining line sections must be evaluated as to whether portions of the load can be interrupted (i.e., shed) and what loads can be energized from the alternative feeder. In this case study, only portions of loads "D" and "E" can be supplied from the alternative feeder. To simplify the calculations, load "D" can be subdivided into two loads, namely, a "transfer load" and an "interruptible load." The reliability indices and cost of interruptions for load point "D" are illustrated in Tables 15.7 and 15.8. The reliability indices and costs of load point interruptions are listed in Table 15.9.

When the substation and feeder capacity levels are restricted, there is a significant increase in the cost of industrial customer interruptions. If there is any significant load growth in the distribution system area, the increased loading can significantly limit the transfer capabilities of the feeder circuits.

TABLE 15.8. Load Point "D" Reliability Indices and Annual Cost of Outages "Interruptible Load"

| Section | Length (miles) | $\lambda$ (failures/year) | $r$ (h) | Average Load (kW) | Cost ($/kW) | Interruption Cost ($/year) |
|---|---|---|---|---|---|---|
| 1 | 6 | 0.12 | 4 | 1439.13 | 18.544 | 3202.47 |
| 2 | 6 | 0.12 | 4 | 1439.13 | 18.544 | 3202.47 |
| 3 | 3 | 0.06 | 1 | 1439.13 | 18.544 | 1601.23 |
| 4 | 2 | 0.04 | 4 | 1439.13 | 18.544 | 1067.49 |
| Total | | 0.34 | 4.0 | | | $9073.66 |

TABLE 15.9. Distribution System Load Point Reliability Indices and Annual Cost of Outages: Case Study 4

| Load Point | $\lambda$ (outages/year) | $U$ (h/year) | $r$ (h/outage) | Interruption Cost ($/year) |
|---|---|---|---|---|
| A | 0.34 | 0.70 | 2.06 | 2092.78 |
| B | 0.34 | 1.06 | 3.12 | 1892.12 |
| C | 0.34 | 1.24 | 3.65 | 12,535.35 |
| D | 0.34 | 0.46/1.36[a] | 1.35/4.00[a] | 17,585.07 |
| E | 0.50 | 0.62/1.58[a] | 1.24/3.16[a] | 8223.68 |
| F | 0.50 | 1.64 | 3.28 | 16,949.09 |
| G | 0.50 | 1.88 | 3.76 | 3521.53 |
| Annual cost of interruptions | | | | $69,835.72 |

[a] Transfer load index/interruptible load index.

In this particular case study, it is assumed that load points “D” and “E” can shed a portion of their loads during line section outages. If these load points cannot shed any load because of their process limitations, for example, then the feeder circuits are radial and the cost of industrial customer interruptions will rise (e.g., see Case Study 1). The other loads on the feeders could not shed load due to the restrictions of both substations.

A summary of the cost of industrial customer interruptions for each case study is provided in Table 15.10.

From an individual industrial customer’s point of view, the difference in the annual cost of interruptions between the different case studies can vary significantly. The cost of

TABLE 15.10. Summary of Load Point Interruption Costs for Each Case Study

| Load Point | Case Study 1 Radial feeders No feeder interties Manual sectionalizing | Case Study 2 Looped radial feeders Manual sectionalizing Unrestricted tie capacity | Case Study 3 Looped radial feeders Automatic sectionalizing Unrestricted tie capacity | Case Study 4 Looped radial feeders Manual sectionalizing Restricted tie capacity |
|---|---|---|---|---|
| A | 2092.78 | 2,092.78 | 1,072.63 | 2,092.78 |
| B | 1892.12 | 1,517.76 | 1,181.47 | 1,892.12 |
| C | 12,535.35 | 8,021.61 | 7,210.65 | 12,535.35 |
| D | 23,866.29 | 13,732.23 | 13,585.07 | 17,585.07 |
| E | 19,804.85 | 12,755.26 | 9,430.52 | 15,259.78 |
| F | 16,949.09 | 11,416.17 | 7,554.96 | 16,949.09 |
| G | 3521.53 | 2,393.51 | 2,241.51 | 3,521.53 |
| Total | $80,622.01 | $51,929.33 | $42,423.97 | $69,835.72 |

TABLE 15.11. Benefit–Cost Ratios for All Case Studies

| Annual Load Growth (%) | Case Study 1<br>Radial feeders<br>Manual sectionalizing<br>No feeder interties | Case Study 2<br>Looped radial feeders<br>Manual sectionalizing<br>Unrestricted tie capacity | Case Study 3<br>Looped radial feeders<br>Automatic sectionalizing<br>Unrestricted tie capacity | Case Study 4<br>Looped radial feeders<br>Manual sectionalizing<br>Restricted tie capacity |
|---|---|---|---|---|
| 0.0 | 1.66 | 1.15 | 0.97 | 1.47 |
| 2.0 | 1.95 | 1.33 | 1.13 | 1.72 |
| 5.0 | 2.57 | 1.73 | 1.45 | 2.25 |

interruptions to some industrial customers does not change when the feeder configurations are operating under different constraints. The greatest impact on the change in cost of interruptions occurs for larger customer loads, particularly when they are close to the end of the radial feeders.

Applying Step 4 to Case Study 4 and assuming a discount rate of 9.47%, an inflation rate of 2.5%, a load growth rate of 5%, the economic life of the project to be 25 years, and the cumulative value of the cost of consumer interruptions is \$1,434,470. The addition of the future feeder, assuming negligible outages and automatic sectionalizers, is assumed for illustration purposes only.

In addition, through load flow studies, the line losses of the new system are significantly less with a cumulative savings based on 1.08 cents per kWh for a total of \$147,208. The total cost benefits to the customer by adding the third feeder would be \$1,581.678. Actual benefits would be slightly lower since the additional feeder would not completely eliminate outages (i.e., these calculations are beyond the scope of this chapter).

Applying Step 5, the cost of building the third feeder circuit (i.e., future expansion) is \$539,000 and assuming the same economical parameters plus 2% of the capital cost for operating and maintenance costs results in a cumulative present value of \$702,879. The benefit–cost ratio = (1,581,678/702,879) = 2.25% or 225%.

A summary of the benefit–cost ratios for each case study is provided in Table 15.11.

It is important to note that load growth has a significant impact on the benefit–cost ratio for any system.

To investigate the impact of varying feeder reliability on the overall system performance, similar computations are performed by varying the 25 kV feeder failure rate (Step 6). The cumulative present value of benefits including loss savings for feeder failure rates of 0.5, 0.75, 1.0, and 2.0 failures per 100 miles per year are presented in Table 15.12.

Examination of Table 15.12 reveals that if distribution feeders have a low failure rate, the benefit–cost ratio is less than 100% and the addition of another feeder cannot be economically justified. If the benefit–cost ratio significantly exceeds 100%, then the system alterations can be justified. It is also clear that when the failure rates of the feeder

TABLE 15.12. Cumulative Present Value of Customer Cost Benefits and Percentage Benefit–Cost Ratio

| Annual Load Growth (%) | 0.5 failures/100 miles/year | 0.75 failures/100 miles/year | 1.0 failures/100 miles/year | 2.0 failures/100 miles/year |
|---|---|---|---|---|
| 0.0 | \$560,091 (80%) | \$641,827 (91%) | \$723,562 (103%) 23 years | \$1,032,282 (147%) 3 years |
| 2.0 | \$645,855 (92%) | \$744,569 (106%) 22 years | \$843,283 (120%) 15 years | \$1,208,865 (172%) 2–3 years |
| 5.0 | \$826,916 (117%) 19–20 years | \$961,473 (137%) 13–14 years | \$1,096,030 (156%) 11 years | \$1,581,678 (225%) 2–3 years |

sections are low and the industrial load is fairly constant over time (i.e., zero load growth), the benefit–cost ratios are less than 1 and a utility cannot economically add the third feeder in this case.

In Table 15.12, the year of crossover point is where the cumulative present value of the industrial customer benefits exceeds the cumulative present value of the revenue requirements of the capital expenditures for the future expansion. It is clear that if the load growth rate and the feeder line section failure rate are high, then the crossover period is several years.

It is important to note that the value-based methodology is clearly very sensitive to a number of parameters, namely, the interruption cost data, the load data, the customer mix in the service area, and the distribution system data such as outage duration, failure rate, operating constraints, and so on.

## 15.5 ILLUSTRATIVE EXAMPLE SYSTEM PROBLEM AND ITS RELIABILITY CALCULATIONS

The load of the distribution system's industrial service area is supplied by two 25 kV distribution feeder circuits, as shown in Fig. 15.3. The 25 kV feeders from substations A and B are operated as radial feeders, although they can be interconnected by a normally open tie point. The disconnects, lateral distributors, step-down transformers, fuses, and the alternative supply are assumed to be 100% available in the analysis to simplify the reliability value-based planning methodology.

The physical lengths of each line section are shown in Table 15.13. The loading conditions at each load point for 2004 are shown in Table 15.14. The interruption cost for

TABLE 15.13. Distribution Feeder Line Section Lengths

| Line section number | 1 | 2 | 3 | 4 | 5 | 6 | 7 | 8 | 9 | 10 | 11 | 12 | 13 | 14 | 15 |
|---|---|---|---|---|---|---|---|---|---|---|---|---|---|---|---|
| Physical length (km) | 2 | 1 | 2 | 3 | 3 | 3 | 2 | 3 | 1 | 5 | 2 | 4 | 3 | 2 | 3 |

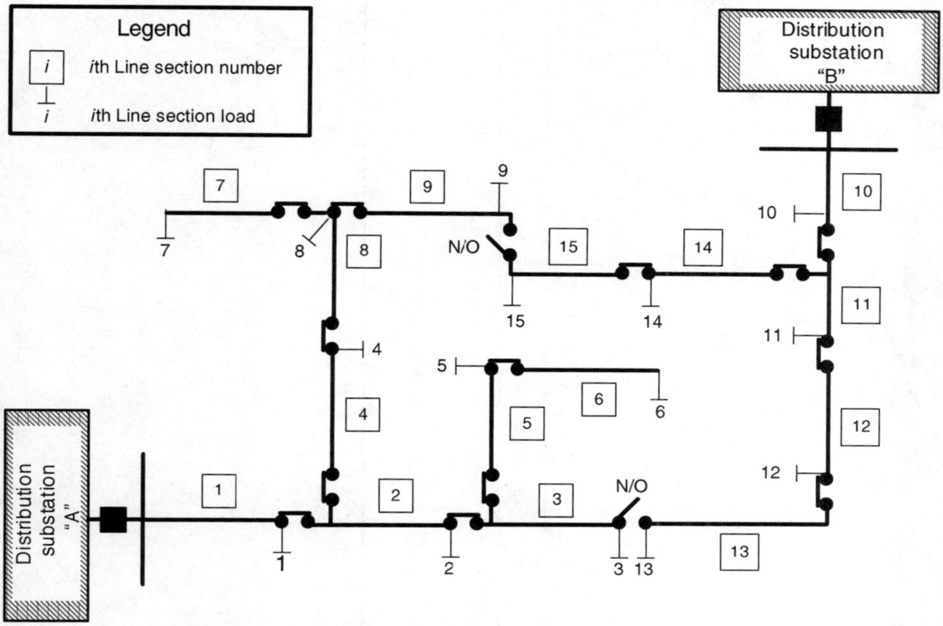


Figure 15.3. Manually sectionalized distribution feeders.

a 1 h interruption to a customer is \$10.00/average kW load interrupted and for a 4 h interruption \$25.00/average kW load interrupted.

## 15.5.1 Operating Procedures

If a fault occurs in any line section, the respective substation breaker is assumed to trip and deenergize the entire feeder. The faulty line section is first isolated, and whether some or all the loads are transferred to the adjacent energized substation depends on the following constrained operating conditions:

1. The normally opened tie switches interconnecting substation A and B will be closed only if the sum of the peak loads of the energized substation plus the sum of the isolated peak loads being transferred to the energized feeder from the deenergized substation do not exceed the rated capacity of the energized substation.
2. If the sum of the peak loads being transferred to the energized feeder and the sum of the peak loads of the energized substation exceed the rated capacity of the substation, the normally opened tie switches will not be closed.

***15.5.1.1 Feeder Characteristics: Looped Radial Feeders—Manual Sectionalizing.*** If both feeders are operated radially and are tied through a normally open tie switch, then any line section outage can be manually isolated and the remaining

TABLE 15.14. Distribution Feeder Loads

| Load Point | 1 | 2 | 3 | 4 | 5 | 6 | 7 | 8 | 9 | 10 | 11 | 12 | 13 | 14 | 15 |
|---|---|---|---|---|---|---|---|---|---|---|---|---|---|---|---|
| Average load (kW) | 800 | 400 | 800 | 1200 | 560 | 1040 | 400 | 800 | 1200 | 1000 | 200 | 400 | 600 | 800 | 600 |
| Peak load (kW) | 1000 | 500 | 1000 | 1500 | 700 | 1300 | 500 | 1000 | 1500 | 1250 | 250 | 500 | 750 | 1000 | 750 |

line sections can be energized from the alternative feeder or from the nearest continuous feeder. The load transfer is possible only when the feeder circuits and substation are unrestricted in capacity and the substations are not overloaded. If the substation is overloaded, then the load transfer will not occur.

#### *15.5.1.2 Reliability Data.*

$\lambda$ (feeder failure rate) = 2 failures per 100 km per year
Repair time = 4 h per feeder section failure
Switching time = 1.0 h (includes one or more switching operations)

The following reliability indices and cost of interruptions for each load point for the three case studies are calculated:

1. $\lambda$: number of feeder outages per year.
2. $U$: total duration of feeder outages in hours per year.
3. $r$: average duration of a feeder outage.
4. Total annual interruption costs per year at each load point and the total annual cost of interruptions for the distribution system.

## Case 15.1

The peak ratings for the 25 kV feeders from substations A and B are 10.0 and 5.0 MVA, respectively, at a power factor of 0.90 lagging.

- Manual feeder sectionalizing

## Case 15.2

The peak ratings for the 25 kV feeders from substations A and B are 15.0 and 10.0 MVA, respectively, at a power factor of 0.90 lagging.

- Manual feeder sectionalizing

## Case 15.3

The peak ratings for the 25 kV feeders from substations A and B are 15.0 and 15.0 MVA, respectively, at a power factor of 0.90 lagging. The OCRs replaced several manual disconnect switches as shown in Fig. 15.4.

*Solution:*

Results for the three case studies are provided in Tables 15.15–15.17

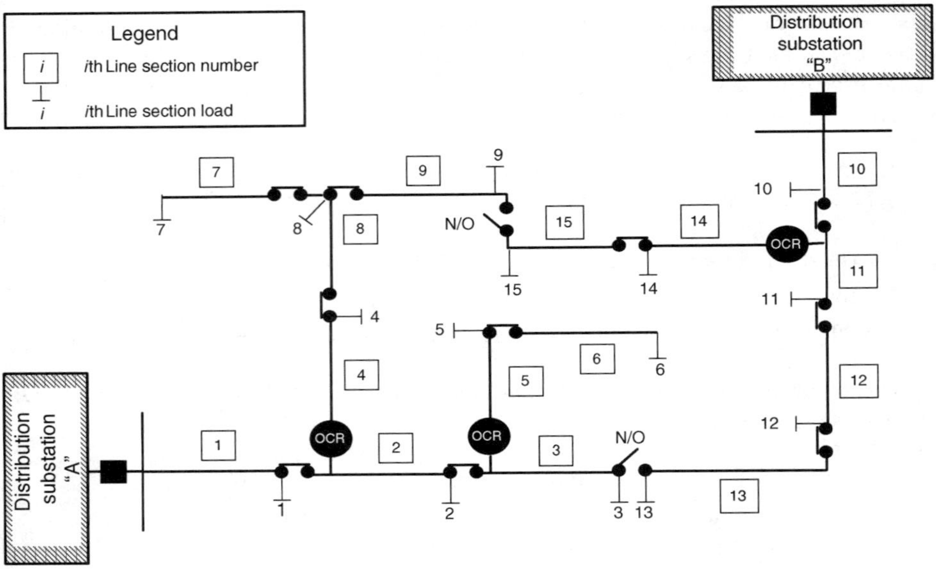


Figure 15.4. Manually sectionalized distribution feeders with OCRs for laterals.

TABLE 15.15. Summary of Results: Case Study 1

| Load Point | $\lambda$ (interruptions/year) | $U$ (h/year) | $r$ (h/interruption) | Interruption Cost ($/year) |
|---|---|---|---|---|
| 1 | 0.40 | 0.52 | 1.30 | 3680.00 |
| 2 | 0.40 | 0.58 | 1.45 | 1960.00 |
| 3 | 0.40 | 0.70 | 1.75 | 4400.00 |
| 4 | 0.40 | 0.76 | 1.90 | 6960.00 |
| 5 | 0.40 | 0.88 | 2.20 | 3584.00 |
| 6 | 0.40 | 1.06 | 2.65 | 7592.00 |
| 7 | 0.40 | 1.06 | 2.65 | 2920.00 |
| 8 | 0.40 | 0.94 | 2.35 | 5360.00 |
| 9 | 0.40 | 1.00 | 2.50 | 8400.00 |
| 10 | 0.38 | 0.68 | 1.79 | 5300.00 |
| 11 | 0.38 | 0.80 | 2.11 | 1180.00 |
| 12 | 0.38 | 1.04 | 2.74 | 2840.00 |
| 13 | 0.38 | 1.22 | 3.21 | 4800.00 |
| 14 | 0.38 | 0.92 | 2.42 | 5200.00 |
| 15 | 0.38 | 1.10 | 2.89 | 4440.00 |
| Annual cost of interruptions for the distribution system | | | | $68,616.00 |

TABLE 15.16. Summary of Results: Case Study 2

| Load Point | $\lambda$ (interruptions/year) | $U$ (h/year) | $r$ (h/interruption) | Interruption Cost ($/year) |
|---|---|---|---|---|
| 1 | 0.40 | 0.52 | 1.30 | 3680.00 |
| 2 | 0.40 | 0.58 | 1.45 | 1960.00 |
| 3 | 0.40 | 0.64 | 1.60 | 4160.00 |
| 4 | 0.40 | 0.64 | 1.60 | 6240.00 |
| 5 | 0.40 | 0.82 | 2.05 | 3416.00 |
| 6 | 0.40 | 1.00 | 2.50 | 7280.00 |
| 7 | 0.40 | 0.76 | 1.90 | 2320.00 |
| 8 | 0.40 | 0.58 | 1.45 | 3920.00 |
| 9 | 0.40 | 0.46 | 1.15 | 5160.00 |
| 10 | 0.38 | 0.68 | 1.79 | 5300.00 |
| 11 | 0.38 | 0.50 | 1.32 | 880.00 |
| 12 | 0.38 | 0.62 | 1.63 | 2000.00 |
| 13 | 0.38 | 0.56 | 1.47 | 2820.00 |
| 14 | 0.38 | 0.50 | 1.32 | 3520.00 |
| 15 | 0.38 | 0.56 | 1.47 | 2820.00 |
| Annual cost of interruptions for the distribution system | | | | $55,476.00 |

TABLE 15.17. Summary of Results: Case Study 3

| Load Point | $\lambda$ (interruptions/year) | $U$ (h/year) | $r$ (h/interruption) | Interruption Cost ($/year) |
|---|---|---|---|---|
| 1 | 0.10 | 0.22 | 2.20 | 1280.00 |
| 2 | 0.10 | 0.16 | 1.60 | 520.00 |
| 3 | 0.10 | 0.22 | 2.20 | 1280.00 |
| 4 | 0.28 | 0.46 | 1.64 | 4440.00 |
| 5 | 0.22 | 0.52 | 2.36 | 2072.00 |
| 6 | 0.22 | 0.70 | 3.18 | 4784.00 |
| 7 | 0.28 | 0.58 | 2.07 | 1720.00 |
| 8 | 0.28 | 0.46 | 1.64 | 2960.00 |
| 9 | 0.28 | 0.34 | 1.21 | 3720.00 |
| 10 | 0.28 | 0.58 | 2.07 | 4300.00 |
| 11 | 0.28 | 0.40 | 1.43 | 680.00 |
| 12 | 0.28 | 0.52 | 1.86 | 1600.00 |
| 13 | 0.28 | 0.46 | 1.64 | 2220.00 |
| 14 | 0.38 | 0.50 | 1.32 | 3520.00 |
| 15 | 0.38 | 0.56 | 1.47 | 2820.00 |
| Annual cost of interruptions for the distribution system | | | | $37,916.00 |

## 15.6 CONCLUSIONS

This chapter has presented the basic concepts involved in using customer interruption cost data to evaluate distribution system reliability worth at individual customer load points. Predictive distribution system reliability indices have been computed and used to estimate the customer interruption costs by considering outages and various feeder constraints in various radial and looped radial feeder circuits within a distribution system area and the benefit–cost ratios obtained for each case study.

The chapter revealed the impact of feeder line section failure rates and load growth on the cost of industrial customer interruptions. The chapter clearly brought out the impact of capacity-restricted distribution feeders that are unable to pick up adjacent feeder loads during line section outages due to insufficient substation capacity levels.

It is important to have a long-term historical database of the feeders in the distribution system area to obtain accurate estimates of the frequency and the duration of industrial customer interruptions. It is also important to have a working knowledge of the operating practices (e.g., switching and restoration activities and procedures) of the distribution feeders and the operational constraints imposed on the feeders.

One basic conclusion of this chapter is that the expansion plans of any distribution system can be optimized in terms of reliability by using an economic criterion in which the sum of both customer interruptions and system costs is minimized.

## REFERENCES

1. R. Billinton and J. E. Billinton, Distribution system reliability indices, *IEEE Transactions on Power Delivery*, **4** (1), 1989, 561–586.
2. Canadian Electrical Association, 1991 Annual Service Continuity Report on Distribution System Performance in Canadian Electric Utilities, May 1992.
3. Canadian Electrical Association, Economic Penalties Including Customer Costs, for Loss of Service Continuity, Research and Development Report No. SD-273, July 1991.
4. L. Goel and R. Billinton, A procedure for evaluating interrupted energy assessment rates in an overall electric power system, *IEEE Transactions on Power Systems*, **8** (3), 1993, 929–936.
5. G. Wacker, R. K. Subramaniam and R. Billinton, Using cost of electric service interruption surveys in the determination of a composite customer damage function, *International Journal of Energy Systems*, **5**, 1985, 100–104.
6. IEEE Standard 493–1990, IEEE Recommended Practice for the Design of Reliable Industrial and Commercial Power Systems (ANSI).
7. IEEE Standard 446-1987, IEEE Recommended Practice for Emergency and Standby Power Systems for Industrial and Commercial Applications, pp. 74–77.
8. Canadian Standards Association Standard ANN3-C235-83, Preferred Voltage Levels for AC Systems, 0 to 50,000 V, 1983, p. 11
9. M. J. Sullivan, B. N. Suddeth, T. Vardell, and A. Vojdani, Interruption Costs, Customer Satisfaction and Expectations for Service Reliability, Presented at the IEEE PES Summer Meeting, Portland, OR, July 23–27, 1995, Paper 95 SM 572-8 PWRS.
10. S. Burns and G. Gross, Value of service reliability, *IEEE Transactions on Power Systems*, **5** (3), 1990, 825–834.
11. M. Munasinghe and M. Gellerson, Economic criteria for optimizing power system reliability levels, *Bell Journal of Economics* Spring, 1979, 353–365.
12. Cost–Benefit Analysis of Power System Reliability: Determination of Interruption Costs—Vol. 1: Measurement Methods and Potential Applications in Reliability Cost–Benefit Analysis, EPRI IL-6791, April 1990.

# 16

# ISOLATION AND RESTORATION PROCEDURES

## 16.1 INTRODUCTION

A recent report on the U.S.–Canada blackout on August 14, 2003 revealed that the duration of restoring the Eastern Interconnect to a normal operating configuration was lengthy and complicated. One of the difficulties in modeling a power system is representing the significant changes in loading patterns that present themselves during the restoration process following a major outage. The capacity of equipment may be adequate during normal operating conditions, although severely compromised during restoration procedures, particularly the restoration of thousands of distribution system feeder circuits. This chapter presents a new restoration methodology for distribution system configurations that maximizes the amount of load that can be restored after a grid blackout, substation outage, and distribution feeder line section outages and evaluates the cost of load point interruptions considering feeder islanding and substation capacity constraints. Several case studies with restoration tables are presented and discussed to clearly reveal the impact of distribution system capacity constraints on load point reliability indices and the cost of load point interruptions.

*Power Distribution System Reliability.* By Ali A. Chowdhury and Don O. Koval

For any network configuration, there are many failure patterns that can cause a major blackout to an entire network or to parts of a network. Control centers monitor only a part of a grid and not the entire grid. When more than "$n - 1$" outages occur in a grid, the impact of these isolated outages in parts of the system and their impact on the entire network as a "whole" cannot be easily assessed or visualized in real time. These findings are particularly relevant to restoring power to deenergized distribution systems containing thousands of feeders and load points that are not monitored in "real time."

The primary operational difficulties during major grid outages are to locate and identify the cause of the outages, isolate the faulty parts, and begin the complicated process of resorting power to the system. When no detailed operational procedures are defined during major outages and/or grid blackouts, a power system network can be fractured into many parts (e.g., islands). The process of restoring power to these isolated islands in a logical operational manner to have a completely operational network again is extremely complex and time consuming when no detailed operational procedures and no "real-time" monitoring are in place particularly for distribution systems.

There are significant research publications on assessing network performance (e.g., voltage stability, voltage collapse, etc.) for various contingencies; however, there are very few, if any, publications on restoring networks that have experienced a major outage (e.g., a grid blackout) to a normal operating configuration, particularly distribution systems. The load points being serviced by distribution systems are often the last to have their electric service restored. When a grid blackout occurs, the electrical characteristics of the distribution system are altered during the blackout period complicating the restoration process.

Prior to any outages, a normally operating network configuration is designed to meet the "diversified" demand of the customers being served. However, when a major outage occurs, the demand on a network is no longer diversified (e.g., a feeder diversified load of 10 MW prior to an outage can easily become an undiversified load of 15 MW following a lengthy outage). This phenomenon, known as "cold load pickup," is strongly correlated to the duration of an outage and the environmental conditions at the time of the outage. Cold load pickup characteristics have a significant impact on all load point reliability indices of a network configuration and the cost of load point interruptions.

One of the difficulties in modeling distribution feeders and network configurations is that every line section outage in the network has a unique impact on the reliability indices of all customers being serviced by the power delivery system. During a major outage (e.g., a substation transformer or bus failure or a feeder section outage), it is necessary to locate and identify the cause of the outage, isolate the faulty equipment, and reconfigure the network to restore power to as many load points as possible, a complicated process to minimize the impact and cost of load point interruptions. Some distribution system line sections permit load transfers but many do not. These physical constraints complicate any restoration procedure.

In many distribution networks, it is not possible to pick up all the feeder loads at once following a major outage (e.g., a grid blackout, a critically located feeder section outage). It is necessary to sectionalize the network and energize the sections in a sequential manner, otherwise cascading outages will reoccur further complicating the isolation and restoration process. Without the detailed restoration procedures in place, load points will be exposed to increased outage durations, a costly issue for society.

The ability of a distribution system configuration to transfer loads during an outage depends on the substation design and operational procedures. Some of the common distribution substation configurations used by utilities are

- Single transmission source: single transformer
- Single transmission source: dual transformers
- Dual transmission sources: single transformer
- Dual transmission sources: dual transformers
- Dual transmission sources: dual transformers with tiebreakers
- Dual transmission sources: dual transformers with three breakers
- Dual transmission sources: dual transformer with ring bus

The switching and lockout complexity of the distribution substation configurations listed above varies from simple to extremely complex.

This chapter presents three case studies revealing the load point restoration transfer procedures defined in tables necessary to restore power from alternative supplies to selected load points on outage from their primary or normal supply. The case studies presented in this chapter are intended to clearly reveal the significant changes in load point reliability indices caused by the following distribution system constraints and operating characteristics:

1. Capacity-limited distribution substations and their inability to pick up "cold loads" following a lengthy outage.
2. Capacity-limited feeder circuit interties.
3. The varying cost of grid blackouts to a distribution system due to restricted and unrestricted substation capacity.
4. The cost of substation outages and their ability or inability to transfer their loads to adjacent substations whose capacity may be restricted or unrestricted.
5. The unique impact of individual line section outages in creating unique islands within a distribution network.

This chapter presents a restoration methodology for distribution system configurations whose power delivery capability has been curtailed in part or in whole that maximizes the amount of load that can be restored after a blackout and/or feeder section outages and minimizes the cost of customer interruptions considering feeder islanding, substation and feeder capacity limitations, and operating constraints. Restoration transfer and switching rules shown in the tables clearly identify load points restored for a given contingency and those still on outage due to system operational and structural constraints are also presented in this chapter.

## 16.2 DISTRIBUTION SYSTEM CHARACTERISTICS

The distribution system configuration analyzed in this chapter is shown in Fig. 16.1. Generally, each distribution substation serves many feeder circuits (e.g., 4–75); however, in this chapter only one circuit from each substation is shown and analyzed. The methodology presented in this chapter can be applied to any number of distribution feeder circuits that are looped radial or strictly radial or combinations of both.

Any distribution circuit can be divided into line sections that can be isolated by either a manual or an automatic switching device. Each line section has a switching device at the end of the line section closest to the normal supply. The dual switching devices often associated with each line section allow for any line section fault to be sectionalized, isolating some load points from their normal supply (e.g., an island). The switching devices associated with the line sections also allow the isolated loads to be transferred to adjacent energized feeder circuits other than their normal supply subject to system constraints depending upon the duration of the outage.

The composite load for each line section is aggregated and placed on a node at the end of the line section. Figure 16.1 shows only manual isolating devices and the normally opened tie switches linking substations A and B. Depending upon which tie switches are opened or closed, the number of unique feeder operating configurations is considerable.

The average and peak composite loads for each line section are shown in Table 16.1. The load factor of the system is 0.80.

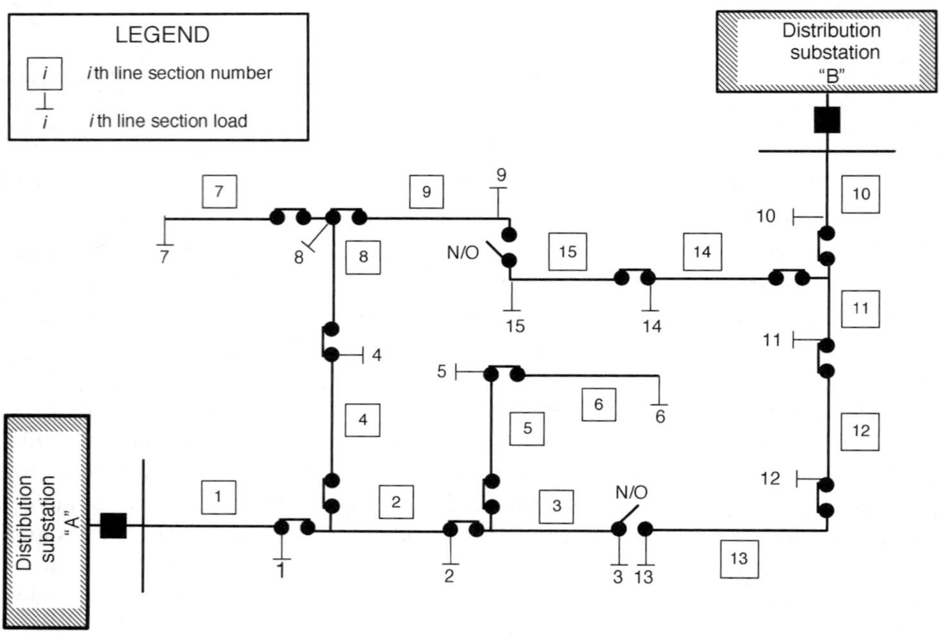


Figure 16.1. Distribution looped feeder: normal operating configuration.

TABLE 16.1. Line Section Composite Loads

| Load Point | Peak Load (kW) | Average Load (kW) |
|---|---|---|
| 1 | 1000 | 800 |
| 2 | 500 | 400 |
| 3 | 1000 | 800 |
| 4 | 1500 | 1200 |
| 5 | 700 | 560 |
| 6 | 1300 | 1040 |
| 7 | 500 | 400 |
| 8 | 1000 | 800 |
| 9 | 1500 | 1200 |
| Substation A total feeder load | 9000.0 | 7200.0 |
| 10 | 1250 | 1000 |
| 11 | 250 | 200 |
| 12 | 500 | 400 |
| 13 | 750 | 600 |
| 14 | 1000 | 800 |
| 15 | 750 | 600 |
| Substation B total feeder load | 4500.0 | 3600.0 |

The physical length of each feeder line section for the distribution feeder circuits of substations A and B is shown in Tables 16.2 and 16.3.

## 16.2.1 Distribution Load Transfer Characteristics

The distribution feeder characteristics of the looped radial feeder circuits are defined as follows. If both feeders are operated radially and are tied through a normally open tie switch, then any line section outage can be *manually* isolated and the remaining line sections can be energized from the alternative feeder or from the nearest continuous feeder. The load transfer is possible only when the feeder circuits and the substation are unrestricted in capacity and the substations are not overloaded. If the substation is overloaded, then the load transfer will not occur.

TABLE 16.2. Distribution Feeder Line Section Lengths: Substation A

| Line section | 1 | 2 | 3 | 4 | 5 | 6 | 7 | 8 | 9 |
|---|---|---|---|---|---|---|---|---|---|
| Physical length (km) | 2 | 1 | 2 | 3 | 3 | 3 | 2 | 3 | 1 |

TABLE 16.3. Distribution Feeder Line Section Lengths: Substation B

| Line section | 10 | 11 | 12 | 13 | 14 | 15 |
|---|---|---|---|---|---|---|
| Physical length (km) | 5 | 2 | 4 | 3 | 2 | 3 |

## 16.2.2 Operating Procedures: Line Section Outages

If a fault occurs in any line section, the respective substation breaker is assumed to trip and deenergize the entire feeder. The faulty line section is first isolated and whether some or all the loads are transferred to the adjacent energized substation depends on the following constrained operating conditions:

1. The normally opened tie switches interconnecting substation A and B will be *closed only* if the sum of the *peak loads* of the energized substation *plus* the sum of the isolated *peak loads* being transferred to the energized feeder(s) from the deenergized feeder(s) *do not exceed* the rated capacity of the energized substation.

$$\sum_{i=1}^{n} L_{\mathrm{pe}_i} + \sum_{j=1}^{m} L_{\mathrm{piso}_j} \leq C_{\mathrm{sub}} \quad \boxed{\text{load transfer occurs}}$$

   where $L_{\mathrm{pe}_i}$ is the $i$th energized feeder peak load, $L_{\mathrm{piso}_j}$ is the $j$th deenergized feeder peak load being transferred to substation $i$ from substation $j$, $C_{\mathrm{sub}}$ is the capacity of the energized substation, $n$ is the number of energized loads, and $m$ is the number of deenergized loads being transferred to the energized substation.

2. If the sum of the peak loads being transferred to the energized feeder plus the sum of the *peak loads* of the energized substation *exceeds* the rated capacity of the substation, the normally opened tie switches will *not be closed* and no load will be transferred.

$$\sum_{i=1}^{n} L_{\mathrm{pe}_i} + \sum_{j=1}^{m} L_{\mathrm{piso}_j} \geq C_{\mathrm{sub}} \quad \boxed{\text{no load transfer occurs}}$$

3. If either or both substations have an outage originating within their system configuration and/or the transmission line outages and/or grid outages, the sum of the peak loads of all the substation's feeder circuits must not exceed the rated capacity of the energized substation. If the capacity of the substation is exceeded, then the feeder circuits must be sectionalized into islands of loads that can be restored in a sequential manner over a period of time.

## 16.2.3 Feeder Circuit Reliability Data

$\lambda$ (feeder failure rate) = 2 failures per 100 km/year

Repair time = 4 h per feeder section failure

Switching time = 1.0 h (includes one or more switching operations)

Grid and substation outage = 1 every 10 years lasting for more than 4 h

### 16.2.4 Cost of Load Point Interruptions

The interruption cost to any load point being supplied by the looped radial feeders is shown in Table 16.4. The values shown in the table are composite values for this specific distribution system.

For a detailed reliability cost analysis, each load point would have a unique composite cost function as a function of the duration of the interruption and is beyond the scope of this chapter.

## 16.3 CASE STUDIES

Three case studies are presented to illustrate the impact of substation capacity constraints in a looped radial feeder configuration. The significance of the duration of the outage and its impact on changing the feeder load characteristics are presented and discussed. Details of the substation ratings and their maximum pickup capacity are shown in the tables. The annual cost of interruptions to grid outages will be calculated for 4 and 8 h grid outages and the results compared with the cost of line section outages.

### 16.3.1 Case Study 1

For Case Study 1, no load transfer or pickup capacity from adjacent substation is possible due to the limited capacity of the substations A and B, as shown in Table 16.5. Under normal operating conditions, the substations can serve their feeder loads adequately but are severely constrained during any outage contingencies.

If an outage to substation A or B caused by a grid or transmission outage exceeds 1 h, the diversified peak load to be restored exceeds the rating of the substation transformer as shown in Table 16.5. To restore service to all the load points, it is necessary to sectionalize the distribution feeders into islands and pick up these islands in a timely sequential manner.

Each line section outage has an impact on the frequency and duration of load point interruptions and the cost of interruptions is illustrated in Table 16.6 for load point 6.

TABLE 16.4. Load Point Interruption Costs

| Duration of Outage (h) | $/Average kW of Load Interrupted |
|---|---|
| 1 | 10.00 |
| 4 | 25.00 |
| 6 | 75.00 |
| 8 | 90.00 |
| 10 | 105.00 |

TABLE 16.5. Substation Ratings and Their Maximum Pickup Capacity—Case Study 1

| | Case Study 1—Capacity of Substation A: 10 MVA, Substation B: 5 MVA | | | | | | | | | |
|---|---|---|---|---|---|---|---|---|---|---|
| | Substation A | | | | Substation B | | | | Maximum Pickup or Reserve Capacity | |
| Outage Duration ($T$) | Peak Load | MVA Rating | PF | MW Rating | Peak Load | MVA Rating | PF | MW Rating | Substation A | Substation B |
| $1 < T < 3$ h | 9 MW | 10 MVA | 0.90 | 9 MW | 4.5 MW | 5 MVA | 0.9 | 4.5 MW | 0 MW | 0 MW |
| $3 > T < 6$ h | 11.7 MW | 10 MVA | 0.90 | 9 MW | 5.85 MW | 5 MVA | 0.9 | 4.5 MW | Substations A and B rating exceeded | |
| $T > 6$ h | 13.5 MW | 10 MVA | 0.90 | 9 MW | 6.75 MW | 5 MVA | 0.9 | 4.5 MW | Substations A and B rating exceeded | |

TABLE 16.6. Detailed Load Point 6 Reliability Indices and Costs of Interruptions—Case Study 1

| Section | Length (km) | λ(failures/ year) | r (h) | U (h/year) | Load Point Interruption | Load Point Annual Cost ($/year) |
|---|---|---|---|---|---|---|
| 1 | 2 | 0.04 | 4.00 | 0.16 | $25.00 | $1040.00 |
| 2 | 1 | 0.02 | 4.00 | 0.08 | $25.00 | $520.00 |
| 3 | 2 | 0.04 | 4.00 | 0.16 | $25.00 | $1040.00 |
| 4 | 3 | 0.06 | 1.00 | 0.06 | $10.00 | $624.00 |
| 5 | 3 | 0.06 | 4.00 | 0.24 | $25.00 | $1560.00 |
| 6 | 3 | 0.06 | 4.00 | 0.24 | $25.00 | $1560.00 |
| 7 | 2 | 0.04 | 1.00 | 0.04 | $10.00 | $416.00 |
| 8 | 3 | 0.06 | 1.00 | 0.06 | $10.00 | $624.00 |
| 9 | 1 | 0.02 | 1.00 | 0.02 | $10.00 | $208.00 |
| Total | | 0.40 | 2.65 | 1.06 | | $7592.00 |

A summary of the load point reliability indices and the cost of interruptions caused by distribution line section outages is shown in Table 16.7 for Case Study 1.

Case Study 1 represents the design of two substations whose capacity to meet their feeder's average and peak loads is adequate during normal operating conditions, although severely constrained during major outages (e.g., grid blackouts and substation

TABLE 16.7. Summary of Load Point Reliability Indices and Cost of Interruptions—Case Study 1

| Load Point | λ(outages/year) | U (h/year) | r (h/outage) | Interruption Cost ($/year) |
|---|---|---|---|---|
| 1 | 0.40 | 1.30 | 0.52 | $3680.00 |
| 2 | 0.40 | 1.45 | 0.58 | $1960.00 |
| 3 | 0.40 | 1.75 | 0.70 | $4400.00 |
| 4 | 0.40 | 1.90 | 0.76 | $6960.00 |
| 5 | 0.40 | 2.20 | 0.88 | $3584.00 |
| 6 | 0.40 | 2.65 | 1.06 | $7592.00 |
| 7 | 0.40 | 2.65 | 1.06 | $2920.00 |
| 8 | 0.40 | 2.35 | 0.94 | $5360.00 |
| 9 | 0.40 | 2.50 | 1.00 | $8400.00 |
| 10 | 0.38 | 1.79 | 0.68 | $5300.00 |
| 11 | 0.38 | 2.11 | 0.80 | $1180.00 |
| 12 | 0.38 | 2.74 | 1.04 | $2840.00 |
| 13 | 0.38 | 3.21 | 1.22 | $4800.00 |
| 14 | 0.38 | 2.42 | 0.92 | $5200.00 |
| 15 | 0.38 | 2.89 | 1.10 | $4440.00 |
| Annual cost of interruptions for the distribution system | | | | $68,616.00 |

outages). The annual cost of line section outages for the entire distribution system shown in Fig. 16.1 is $68,616.00 for case Study 1.

For a *grid blackout* of the distribution system for 4 h, the capacity of both substations is unable to pick up the entire feeder loads. Therefore, it is necessary to shed loads because of the cold load pickup and pick them up 2 h later resulting in a total outage duration of 6 h for these loads.

The average annual cost of the interruptions to load points being serviced by substation A is

$$(2700 \text{ kWp shed } \times 0.8 \times \$75/\text{kW}_{\text{av}} \times 0.10) + \cdots$$
$$+ (9000 \text{ kWp energized} \times 0.8 \times \$25/\text{kW}_{\text{av}} \times 0.10) = \$34,200$$

If the grid blackout is 8 h, then the average annual cost of interruptions to load points being serviced by substation A would be $102,600.

The average annual cost of the interruptions to load points being serviced by substation B is

$$(1350 \text{ kWp shed} \times 0.8 \times \$75/\text{kW}_{\text{av}} \times 0.10) + \cdots$$
$$+ (4500 \text{ kWp energized} \times 0.8 \times \$25/\text{kW}_{\text{av}} \times 0.10) = \$17,100$$

If the grid blackout is 8 h, then the average annual cost of interruptions to loads points being serviced by substation A would be $51,300.

The average annual cost of the interruptions to load points being serviced by the entire distribution system is $51,300 for a 4 h grid outage and $153,900 for an 8 h grid outage.

A *single event cost* of a grid or substation outage to the load points being serviced by the distribution system would be $513,000 for a 4 h grid outage and $1,539,000 for an 8 h grid outage, a significant impact on society.

It is important to note that a single grid blackout event is extremely costly in terms of load point interruptions and the average annual cost of interruptions to load points is significant. If the frequency of blackout increases, then the average annual cost of load point interruptions will increase significantly.

These high interruption costs for Case Study 1 are due to the limited capacity of substations A and B and the necessity to shed load extending the interruption duration of the load being shed. When the capacities of the substations A and B are increased, the possibility of load transfer between the substations is possible. In Case Studies 2 and 3, the transfer capabilities depend on the duration of the outage and decrease as the duration of the outage increases.

### 16.3.2 Case Study 2

As shown in Table 16.8 for Case Study 2, the limiting capacity of substation A is 15 MVA or 13.5 MW with a total peak load of 9000 kW with 4500 kW capacity in reserve. The limiting capacity of substation B is 10 MVA or 9.0 kW with a total peak load of 4500 kW and a 4500 kW capacity in reserve for short-term outages. The reserve capacity not only provides the alternative energy supply for load points in the adjacent substation

TABLE 16.8. Substation Ratings and Their Maximum Pickup Capacity—Case Study 2

| Outage Duration ($T$) | Case Study 2—Capacity of Substation A: 15 MVA, Substation B: 10 MVA | | | | | | | | | |
|---|---|---|---|---|---|---|---|---|---|---|
| | Substation A | | | | Substation B | | | | Maximum Pickup or Reserve Capacity | |
| | Peak Load | MVA Rating | PF | MW Rating | Peak Load | MVA Rating | PF | MW Rating | Substation A | Substation B |
| $1<T<3$ h | 9 MW | 15 MVA | 0.90 | 13.5 MW | 4.5 MW | 10 MVA | 0.9 | 9 MW | 4.5 MW | 4.5 MW |
| $3>T<6$ h | 11.7 MW | 15 MVA | 0.90 | 13.5 MW | 5.85 MW | 10 MVA | 0.9 | 9 MW | 1.8 MW | 3.15 MW |
| $T>6$ h | 13.5 MW | 15 MVA | 0.90 | 13.5 MW | 6.75 MW | 10 MVA | 0.9 | 9 MW | 0 MW | 2.25 MW |

network's feeder line sections in the event of a substation outage or a line section outage but also changes depending upon the duration of the outage. Restoration tables for the distribution looped feeder (normal operating configuration) are shown in Tables 16.9 and 16.10 for Case Study 2.

It can be seen from Table 16.9 that for a fault in feeder section 1, load points 1, 2, 3, 5, and 6 cannot be transferred to the alternative substation feed. The duration of interruptions to these load points depends on the repair time of faulted section 1. However, load points 4, 7, 8, and 9 can be transferred to substation B feeder and the duration of interruptions to this set of load points depends on the duration of switching activities.

TABLE 16.9. Restoration Table: Feeder Circuit—Substation A—Case Study 2

**LEGEND**

i - Load Point "I" energized from normal supply

- Load interrupted

- Load transfered to alternate supply

| Faulted Section | Sub A Loads Restored Following Isolation of Fault | | | | | | | | | Sub A Loads Transferred to Sub B | | | | | | | | | Total kW Transfered |
|---|---|---|---|---|---|---|---|---|---|---|---|---|---|---|---|---|---|---|---|
| | 1 | 2 | 3 | 4 | 5 | 6 | 7 | 8 | 9 | 1 | 2 | 3 | 4 | 5 | 6 | 7 | 8 | 9 | |
| 1 | | | | | | | | | | | | | 4<br>1,500 | | | 7<br>500 | 8<br>1,000 | 9<br>1,500 | 4,500 |
| 2 | 1 | | | | | | | | | | | | 4<br>1,500 | | | 7<br>500 | 8<br>1,000 | 9<br>1,500 | 4,500 |
| 3 | 1 | 2 | | 4 | | | 7 | 8 | 9 | | | | | | | | | | 0 |
| 4 | 1 | 2 | 3 | | 5 | 6 | | | | | | | | | | 7<br>500 | 8<br>1,000 | 9<br>1,500 | 3,000 |
| 5 | 1 | 2 | 3 | 4 | | | 7 | 8 | 9 | | | | | | | | | | 0 |
| 6 | 1 | 2 | 3 | 4 | 5 | | 7 | 8 | 9 | | | | | | | | | | 0 |
| 7 | 1 | 2 | 3 | 4 | 5 | 6 | | 8 | 9 | | | | | | | | | | 0 |
| 8 | 1 | 2 | 3 | 4 | 5 | 6 | | | | | | | | | | | | 9<br>1,500 | 1,500 |
| 9 | 1 | 2 | 3 | 4 | 5 | 6 | 7 | 8 | | | | | | | | | | | 0 |

TABLE 16.10. Restoration Table: Feeder Circuit—Substation B—Case Study 2

| Faulted Section | Sub B Loads Restored Following Isolation of Fault | | | | | | Sub B Loads Transferred to Sub A | | | | | | Total kW Transfered |
|---|---|---|---|---|---|---|---|---|---|---|---|---|---|
| | 10 | 11 | 12 | 13 | 14 | 15 | 10 | 11 | 12 | 13 | 14 | 15 | |
| 10 | | | | | | | | 250 11 | 500 12 | 750 13 | 1,000 14 | 750 15 | 3,250 |
| 11 | 10 | | | | | | | | 500 12 | 750 13 | 1,000 14 | 750 15 | 3,000 |
| 12 | 10 | 11 | | | 14 | 15 | | | | 750 13 | | | 750 |
| 13 | 10 | 11 | 12 | | 14 | 15 | | | | | | | 0 |
| 14 | 10 | 11 | 12 | 13 | | | | | | | | 750 15 | 750 |
| 15 | 10 | 11 | 12 | 13 | 14 | | | | | | | | 0 |

It can be seen from Table 16.10 that for a fault in feeder section 10, load points 11–15 can be transferred to substation A. Each line section outage will result in a unique operating configuration following the outage and the amount of load transferred to substation A.

Based on the restoration and the detailed line section failure tables, the load point frequency and duration of each load point and the cost of each load point interruption can be calculated as summarized in Table 16.11.

The annual cost of line section outages for the entire distribution system shown in Fig. 16.1 is $55,716.00 for Case Study 2.

For a *grid blackout* of the distribution system for 4 h, the capacity of both substations is able to pick up the entire feeder loads following the blackout. Therefore, it is not necessary to shed load. The average annual cost of the interruptions to load points being serviced by both substations A and B is $35,100.

If the grid blackout is 8 h, then the average annual cost of interruptions to loads points being serviced by substations A and B would be $145,800.

A *single event cost* of a grid or substation outage to the load points being served by the distribution system would be $351,000 for a 4 h grid outage and $1,458,000 for an 8 h grid outage.

TABLE 16.11. Summary of Load Point Reliability Indices and Cost of Interruptions—Case Study 2

| Load Point | $\lambda$(outages/year) | $U$ (h/year) | $r$ (h/outage) | Interruption Cost ($/year) |
|---|---|---|---|---|
| 1 | 0.40 | 1.30 | 0.52 | $3680.00 |
| 2 | 0.40 | 1.45 | 0.58 | $1960.00 |
| 3 | 0.40 | 1.75 | 0.70 | $4400.00 |
| 4 | 0.40 | 1.45 | 0.58 | $5880.00 |
| 5 | 0.40 | 2.20 | 0.88 | $3584.00 |
| 6 | 0.40 | 2.65 | 1.06 | $7592.00 |
| 7 | 0.40 | 1.75 | 0.70 | $2200.00 |
| 8 | 0.40 | 1.45 | 0.58 | $3920.00 |
| 9 | 0.40 | 1.15 | 0.46 | $5160.00 |
| 10 | 0.38 | 1.79 | 0.68 | $5300.00 |
| 11 | 0.38 | 1.32 | 0.50 | $880.00 |
| 12 | 0.38 | 1.63 | 0.62 | $2000.00 |
| 13 | 0.38 | 1.47 | 0.56 | $2820.00 |
| 14 | 0.38 | 1.32 | 0.50 | $3520.00 |
| 15 | 0.38 | 1.47 | 0.56 | $2820.00 |
| Annual cost of interruptions for the distribution system | | | | $55,716.00 |

A summary of the load point reliability indices and cost of interruptions caused by line section outages is shown in Table 16.10 for Case Study 2.

### 16.3.3 Case Study 3

A summary of the load point reliability indices and cost of interruptions caused by line section outages is shown in Table 16.12 for Case Study 3. As shown in Table 16.13 for Case Study 3, the limiting capacity of substation A is 15 MVA or 13.5 MW with a total peak load of 9000 kW with a 4500 kW capacity in reserve. The limiting capacity of substation B is 15 MVA or 9.0 MW with a total peak load of 4500 kW and a 9000 kW capacity in reserve.

No restoration tables for Case Study 3 are presented as the substations can pick up the entire load of the adjacent substation in the event of it being out of service due to an outage.

The annual cost of line section outages for the entire distribution system shown in Fig. 16.1 is $37,916.00 for Case Study 2. There is sufficient substation capacity of both substations to pick up all the feeder loads following a 4 or 8 h grid outage.

For a *grid blackout* of the distribution system for 4 h, the capacity of both substations is able to pick up the entire feeder loads following the blackout. Therefore, it is not necessary to shed load. The average annual cost of the interruptions to load points being serviced by both substations A and B is $35,100.

If the grid blackout is for 8 h, then the average annual cost of interruptions to loads points being serviced by substations A and B would be $145,800.

TABLE 16.12. Summary of Load Point Reliability Indices and Cost of Interruptions—Case Study 3

| Load Point | $\lambda$(outages/year) | $U$ (h/year) | $r$ (h/outage) | Interruption Cost ($/year) |
|---|---|---|---|---|
| 1 | 0.10 | 2.20 | 0.22 | $1280.00 |
| 2 | 0.10 | 1.60 | 0.16 | $520.00 |
| 3 | 0.10 | 2.20 | 0.22 | $1280.00 |
| 4 | 0.28 | 1.64 | 0.46 | $4440.00 |
| 5 | 0.22 | 2.36 | 0.52 | $2072.00 |
| 6 | 0.22 | 3.18 | 0.70 | $4784.00 |
| 7 | 0.28 | 2.07 | 0.58 | $1720.00 |
| 8 | 0.28 | 1.64 | 0.46 | $2960.00 |
| 9 | 0.28 | 1.21 | 0.34 | $3720.00 |
| 10 | 0.28 | 2.07 | 0.58 | $4300.00 |
| 11 | 0.28 | 1.43 | 0.40 | $680.00 |
| 12 | 0.28 | 1.86 | 0.52 | $1600.00 |
| 13 | 0.28 | 1.64 | 0.46 | $2220.00 |
| 14 | 0.38 | 1.32 | 0.50 | $3520.00 |
| 15 | 0.38 | 1.47 | 0.56 | $2820.00 |
| Annual cost of interruptions for the distribution system | | | | $37,916.00 |

## 16.4 MAJOR SUBSTATION OUTAGES

If substation A is out of service for a period of 4 h, then the primary question that must be addressed is how much of the substation's load can be transferred to substation B. For Case Study 1, no transfer is possible due to the limited capacity of substation B. For Case Study 2, some of the substation load can be transferred to substation B after some switching and transfer activities and some of the load will remain interrupted until substation A is restored to full service. For Case Study 3, substation B can pick up the entire load curtailed by substation A. Details of the switching activities are beyond the scope of this chapter. A similar scenario can be constructed for a 4 or 8 h outage to substation B and a decision on whether load can be transferred to substation A can be taken.

It is important to note that following any substation outage of lengthy duration, it is necessary to know the change in feeder loads from a diversified load to a nondiversified load. This change in load characteristics has a significant impact on whether the load from a substation on outage can be transferred to an adjacent substation. The capacity of the substation is critical to determining the amount of load that can be transferred and the amount of load that has to be shed and therefore subject to the duration of the substation outage. Whether it is a substation outage or a grid blackout, the capacity of the substation determines the restoration procedures that can vary from very simple to very complex.

TABLE 16.13. Substation Ratings and Their Maximum Pickup Capacity—Case Study 3

| Outage Duration ($T$) | Case Study 3—Capacity of Substation A: 15 MVA, Substation B: 15 MVA | | | | | | | | | |
|---|---|---|---|---|---|---|---|---|---|---|
| | Substation A | | | | Substation B | | | | Maximum Pickup or Reserve Capacity | |
| | Peak Load | MVA Rating | PF | MW Rating | Peak Load | MVA Rating | PF | MW Rating | Substation A | Substation B |
| $1 < T < 3$ h | 9 MW | 15 MVA | 0.90 | 13.5 MW | 4.5 MW | 15 MVA | 0.9 | 13.5 MW | 4.5 MW | 9 MW |
| $3 > T < 6$ h | 11.7 MW | 15 MVA | 0.90 | 13.5 MW | 5.85 MW | 15 MVA | 0.9 | 13.5 MW | 1.8 MW | 7.65 MW |
| $T > 6$ h | 13.5 MW | 15 MVA | 0.90 | 13.5 MW | 6.75 MW | 15 MVA | 0.9 | 13.5 MW | 0 MW | 6.75 MW |

The average annual costs of load point interruptions for substations A and B outages lasting 4 and 8 h are presented in Table 16.14. It is important to note that in Case Study 1 the substation capacities are restricted, resulting in a high cost of load point interruptions due to the system's inability to transfer loads to an adjacent substation. Case Study 1 was chosen as a typical distribution system whose capacity to meet the demand of the feeder loads is adequate, but is severely constrained in its ability to pick up loads during major outages, that is, grid outage or a substation outage.

## 16.5 SUMMARY OF LOAD POINT INTERRUPTION COSTS

To understand the restoration processes of distribution systems, it is necessary to obtain a perspective on the cost of load point interruptions due to the various types of outages, that is, line section, substation, and grid. A summary of the costs of interruptions for various outage scenarios for Case Studies 1 to 3 is shown in Table 16.14. Based on these customer costs, a reliability cost–reliability worth analysis can be performed to determine what mitigating actions can be undertaken to improve the reliability seen at the various load points and reduce the cost of load point interruptions. It is also important to note that each outage contingency generates a unique restoration table that enables a visual and mathematical view of the isolation and restoration procedures necessary to cope with that given contingency.

It is clear that the capacity of distribution substations has a significant impact on the cost of interruptions to consumers. It is also clear that the substation capacity has less

TABLE 16.14. Average Annual Costs of Various Types of Outages in the Distribution System

| Type of Outage | Case Study 1 | Case Study 2 | Case Study 3 |
|---|---|---|---|
| Line section outages | \$68,616 | \$55,716 | \$37,916 |
| Substation A<br>4 h outage | \$34,200 | \$18,000 | \$12,600 |
| Substation B<br>4 h outage | \$17,100 | \$3600 | \$3600 |
| Grid blackout<br>4 h outage | \$51,300 | \$35,100 | \$35,100 |
| Substation A<br>8 h outage | \$102,600 | \$68,400 | \$39,600 |
| Substation B<br>8 h outage | \$51,300 | \$3600 | \$3600 |
| Grid blackout<br>8 h outage | \$153,900 | \$145,800 | \$145,800 |

*Note:* Case Study 1, restricted substation capacity; Case Study 2, partial restricted substation capacity; Case Study 3, unrestricted substation capacity.

impact on the cost of load point interruptions for grid outages that deenergize both substations in this case. However, the capacity of the substations will significantly reduce the restoration time if their capacity is unrestricted and can pick up the additional feeder loads due to cold load pickup and enable the transfer of loads from one substation to another. Detailed studies of the restoration procedures of these events are beyond the scope of this chapter; however, the methodology presented in the chapter is directly applicable.

The impact of line section outages and the cost of load point interruptions depend on the transfer capacity of both substations. If the substation is severely in its ability to pick up loads from adjacent substations, then the cost of load interruptions will be high compared to substations that have adequate capacity for load transfer (e.g., annual cost $68,616 for Case Study 1 compared to $37,916 for Case Study 3).

## 16.6 CONCLUSIONS

This chapter has presented three case studies analyzing the frequency and duration of load point interruptions and the annual cost of load point interruptions subject to substation capacity constraints including grid outages. Several restoration tables were presented to illustrate the complexity of restoring service to deenergized load points from an alternative substation supply when a given distribution feeder outage occurs.

One of the objectives of this chapter was to emphasize the fact that distribution loads that have been interrupted change in magnitude significantly immediately after an outage depending upon the duration of the interruption. These changes can significantly compromise restoration procedures, particularly after a lengthy outage (e.g., a grid blackout).

The cost of interruptions to load points for various types of distribution system outages was presented. First distribution line section outages had the highest average annual cost of interruptions compared to substation and grid outages. The cost of line section outages to various load points decreased as the capacity of the primary and alternative substations was changed from restricted to unrestricted capacity limits.

Without a detailed working knowledge of a distribution system's operating practices and load characteristics, reliability analysis of load point interruptions and the cost of these interruptions can be severely compromised possibly resulting in incorrect analytical and mitigating solutions. This is particularly true of computer programs that generate reliability indices without considering the complex issues associated with distribution systems. Some of the critical variables necessary for distribution system analysis are

- Line section loading limitations
- Line equipment (e.g., sectionalizing switches) loading and operating characteristics

- Automatic or manual sectionalizing
- Load point load characteristics as a function of time (e.g., average, peak, and median values)
- Degree of "real-time" monitoring points in the distribution system
- Number of remotely controlled isolating and load transfer switches in the distribution system
- Existing restoration and isolation procedures
- Number of line crew available at the time of the outage
- Equipment failure rates and the duration of repair activities specific to the distribution system
- Duration of restoration activities (e.g., switching, isolating, identifying cause of outage, notification time, trouble crew assembly time, traveling time, etc.)
- Protection coordination characteristics of the protective devices at the substation and in the feeder circuits
- Composition of loads (e.g., mix of residential, commercial, industrial, institutional, etc.) and the loading patterns
- Interruption costs for individual load types and for composite loads; and so on.

## REFERENCES

1. J. E. Propst, Calculating electrical risk and reliability, *IEEE Transactions on Industry Applications*, **31** (5), 1995, PCIC Conference Record 94-2.
2. D. O. Koval, Zone-branch reliability methodology for analyzing industrial power systems, *IEEE Transactions on Industry Applications*, **36** (5), 2000, 1212–1219.
3. D. O. Koval, Evaluation of distribution system reliability, Ph.D. Thesis, University of Saskatchewan, Saskatoon, Saskatchewan, June 1978.
4. IEEE Standard 493-1997, IEEE Recommended Practice for Design of Reliable Industrial and Commercial Power Systems (IEEE Gold Book), The Institute of Electrical and Electronics Engineers, Inc., New York, 1997.
5. D. O. Koval, X. Zhang, J. E. Propst, T. Coyle, R. G. Arno, and R. S. Hale, Jr, Reliability methodologies applied to The IEEE Gold Book Standard Network, *IEEE Industry Applications Magazine*, **9** (1), 2003, 32–41.
6. J. E. Propst and D. R. Doan, Improvements in modeling and evaluation of electrical power system reliability, *IEEE Transactions on Industry Applications*, **37** (5), 2001, 1413–1422.
7. S. Peyton, R. S. Hale, Jr, R. G. Arno, and D. O. Koval, Analysis techniques for electrical and mechanical power systems, 2001 IEEE Industry and Commercial Power Systems Technical Conference Record 01CH37226, 2001, pp. 61–65.
8. J. Prost, PCIC 2000-2 Reliability Software and Documentation, Reliability Model Spreadsheet Software, available from Equilon Enterprises LLC and Web: www.ieee-pcic.org
9. D. O. Koval, L. Jiao, R. G. Arno, and R. S. Hale, Jr, Zone-branch reliability methodology applied to Gold Book Standard Network, IEEE IAS Annual Conference, Chicago, November/December 2001.

10. R. E. Brown and J. J. Burke, Managing the risk of performance based rates, *IEEE Transactions on Power Systems*, **15** (2), 2000, 893–898.
11. Ontario Energy Board Electricity Distribution Rate Handbook, March 2000.
12. Canadian Electricity Association, 2001 Annual Service Continuity Report on Distribution System Performance in Electrical Utilities, May 2002.
13. C. A. Warren and M. J. Adams, Reliability on the regulatory horizon, IEEE T&D Conference and Exposition, Atlanta, GA, October 2001 (CDROM).
14. G. Tollefson, R. Billinton, and G. Wacker, Comprehensive bibliography on reliability worth and electrical service interruption costs: 1980–1990, *IEEE Transactions on Power System*, **6** (4), 1991, 1508–1514.
15. Canadian Electricity Association, Economic Penalties Including Customer Costs for Loss of Service Continuity, Research and Development Report No. SD-273, July 1991.
16. M. J. Sullivan, B. N. Suddeth, T. Vardell, and A. Vojdani, Interruption costs, customer satisfaction and expectations for service reliability, *IEEE Transactions on Power Systems*, **11** (2), 1996, 989–995.
17. A. A. Chowdhury and D. O. Koval, Application of customer interruption costs in transmission network reliability planning, *IEEE Transactions on Industry Applications*, **37** (6), 2001, 1590–1596.
18. A. A. Chowdhury and D. O. Koval, Value-based distribution system reliability planning, *IEEE/IAS Transactions*, **34** (1), 1998, 23–29.
19. Electric Power Research Institute, Cost–Benefit Analysis of Power System Reliability: Determination of Interruption Costs. Vol. I: Measurement Methods and Potential Applications of Reliability Cost–Benefit Analysis, EPRI EL-6791, Palo Alto, CA, April 1990.
20. M. Munasinghe and M. Gellerson, Economic criteria for optimizing power system reliability levels, *Bell Journal of Economics*, Spring, 1979, 353–365.

# 17

# MESHED DISTRIBUTION SYSTEM RELIABILITY

## 17.1 INTRODUCTION

Achieving high-distribution reliability levels and concurrently minimizing capital costs can be viewed as a problem of optimization. Assuming given outage rates and repair times, distribution system design is the remaining factor in determining customer reliability. Including customer value of reliability in an economic analysis allows optimization of the major components of distribution system design. By using mathematical models and simulations, a comparison of design concepts can be performed to compute the optimal feeder section length, feeder loading level, and distribution substation transformer loading level. The number of feeder ties and feeder tie placement are also optimized through the models. The overall outcome of this analysis is that capital costs can then be directed toward system improvements that will be most cost-effective in improving distribution system reliability. This chapter presents a value-based probabilistic approach to designing urban or meshed distribution systems. The inclusion of the customer value of service reliability in an economic analysis permits the optimization of the major components of distribution system design. Using mathematical models and simulations, a comparison of different design

*Power Distribution System Reliability.* By Ali A. Chowdhury and Don O. Koval

concepts can be performed to compute the optimal feeder section length, the feeder loading level, and the distribution substation transformer loading level. The number of feeder ties and feeder tie placement can also be optimized through the probabilistic models. The overall outcome of this analysis is that capital costs can then be directed towards system improvements that will be most cost-effective in improving distribution system reliability. The inclusion of the customer value of service reliability in an economic analysis permits the optimization of the major components of distribution system design. Using mathematical models and simulations, a comparison of different design concepts can be performed to compute the optimal feeder section length, the feeder loading level, and the distribution substation transformer loading level. The number of feeder ties and feeder tie placement can also be optimized through the probabilistic models. The overall outcome of this analysis is that capital costs can then be directed towards system improvements that will be most cost-effective in improving distribution system reliability. The value-based reliability methodology is illustrated by using a practical urban distribution system of a Canadian utility. A commercial computing tool is used for distribution reliability indices.

The importance of electricity supply reliability, which influences customer purchasing decisions is being recognized by electric utilities as the electricity industry is moving toward deregulation and customer choice. The distribution system is an important part of the total electric supply system as it provides the final link between a utility's bulk transmission system and its ultimate customers. As was stated earlier in this book, it has been reported in many technical publications that over 80% of all customer interruptions occur due to failures in the distribution systems.

In the past, the distribution segment of a power system received less attention dedicated to reliability planning than the generation and transmission segments. The distribution segment has been the weakest link between the source of supply and the customer load point. This is because generation and transmission segments are very capital intensive, and outages in these segments can have widespread catastrophic economic consequences to both utilities and customers, as was in the case of the US–Canada blackout on August 14, 2003. Though a distribution system reinforcement scheme is relatively inexpensive compared to a generation or a transmission improvement plan, a utility routinely spends a large sum of money collectively on a number of distribution improvement projects.

## 17.2 VALUE-BASED RELIABILITY ASSESSMENT IN A DEREGULATED ENVIRONMENT

At present, deregulation of the electric energy industry is forcing electric utilities into unfamiliar territories. For the first time in the electricity supply history, the customers are having opportunities to look for value-added services from their suppliers or they will start to shop around. It is a foregone conclusion that failure to recognize customer preferences in a competitive market would bring disastrous results to many utilities. The emerging competitive energy market will be characterized by intense price competition. Utilities will be faced with new challenges of budget constraints, safety, environment, lower load growth, need for more involvement of different interest groups in the planning and designing process, and more competitive independent distributed generators. Moreover, electric utilities will be under conflicting pressures for providing even higher standards of service reliability and hold the line on rates.

It is apparent that modern society is increasingly becoming dependent on cost-effective reliable electric power supply, and unreliable electric power supplies can be extremely costly to both utilities and customers. It has also been recognized that rules of thumb and implicit criteria cannot be used in a consistent manner when a very large number of capital and operation and maintenance (O&M) investments are routinely being made. There is therefore a growing interest in economic optimization approaches to distribution system planning and expansion. For a rational decision making on the requirements of changing the supply reliability levels experienced by customers, utility costs and the costs incurred by customers associated with interruption of service must be incorporated in the distribution system planning practices.

A value-based distribution system reliability planning approach attempts to locate the minimum cost solution where the total cost includes the utility investment costs plus the operating costs plus the customer interruption costs as shown in Fig. 11.2. Value-based distribution system reliability planning, therefore, becomes an invaluable tool using which a proactive, customer-responsive utility will be able to retain its existing customer-base and win new customers. This chapter illustrates the use of a value-based reliability method in the optimal design of urban distribution systems that benefit both electricity suppliers and customers. The value-based planning approach is illustrated by using a practical urban distribution system.

## 17.3 THE CHARACTERISTICS OF THE ILLUSTRATIVE URBAN DISTRIBUTION SYSTEM

Although normally operated radially, urban areas with high-density commercial, industrial, residential, government, and institutional loads are supplied from a number of meshed distribution supply systems such as primary selective systems, primary loop systems, and secondary grid networks. On the contrary, the sparsely populated rural service areas with a mix of commercial and residential customers are normally serviced by overhead radial distribution systems.

Although variations exist among urban feeders across the MidAmerican service territory, the majority of the urban feeders have principally similar characteristics. To broadly apply conclusions reached in the study, the features of a typical urban distribution feeder and substation were agreed upon and used in the study. These typical features of the MidAmerican urban distribution system are as follows:

1. the load density is 2.5 MW per mainline feeder mile;
2. the feeder normal rating is 10 MVA, and the emergency rating is 11.6 MVA;
3. the feeder length and conductors used in the urban system prevent voltage from limiting backup capability;

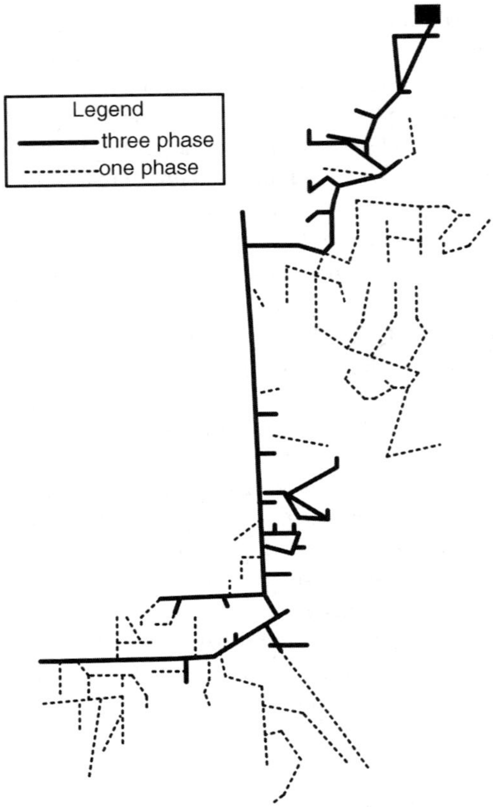


Figure 17.1. Illustrative urban distribution system.

4. the typical urban distribution substation consists of dual 33 MVA transformers each with an emergency rating of 48 MVA.

On the basis of these criteria, an existing MidAmerican urban circuit was selected for use in the study. The selected circuit serves a peak load of 8.2 MW and consists of 2.9 miles of three-phase mainline, 2.3 miles of fused three-phase taps, and 7.4 miles of fused single-phase taps. The circuit is shown in Fig. 17.1.

The greatest impact on customer reliability from a design standpoint can be obtained from improving mainline reliability. So for study purposes, the circuit was reduced to the three-phase mainline. Loads and customers on fused taps were lumped back to the mainline section serving the tap. The study was performed using equipment failure rates and repair times, which were based upon past performance of the Canadian utility's urban distribution system and industry averages. A listing of the failure rates and repair times used in the study is provided in Table 17.1.

The assumed switching time used in the study was 60 min to isolate a failure, and an additional 10 min to close feeder ties when available. The average customer interruption

TABLE 17.1. Component Failure Rates and Repair Times

| Component | Failure Rate (F) | Repair Time (RT) |
|---|---|---|
| Substation transformer | 0.07/year | 24 h |
| Bus/switchgear | 0.001/year | 15 h |
| Circuit breaker | 0.0036/year | 32 h |
| Three-phase UG | 0.35/miles year | 18 h |
| Three-phase OH | 0.8/miles year | 4 h |
| Switch | 0.001/year | 5.5 h |

cost figure used in the cost–benefit analyses is $10.76/kWh. The unserved energy in kilowatt hour can be converted to a cost based upon customer value of reliability. In this manner, system modification costs can be compared against the associated reliability benefits to determine if the improvements are economically justifiable. Various system layouts, including reliability benefits, were compared using economic analysis to arrive at the optimal layout for a distribution system. A number of urban distribution system design criteria were optimized and are described in the following sections. A brief description of the items optimized and the process used is provided in the following.

Optimal feeder mainline section length was calculated mathematically by deriving an equation for the reduction of unserved energy costs associated with adding a switch at the midpoint in a single mainline section. The optimal section length was then calculated by setting reduction in unserved energy costs equal to the cost of the switch installation. Equations are included in Section 17.10.

Optimal feeder loading and transformer loading were determined by performing simulations to calculate the unserved energy costs for the total load served by an urban distribution substation. Simulations were run at different system load levels (100%, 80%, 70%, etc.) and annual unserved energy cost was calculated by weighting the results of each simulation by the percentage of the year that each load level is present. The five-step load model used in the study is presented in Table 17.2.

The annual unserved energy cost for a particular transformer and feeder-loading scenario was then added to a charge for any unused transformer and feeder capacity to create the total annual cost of the loading scenario. Annual costs of each loading scenario were then compared to determine the most cost-effective solution. Equations are included in Section 17.11.

TABLE 17.2. Five-Step Load Duration Curve Approximation

| Load Level (%) | Probability |
|---|---|
| 100 (Peak) | 0.001 |
| 80–90 | 0.025 |
| 70–80 | 0.040 |
| 60–70 | 0.097 |
| <60 | 0.837 |

The optimal number of feeder ties was determined by performing simulations to calculate the unserved energy for a single feeder while varying the number of feeder ties. The incremental reduction in unserved energy costs associated with each feeder tie addition is then compared against the incremental cost of installing the tie. Differing level benefits are obtained when adding feeder ties depending on where the tie is located.

## 17.4 DISCUSSION OF RESULTS

The optimal section length can be mathematically determined independent of the other distribution system design issues if two assumptions are made. The first assumption is that once a faulted section is isolated, there is enough feeder tie capacity to serve the remaining sections. This is a safe assumption taking into account the Canadian utility's load duration curve (see Table 17.2). Over 93% of the hours fall at 60% load level or less. At 60% loading, even one tie should be able to cover the circuit assuming that voltage is not a limiting issue. Also, for most faults, the feeder ties will not be required to pick up the entire feeder because a portion of the feeder will be served by the normal source, unless one of the first sections is faulted.

The second assumption is that load along the feeder is uniformly distributed. This may not be the case, but the majority of the benefit of adding sectionalizing switches comes from shortening section length, no matter what the load distribution. If there is large spot load, additional switches can be put on either side or both sides to protect it.

Once these two assumptions are made, a mathematical analysis can be carried out for the addition of switches. The benefit in unserved energy cost reduction can be calculated, and since the cost of the switch and installation is also known, the analysis can be used to determine optimal section length with a cost–benefit analysis. A detailed derivation of the equations can be found in Section 17.10. Table 17.3 shows the results of the cost–benefit analysis taking into account customer reliability benefits and using different years to payback. It should be noted that the calculation used equipment and installation costs of overhead distribution switches, and therefore the resulting section length is applicable to overhead portions of the three-phase feeder.

The results in Table 17.3 indicate a rather short section length is beneficial even using a 1-year payback period. By using the results from this analysis, it was decided to modify the feeder to have a section length of around 1000 ft. The circuit originally had

TABLE 17.3. Optimum Section Lengths

| Year to Payback | Length (miles) | Length (ft) |
|---|---|---|
| 1 | 0.26 | 1396 |
| 2 | 0.19 | 1012 |
| 3 | 0.16 | 846 |
| 4 | 0.14 | 751 |
| 5 | 0.13 | 687 |

5 sectionalizing switches over a length of about 2.9 miles; therefore, 11 more switches were added to bring the section length to around 1000 ft.

## 17.5 FEEDER AND TRANSFORMER LOADING LEVELS

Feeder loading and transformer loading also need to be addressed to set up the base case. To perform this part of the analysis, the circuit was modified to reflect different feeder load levels. Additional sections were added in specific cases to represent the circuit configuration for a higher loaded feeder. The unserved energy costs associated with different feeder loading levels were determined.

The unserved energy costs decrease as feeder load levels are lowered; however, to allow a comparison of alternatives, a charge was applied for unused feeder and transformer capacity. To fairly incorporate the transformer loading, the approach used was to determine the unserved energy cost and unused capacity charge associated with serving 33 MVA of load. That way, each alternative case could be directly compared against a base case that was selected to be a completely loaded transformer with three 11 MVA feeders. A cost–benefit analysis was performed on each of the cases, where the benefit was the difference between the unserved energy costs of the base case and the unserved energy costs of the alternative case. The cost for the alternative case was the capacity charge associated with operating the system below full nameplate capacity levels. The values used for capacity charges were those agreed upon for alternative source calculations and are listed in Section 17.11.

Each alternative case was run by using two feeder ties and a bus tie, because earlier runs indicated that was the optimal way to operate the system no matter what load level was chosen. Table 17.4 shows the calculated results. In this chapter, UE and $\Delta$UE denote unserved energy and marginal unserved energy reduction respectively.

A utility cost–benefit analysis model was used to calculate the years to payback for each alternative. The option with the shortest payback period indicates that it is the most economical way to serve the load taking into account the customer value of reliability. Since the transformer capacity charge is much larger than the feeder capacity costs, the results show that loading the transformer closer to nameplate rating

TABLE 17.4. Feeder and Transformer Loading Comparison

| MVA | Feeders/ xfmr | Unserved Energy Cost ($) | UE Cost for 33 MVA xfmr ($) | Unused Capacity Cost | Cost for xfmr Load Losses ($) | Annual Benefit ($) | Years to Payback |
|---|---|---|---|---|---|---|---|
| 11 | 3 | 226,468 | 679,403 | 0 | 25,826 | 0 | – |
| 10 | 3 | 190,744 | 629,455 | 88,896 | 22,155 | 53,620 | 3 |
| 9 | 3 | 158,484 | 581,110 | 205,230 | 18,833 | 105,2887 | 3 |
| 8 | 4 | 129,799 | 535,423 | 102,800 | 24,563 | 145,243 | 1 |
| 7 | 4 | 105,460 | 497,168 | 257,912 | 19,901 | 188,160 | 2 |
| 6 | 4 | 84,420 | 464,310 | 413,024 | 15,860 | 225,058 | 3 |

is advantageous. Also, because most of the unserved energy costs come from faults in the line, increasing the transformer loading only incrementally increases unserved energy costs.

## 17.6 BUS AND FEEDER TIE ANALYSIS

Once the base case was determined, it was used to verify how many feeder ties are cost-beneficial taking into account the customer value of reliability. By using the results from Tables 17.3 and 17.4, the base case was chosen as an 8 MVA circuit with a section length of approximately 1000 ft. Starting at a base case with no ties, ties were added one at a time and the utility cost–benefit analysis model was used to determine if that incremental change was beneficial. The calculated results for each addition are shown in Table 17.5.

### 17.6.1 Tie Costs and Descriptions

The bus tie was simulated in the analyses as a backup source to the distribution substation switchgear in the case of a transformer failure. The cost of bus tie was estimated at $100,000 and consisted of two breakers and associated cable work for the bus tie installation. The first feeder tie was simulated as a backup source at the end of the feeder because typically as the distribution system expands outward a normally closed switch turns into a normally open tie point. Therefore, the cost of this improvement was set at zero. The second and third feeder ties were simulated as backup sources near the midpoint and quarter-point of the feeder respectively. The cost of these improvements consisted of the switch and three-phase construction required to connect the test feeder to an assumed adjacent feeder. Section 17.12 illustrates how the three-phase construction cost was calculated. Since the test feeder and the adjacent feeder each receive a benefit from a feeder tie, the three-phase construction cost was cut in half for the cost–benefit analysis.

The results shown in Table 17.5 indicate large benefits associated with adding the bus tie and adding the first feeder tie. However, the benefit gained from adding the second and third feeder ties is very small.

TABLE 17.5. Cost–Benefit Analysis of Adding Bus Tie

| Tie | Incremental Cost ($) | Incremental Benefit (annual) ($) | Years to Payback |
|---|---|---|---|
| Bus tie | 100,000 | 79,789 | 2 |
| Feeder tie 1 | 0 | 152,367 | $\leq 1$ |
| Feeder tie 2 | 13,384 | 302 | $>30$ |
| Feeder tie 3 | 6404 | 3 | $>30$ |

## 17.7 MAINTENANCE

The simulation results indicate that in many cases only one tie is cost-beneficial. While having only one tie may be economically advantageous, switching options are greatly reduced, especially when one circuit is restricted due to maintenance or construction projects such as road widening. If a fault occurred when the tie was unavailable, it would lead to significant customer outage duration. To include this aspect, two different scenarios, one *with* the second tie and one *without* the second tie, can be combined to represent a system with feeder tie availability considered. The only requirement is to determine what percentage of a year the tie typically is unavailable. Section 17.13 illustrates the computation model. Table 17.6 summarizes the results with maintenance included.

Table 17.6 illustrates that when maintenance is included, the benefit of the second tie is more apparent. As the maintenance becomes less frequent, meaning the tie is available for a higher percentage of the year, the benefit of adding the second tie decreases considerably.

### 17.7.1 Single Transformer

Cases were also run to see if the results differ for a single-transformer substation. In these cases, the bus tie was removed, and all ties were assumed to be from circuits served from the same transformer, excluding the tie at the end, which was assumed to be from a circuit served from another substation. Because the bus tie is used only for transformer faults, which is a small percentage of all faults, these results lead to the same conclusions as a two-transformer substation. Even without the bus tie, the feeder tie at the end of the circuit will be able to pick up a large portion of an 8 MVA feeder during many hours of the year, considering utility's load duration curve. For the feeder ties to pick up significant load during a transformer failure, there needs to be an emergency rating on the transformers at the surrounding substations.

### 17.7.2 Conductor Sizing

In most areas, the distribution system can be classified as a 600 A system, and in most instances, conductor sizes can be chosen in accordance with that concept. However, in

TABLE 17.6. Cost–Benefit Analysis of Second Tie Including Maintenance Considerations

| Frequency of Maintenance | Benefit of Second Tie ($) | Cost ($) | Years to Payback |
|---|---|---|---|
| 4 weeks every 1 year | 11,990 | 13,384 | 2 |
| 4 weeks every 3 years | 4198 | 13,384 | 6 |
| 4 weeks every 5 years | 2639 | 13,384 | 12 |
| 4 weeks every 10 years | 1470 | 13,384 | >30 |

TABLE 17.7. Benefit of Additional Tie for Three-Phase Branches

| Branch Length (miles) | ΔUE Sections (kWh) | ΔUE Switches (kWh) | ΔUE Total (kWh) | Benefit ($) | Feeder Tie Investment Allowed for a 5-Year Payback ($) |
|---|---|---|---|---|---|
| 0.2 | 0.0 | 0.0 | 0.0 | 0 | 0 |
| 0.4 | 226.7 | 2.2 | 228.8 | 2462 | 6298 |
| 0.6 | 680.0 | 4.3 | 684.3 | 7363 | 18,833 |
| 0.8 | 1360.0 | 6.5 | 1366.5 | 14,704 | 37,610 |
| 1 | 2266.7 | 8.7 | 2275.3 | 24,483 | 62,623 |
| 1.2 | 3400.0 | 10.8 | 3410.8 | 36,701 | 93,874 |
| 1.4 | 4760.0 | 13.0 | 4773.0 | 51,357 | 131,362 |

areas where voltage limitations are a concern, a larger conductor has been used to help return the system to an ampacity-limited rather than to a voltage-limited one. Designing the system by using a larger conductor can be an economical solution for systems where voltage limitations exist. However, in an ampacity-limited system, selecting a larger conductor size will not provide significant benefit in terms of reliability. To illustrate this point, additional scenarios were run to compare the benefit in reliability from increased available tie capacity. The results of the simulation show that installing a larger sized conductor strictly for increasing tie capacity is not cost justified when taking into account the customer value of improved reliability.

## 17.8 FEEDERS WITH NONFUSED (LATERAL) THREE-PHASE BRANCHES

Another issue to address is adding ties for three-phase nonfused lateral (see Fig. 17.1) branches on a circuit. For long branches of the main circuit, considerable unserved energy costs can be observed if there is no backup tie. An equation representing the benefit of adding a tie for branches of different lengths is provided in Section 17.14. Table 17.7 summarizes the reduction of unserved energy costs for various branch lengths.

It is apparent that as the length/load of the three-phase branch increases, the benefit of constructing an additional feeder tie at the end of the branch also increases. Assuming that a project with a 5-year payback period is economically justifiable, the last column in the table indicates the allowable costs associated with constructing the additional feeder tie for a generic feeder.

## 17.9 FEEDER TIE PLACEMENT

Feeder tie points should be spread evenly across the circuit taking into account customer loading. Considering a feeder tie as an additional source verifies this conclusion. Spreading out the sources allows more customers to be picked up following an outage.

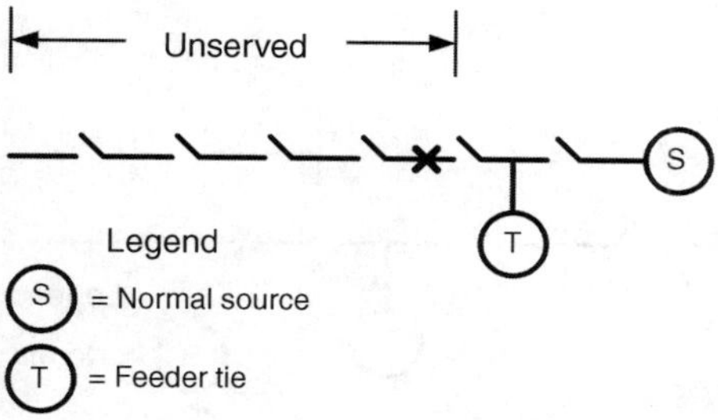


Figure 17.2. Tie location near source.

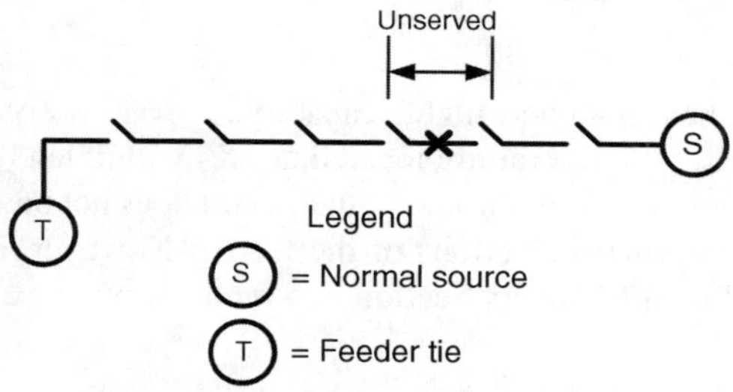


Figure 17.3. End of feeder tie location.

Figure 17.2 shows an example of this on a system with only one feeder tie. A failure occurring at the spot marked "**X**" will result in the majority of the customers remaining unserved until the failure is resolved. The tie as positioned in Fig. 17.2 serves no purpose for failures beyond the first switch. However, if the tie is placed at the end of the circuit as illustrated in Fig. 17.3, it gives a far greater benefit to reducing unserved energy.

The same argument illustrated in Figs. 17.2 and 17.3 for one feeder tie, also applies to circuits with more than one tie (when there is a limited capacity on each tie). An example of this for a circuit with two feeder ties is illustrated in Fig. 17.4. Assume for this

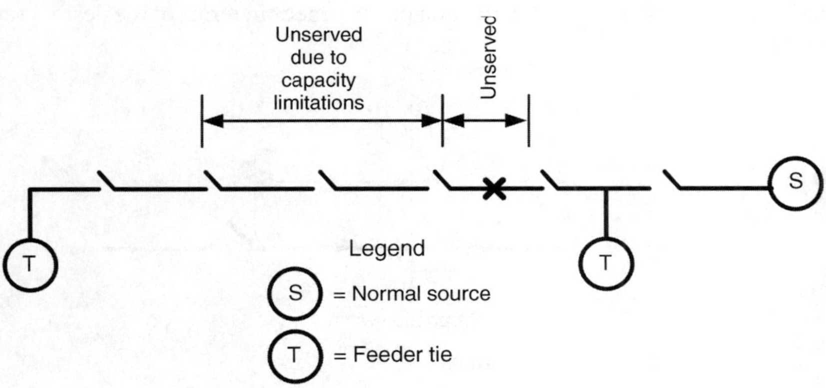


Figure 17.4. Two feeder ties unevenly spaced.

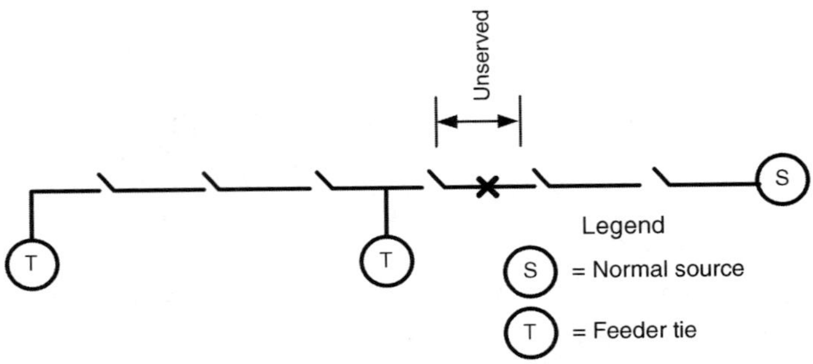


Figure 17.5. Two feeder ties evenly spaced.

example that each tie has enough available capacity to serve two sections of the original feeder. As shown in Fig. 17.4, a failure located at "**X**" would lead to a large amount of unserved energy because the tie at the end of the circuit does not have enough capacity to pick up more than the last two sections of the feeder. However, if the ties are evenly spaced as shown in Fig. 17.5, every section can retain service except the one directly affected by the fault.

## 17.10 FINDING OPTIMUM SECTION LENGTH

Assumptions:

1. Feeder load is uniformly distributed at 2500 kW per mile of three-phase line.
2. All load can be picked up once failure is isolated.

The illustration in Fig. 17.6 shows a section where a switch will be added to form two sections each of length "*X*".

The annual benefit of this improvement is the reduction in unserved energy costs. The following is a mathematical derivation of the benefit in terms of the desired section length, "*X*."

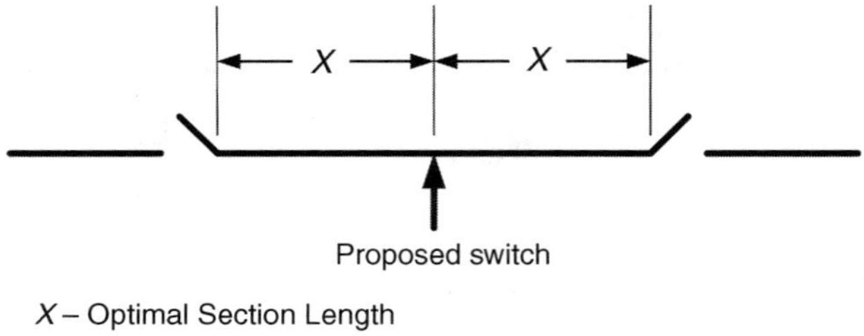


Figure 17.6. Illustration of switch addition.

### 17.10.1 Definition of Terms

$F_{line}$ failure rate of line (events/mile year)
$F_{switch}$ failure rate of line switch (events/year)
$Load_{total}$ total feeder load (kW)

$RT_{line}$ repair time of line (h)
$RT_{switch}$ repair time of switch (h)
$T_{switching}$ time required to isolate failure and close ties in hours ($T_{switching}$ 70 min = 1.166 h)
UE unserved energy (kW/h)
$x$ section length (miles)

$$\text{Benefit} = \Delta\text{UE} \times \text{Cost/kWh} = (\text{UE}_{old} - \text{UE}_{new}) \times \text{Cost/kWh} \tag{17.1}$$

$$\text{UE}_{old} = (2x \times \text{F}_{line})(2x \times \text{kW/mile}) \times \text{RT}_{line} \tag{17.2}$$

$$\begin{aligned} \text{UE}_{new} = {} & (x \times F_{line})(x \times \text{kW/mile}) \times \text{RT}_{line} + (x \times F_{line})(x \times \text{kW/mile}) \\ & \times T_{switching} + (x \times F_{line})(x \times \text{kW/mile}) \times \text{RT}_{line} \\ & + (x \times F_{line})(x \times \text{kW/mile}) \times T_{switching} + (F_{switch})(2x \times \text{kW/mile}) \\ & \times \text{RT}_{switch} + (F_{switch})(\text{Load}_{total}) \times T_{switching} \end{aligned} \tag{17.3}$$

The failure rates and repair times listed in Table 17.8 can be used in the calculations. Substituting the failure rates and repair times along with assumed constants into the previous equations results in the following:

$$\begin{aligned} \text{UE}_{old} &= (2x \times 0.8)(2x \times 2500) \times 4 \\ &= 32,000x^2 \text{ kWh} \end{aligned}$$

$$\begin{aligned} \text{UE}_{new} &= (x \times 0.8)(x \times 2500) \times 4 + (x \times 0.8)(x \times 2500) \times 1.166 \\ &\quad + (x \times 0.8)(x \times 2500) \times 4 + (x \times 0.8)(x \times 2500) \times 1.166 \\ &\quad + (0.001)(2x \times 2500) \times 5.5 + (0.001)(10000) \times 1.166 \\ &= 20666.7x^2 + 27.5x + 11.67 \text{ kWh} \end{aligned}$$

$$\begin{aligned} \Delta\text{UE} &= 32000x^2 - (20666.7x^2 + 27.5x + 11.67) \text{ kWh} \\ &= 11333.3x^2 - 27.5x - 11.67 \text{ kWh} \end{aligned}$$

$$\Delta\text{UE} \times \$10.76/\text{kWh} = 121,946.7x^2 - 295.9x - 125.5$$

TABLE 17.8. Component Failure Rates and Repair Times

| Component | Failure Rate (F) | Repair Time (RT) |
|---|---|---|
| Three-phase OH | 0.8/miles year | 4 h |
| Switch | 0.001/year | 5.5 h |

The cost of a switch and installation was estimated to be $5000. Using UCBM, the 30-year present worth of that investment was calculated to be $8481. Setting the cost equal to the annual reliability benefit to customers and solving for $x$, will give the desired section length for a 1-year payback as shown in Equation 17.4 The terms can also be multiplied by the appropriate factors to obtain the desired section length for a 2-year payback, and so on.

$$121,946.7x^2 - 295.9x - 125.5 = 8481 \tag{17.4}$$

## 17.11 FEEDER AND TRANSFORMER LOADING

Feeder tie capacity and bus tie capacity used for these simulations were calculated using Equations 17.5 and 17.6:

$$\text{Feeder tie capacity} = \text{feeder emergency rating} - \text{feeder peak load} \tag{17.5}$$

$$\text{Bust tie capacity} = \begin{array}{l}(\text{xfmr emergency rating} - \text{xfmr load level})/\\ \text{number of feeders/xfmr}\end{array} \tag{17.6}$$

The feeder tie capacity is equal to the emergency rating minus the feeder peak load level, assuming an ampacity-limited feeder. For a typical feeder, the normal rating is 10 MVA and the emergency rating is 11.6 MVA. The emergency rating for a new 33 MVA transformer was calculated by using PT load. Results showed that the transformer could be loaded to 48 MVA, assuming a typical MidAmerican daily load curve. The PT load studies were performed for a 24 h period, which was taken to be the time required for a mobile substation to be installed.

In simulating a transformer failure, it was assumed that the bus tie capacity was split up evenly among the circuits fed from a transformer. This was necessary because simulations were run on a per feeder basis, so assigning the total bus tie capacity to the feeder being simulated would underestimate the total unserved energy associated with a transformer failure. Splitting the bus tie capacity evenly represents the fact that for a transformer failure, the bus tie to the other transformer is able to pick up only a percentage of the total transformer load. Table 17.9 shows the results of the simulation runs and the cost–benefit calculations.

$$\text{Circuit length} = 2.25 + (\text{feederMVA} - 6) \times .2 \tag{17.7}$$

$$\text{xfmr load} = \text{number of feeders} \times \text{feederMVA}$$

$$\text{UE for 33MV}A = \text{number of feeders} \times \left[\text{UE}_{\text{per feeder}} + \frac{(33 - \text{xfmr load})}{\text{feederMVA}}\right] \tag{17.8}$$

$$\text{Annual benefit} = \text{UE}_{11\text{ MVA system}} - \text{UE}_{\text{alternative system}} \tag{17.9}$$

TABLE 17.9. Feeder and Transformer Loading Comparison

| MVA | Circuit Length (miles) | No. of Feeders/ xfmr | Unserved Energy Cost ($) | UE Cost for 33 MVA xfmr ($) | Capacity Cost ($) | Benefit (annual, $) | Years to Payback |
|---|---|---|---|---|---|---|---|
| 11 | 3.25 | 3 | 150,441 | 451,324 | 0 | 0 | – |
| 10 | 3.05 | 3 | 125,258 | 413,352 | 88,896 | 37,972 | 4 |
| 9 | 2.85 | 3 | 102,873 | 377,201 | 205,230 | 74,123 | 5 |
| 8 | 2.65 | 4 | 84,425 | 348,254 | 102,800 | 103,070 | 2 |
| 7 | 2.45 | 4 | 68,632 | 323,552 | 257,912 | 127,772 | 3 |
| 6 | 2.25 | 4 | 55,529 | 305,411 | 413,024 | 145,913 | 5 |

The capacity charges used were the same ones used for second source calculations. They are listed below.

$$\begin{aligned} \text{xfmr capacity cost} &= \$29,632/\text{MVA} \\ \text{Feeder capacity cost} &= \$9146/\text{MVA} \end{aligned} \tag{17.10}$$

The equations used to calculate the capacity charge for the total system are as follows for feeder capacities of 10 MVA and below:

$$\begin{aligned} \text{Feeder capacity cost} &= \text{number of feeders} \times (10\text{MVA} - \text{feederMVA}) \times \$9146 \\ \text{xfmr capacity cost} &= (33 - (\text{number of feeders} \times \text{feeder MVA})) \times \$29,632 \\ \text{Unused capacity cost} &= \text{feeder capacity cost} + \text{xfmr capacity cost} \end{aligned} \tag{17.11}$$

Using the "annual benefit" and the "unused capacity cost," a cost–benefit analysis can be performed on each scenario that takes into account the customer value of reliability. The option with the shortest payback period is the desired alternative. The results in Table 17.4 in Section 17.5 show that four 8 MVA circuits are the alternative with the shortest payback period.

## 17.12 FEEDER TIE COST CALCULATION

To find the construction cost of the three-phase line that is required for a feeder tie, the distance between the two circuits must be calculated. That number can then be multiplied by the cost per mile of three-phase construction. Assuming a two-transformer substation with four feeders per transformer, area served will be split into 1/8th, as indicated in Fig. 17.7. Each line emanating from the center represents a feeder.

Assuming a 1.75-mile radius for the substation, the following figures represent a tie placed at the midpoint of a feeder (Fig. 17.8).

$$\sin 22.5^\circ = \frac{X/2}{0.85}$$

$$X = 2 \times 0.85 \times \sin 22.5^\circ$$
$$X = 0.55 \text{ miles}$$

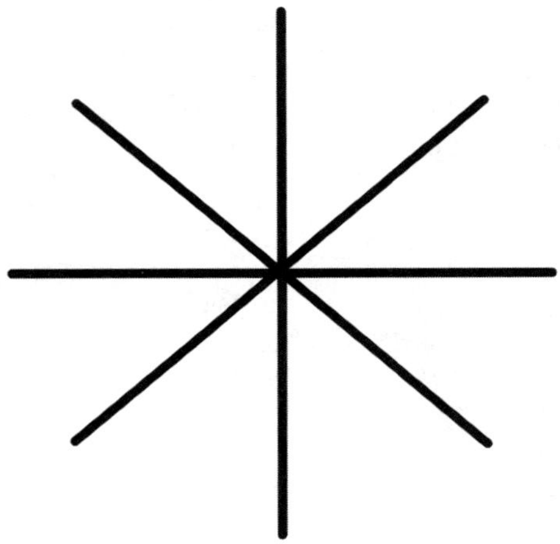

Figure 17.7. Service area split into 1/8th.

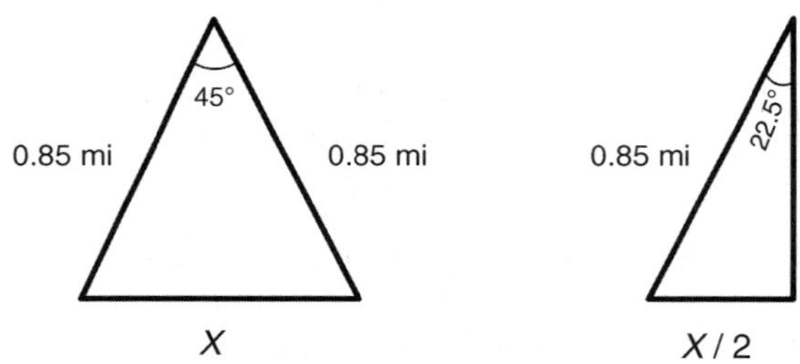


Figure 17.8. Illustration of tie placement.

$$\text{Cost} = \$30,487/\text{mile} \times 0.55 \text{ miles}$$
$$\text{Cost} = \$16,768$$
$$\text{Cost per feeder} = \$16,768/2 = \$8384$$

## 17.13 EFFECTS OF TIE MAINTENANCE

To model a more accurate representation of adding a second feeder tie to the circuit, the effect of the feeder tie maintenance had to be included. Different frequencies of tie maintenance were selected, which allowed the tie availability to be calculated. The tie availability was used to calculate a new unserved energy cost for Case 2 (single-feeder tie). If the feeder tie is available, the system remains the same as Case 2, which has an unserved energy cost of \$96,561. If the feeder tie is unavailable, the system has an unserved energy cost of \$272,489. Multiplying each of these by the correct factor and

TABLE 17.10. Frequencies of Maintenance

| Maintenance Duration (h) | Frequency (once every *N* years) | Tie Availability (%) | New UE Cost of Case 2 ($) |
|---|---|---|---|
| 672 | 1 | 92.3 | 158,879 |
| 672 | 3 | 97.4 | 151,097 |
| 672 | 5 | 98.5 | 149,528 |
| 672 | 10 | 99.2 | 148,360 |
| 672 | 15 | 99.5 | 147,970 |

then adding the results will give the modified unserved energy cost for Case 2 with maintenance effects included. The equation used to arrive at the modified unserved energy cost is listed below.

$$\mathrm{UE}_{\mathrm{new}} = \mathrm{TA} \times \$96,561 + (1 - \mathrm{TA}) \times \$272,489 \tag{17.12}$$

where TA denotes tie availability.

Table 17.10 lists various frequencies of maintenance. The maintenance duration of 672 h corresponds to 4 weeks. The tie availability and corresponding UE costs are also listed.

The new unserved energy cost of Case 2 (single-feeder tie) can then be compared against the unserved energy cost of the case with two feeder ties ($96,303) to determine the benefit of adding the second tie. Table 17.6 in Section 17.7 shows the yearly benefit associated with each maintenance frequency for the addition of a second feeder tie. These values were used in UCBM to determine the payback period for the incremental investment taking into account customer value of improved reliability.

## 17.14 ADDITIONAL TIES FOR FEEDERS WITH THREE-PHASE BRANCHES

The benefit of adding a tie at the end of a three-phase branch is that for line or switch failures; only the section that failed will be unserved during repair instead of all downstream sections in the case of no backup tie. Figure 17.9 illustrates this concept.

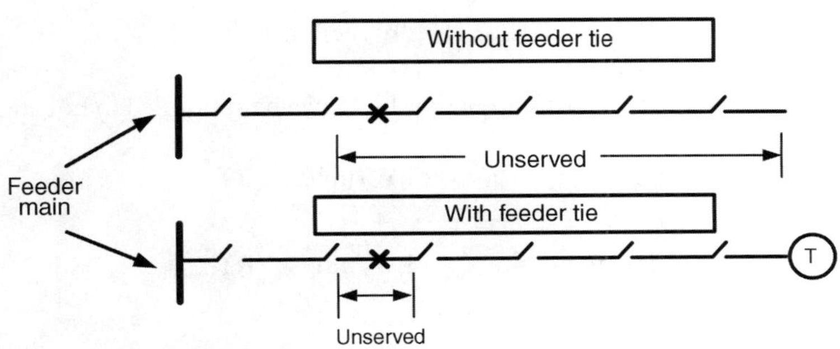


Figure 17.9. Benefit of tie for three-phase branch.

If following two assumptions are made, an equation can be developed for the benefit of installing the additional feeder tie:

1. Feeder load is uniformly distributed at 2500 kW/mile of three-phase.
2. All load can be picked up once failure is isolated.

The method used in developing the equation was to calculate the reduction in unserved energy for a fault on each section and each switch on the branch. For example, for a fault on the second section of the branch as depicted in Fig. 17.9, four sections will be unserved if there is no feeder tie and only one section will be unserved if there is a feeder tie. The three sections that are served in the case with the feeder tie are not picked up back instantaneously; instead, they are out for the time it takes to isolate the failure and close the backup tie. Therefore, the load on those three sections will be unserved for a smaller amount of time if the backup tie is available.

To apply this to a branch with "$n$" sections, the change in unserved energy for a fault on each individual section and switch needs to be calculated. Equation 17.13 calculates the change in unserved energy for faults on each section.

### 17.14.1 Definition of Terms

$F_{\text{line}}$ failure rate of line (events/mile year)
$F_{\text{switch}}$ failure rate of line switch (events/year)
$n$ number of sections
$\text{RT}_{\text{line}}$ repair time of line (h)
$\text{RT}_{\text{switch}}$ repair time of switch (h)
$T_{\text{switching}}$ time required to isolate failure and close ties (h)
UE unserved energy (kW h)

$$\begin{aligned}\Delta\text{UE} = {} & \text{section length} \times F_{\text{line}} \times \text{kW/mile} \times \text{section length} \times (n-1) \\ & \times(\text{RT}_{\text{line}} - T_{\text{switching}}) \quad \leftarrow \text{1st section} + \text{section length} \times F_{\text{line}} \\ & \times \text{kW/mile} \times \text{section length} \times (n-2) \times (\text{RT}_{\text{line}} - T_{\text{switching}}) \\ & \leftarrow \text{2nd section} + \text{section length} \times F_{\text{line}} \times \text{kW/mile} \times \text{section length} \\ & \times (n-3) \times (\text{RT}_{\text{line}} - T_{\text{switching}}) \leftarrow \text{3rd section}\end{aligned} \tag{17.13}$$

Factoring out some like-terms, Equation 17.12 can be simplified to Equation 17.13.

$$\begin{aligned}\Delta\text{UE} &= \text{section length}^2 \times F_{\text{line}} \times \text{kW/mile} \times (\text{RT}_{\text{switch}} - T_{\text{switching}}) \\ &\quad [n-1+n-2+n-3+\cdots+1] \\ \Delta\text{UE}_{\text{sections}} &= \text{section length}^2 \times F_{\text{line}} \times \text{kW/mile} \times (\text{RT}_{\text{switch}} - T_{\text{switching}}) \\ &\quad \times \sum_{i=0}^{n-1} n-(i+1)\end{aligned} \tag{17.14}$$

A similar approach is used to determine the change in unserved energy associated with switch failures, and Equation 17.15 below is arrived at.

$$\Delta \mathrm{UE}_{\mathrm{switches}} = F_{\mathrm{switch}} \times \mathrm{kW/mile} \times \text{section length} \times (\mathrm{RT}_{\mathrm{switch}} - T_{\mathrm{switching}}) \times \sum_{i=0}^{n-1} n - (i+1) \tag{17.15}$$

To obtain the total change in unserved energy from adding the switch, the previous two equations can be added. Table 17.7 in Section 17.5 summarizes the benefits associated with adding a feeder tie at the end of a three-phase branch for various branch lengths.

## 17.15 CONCLUSIONS

This chapter presented value-based probabilistic urban distribution system planning models for determining optimal section length for switch placement on the main feeder, number and placement of feeder ties, and feeder and transformer loadings. The following conclusions were reached on the basis of the assumed failure rates, repair times, switching time, and customer value of reliability used in the analyses. A sectionalizing switch should be placed every 0.7 MW of feeder load or approximately every one-fourth mile. Two feeder ties should be installed on a radial feeder with no three-phase branches and with no voltage constraints. The most essential tie, in terms of reliability, is the tie located at the end of the feeder. This tie allows the most flexibility because it can provide backup for a failure anywhere along the feeder. Available transformer capacity (top nameplate rating) should be used for normal loading conditions. Feeders should be loaded to approximately 8 MVA, leaving 3–4 MVA available in emergency. Having sufficient feeder tie capacity on adjacent feeders is essential for providing backup capacity following a transformer outage and, more importantly, for backup following feeder outages, which occur more frequently.

## REFERENCES


1. A. A. Chowdhury, S. K. Agarwal, and D. O. Koval, Reliability modeling of distributed generation in conventional distribution systems planning and analysis, *IEEE Transactions on Industrial Applications*, **39** (5), 2003, 1493–1498.
2. A. A. Chowdhury and D. O. Koval, Value-Based distribution system reliability planning, *IEEE/IAS Transactions*, **34** (1), 23–29, 1998.
3. A. A. Chowdhury and D. O. Koval, Application of customer interruption costs in transmission network reliability planning, *IEEE Transactions on Industry Applications*, **37** (6), 2001, 1590–1596.
4. R. Billinton and P. Wang, Distribution reliability cost/worth analysis using analytical and sequential simulation techniques, *IEEE Transactions on Power Systems*, **13** (4), 1998, 1245–1250.

5. R. E. Brown, S Gupta, R. D. Christie, S. S. Venkata, and R. Fletcher, Automated primary distribution system design: reliability and cost optimisation, *IEEE Transactions on Power Delivery*, **12** (2), 1997, 1017–1022.
6. A. A. Chowdhury and D. O. Koval, Current practices and customer value-based distribution system reliability planning, *Transactions of the IEEE/IAS Annual Meeting*, **2**, 2000, 909–916.
7. R. E. Brown and M. M. Marshall, Budget constrained planning to optimize power system reliability, *IEEE Transactions on Power Systems*, **15** (2), 2000, 887–892.
8. R. E. Brown The impact of heuristic initialization on distribution system reliability optimisation, *International Journal of Engineering Intelligent Systems for Electrical Engineering and Communications*, **8** (1), 2000, 45–52.
9. Electric Power Research Institute Utility Cost Benefit Analysis Model (UCBM), EPRI RP1678–15, March, 1987.

# 18

# RADIAL FEEDER RECONFIGURATION ANALYSIS

## 18.1 INTRODUCTION

As stated earlier in this book, historical distribution feeder reliability assessment generally summarizes discrete interruption events occurring at specific locations over specific periods, whereas predictive assessment estimates the long-run behavior of systems by combining component failure rates and repair (restoration) times that describe the central tendency of an entire distribution of possible values with feeder configurations. The outage time due to component failures can substantially be reduced by protection and sectionalizing schemes. The time required to isolate a faulted component by isolation and switching action is known as switching or restoration time. An alternative supply in radial networks normally enhances the load point reliability. Fuses usually protect the lateral distributors connected to the customers. This chapter presents a reliability methodology to improve the radial distribution feeder reliability performance using a simple illustrative feeder configuration.

It is a well-known fact that the distribution system is an important link between the bulk transmission system and its customers, and in many cases these links are radial in nature, which makes them vulnerable to outages due to a single event. A radial distribution circuit generally uses main feeders and lateral distributors to supply customer energy requirements.

*Power Distribution System Reliability.* By Ali A. Chowdhury and Don O. Koval

The basic activities associated with distribution system reliability assessment can be divided into two fundamental segments of measuring past performance and predicting future performance. Service continuity data are being routinely published by many organizations and utilities. The calculated statistics aid in planning, operating, and maintaining the distribution system.

The customer outage time due to component failures can significantly be reduced by protection and sectionalizing schemes. The time required to isolate a faulted component by isolation and switching action is defined as switching or restoration time. Options for an alternative supply to radial feeders normally increases customer load point reliability levels.

At most electric utilities, the determination of acceptable levels of service continuity is currently achieved by comparing the actual interruption frequency and duration indices with arbitrary targets. For example, monthly reports on service continuity statistics produced many utilities that contain the arbitrary targets of system reliability indices for performance comparison purposes. It has been recognized that rules of thumb and implicit criteria cannot be used in a consistent manner to the very large number of capital investment and operating decisions that are routinely being made.

This chapter is concerned with the aspects of predictive assessment, and it presents the basic concepts involved in the use of customer interruption cost data to evaluate the worth of distribution system reliability at the individual customer load points. Predictive distribution system reliability indices were computed and used to estimate the customer interruption costs by considering outages in the radial distribution networks associated with an illustrative rural feeder configuration. The expansion plan of a distribution feeder can be optimized in terms of reliability by using an economic criterion in which the sum of both customer interruption and system costs is minimized. This chapter presents a reliability cost/reliability worth methodology to improve the distribution feeder reliability performance by using a simple illustrative rural feeder configuration.

## 18.2 PREDICTIVE FEEDER RELIABILITY ANALYSIS

To render a rational means of decision making for changing service continuity levels experienced by customers, it is necessary to include system costs and the costs incurred upon customers associated with an interruption in current planning and operating practices. A value-based reliability planning approach therefore attempts to locate the minimum cost solution, where the total cost includes investment cost plus operating cost plus customer interruption cost. Using the concept of value-based reliability planning, a given level of service reliability can be examined in terms of the cost and the worth to the customer of providing the electric service.

Value-based distribution system reliability planning, therefore, becomes an invaluable tool using which a proactive, customer-responsive utility will be able to retain its existing customer base and win new customers. The value of service, that is, the worth of reliability expressed in terms of costs of customer interruptions can be established on the basis of actual surveys of the customer perception regarding the level of reliability they are willing to pay for. By establishing a method of giving a dollar value to various levels of reliability, it is possible to ascertain the balance where reliability and cost are best matched. The data compiled from customer surveys lead to the creation of sector damage functions. The cost of interruption at a single-customer load point depends entirely on the cost characteristics of that customer. The sector damage function presents the sector interruption costs as a function of interruption duration. The customer costs associated with an outage at any load point in the system involve the combination of costs associated with all customer types affected by that interruption. This combination leads to the development of a composite customer damage function. For the purposes of this chapter, a generic cost of interruption value of \$10.76/kWh has been used in these studies to illustrate the application of the value-based reliability methodology in distribution feeder reliability assessment. The terms "unserved energy cost," "customer interruption cost," and "customer outage cost" are used interchangeably in this chapter.

Predictive reliability performance is normally concerned with the supply adequacy at the customer load points. The basic indices used are the load point average failure rate ($\lambda$), average outage duration ($r$), and average annual outage time ($U$). The mathematical models for the basic reliability indices for series and parallel systems are given in the following.

For a radial system, the indices at each load point "p" are

$$\lambda_{\mathrm{p}} = \sum_{i=1}^{N} \lambda_i \quad \text{failures/year} \tag{18.1}$$

$$U_{\mathrm{p}} = \sum_{i=1}^{N} \lambda_i r_i \quad \text{h/year} \tag{18.2}$$

$$r_{\mathrm{p}} = \frac{U_{\mathrm{p}}}{\lambda_{\mathrm{p}}} \quad \text{h/failure} \tag{18.3}$$

For a two-component parallel system, the indices at each load point "p" are

$$\lambda_{\mathrm{p}} = \frac{\lambda_1 \lambda_2 (r_1 + r_2)}{1 + \lambda_1 r_1 + \lambda_2 r_2} \quad \text{failures/year} \tag{18.4}$$

$$r_{\mathrm{p}} = \frac{r_1 r_2}{r_1 + r_2} \quad \text{h/failure} \tag{18.5}$$

$$U_{\mathrm{p}} = \frac{\lambda_1 \lambda_2 r_1 r_2}{1 + \lambda_1 r_1 \lambda_2 r_2} \quad \text{h/year} \tag{18.6}$$

where "$N$" denotes the number of outage events affecting load point "p."

A distribution reliability program has been used in this chapter to analyze the reliability improvement options for the illustrative distribution feeder. The program is designed to aid electric utility and industrial/commercial customers with predictive reliability assessment of a distribution network. The customer-responsive utility would address reliability problems by selecting project alternatives that have the highest internal and external benefits. Customers may be willing to share the costs when approached with quantifiable plans. In addition, it can assist in developing reliability guidelines and service-based pricing by quantifying the system reliability. The computer program used computes a set of reliability indices including System Average Interruption Frequency Index (SAIFI), System Average Interruption Duration Index (SAIDI), Average Service Availability Index (ASAI), load/energy curtailed and the cost of outages based on the component outage data, and the cost of interruption to a customer. The program models time-sequenced switching actions taken by an operator/repair person following an outage. It can also be used to quantify benefits of automating distribution systems and feeder reconfiguration and to compare various competing projects by using cost of outages and utility benefits.

## 18.3 RELIABILITY DATA AND ASSUMPTIONS

The input data and assumptions used to assess the feeder reliability improvements are presented in Table 18.1. As shown in Table 18.1, the three-phase (PH) overhead (OH) distribution feeder failure rate seems to be higher than that of the industry average values. The rate of 0.8 failures/mile year reflects the actual failure performance over the years of the rural distribution feeder analyzed in this chapter. This specific feeder is subject to higher failures due to its geographical location, terrain, proximity to wildlife and public interference, weather conditions, and feeder design considerations. The repair time figures presented in Table 18.1 include crew callout, fault isolation, and actual repair or replacement times for a piece of distribution component.

A utility cost–benefit analysis model was used to calculate the years to payback for each alternative. The alternative with the shortest payback period indicates reliability improvements that it is the most economical way to serve the load taking into account customer value of reliability. The cumulative present values (CPV) in 2003 dollars for the facility costs and the reliability benefits have been computed by considering a 35-year project life, a discount rate of 8.34%, an inflation rate of 2.5%, and an unserved energy cost of $10.76/kWh. The unserved energy cost of $10.76/kWh was derived from published industry data by weighting the customer composition mix served by the

TABLE 18.1. Distribution Component Failure Data

| Component | Failure Rate | Repair Time (h) |
|---|---|---|
| 3 PH OH | 0.8/miles year | 4 |
| Switch | 0.001/year | 5.5 |
| Transformer | 0.07/year | 24 |
| Regulator | 0.0036/year | 15 |
| Recloser/breaker | 0.0036/year | 32 |
| Switchgear/bus | 0.001/year | 15 |

TABLE 18.2. Five-Step Load Duration Curve Approximation

| Load Level | Probability |
|---|---|
| 100% (Peak) | 0.001 |
| 80–90% | 0.025 |
| 70–80% | 0.040 |
| 60–70% | 0.097 |
| <60% | 0.837 |

feeder studied in this chapter. Section 18.4 presents the results of the analyses for the illustrative distribution feeder. Simulations were run at different system load levels (100, 80, 70%, etc.), and the annual unserved energy cost was calculated by weighting the results of each simulation by the percentage of the year that each load level is present. A five-step approximation of the feeder load duration curve is used in computing the annual expected unserved energy for component outages. The five-step load model with corresponding exposure probability is presented in Table 18.2.

## 18.4 RELIABILITY ASSESSMENT FOR AN ILLUSTRATIVE DISTRIBUTION FEEDER

The basic objective of the study is to ascertain the different reliability improvement options for the rural illustrative distribution feeder. The following sections present the reliability improvement options for the feeder of study.

### 18.4.1 Base Case Circuit Description

The feeder studied in this chapter serves mainly rural area load. The feeder has a peak load of 11.5 MW and consists of approximately 28 miles of three-phase overhead circuit. This is a considerable amount of exposure for a single distribution feeder; however, the feeder does have five reclosers to help reduce the exposure to any particular customer. The feeder load is mainly concentrated on the north of the substation AA; that is, about 10 MW of load is served from the north branch. The feeder presently has six feeder ties to four surrounding substations. Figure 18.1 shows the current feeder layout. Some of the ties are fairly weak due to the distance from the surrounding substations; but each tie is able to provide backup for at least one branch of the circuit, so there is sufficient backup capacity available. A brief description of the existing ties and their capabilities is presented in the following.

### 18.4.2 Circuit Tie 47-2

This tie circuit serves 3 MW of load at peak, which would leave a significant amount of backup capacity, but the tie point to the circuit is 7 miles away from Sub 47. Therefore, tie capability is limited to 2 MW at peak due to voltage constraints. At lower load levels, the

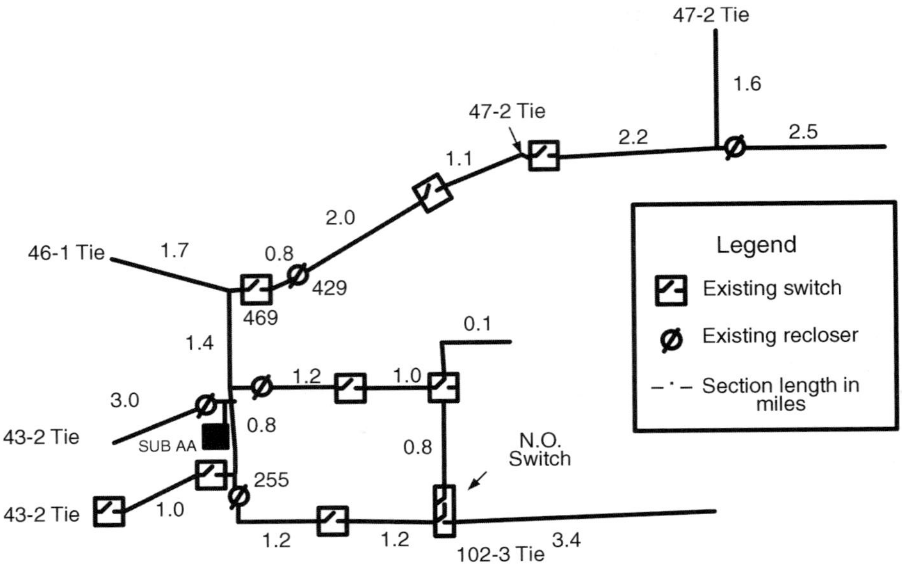


Figure 18.1. The illustrative feeder layout.

tie capacity would increase somewhat, but it is doubtful that it would be possible to serve past Switch 469 (see Fig. 18.1 for map of the circuit) because of voltage constraints. There are two ties to the feeder, which are relatively close to each other. For certain outages, it is possible to close both ties and open a feeder switch between them, which helps split the load, reducing voltage drop along the lines. The assumed tie capacity used in the simulation is listed in Table 18.3. Note that this tie was not allowed to serve past Switch 469 in the simulation.

### 18.4.3 Circuit Tie 46-1

This tie circuit has 8 MW of load and is served by a 12.5 MVA transformer. The tie point is relatively close to Sub 46 (about 0.9 miles), so voltage drop is not much of a concern; however, there are no sectionalizing switches between this tie and the source for the feeder of concern. This makes the tie very ineffective for providing backup for failures along the feeder. It is only useful for a failure of the transformer, breaker, or feeder exit. The assumed tie capacity used in the simulation model is listed in Table 18.4.

TABLE 18.3. Illustrative Circuit Tie 47-2 Capacity

| Load (% of Peak) | Tie Capacity (MW) |
|---|---|
| 100 | 2 |
| 80 | 2.6 |
| 70 | 2.9 |
| 60 | 3.2 |
| 55 | 3.6 |

TABLE 18.4. Circuit Tie 46-1 Capacity

| Load (% of Peak) | Tie Capacity (MW) |
|---|---|
| 100 | 4.5 |
| 80 | 6.1 |
| 70 | 6.9 |
| 60 | 7.7 |
| 55 | 8.1 |

### 18.4.4 Circuit Tie 43-2

This tie circuit serves 5 MW of load and has two tie points to the feeder studied in the chapter. The source at Sub 43 is approximately 5.5 miles away from the tie points. This again limits the tie capability to roughly 2 MW at peak because of voltage constraints. The branches that these ties backup are very lightly loaded and there is sufficient capacity available to cover them. However, one of the branches has no sectionalizing switches between the tie point and the recloser serving that particular branch, so the usefulness of the tie is limited. The assumed tie capacity used in the simulation model is listed in Table 18.5.

Note that this capacity is split equally between the two ties.

### 18.4.5 Circuit Tie 102-3

This tie circuit serves 2 MW of load, but the tie point is 7 miles from the source at Sub 102. Tie capability is therefore limited to 2 MW at peak due to voltage constraints. Also due to existing circuit layout, this tie provides little benefit because there is normally an open switch connecting the two branches of the feeder very close to this tie. The normally open switch serves the same function as a feeder tie, even though it is to the same circuit. The tie to Circuit 102-3 serves no additional benefit for normal line failures, only for a transformer, breaker, or feeder exit failure. The assumed tie capacity used in the simulation model is listed in Table 18.6.

### 18.4.6 Base Case Reliability

Using outage rates, switching times, and repair times, an expected unserved energy figure is computed by the simulation runs. This figure can be converted to an unserved

TABLE 18.5. Circuit Tie 43-2 Capacity

| Load (% of Peak) | Tie Capacity (MW) |
|---|---|
| 100 | 2 |
| 80 | 3 |
| 70 | 3.5 |
| 60 | 4 |
| 55 | 4.25 |

TABLE 18.6. Circuit Tie 102-3 Capacity

| Load (% of Peak) | Tie Capacity (MW) |
|---|---|
| 100 | 2 |
| 80 | 2.4 |
| 70 | 2.6 |
| 60 | 2.8 |
| 55 | 3 |

TABLE 18.7. Base Case Results

| | Unserved Energy (kWh) | Probability | Expected Unserved Energy (kWh) |
|---|---|---|---|
| 90–100% | 221,796 | 0.001 | 221.8 |
| 80–90% | 154,342 | 0.025 | 3,858.6 |
| 70–80% | 125,883 | 0.04 | 5,035.3 |
| 60–70% | 107,060 | 0.097 | 10,384.8 |
| <60% (55%) | 98,138 | 0.837 | 82,141.5 |
| | | Annual total | 101,642.0 |

energy cost by multiplying by the average customer cost of unserved energy. Table 18.7 summarizes the expected annual unserved energy for the base case situation.

As shown in Table 18.7, the annual expected total unserved energy is 101,642.0 kWh and the expected customer outage cost is $1,093,668.

## 18.5 ALTERNATIVE IMPROVEMENT OPTIONS ANALYSIS

The different reliability improvement alternative options considered in the studies are shown in Fig. 18.2. The reliability improvements were simulated in the computer model used to determine the reduction in customer outage costs, taking into account the customer value of reliability. This allows cost–benefit analysis to be performed on feeder improvements by comparing the benefits in reduced customer outage costs against the capital cost required for the improvement. A total of four incremental improvement options were considered for increasing the feeder reliability. As each incremental improvement was analyzed, it was added to the previous improvements to determine the benefit that related strictly to each particular addition.

### 18.5.1 Incremental Improvement Alternative 1: Add Distribution Automation Switch

At present, the breaker will clear a line fault on any section along the north branch of the feeder before Recloser 429. This adds 3.1 miles of exposure to the entire feeder load. Adding a distribution automation (DA) switch would remove this exposure and improve

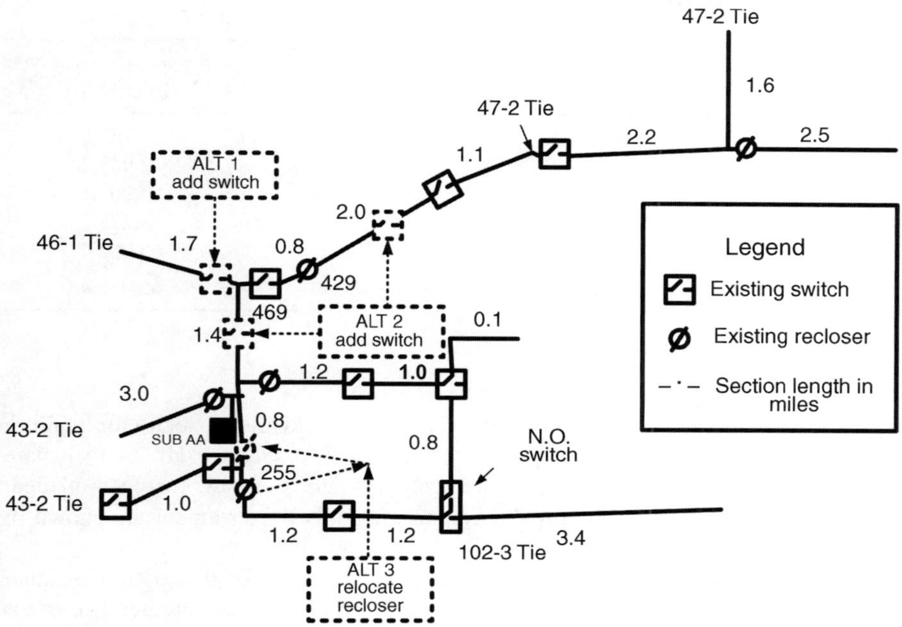


Figure 18.2. The illustrative feeder improvement alternatives.

the reliability of the feeder. This improvement was modeled, and the results are listed in Table 18.8.

The annual expected total unserved energy is 87,249.6 kWh and the calculated customer outage cost is $938,806. For this improvement option, the annual incremental benefit is $154,862, the incremental facility cost is $25,000 for the DA switch and installation, and the payback period for the improvement project is less than 1 year.

### 18.5.2 Incremental Improvement Alternative 2: Add Sectionalizing Switch

As discussed earlier, the usefulness of the tie to the Tie Circuit 46-1 is restricted due to the fact that there is no sectionalizing switch between the tie and the normal source for the

TABLE 18.8. Incremental Improvement Alternative 1 Results

| | Unserved Energy (kWh) | Probability | Expected Unserved Energy (kWh) |
|---|---|---|---|
| 90–100% | 181,793 | 0.001 | 181.8 |
| 80–90% | 129,420 | 0.025 | 3,235.5 |
| 70–80% | 107,570 | 0.04 | 4,302.8 |
| 60–70% | 92,021 | 0.097 | 8,926.0 |
| <60% (55%) | 84,353 | 0.837 | 70,603.5 |
| | | Annual total | 87,249.6 |

TABLE 18.9. Incremental Improvement Alternative 2 Results

| | Unserved Energy (kWh) | Probability | Expected Unserved Energy (kWh) |
|---|---|---|---|
| 90–100% | 162,583 | 0.001 | 162.6 |
| 80–90% | 117,806 | 0.025 | 2,945.2 |
| 70–80% | 99,177 | 0.04 | 3,967.1 |
| 60–70% | 84,827 | 0.097 | 8,228.2 |
| <60% (55%) | 77,758 | 0.837 | 65,083.4 |
| | | Annual total | 65,083.4 |

feeder of study. Placing a switch as shown in Fig. 18.2 would allow backup for a failure along the section north of the substation AA. It also provides the ability to isolate the faulted section for a failure along the section where the new switch is placed. The computer simulation results for the system after this improvement are shown in Table 18.9.

The expected annual customer outage cost is \$864,959 for this option. Annual incremental benefit for this improvement option is \$73,847, and the incremental cost for the sectionalizing switch is \$5000 (cost of switch and installation). The investment payback period is less than 1 year.

### 18.5.3 Incremental Alternative 3: Relocate Recloser 255

If Recloser 255 was removed from its current location and placed just south of the substation, it would reduce exposure of the north section of the feeder by almost 2 miles. It must be verified that coordination between the breaker and the recloser can still be attained if the recloser is moved. Since the majority of the load is on the north section of the feeder, reducing its exposure is desirable. This improvement was modeled and simulated in the computer model used, producing the following results shown in Table 18.10.

TABLE 18.10. Incremental Improvement Alternative 3 Results

| | Unserved Energy (kWh) | Probability | Expected Unserved Energy (kWh) |
|---|---|---|---|
| 90–100% | 142,435 | 0.001 | 142.4 |
| 80–90% | 104,632 | 0.025 | 2,615.8 |
| 70–80% | 89,487 | 0.04 | 3,579.5 |
| 60–70% | 76,703 | 0.097 | 7,440.2 |
| <60% (55%) | 70,311 | 0.837 | 58,850.3 |
| | | Annual total | 72,628.2 |

TABLE 18.11. Incremental Alternative 4 Results

| | Unserved Energy (kWh) | Probability | Expected Unserved Energy (kWh) |
|---|---|---|---|
| 90–100% | 137,400 | 0.001 | 137.4 |
| 80–90% | 99,388 | 0.025 | 2,484.7 |
| 70–80% | 84,896 | 0.04 | 3,395.8 |
| 60–70% | 72,768 | 0.097 | 7,058.5 |
| <60% (55%) | 66,704 | 0.837 | 55,831.2 |
| | | Annual total | 68,907.7 |

The expected annual customer outage cost for this improvement option is $781,480. The annual incremental benefit is $83,479. The incremental recloser relocation cost is $6000. This includes the cost of removal and reinstallation of existing unit, along with a group-operated bypass switch. This estimate assumes that a pole changeout is not required; this would add approximately $2000 to the estimate. The investment payback period is less than 1 year.

### 18.5.4 Incremental Improvement Alternative 4: Place 2 New Switches

The existing feeder has some lengthy, highly loaded sections. Prior study on this feeder indicated that adding sectionalizing switches to break up sections serving more than 0.75 MW of load can be cost justified by the reliability improvements associated with the addition. Two existing sections were selected to place a new switch. These switches (shown in Fig. 18.2) were modeled, and the computer simulation results with this improvement included are shown in Table 18.11.

The expected annual customer outage cost is $741,447 for this option. The annual incremental benefit is $40,333. The incremental facility cost for this option is $10,000 (cost of two switches and installation). The investment payback period is less than 1 year.

## 18.6 SUMMARY OF THE ILLUSTRATIVE FEEDER RELIABILITY PERFORMANCE IMPROVEMENT ALTERNATIVES

The computed results show that the improvements considered can be justified in a cost–benefit analysis including customer value of reliability. Combining all four improvements results in a total annual benefit of $352,221 against a total capital cost of $46,000. It should be noted that an alternative project was considered to add switchgear to break up this feeder into three feeders. The switchgear addition would replace the need for incremental improvement alternatives 1 and 3. This switchgear addition project would provide nearly identical results, depending on the final configuration, of the newly created circuits. The switchgear addition project was estimated at $478,000, in comparison to $31,000 for the combination of alternatives 1 and 3. It therefore can be concluded that these improvement alternatives of 1 and 3 considered in this chapter provide the higher benefit–cost ratio than that of the switchgear addition project.

## 18.7 CONCLUSIONS

This chapter presents the basic concepts involved in using customer interruption costs data to evaluate distribution feeder reliability worth at the individual customer load points. Predictive distribution feeder reliability indices were calculated and used to estimate the customer interruption costs by considering outages in a rural distribution feeder. One basic conclusion of this chapter is that the reliability improvement plan of a distribution feeder may be optimized in terms of reliability by using an economic criterion in which the sum of both customer interruption and system costs is minimized.

## REFERENCES


1. Canadian Electricity Association, 2002 Annual Service Continuity Report on Distribution System Performance in Electrical Utilities, May 2003.
2. A. A. Chowdhury and D. O. Koval, Value-based distribution system reliability planning, *IEEE/IAS Transactions*, **34** (1), 1998, 23–29.
3. A. A. Chowdhury and D. O. Koval, Application of customer interruption costs in transmission network reliability planning, *IEEE Transactions on Industry Applications*, **37** (6), 2001, 1590–1596.
4. R. Billinton and P. Wang, Distribution reliability cost/worth analysis using analytical and sequential simulation techniques, *IEEE Transactions on Power Systems*, **13** (4), 1998, 1245–1250.
5. R. E. Brown, S. Gupta, R. D. Christie, S. S. Venkata, and R. Fletcher, Automated primary distribution system design: reliability and cost optimization, *IEEE Transactions on Power Delivery*, **12** (2), 1997, 1017–1022.
6. A. A. Chowdhury and D. O. Koval, Current practices and customer value-based distribution system reliability planning, *IEEE/IAS Transactions*, **40** (5), 2004, 1174–1182.
7. R. E. Brown and M. M. Marshall, Budget constrained planning to optimize power system reliability, *IEEE Transactions on Power Systems*, **15** (2), 2000, 887–892.
8. R. E. Brown, The impact of heuristic initialization on distribution system reliability optimisation, *International Journal of Engineering Intelligent Systems for Electrical Engineering and Communications*, **8** (1), 2000, 45–52.
9. A. A. Chowdhury, S. K. Agarwal, and D. O. Koval, Reliability modeling of distributed generation in conventional distribution systems planning and analysis, *IEEE Transactions on Industrial Applications*, **39** (5), 2003, 1493–1498.
10. Electric Power Research Institute, Utility Cost Benefit Analysis Model (UCBM), EPRI RP1678-15, March 1987.

# 19

# DISTRIBUTED GENERATION

## 19.1 INTRODUCTION

The primary objective of any electric utility company in the new competitive environment is not only to increase the market value of the services by providing the right amount of reliability but also to lower the costs of operation, maintenance, and construction of new facilities so as to provide customers electricity at lower rates. An electric utility company will strive to achieve this objective through various means, one of which is to defer the capital distribution facility requirements in favor of a distributed generation (DG) solution by an independent power producer (IPP) to meet the growing customer load demand. In this case, the distribution capital investment deferral credit received by the IPP will depend on the incremental system reliability improvement rendered by the DG solution. In other words, the size, location, and reliability of the DG will be based on the comparable incremental reliability provided by the distribution solution under consideration. This chapter presents a reliability model for determining the DG equivalence to a distribution facility for use in distribution system planning studies in the new competitive environment.

*Power Distribution System Reliability.* By Ali A. Chowdhury and Don O. Koval

At present, the electric power industry is undergoing considerable change with respect to structure, operation, and regulation. The various electric utility "Acts" introduced in different countries have initiated the restructuring process, and the traditional, vertically integrated utility structure consisting of generation, transmission, and distribution functions has been dismantled. Instead, distinct generation, transmission, and distribution companies have been established in which each company performs a single function in the overall electricity supply task. As a result, the overall responsibility of serving the individual customer needs does not reside in a single electric utility, as has been the case in the vertically integrated utility structure.

To appreciate the reliability issues arising in the present electric power industry environment, it is necessary to recognize the various forces and actions that are shaping the environment. The deregulation legislations establish many new entities to facilitate system operations and market functions independent of the owners of facilities. In the new competitive environment, power generation is no longer a natural monopoly. Generation expansion will be decided by the market forces and new players such as IPP and cogenerators who will make their presence felt in the generation arena.

As customers will increasingly demand lower rates and higher reliability in the new competitive environment, the challenging task for an electric utility company will be to minimize the capital investments and operation and maintenance expenditures to hold down electricity rates. If, however, the cost is cut too far, it may jeopardize the system's ability to provide a reliable power supply to its customers. The movement toward deregulation will therefore introduce a wide range of reliability issues that require system reliability criteria and tools that can incorporate the residual risks and uncertainties in distribution system planning and operation. Probabilistic techniques offer a rational response to these conflicting new requirements. This chapter illustrates a probabilistic reliability-based distribution system expansion and investment model to satisfy increasing customer demands of lower rates and higher service reliability in the competitive market.

## 19.2 PROBLEM DEFINITION

Distribution system reliability is an important issue in system planning and operations. In the past, electric utilities were continuously adding more facilities to their systems to satisfy the increasing customer load requirements. An electric utility company has traditionally relied on a set of deterministic criteria to guide distribution planning. Such criteria specified the outage conditions under which the system must meet future load forecasts. In most cases, the systems were overbuilt resulting in higher electricity rates for customers. As customers become more cost and service sensitive in the emerging competitive market, it will be extremely difficult for distribution companies to rationalize capital expenditures on the basis of their deterministic criteria. The distribution companies will be forced to look for different means to avert the risk of overinvestment in providing competitive rates and acceptable reliability levels to customers. As load increases, the distribution system has to be expanded to satisfy increased customer load requirements. For example, due to the increased load growth to a specific area of a distribution system, the local area distribution network is deemed inadequate and requires expansion. The distribution system planners would come up with a number of local area distribution improvement solutions such as adding a distribution feeder, adding a reactive compensation to the area, or adding a distribution substation to meet the growing customer loads. The cost of capital will be added to the rate base and will be reflected in the electricity rates.

To remain competitive, the electric utility company will look for ways to reduce costs and still provide the acceptable level of reliability required by the customers. One solution is to add smaller and environmentally friendlier distributed generation that can now be built economically by independent generators. Distributed generation consists of small generators typically ranging in capacity from 15 to 10,000 kW connected to the electric distribution system. DG can be installed at the utility or at the customer sites. Distributed generation technologies include conventional and nonconventional energy solutions such as diesel engine driven generators, wind turbines, fuel cells, and microturbines. Recent technical advances have significantly reduced the cost of DG and could eventually compete with gas turbines. A recent study on DG market potential notes that a generator selling into the real-time market could have made more than $3068/MWh during just 5 h on a particular day and would have made more than twice as much money if it could have earned the real-time price on days when the real-time price averaged more than $35/MWh. Another similar study on DG indicates that in the next 10–15 years, DG would capture 10–15% of new generating capacity in the United States. The growing demand for power could reach 60,000–120,000 MW of generation over the next 10–15 years of which DG will be an increasing component. This could amount to 6000–12,000 MW of DG over 10–15 years. An EPRI study provides a probabilistic area investment model for the determination of whether or not DG is an economic option in the overall distribution system expansion planning.

Another recent DG study states that many DG technologies are expected to see 25–40% decrease in capital costs and 10–15% increase in efficiency. In addition, many studies on DG potential predict that over the next 10 years, DG will emerge worldwide in many different shapes and sizes, possibly accounting for 8–14% of all additions.

In the light of the above discussions, one prudent investment decision by an electric utility company in the competitive market would be to issue a request for the proposal of distributed generation addition by an IPP to mitigate the distribution deficiency in the system. In this case, the distribution requirements can be met by a generation solution and significant savings through capital deferral by the electric utility company can be achieved thus enabling them to hold the line on rates. The IPP would receive incentives in the form of capital deferral credit from the electric utility company for replacing a distribution facility requirement. The amount of the capital deferral credit received by the IPP would be negotiated between the electric utility company and the IPP based on the size of the generator, the amount of must-run capacity from the unit to satisfy distribution requirements, and the comparable reliability improvement to the area where the generator will be located. This chapter illustrates a reliability model to determine the DG equivalence to a distribution facility based on comparable reliability rendered by distribution and generation solutions using a small illustrative distribution system.

## 19.3 ILLUSTRATIVE DISTRIBUTION SYSTEM CONFIGURATION CHARACTERISTICS

The basic objective of this chapter is to present a reliability model to determine distributed generation equivalence to a distribution facility in an attempt to improve the distribution system reliability while meeting increasing customer load requirements. This chapter considers a simple illustrative distribution system's loading conditions and needed reinforcements. The practical distribution system used to illustrate the DG modeling is the same as has been used in Chapters 11 and 15 and is repeated in this chapter for convenience and easy understanding. The load of the distribution system is supplied by two 13 kV distribution feeder circuits, as shown in Fig. 19.1.

The 13 kV feeders from substations A and B are operated as radial feeders, but they can be interconnected by a normally open tie point. The disconnects, lateral distributors, step-down transformers, and fuses and the alternative supply are assumed to be 100% available in the analysis to illustrate the reliability model.

The load factor for the service area is assumed to be 77%. The loading conditions at each load point are shown in Fig. 19.1. The peak rating for the 13 kV feeders from substations A and B are 12.00 and 10.5 MVA, respectively, at a power factor of 0.90 lagging. To evaluate the load point reliability levels of the distribution system, it is essential to have a working knowledge base of the operation of the feeder circuits and their operational constraints. The feeders can supply their respective loads when operated radially. For a line section outage on either feeder, the healthy feeder cannot supply the entire load of the faulted feeder due to the fact that the feeders are thermally limited. In this case, if both the feeders are operated radially and are tied through a normally open tie switch, then any line section outage can be manually isolated and the

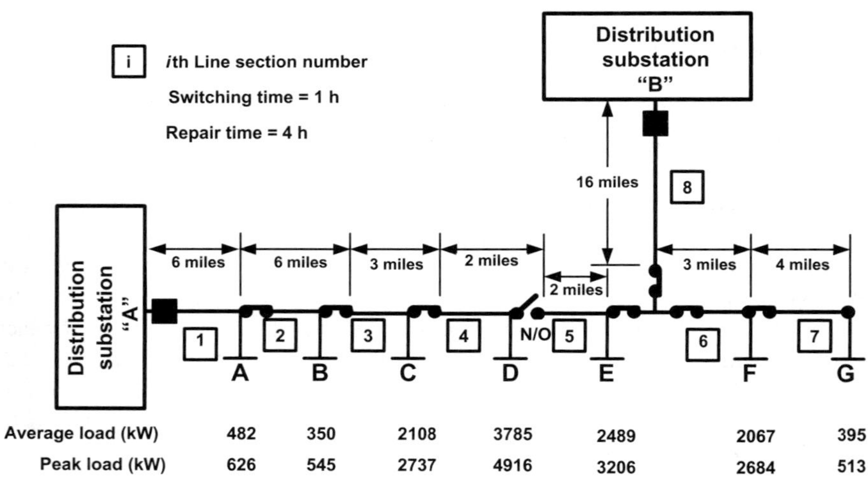


Figure 19.1. Distribution radial network configuration showing peak and average loads at each load point.

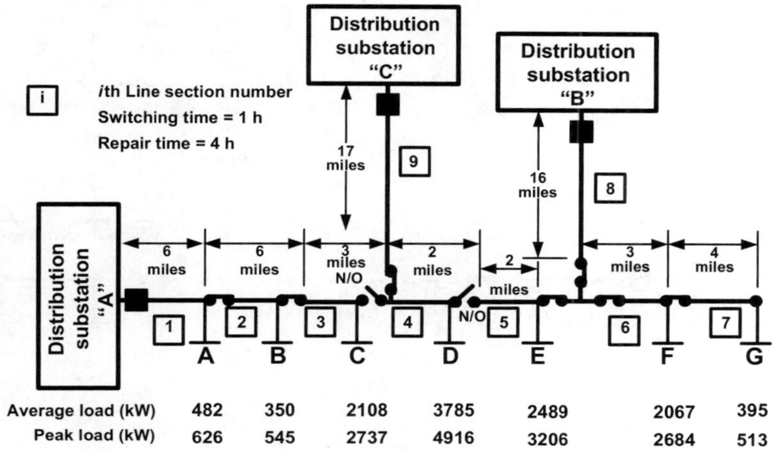


Figure 19.2. Distribution radial network configuration showing a third feeder from substation C to the area.

load on the remaining line sections must be evaluated as to whether portions of the load can be interrupted, that is, shed and what loads can be energized from the alternative feeder. In this example, only portions of the loads "D" and "E" can be supplied from the alternative feeder.

To address the feeder limitation issues, a third feeder from an adjacent substation C to the area has been added. The feeder rating is 12.0 MVA, similar to the feeder from substation A. The length of this feeder is 17 miles as shown in Fig. 19.2. In normal operation of the local distribution system, it is assumed that the feeders from substations A and B will be offloaded by transferring loads D and E to the third feeder. For simplicity, it will be assumed that the average duration to repair any line section is 4.0 h, and the duration to perform the necessary isolation, switching, and load transfer activities is 1.0 h. In this chapter, the emphasis is placed on illustrating the reliability-based determination of DG equivalence to a distribution facility.

Before proceeding with this third feeder solution to solve the capacity problem, the electric utility company should also explore alternative proposals for distributed generation or other solutions that adequately expand the distribution capacity in the area. In this case, the capital cost of the third feeder could be avoided or deferred, thereby holding the line on customer rates. The DG solution is illustrated in Fig. 19.3.

Although the DG solution is an expensive solution compared to the distribution solution, it has the benefit of providing much needed voltage control, and the cost borne by the IPP would be much less, as the IPP would receive the distribution capacity deferral credit, which is a percentage of the annual revenue requirements of the distribution solution. In the request for proposals, the electric utility company would identify the minimum capacity of the unit based on the incremental reliability provided by the distribution solution. The following section describes the probabilistic reliability technique for determining the equivalent capacity for a distributed generating unit

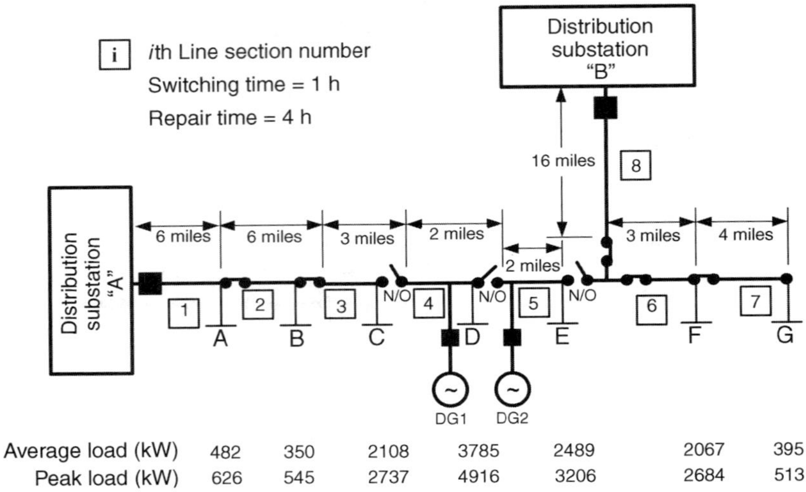


Figure 19.3. Distribution radial network configuration showing two DG additions at load points D and E.

(s) that would replace the requirements of the third feeder from the substation C to the area.

## 19.4 RELIABILITY ASSESSMENT MODEL

Reliability analysis of an electric distribution system is conventionally done by using either the analytical method based on the contingency enumeration approach or the Monte Carlo simulation. The analytical approach based on contingency enumeration can identify low voltage and voltage collapse problems in addition to thermal overloads. The enumeration method, however, cannot model a wide range of operating conditions and is therefore subject to different simplifying assumptions. Monte Carlo simulation, on the contrary, is capable of modeling the full range of operating conditions. One disadvantage of this model is that computer resource limitations limit the solution precision to DC power flow problems. In this case, the simulated performance indices reflect only system overload problems. The important but extremely low-probability transmission outages as well as low voltage and voltage collapse problems cannot be modeled in this method.

A commercial grade computer model is used in this chapter to determine DG equivalence to the third feeder addition to the area. The program is designed to aid electric utility and industrial/commercial customers with predictive reliability assessment of a distribution network. The customer-responsive utility would address reliability problems by selecting project alternatives that have the highest internal and external benefits. Customers may be willing to share the costs when approached with quantifiable plans. The computer model computes a set of reliability indices including System

Average Interruption Frequency Index (SAIFI), System Average Interruption Duration Index (SAIDI), Average Service Availability Index (ASAI), load/energy curtailed, and the cost of outages based on the component outage data and the cost of interruption to a customer. The program models time-sequenced switching actions taken by an operator/repair person following an outage. It can also be used to quantify benefits of automating distribution systems, feeder reconfiguration, and compare various competing projects using cost of outages and utility benefits.

### 19.4.1 Reliability Indices

The computer program computes a set of reliability indices that has been recommended in various publications. Some of the load point indices computed are as follows:

1. Frequency of load interruptions (occurrences/year)
2. Duration of load interruptions (h/occurrence)
3. Duration of load interruptions (h/year)
4. Frequency of customer interruptions (customer interruptions/year)
5. Duration of customer interruptions (customer hours/year)
6. Expected unsupplied energy (EUE (kWh/year)
7. Expected outage cost ($).

The computer model also computes indices for the system under study. A list of system indices is as follows:

1. System Average Interruption Frequency Index
2. System Average Interruption Duration Index
3. Customer Average Interruption Frequency Index (CAIFI)
4. Average Service Availability Index
5. Average Service Unavailability Index (ASUI)
6. Expected unsupplied energy (kWh/year)
7. Expected outage cost ($).

### 19.4.2 Reliability Data

The input data used in quantifying the reliability improvements due to the distribution solution and different distributed generator alternatives to match the equivalent reliability enhancements to the distribution system shown in Fig. 19.1 are listed in Table 19.1.

## 19.5 DISCUSSION OF RESULTS

The study begins by first determining the reliability of the existing system. Next, the reliability indices are calculated after adding a third feeder from substation C to the area

TABLE 19.1. Distribution Network Generation and Feeder Reliability Data

| Component | Failure Rate (occurrences/year) | Repair Time (h) | Switching Time (min) | Stuck Probability |
|---|---|---|---|---|
| Substation A, B, or C | 0.02 | 4.0 | 60.0 | |
| DG1 at D | 5.00 | 50.0 | 60.0 | |
| DG2 at E | 5.00 | 50.0 | 60.0 | |
| Section 1 or 2 | 0.12 | 4.0 | 60.0 | |
| Section 3 or 6 | 0.06 | 4.0 | 60.0 | |
| Section 4 or 5 | 0.04 | 4.0 | 60.0 | |
| Section 7 | 0.08 | 4.0 | 60.0 | |
| Third feeder | 0.32 | 4.0 | 60.0 | |
| Breaker A, B, or C | 0.0036 | 12.0 | 60.0 | 0.01 |

served by the distribution system shown in Fig. 19.1. The last step is to determine the size of a distributed generator or a combination of smaller distributed units by adding to the existing system that would provide the similar reliability level for the area.

As mentioned earlier, the distribution reinforcement to the area considered is a 17-mile long 13 kV feeder from substation C to the area of concern served by the distribution system shown in Fig. 19.1.

### 19.5.1 Equivalent Distributed Generation Reinforcement Alternative

To compute the amount of distributed generation capacity providing the reliability enhancement identical to that of the 13 kV feeder, a range of capacities from 1 to 6 MW is considered in the studies. The computed different load points frequency and duration values, as well as the overall system expected energy not supplied (EENS) figures for various options considered in this study are listed in Tables 19.2–19.4. Adding a third feeder or a DG has improved the overall feeder reliability. The improvement in load point (customer) indices depends on the selected option as seen from the frequency and duration results listed in Tables 19.2 and 19.3. If the objective is to improve indices for a set of customers, then it is important to focus on individual load point indices in selecting the most optimal alternative.

In this study, the main focus is to improve the overall feeder reliability. The reliability index chosen for sizing an equivalent generator(s) is EENS, which is expressed in kWh/year. EENS is calculated based on the frequency, the duration, and the amount of load interruptions. The computed EENS indices for the existing configuration, the distribution reinforcement, and different distributed generation reinforcements are summarized in Table 19.4.

The EENS results listed in Table 19.4 indicate that adding a third feeder greatly improves the reliability of the existing system. The EENS reduces to almost one-fourth when a third feeder is added. To get the same reduction in EENS by adding DGs, a number of DG combinations were considered. Results are provided for adding

TABLE 19.2. Frequency of Interruptions Per Year for Existing Configuration, Third Feeder Addition, and Distributed Generation Reinforcements for the Distribution Network Considered

| Load ID | Existing Distribution Configuration | Distribution Feeder Reinforcement Third Feeder Addition | Distributed Generation Reinforcement | | | | |
|---|---|---|---|---|---|---|---|
| | | | One 1 MW DG at "D" and one 1 MW DG at "E" | One 2 MW DG at "D" and one 2 MW DG at "E" | One 3 MW DG at "D" and one 3 MWDG at "E" | One 6 MW DG at "D" | One 6 MW DG at "E" |
| Load A | 0.55886 | 0.40967 | 0.42912 | 0.41714 | 0.40917 | 0.40945 | 0.40945 |
| Load B | 0.55886 | 0.40967 | 0.42912 | 0.41714 | 0.40917 | 0.40945 | 0.40945 |
| Load C | 0.55886 | 0.18985 | 0.21873 | 0.21873 | 0.18961 | 0.18975 | 0.18975 |
| Load D | 0.55886 | 0.18985 | 0.18961 | 0.18961 | 0.18961 | 0.18975 | 0.18975 |
| Load E | 0.71854 | 0.56954 | 0.62861 | 0.60861 | 0.56940 | 0.56924 | 0.51722 |
| Load F | 0.71854 | 0.56954 | 0.67857 | 0.62861 | 0.56940 | 0.56924 | 0.56924 |
| Load G | 0.71854 | 0.56954 | 0.67857 | 0.62857 | 0.56940 | 0.56924 | 0.56924 |

TABLE 19.3. Duration of Interruptions (h/year) for Existing Configuration, Third Feeder Addition, and Distributed Generation Reinforcements for the Distribution Network Considered

| Load ID | Existing Distribution Configuration | Distribution Feeder Reinforcement Third Feeder Addition | Distributed Generation Reinforcement | | | | |
|---|---|---|---|---|---|---|---|
| | | | One 1 MW DG at "D" and one 1 MW DG at "E" | One 2 MW DG at "D" and one 2 MW DG at "E" | One 3 MW DG at "D" and one 3 MW DG at "E" | One 6 MW DG at "D" | One 6 MW DG at "E" |
| Load A | 1.879 | 0.410 | 1.836 | 1.836 | 0.409 | 0.409 | 0.409 |
| Load B | 2.239 | 0.410 | 2.196 | 2.196 | 0.409 | 0.409 | 0.409 |
| Load C | 2.418 | 0.190 | 2.365 | 2.122 | 0.190 | 0.190 | 0.190 |
| Load D | 2.538 | 0.190 | 0.190 | 0.190 | 0.190 | 0.190 | 0.190 |
| Load E | 2.638 | 0.570 | 0.679 | 0.679 | 0.679 | 0.569 | 0.569 |
| Load F | 2.818 | 0.749 | 1.814 | 0.858 | 0.858 | 0.749 | 0.749 |
| Load G | 5.147 | 0.989 | 3.124 | 1.738 | 1.098 | 0.989 | 0.989 |

TABLE 19.4. Expected Energy Not Supplied (kWh/year) for Existing Configuration, Third Feeder Addition, and Distributed Generation Reinforcements for the Distribution Network Considered

| | | Distributed Generation Reinforcement | | | | |
|---|---|---|---|---|---|---|
| Existing Distribution Configuration | Distribution Feeder Reinforcement Third Feeder Addition | One 1 MW DG at "D" and one 1 MW DG at "E" | One 2 MW DG at "D" and one 2 MW DG at "E" | One 3 MW DG at "D" and one 3 MW DG at "E" | One 6 MW DG at "D" | One 6 MW DG at "E" |
| 17,416 | 4660 | 11,426 | 6639 | 4628 | 4630 | 4630 |

DGs of various sizes. One 6 MW DG or two smaller 3 MW DGs yield almost similar reliability improvement of distribution reinforcement to the distribution system. However, it is preferred to connect two smaller units as they will provide higher reliability. The difference in EENS is more pronounced if higher level outages are also considered. In this example, the location of the unit is not making much difference to reliability, but in real life it is important to include location of the unit in comparing various options.

The probabilistic method illustrated in this chapter helps in identifying the best location for the units in the local area and in determining the output requirements of the distributed generator(s). The computation of the reliability-based equivalent distributed generation capacity to replace a distribution reinforcement requirements will also provide important input to economic feasibility studies performed by the IPP willing to penetrate the new generation market. It is a well-known fact that smaller, distributed, and environment friendly distributed generators hold much promise in the generation of future electric energy as opposed to large and centralized coal- and nuclear-fired units. In addition, smaller units are more suited to replace distribution capacity requirements and the smaller units have the economic advantage of receiving distribution capital deferral credit by replacing distribution requirements.

## 19.6 CONCLUSIONS

The concepts and applications of a probabilistic reliability model for computing distributed generation equivalence to a distribution facility in the deregulated electric utility environment are presented in this chapter. Local area distribution reliability planning is a powerful methodology especially when the area capacity improvement options are disparate. One important conclusion of this chapter is that while the distribution generation addition may be the most expensive alternative, with the right generator size determined by using the reliability techniques and the distribution capital deferral credit obtained from the utility company, the distributed generation option could become a cost-effective solution to the energy supply problem of the future benefiting both the energy suppliers and the energy consumers. Finally, the methodology can be effectively used in the emerging competitive electric energy market to evaluate a wide range of power supply problems.

## REFERENCES

1. S. V. Vadari and J. D. Hammerly, New faces and functions in a competitive market, *IEEE Computer Applications in Power*, **10** (1), 1997, 47–52.
2. T. J. Overbye, G. Gross, M. J. Laufenberg and P. W. Sauer, Visualizing power system operations in an open market, *IEEE Computer Applications in Power*, **10** (1), 1997, 53–58.
3. R. Billinton, L. Salvaderi, J. D. McCalley, H. Chao, Th. Seitz, R. N. Allan, J. Odom, and C. Fallon, Reliability issues in today's electric power utility environment, *IEEE Transactions on Power Systems*, **12** (4), 1997, 1708–1714.

4. A. A. Chowdhury and D. O. Koval, A probabilistic approach to determining generation equivalence to a transmission facility, *IEEE/IAS Transactions*, **35** (6), 1999, 1260–1264.
5. H. L. Willis and W. G. Scott, Distributed Power Generation: Planning and Evaluation, Marcel Dekker, 2000.
6. S. M. Harvey Source distributed generation: market rules and market penetration, in Proceedings of Distributed Generation Conference, Denver, CO, September 1999.
7. Electric Power Research Institute, Strategic Market Assessment of Distributed Resources, Palo Alto, CA, December 1995.
8. Electric Power Research Institute, Applications Guide: Distribution Capacity Planning with Distributed Resources, Palo Alto, CA, January 2000.
9. J.-L. Poirier, Domestic and international markets for distributed generation, in Proceedings of Distributed Generation Conference, Denver, CO, September 1999.

# 20

# MODELS FOR SPARE EQUIPMENT

## 20.1 INTRODUCTION

This chapter presents probabilistic models developed based on Poisson probability distribution for determining the optimal number of transformer spares for distribution transformer systems. The outage of a transformer is a random event and the probability mathematics can best describe this type of failure process. The developed models will be described using an illustrative 72 kV distribution transformer system. Industry average catastrophic transformer failure rate and a 1-year transformer repair or procurement time have been used in the examples considered in this chapter. Among the models developed for determining the optimum number of transformer spares, the statistical economic model provides the best result as it attempts to minimize the total system cost including the cost of spares carried in the system.

It is recognized that some equipment failures in an electrical system are unavoidable. An electric utility delivery system must be designed to withstand occasional equipment failures by including redundant or standby equipment into the overall system operation plan. The question what would be the optimal number of spare equipment for a given system is the subject matter of the analysis performed in this chapter.

This chapter attempts to answer the above question and deals with the use of distribution substation transformer spares, particularly for determining the required number of transformer spares to maintain an acceptable level of system reliability. The question of how many spares should be carried in a system depends on the system reliability requirements and cost of having that reliability level. As the number of spares carried is increased, the capital and operation and maintenance (O&M) costs of the system also increase. To ascertain the optimal number of spares, it is necessary to perform an economic comparison between the increased economic return due to increased system reliability and the required capital and O&M investment to achieve the increased reliability.

Considerable efforts have been made to power system probabilistic planning and design for the past four decades. However, very little attention has been paid from a viewpoint of probabilistic methods to power equipment spare planning. This chapter presents probabilistic models for computing the optimal number of transformer spares for electric distribution transformer systems. Although the models are developed using probability mathematics, the reader will require no background in probability mathematics to be able to use these models to determine the number of electric equipment types and situations encountered in electric power delivery systems.

## 20.2 DEVELOPMENT OF PROBABILISTIC MODELS FOR DETERMINING OPTIMAL NUMBER OF TRANSFORMER SPARES

A particular unit of equipment, such as transformer, line, and breaker, has a failure rate that varies over the life of that unit. It is recognized that populations of equipment normally tend to attain equipment static age distributions, thereby allowing failures to be modeled as stationary random processes. This permits the construction of statistical models to predict system performance and to build an appropriate level of system redundancy. The general concepts associated with failure probabilities will be presented in detail for three probabilistic models.

### 20.2.1 Reliability Criterion Model for Determining the Optimal Number of Transformer Spares

The Poisson probability distribution represents the probability of an isolated event occurring a specified number of times in a given interval of time or space when the rate of occurrence, hazard rate in a continuum of time or space, is constant. In such situations, the hazard rate is normally termed as the failure rate.

The failure of a single unit of equipment is a random event and must be affected by chance alone. The parameter $\lambda$ is the failure rate of a given type of equipment, defined as the mean number of failures per unit year in service. Let $dt$ be a sufficiently small interval of time such that the probability of more than one failure occurring during this interval is negligible and can be neglected.

$\lambda dt$ = probability of failure in the interval $dt$, that is, in the period $(t, t+dt)$. Let $P_x(t)$ be the probability of failure occurring $x$ times in the interval $(0, t)$, then probability of zero failures in the interval $(0, t + \mathrm{d}t)$ equals probability of zero failures in the interval $(0, t)$ times probability of zero failures in the interval $(t, t + \mathrm{d}t)$. Then, $P_0(t + \mathrm{d}t) = P_0(t) \times (1 - \lambda \mathrm{d}t)$, assuming event independence, $\{P_0(t + \mathrm{d}t) - P_0(t)\}/\mathrm{d}t = -\lambda P_0(t)$. As $\mathrm{d}t$ approaches zero or becomes incrementally small, $\mathrm{d}P_0(t)/\mathrm{d}t = -\lambda P_0(t)$, which by integrating becomes $\ln P_0(t) = -\lambda P_0(t) + C$.

At $t = 0$, the equipment is known to be operating. Therefore at $t = 0$, $P_0(0) = 1$, $\ln P_0(t) = 0$, and $C = 0$, giving $P_0(t) = \mathrm{e}^{-\lambda t}$. This expression provides the probability of zero failures in a specified time period $t$. If $\lambda(t) = \lambda$, a constant, then for zero failures, $R(t) = \mathrm{e}^{-\lambda t}$. This is the first term of the Poisson probability distribution and is widely used to calculate the reliability of a system.

From the equations deduced earlier, it can be concluded that the probability that a given equipment will survive for a period of $t$ years in service, $P_x(t)$, is determined by the exponential decay function $P_x(t) = \mathrm{e}^{-\lambda t}$. For a population of $N$ such equipment, the mean number of failures per year is equal to $N\lambda$. If failures are statistically independent, the probability of exactly $x$ failures occurring over the period $t$ years, $P_x(t)$, is given by the Poisson probability distribution

$$P_x(t) = (\mathrm{e}^{-N\lambda t} N\lambda t^x)/x! \tag{20.1}$$

This Poisson reliability model can be used for calculating reliability of a system with $n$ spares as given in the following:

$$R(t) = \mathrm{e}^{-\lambda t}[1 + N\lambda t + (N\lambda t)^2/2! + (N\lambda t)^3/3! + (N\lambda t)^4/4! + \cdots + (N\lambda t)^n/n!] \tag{20.2}$$

### 20.2.2 Mean Time Between Failures ($\mathrm{MTBF_u}$) Criterion Model for Determining the Optimal Number of Transformer Spares

In Equation (20.2), the system reliability is given by the sum of the first $n$ terms of the Poisson probability distribution. Poisson statistics can therefore be effectively utilized to determine the equipment unavailability in a mature population of $N$ units in service including a number of $n$ spare units. The mean number of units entering the repair status per year equals the mean number of failures per year, which is $N\lambda$. Let MTTR be the mean time to repair a unit or to procure a new unit if the failed unit is scrapped. The mean number of units in the under-repair status at any given instant in time, $\mu_r$, is the mean number of units entering the under-repair status in the time interval MTTR. Therefore,

$$\mu_r = N\lambda \mathrm{MTTR} \tag{20.3}$$

Since the probability of exactly $n$ units entering the under-repair status in any time interval MTTR is determined by the Poisson probability distribution, the probability of exactly $n$ units existing in the under-repair status, $P_x(t)$, is given by the same Poisson probability distribution,

$$P_x(t) = (\mathrm{e}^{-\mu_r} \mu_r^x)/x! \tag{20.4}$$

If there are $n$ units assigned as spare units, the probability that all $n$ units are depleted at any instant in time, $P_u$, equals the sum of probabilities in Equation (20.4) for $x \geq n$.

$$P_u = P_x(x \geq n) = 1 - P_x(x < n) = 1 - \sum_{x=0}^{n-1} \frac{e^{-\mu_r} \mu_r^x}{x!} \tag{20.5}$$

Let $MTBF_u$ be the mean time between failures on the system when all spares have been depleted. This time interval is the mean time between equipment unavailabilities.

$$MTBF_u = 1/(N\lambda P_u) \tag{20.6}$$

Let $\mu_u$ be the mean number of units in the population at any instant in time, which are unavailable due to the fact that $n$ spare units were depleted from prior failures.

$$\mu_u = \sum_{x=n}^{N+n} \frac{(x-n)e^{-\mu_r} \mu_r^x}{x!} \approx \mu_r - n + \sum_{x=0}^{n-1} \frac{(n-x)e^{-\mu_r} \mu_r^x}{x!} \tag{20.7}$$

Let $MTTR_u$ be the mean time that a unit unavailable for service will remain out of service until the first under-repair unit is repaired.

$$MTTR_u = \mu_u MTBF_u \tag{20.8}$$

### 20.2.3 Determination of Optimal Transformer Spares Based on the Model of Statistical Economics

It is important to note that as the number of spares carried in a system is increased, the capital and O&M costs of the system also increase. To determine the optimum number of spares in a system, it is necessary to make an economic comparison between the cost and the benefit of carrying a certain number of spares. Economic aspects become an important factor in deciding the required level of system reliability.

Economic analyses including customer outage costs can also be used to determine the optimum number of spares. Let $C_u$ be the cost increase due to a unit of equipment being unavailable for service, expressed in terms of dollars per unit year out of service. Since $\mu_u$ is the average number of units unavailable at any instant in time, the expected annual cost of having units unavailable is $\mu_u \times C_u$. Let $C_s$ be the cost of owning and maintaining one spare unit of equipment, expressed as carrying charge in dollars per unit year. If there are $n$ such units in the inventory, the total cost of maintaining the inventory is $n \times C_s$. The optimal number of units in the inventory is the figure of $n$ such that the total cost $\mu_u \times C_u + n \times C_s$ is minimized.

The probability models presented in Sections 20.2.1–20.2.3 are utilized in Section 20.3 for determining optimum spare transformer requirements for illustrative distribution transformer systems.

## 20.3 OPTIMAL TRANSFORMER SPARES FOR ILLUSTRATIVE 72 kV DISTRIBUTION TRANSFORMER SYSTEMS

The $N-1$ reliability criterion is widely used in the substation transformer planning and design. Each substation is normally designed to have two or more transformers in parallel so that the peak load can be supplied when one transformer fails. This is a reliable but expensive substation design criterion. Utilization of common spare transformers has been a common practice in many utilities in planning and design of substations. The policy of using common spares can be extended to multiple transformer substations and new single-transformer substation design. For example, for a two-transformer substation, where either one will not be able to meet the peak load due to load growth, the substation can become a member of the substation group with the same class of transformers to share common spares rather than having a third transformer added to the individual substation. For noncritical loads, substations instead of two transformers in parallel can be considered with shared common spares. Compared to the $N-1$ contingency design principle in each substation, common spare transformers shared by multiple substations can avoid significant capital and O&M expenditure, and still provide adequate service reliability.

There are two failure modes for a transformer: field repairable and nonfield repairable or catastrophic failure. The installation time for a spare transformer is 1–5 days, which is comparable with the field repairable time of 1–10 days for a random repairable failure and much shorter than the replacement or procurement time of 1.0–1.5 years for buying or rebuilding a transformer in the case of a catastrophic failure and no spare available.

The number of transformers requiring spare backup for the illustrative 72 kV distribution systems considered in this chapter for different transformer MVA ratings is summarized in Table 20.1.

The catastrophic failure rate for distribution transformers of 0.011 failures per transformer per year will be used in the analysis performed in this chapter. Canadian Electricity Association publishes transformer failure statistics for all Canadian utilities annually. Using the catastrophic failure rate of 0.011 failures per transformer per year for the transformers of all MVA ratings in Table 20.1, the 72 kV transformer system failure rates per year have been calculated and are presented in Table 20.2.

TABLE 20.1. Number of Distribution Substation Transformers Requiring Spare Backup for the N − 1 or First Contingency (72 kV Primary and 25 kV Secondary)

| 72 kV $\leq$ 7 MVA | 72 kV 7.5$\leq$16 MVA | 72 kV $\geq$ 16 MVA |
|---|---|---|
| 67 | 47 | 132 |

TABLE 20.2. The Calculated Catastrophic Transformer Failure Rate Per Year by Transformer MVA Ratings for 72 kV Systems

| 72 kV ≤ 7 MVA | 72 kV 7.5≤16 MVA | 72 kV ≥ 16 MVA |
|---|---|---|
| 0.737 occurrences/year | 0.517 occurrences/year | 1.452 occurrences/year |

Transformer outages in general are somewhat different because while they are infrequent compared to transmission line outages, restoration/repair times are normally long. This is especially true for catastrophic transformer failures. For a catastrophic failure, the damaged transformer or its replacement may be unavailable for up to a year. Extended repair times normally result from problems with windings, core, leads, bushings, and load tap changer that require untanking and/or a trip to the repair facility. If the damage to the failed transformer is such that the transformer is beyond repair and is scrapped, then a new transformer is required to be purchased, which might normally take a year or more. In the analysis for determining the optimal number of spares, it is considered that the repair/restoration time of the failed transformer or the procurement time of a new transformer would be 1 year.

Three different probabilistic models are used in this chapter for determining optimal transformer spares, such as satisfying the minimum mean time between failures (MTBF) requirements for the system, satisfying the minimum reliability requirements for the system, and satisfying the minimum statistical economics criterion for the system. In the following subsections, three example calculations for determining optimal transformer spares using the developed probabilistic models are illustrated. The transformers used in the example calculations are for 72 kV transformers of MVA rating equal to and greater than 16 MVA as shown in Table 20.1. Similar mathematical models can be used for other MVA ratings.

### 20.3.1 Determination of Optimal Transformer Spares Based on the Minimum Reliability Criterion

The example of 72 kV transformers with MVA rating of ≥16 MVA is utilized in this subsection to illustrate the optimum transformer spare calculation methodology using the minimum reliability criterion. Equation (20.2) is used for calculating the system reliability for different spare levels. In this example, there are 132 transformers (see Table 20.1). The system failure rate is 1.452 failures per year (see Table 20.2). Consider that the system failure occurs when any one transformer fails and the repair time of the failed unit or the procurement time for a replacement transformer is 1 year. If the system is to have a minimum reliability of 0.9950, a number typically used in the electric utility industry, what is the minimum number of spares that must be carried as immediate replacements? The calculated reliability figures and corresponding number of spare requirements are presented in Table 20.3.

It can be seen from Table 20.3 that the number of spares should be five to achieve a minimum reliability of 0.9950. The results for remaining MVA ratings using the reliability criterion of 0.9950 are summarized in Table 20.4.

TABLE 20.3. Reliability that $N$ Spares Will be Depleted in a System with 132 Transformers with $\geq$16 MVA Rating Using the System Failure Rate of 1.452 failures/year and a Transformer Repair/Procurement Time of 1 Year

| Number of Spares, $n$ | System Reliability |
|---|---|
| 0 | 0.2341 |
| 1 | 0.5740 |
| 2 | 0.8208 |
| 3 | 0.9402 |
| 4 | 0.9836 |
| **5** | **0.9962** |
| 6 | 0.9992 |

### 20.3.2 Determination of Optimal Transformer Spares Based on the Minimum $MTBF_u$ Criterion

The example transformer system considered has 132 (72–25 kV) transformers with MVA rating equal to and greater than 16 MVA (see Table 20.1). The failure rate for the transformer system is 1.452 failures per year and the mean time to repair (MTTR), which is the replacement or the procurement time for the failed transformer unit, is 1 year. How many spares, $n$, will be needed to assure that the mean time between prolonged outages on the system is greater than 35 years, which is the average useful life of a transformer?

From Equation 20.3: $$\mu_r = N\lambda \text{MTTR} = 132 \times 0.011 \times 1 = 1.452 \text{ units under repair}$$

where $\mu_r$ is the mean number of transformers entering the under-repair status in the time interval MTTR, $N$ is the population of the transformers, and $\lambda$ is the catastrophic failure rate of a transformer.

From Equation 20.6: $$\text{MTBF}_u = 1/(N\lambda P_u) > 35 \text{ years}$$

$$P_u < 1/1.452 \times 35 = 0.0197$$

where $MTBF_u$ is the mean time between failures on the system when all spares have been depleted and $P_u$ is the probability that all $n$ units are depleted at any instant in time.

TABLE 20.4. Optimum Number of Transformer Spares for Different MVA Ratings Using the Reliability Criterion of 0.9950

| Optimum Number of Spares | 72 kV $\leq$ 7 MVA | 72 kV 7.5$\leq$16 MVA | 72 kV $\geq$ 16 MVA |
|---|---|---|---|
| $n$ | 3 | 3 | 5 |

TABLE 20.5. Results for Spares Analysis Using $MTBF_u$ Criterion

| | |
|---|---|
| $n=0$ | $P_u=1.0000>0.0197$ |
| $n=1$ | $P_u=0.7659>0.0197$ |
| $n=2$ | $P_u=0.4260>0.0197$ |
| $n=3$ | $P_u=0.1792>0.0197$ |
| $n=4$ | $P_u=0.0598>0.0197$ |
| $\mathbf{n=5}$ | $\mathbf{P_u=0.0164<0.0197}$ |
| $n=6$ | $P_u=0.0038<0.0197$ |

Using Equation 20.5 to determine $P_u$ as a function of $n$, the results are computed and presented in Table 20.5.

Therefore, $n=5$ spares will satisfy the condition that $MTBF_u > 35$ years. With $n=5$, using Equation 20.6, $MTBF_u$ is 42 years.

Using the similar approach, the optimum number of transformer spares for remaining MVA ratings has been computed and is summarized in Table 20.6.

The results are identical to those obtained by using the reliability criterion method.

### 20.3.3 Determination of Optimal Transformer Spares Based on the Criterion of Statistical Economics

The example system used in Sections 20.3.1 and 20.3.2 containing 132 (72 kV) transformers of ≥16 MVA rating is used to illustrate the statistical economics methodology for determining the optimal number of spare transformers for the illustrative distribution systems. In this case, a typical 72–25 kV transformer of 12/16/20/22.4 MVA rating is considered for the calculations of the cost increase in terms of kilowatt hour loss, the revenue lost cost, the customer outage cost, and the capital cost for the spare transformer. As indicated earlier, the 132-transformer system failure rate is 1.452 failures per year. The repair time/replacement time/procurement time for a new unit (if the failed unit is beyond repair) is considered to be 1 year. The loss of a transformer would normally increase system power losses. The estimated system O&M cost increase due to the catastrophic failure of a 12/16/20/22.4 MVA 72 kV transformer is $7160 per unit year. This cost is derived assuming loss cost of 1.73 cents per kWh, system load factor of 0.5241, and loss factor of 45%.

For the catastrophic failure of a 12/16/20/22.4 MVA 72 kV transformer, the average power not supplied at 0.87 power factor and 0.5241 load factor if the transformer was supplying a peak load of 16 MVA is 7.3 MW. The energy lost for the downtime of 1 year is 63,909 MWh per year. The computed annual revenue lost at assumed 6.25 cents per kWh

TABLE 20.6. Optimum Number of Transformer Spares for Different MVA Ratings Using the $MTBF_u$ Criterion of 35 years

| Optimum Number of Spares | 72 kV ≤7 MVA | 72 kV 7.5≤16 MVA | 72 kV ≥16 MVA |
|---|---|---|---|
| $n$ | 3 | 3 | 5 |

TABLE 20.7. The Optimum Number of Transformer Spares Computed Using the Statistical Economics Criterion for 72 kV Transformers with ≥16 MVA Rating

| $n$ | $\mu_u$ | $\mu_u \times C_u$ ($/year) | $n \times C_s$ ($/year) | Total Cost ($/year) |
|---|---|---|---|---|
| 1 | 0.6861 | $474,544,573 | $52,500 | $474,597,073 |
| 2 | 0.2601 | $179,899,495 | $105,000 | $180,004,495 |
| 3 | 0.0809 | $55,954,899 | $157,500 | $56,112,399 |
| 4 | 0.0212 | $14,663,088 | $210,000 | $14,873,088 |
| 5 | 0.0047 | $3,250,779 | $262,500 | $3,513,279 |
| 6 | 0.0009 | $622,490 | $315,000 | $937,490 |
| 7 | 0.00016 | $110,665 | $367,500 | $478,165 |
| **8** | **0.0000251** | **$17,361** | **$420,000** | **$437,361** |
| 9 | 0.00000355 | $2455 | $472,500 | $474,955 |
| 10 | 0.000000457 | $316 | $525,000 | $525,316 |

would be $3,994,271 per unit year. The calculated customer outages cost at assumed $10.76 per kWh not supplied would be $687,653,682 per unit year. Therefore, $C_u$ for a transformer out of service is ($7160 + $3,994,271 + $687,653,682) $691,655,113 per year. If the estimated capital cost of a 12/16/20/22.4 MVA 72 kV transformer spare is $350,000, then the carrying charge for the spare (@ 15%) would be $C_s = \$52,500$ per year. Again, the MTTR to fix the failed transformer, to get a replacement transformer, or to procure a new transformer is considered to be 1 year in the reliability cost–reliability benefit computation process. The question now is: How many transformers should be stocked as spares?

From Equation 20.3: $\mu_r = 132 \times 0.0110 \times 1 = 1.452$ units on order

From Equation 20.7, $\mu_u$ as a function of $n$ can be computed. The results are presented in Table 20.7, which indicate that the optimal number of transformer spares is $n = 8$, since this number minimizes the total cost.

It is worth noting that the statistical economics model yields the higher number of spares for all transformer categories. However, if the appropriate minimum $MTBF_u$ and the reliability criteria are chosen, three probabilistic models would render identical results. Reliability cost–reliability benefit analyses were performed for calculated optimum number of spares for different transformer MVA ratings as shown in Table 20.8. The cumulative present values (CPVs) in 2004 dollars for the spare costs and the reliability benefits have been computed considering a 30-year project life, a 7.7%

TABLE 20.8. Optimum Number of Transformer Spares for Different MVA Ratings Using the Statistical Economics Criterion

| Optimum Number of Spares | 72 kV ≤7 MVA | 72 kV 7.5≤16 MVA | 72 kV ≥16 MVA |
|---|---|---|---|
| $n$ | 6 | 6 | 8 |

TABLE 20.9. Cumulative Present Value of Reliability Cost–Benefits for Calculated Optimum Spare Transformers

| kV Ratings | Optimum Number of Spares | CPV of the Revenue Requirement of Spare in 2004 Dollars | CPV of the Reliability Benefits in 2004 Dollars | Year of Payback |
|---|---|---|---|---|
| 72 kV ≤ 7 MVA | 6 | $234,633 | $1,556,810 | 2 |
| 72 kV 7.5≤16 MVA | 6 | $312,844 | $1,148,639 | 4 |
| 72 kV ≥ 16 MVA | 8 | $547,478 | $1,915,603 | 5 |

discount rate, and a 2.5% inflation rate. The results of the cost–benefit analyses are summarized in Table 20.9.

It is worth noting that the three probabilistic models did not yield the same number of spares for all transformer categories. This is due to the fact that if the $MTBF_u$ and the reliability criterion used were different from those used in the calculations, the calculated optimum spares provided by the three models could have been the same. Reliability cost–reliability benefit analyses were performed for calculated optimum number of spares for the same transformer MVA ratings as shown in Table 20.8. The cumulative present values in 2004 dollars for the spare costs and the reliability benefits have been computed considering a 30-year project life, a 7.7% discount rate, and a 2.5% inflation rate. The results of the cost–benefit analyses are summarized in Table 20.9.

As shown in Table 20.9, the CPV of the reliability benefits exceeds the CPV of the revenue requirements of the capital expenditures required for stocking the optimum number of spares in the system.

## 20.4 CONCLUSIONS

This chapter presents three probabilistic models developed based on Poisson probability distribution for determining the optimal number of spare distribution transformers. The outage of a transformer is a random event and the probability mathematics can best describe this type of failure process. Three developed models have been illustrated using 72 kV transformer systems. Industry average catastrophic transformer failure rate and a 1-year transformer repair or procurement time have been used in the examples considered in distribution transformer analyses. Among the three models considered for determining the optimum number of transformer spares, the statistical economics model provides the best result as it attempts to minimize the total system cost including the cost of spares carried in the system.

It is important to note that the developed probabilistic models use the failure rates, repair times, and transformer inventory information as input to the models. The optimum number of transformer spares for a system is a function of the aforementioned input parameters. The number of spare requirements for a system also depends on the reliability level demanded from the system. Substation design and operation policies can greatly impact the number of spare requirements. For example, for multiple transformer substations, for a $N - 1$ or first contingency, if the remaining transformer(s) can carry the substation peak demand, then there is no need for a spare transformer to back up the next $N - 2$ or second transformer contingency. In the case of a single-transformer substation, if the adjacent substations and the local distribution network can absorb the substation load for the time when the failed transformer is being fixed, then there is no need for a spare transformer to be brought in to the substation. The overall system capacity to ride through the $N - 1$ or first contingency would greatly reduce the number of transformer inventory, which in turn would reduce the number of spares to be stocked in the system.

A uniform substation design and consistent operating voltages across the system also greatly reduce the number of spare transformer requirements. In addition, spares with dual primary and secondary voltages could also reduce the number of spare requirements in the system. It is important to note that each of the aforementioned system design and operating policies would greatly reduce the number of spare requirements, which in turn would reduce the system capital and O&M costs without compromising the service reliability.

It is a well-known fact that a population of transformers with a high failure rate and a long repair or replacement time will have a high system unavailability until a large number of spares are maintained in the system. Spare inventories incur huge costs that warrant serious considerations in system planning, design, operation, and maintenance activities so that the cost of a utility system would not become prohibitively high. It is important to note that the higher the equipment failure rate and the repair time, the higher would be the number of spare requirements for maintaining an adequate level of system reliability.

## REFERENCES


1. R. N. Allan, R. Billinton, A. M. Breipohl, and C. H. Grigg, Bibliography on the application of probability methods in power system reliability evaluation: 1967–1991, *IEEE Transactions on Power Systems*, **9** (1), 1994, 41–49.
2. J. Endrenyi, *Reliability Modeling in Electric Power Systems*, John Wiley & Sons, Ltd., Chichester, 1978.
3. R. Billinton and R. N. Allan, *Reliability Evaluation of Power Systems*, Plenum Press, New York, 1996.
4. R. Billinton and R. N. Allan, *Reliability Evaluation of Engineering Systems*, Plenum Press, New York, 1992.
5. R. Sahu, Using transformer failure data to set spare equipment inventories, *Electric Light and Power*, October 1980, 41–42.
6. V. I. Kogan, J. A. Fleeman, J. H. Provanzana, and C. H. Shih, Failure analysis of EHV transformers, *IEEE Transactions on Power Delivery*, **3** (2), 1988, 672–683.
7. G. E. Henry III, A method for economic evaluation of field failures such as low voltage side lightning surge failure of distribution transformers, *IEEE Transactions on Power Delivery*, **3** (2), 1988, 813–818.

8. IEEE Working Group D1 on Operating and Reliability Aspects for Substation Design of the Distribution Substation Subcommittee, Substation Subcommittee of the Power Engineering Society, Bibliography for reliability and availability engineering for substations, *IEEE Transactions on Power Delivery*, **4** (3), 1989, 1689–1694.
9. W. J. McNutt and G. H. Kaufman, Evaluation of a functional life model for power transformers, *IEEE Transactions on Power Apparatus and Systems*, **102**, 1983, 1151–1162.
10. Canadian Electricity Association, Forced Outage Performance of Transmission Equipment for the Period January 1, 1995 to December 31, 1999, 2000.

# 21

# VOLTAGE SAGS AND SURGES AT INDUSTRIAL AND COMMERCIAL SITES

## 21.1 INTRODUCTION

The previous chapters primarily focused on the impact of permanent equipment outages on service interruptions. Voltage sags and surges occurring at industrial and commercial sites can also disrupt computerized equipment resulting in customer interruptions. This chapter presents some of the known origins of power supply disruptions and some generalized susceptibility characteristics of electronic equipment. The chapter also answers many questions concerning origins and the frequency and duration of voltage sags and surges that occur at commercial and industrial sites. Many of the answers are based on the results of a cross-Canada power quality survey conducted by the Canadian Electricity Association. A methodology for estimating the frequency of voltage sags at a particular site considering the impact of distribution feeder outages will be presented.

A voltage dip or voltage sag may be caused by a switching operation involving heavy currents or by the operation of protective devices (including autoreclosers) resulting from faults. These events may originate from the consumers' systems or from the public supply network. Voltage dips and short supply interruptions may disturb the electronic

*Power Distribution System Reliability.* By Ali A. Chowdhury and Don O. Koval

and electrical equipment connected to the supply network. Some of the typical incorrect equipment operations are

- Extinction of discharge lamps
- Incorrect operation of regulation devices
- Speed variation or stopping of motors
- Tripping of contactors
- Failures and computation errors for computers or measuring instruments equipped with electronic devices
- Loss of synchronism of synchronous motors and generators
- Commutation failure in thyristor bridges operating in the inverter mode.

Some of the inconveniences mentioned above are made worse by the fact that restarting a machine may take from a few minutes to a few hours.

## 21.2 ANSI/IEEE STANDARD 446—IEEE ORANGE BOOK

IEEE Recommended Practice for Emergency and Standby Power Systems for Industrial and Commercial Applications.

### 21.2.1 Typical Range for Input Power Quality and Load Parameters of Major Computer Manufacturers

By the proper selection of an electric supply system, the power needs associated with data processing with a computer can be met, namely, a reliable source of noise-free electric power at all times and of a much higher quality than previously demanded by most devices. In the early stages of data processing equipment and computer equipment development, it was not unusual to experience problems with hardware and software when power disturbances of microseconds in duration were experienced. Most equipment built in that era was extremely vulnerable to short-time disturbances. The goal of most manufacturers in today's technology is to build from 4 ms to 1 cycle of carryover, or ride-through, time into their equipment.

Table 21.2 shows computer input power quality parameters for several manufacturers. The user should consider Table 21.1 only as a source of some examples since computer designs vary with the size of computers, their processing power, and the technology available when the design was created. They are continually changing and the parameters of power needs are changing rapidly with the designs.

### 21.2.2 Typical Design Goals of Power Conscious Computer Manufacturers (Often Called the CBEMA Curve)

Figure 21.1 shows an envelope of voltage tolerances that is representative of the present design goal of a cross section of the electronic equipment manufacturing industry.

TABLE 21.1. Typical Range of Input Power Quality and Load Parameters of Major Computer Manufacturers

| Parameters | Range or Maximum |
|---|---|
| 1. Voltage regulation, steady state | +5%, – 10% to +10%, – 15% (ANSI C84.1–1970) +6%, – 13% |
| 2. Voltage disturbances, momentary undervoltage, transient overvoltage | – 25% to – 30% for less than 0.5 ms with – 100% acceptable for 4–20 ms + 150–200% for less than 0.2 ms |
| 3. Voltage harmonic distortion[a] | 1–5% (with linear load) |
| 4. Noise | No standard |
| 5. Frequency variation | $60 \pm 0.5$ Hz |
| 6. Frequency rate of change | 1 Hz/s (slew rate) |
| 7. Three-phase, phase voltage unbalance[b] | 2.5–5% |
| 8. Three-phase, load unbalance[c] | 5–20% maximum for any one phase |
| 9. Power factor | 0.8–0.9 |
| 10. Load demand | 0.75–0.85 of connected demand |

*Note*: Prameters 1, 2, 5, and 6 depend on the power source, while parameters 3, 4, and 7 are the product of an interaction between source and load, and parameters 8, 9, and 10 depend on the computer load alone.

[a]Computed as the sum of all harmonic voltages added vectorally.

[b]Computed as follows: % phase voltage unbalance $= [3(V_{\min} - V_{\min})/(V_a + V_b + V_c)] \times 100$.

[c]Computed as difference from average single-phase load.

Shorter duration overvoltages have higher voltage limits. Some computer manufacturers specify a maximum allowable limit for volt-seconds, typically 30% of the nominal volt-seconds (areas under the sine wave). The CBEMA curve presented here is just an illustration to show the susceptibility patterns of electronic equipment.

## 21.3 IEEE STANDARD 493-2007—IEEE GOLD BOOK

*IEEE Gold Book* is the recommended practice for designing reliable industrial and commercial power systems.

### 21.3.1 Background

A voltage sag can be characterized by its magnitude and duration. The voltage sag magnitude, in this chapter, is the net rms voltage in percentage or per unit of the system nominal voltage. Sag duration is the time when the voltage is low, usually less than 1 s. Fault clearing within the electrical system has been reported as a major cause of voltage sags. Some studies found that nearly all disruptive voltage sags were caused by current flowing to short circuits either within the plant or on utility lines in the electrical neighborhood. Starting motor and welders can also cause voltage sags with predictable characteristics.

This chapter concentrates on voltage sags associated with short circuits (i.e., faults) on the electrical supply system. The principal voltage drop only occurs while short-circuit current flows. The voltage increases as soon as a fault clearing device interrupts

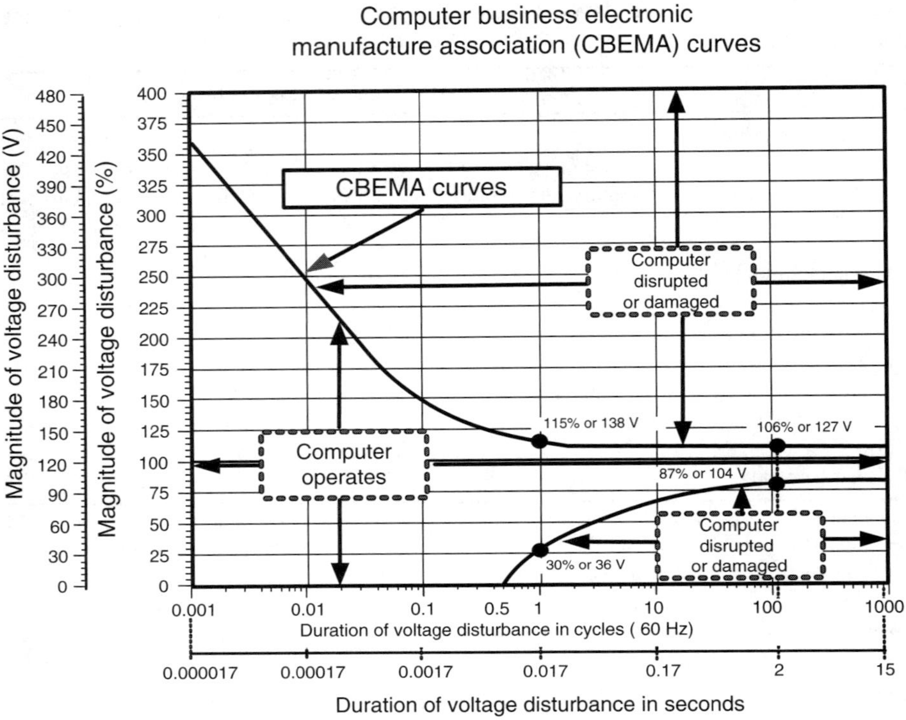


Figure 21.1. Typical design goals of power conscious computer manufacturers.

the flow of current. These faults may be at a distance of kilometer from the interrupted process, but close enough to cause disruptive problems. A clear understanding of voltage drop during faults and the fault clearing process is necessary before making accurate voltage sag predictions.

Voltage sags associated with fault clearing have many predictable characteristics. It is possible to predict the voltage sag magnitude for individual faults by calculating the voltage drop at the critical load. Predicting how long the voltage sag will last requires an estimate of the total clearing time of the overcurrent protective devices. The waveform of voltage sags is somewhat predictable from the analysis of recorded voltage sag data available and with the aid of transient network analysis. However, it is very important to estimate how often voltage sags will upset sensitive electronic equipment.

Predicting characteristics for just one sag caused by a specific fault at a specific location is straightforward. Prepare an accurate electrical model of the system, apply the fault, and calculate the voltage sag magnitude at the specific location. Use the protective device characteristics to estimate sag duration and compare the sag characteristics with the sensitive equipment susceptibility to determine if the process will have an outage.

Predicting the voltage sag characteristics that a sensitive load will see during several years of operation requires a probabilistic approach. It is impossible to predict exactly where each fault will occur, but it is reasonable to assume that many faults will occur. The most accurate predictions require voltage sag calculations for every possible fault in the electrical system and estimating each fault's frequency of occurrence. The overall voltage sag frequency at a sensitive load is the sum of the individual frequencies of the faults that occurred. A practical approach is to locate boundaries on the electrical system where specific sag magnitudes are possible and then to estimate the fault frequency in the boundary.

To understand how electrical faults on the utility system create voltage sags, a simplified distribution system shown in Fig. 21.2 will be studied in some detail.

Each feeder has a circuit breaker with protective relays to detect and clear faults. Point "C" is an industrial customer supplied by 600 Y/347 V from a distribution transformer. The lower half of Fig. 21.2 shows what happens to the rms voltage when

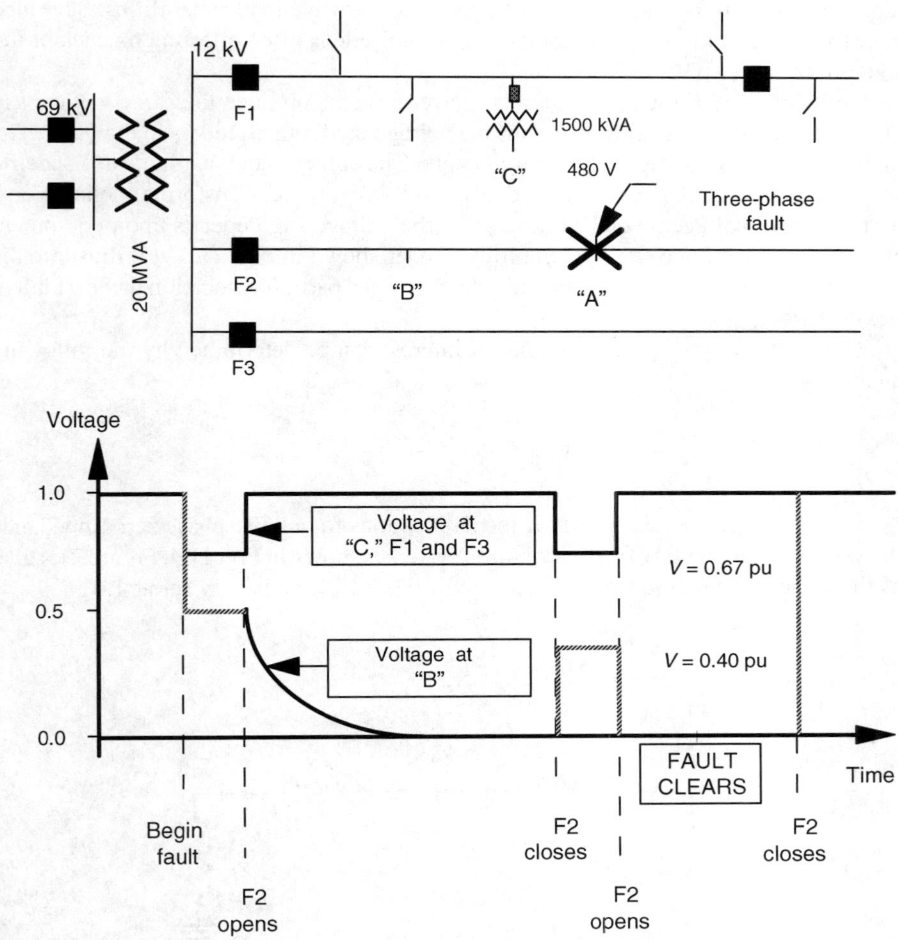


Figure 21.2. Voltage sags from faults and fault clearing.

a temporary three-phase fault occurs at "A" on feeder F2. The dashed lines show the rms voltage at point "B" and the solid line shows the rms voltage on feeders F1 and F3 during the same fault. The load at "C" will also see the voltage represented by the solid line. A timeline shows the sequence of events after the initiation of the fault. Note that F2 uses reclosing relays. Reclosing can cause several sags for one permanent fault. Also, the voltage decay on the first interruption represents motor voltage decay. The motors trip off before reclosure.

All loads on F2 including "B" suffer a complete interruption when breaker F2 clears the fault. All loads on F1 and F3 see two voltage sags. The first sag begins at the initiation of the fault. The second sag begins when breaker F2 recloses. Sags occur whenever fault current flows through the impedance to a fault. Voltage returns to normal on feeders F1 and F3 once the breaker on F2 interrupts the flow of current. Unfortunately, sensitive loads on F1 and F3 experience a production outage if the sag magnitude and duration are more severe than the sensitive load capabilities. Sags also occur for single- and two-phase faults. The magnitude is often different on each of the three phases.

Faults in industrial and commercial power systems produce the same voltage sag phenomena. A fault in one feeder drops the voltage in all other feeders in the plant. The voltage sag even shows up in the utility system. The voltage sag magnitude at a specific location depends on the system impedance, fault impedance, transformer connections, and the presag voltage level. The impact of the voltage sag depends upon equipment susceptibility to voltage sags. A simplified feeder shown in Fig. 21.3 will illustrate the basic methodology for determining the voltage sag at a particular location when a fault in the system occurs.

The voltage sag ($V_{sag}$) at the point of interest can be determined by the following equation:

$$V_{sag} = \frac{Z_2 + Z_f}{Z_1 + Z_2 + Z_f} \times V_{source}$$

The voltage sag levels at various feeder locations for a three-phase zero impedance fault ($Z_f = 0$) in feeder F2 of the distribution system is shown in Fig. 21.4. Figure 21.4 also has the impedance diagram for feeder circuit F2. $V_{source}$ is usually set at 1.0 pu.

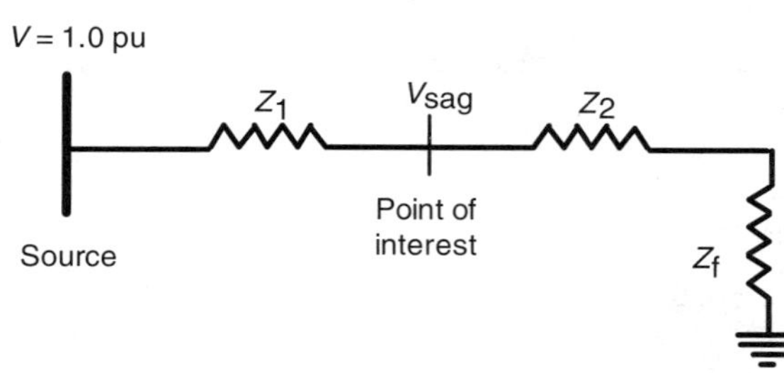


Figure 21.3. Basic feeder circuit.

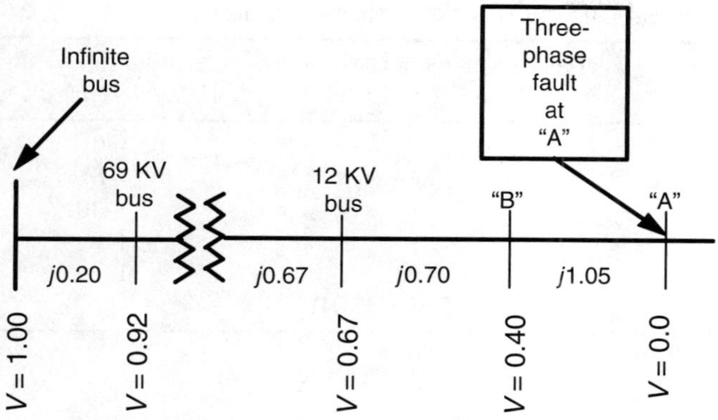


Figure 21.4. Impedance diagram for feeder F2.

While the fault current is flowing from the infinite bus to "A," the magnitude of the voltage at "B" bus is

$$\begin{aligned} V_{\mathrm{B}} &= \frac{j1.05}{j0.20 + j0.67 + j0.70 + j1.05} \times 1.0 \\ &= 0.40\ \mathrm{pu} \end{aligned}$$

The magnitude of the voltage on the 12 kV bus and all loads on feeders F1 and F3 including bus "C" (see Fig. 21.4) is

$$\begin{aligned} V_{\mathrm{B}} &= \frac{j0.70 + j1.05}{j0.20 + j0.67 + j0.70 + j1.05} \times 1.0 \\ &= 0.67\ \mathrm{pu} \end{aligned}$$

The magnitude of the voltage on the 69 kV bus is

$$\begin{aligned} V_{\mathrm{B}} &= \frac{j0.67 + j0.70 + j1.05}{j0.20 + j0.67 + j0.70 + j1.05} \times 1.0 \\ &= 0.92\ \mathrm{pu} \end{aligned}$$

The typical clearing times of commonly used devices in practice are listed in Table 21.2.

### 21.3.2 Case Study: Radial Distribution System

A distribution system consists of two feeders as shown in Fig. 21.5. An industrial load on feeder F1 is sensitive to voltage sags. The industrial customer wants to know how many sags can be expected from faults in feeder F2. For this example, consider all faults to be bolted three phase only. The prefault voltage is assumed to be 1.0 per unit. The source reactance to the feeder bus is $+j0.50$. Feeder F2 is 12 km long with a reactance

TABLE 21.2. Typical Clearing Times of Protective Devices

| Type of Fault Clearing Device | Typical Minimum Clearing Time in Cycles | Typical Time Delay in Cycles | Number of Retries |
|---|---|---|---|
| Expulsion fuse | 0.50 | 0.5–60 | None |
| Current limiting fuse | 0.25 or less | 0.25–6 | None |
| Electronic recloser | 3 | 1–30 | 0–4 |
| Oil circuit breaker | 5 | 1–60 | 0–4 |
| SF6 or vacuum breaker | 3 (actual) (5 by C37.04) | 1–60 | 0–4 |

*Note*: C37.04 CIGRE Working Group.

of $+j0.40$/unit/km. The average number of the three-phase faults is 0.15 faults/km/year. Results for voltage sags for the radial distribution system of Fig. 21.5 are shown in Table 21.3.

Any fault closer to the feeder bus can cause voltage sags worse than those at the point of interest. For example, three-phase faults between the feeder bus and 5 km out on the feeder F2 will cause at least a voltage sag of 0.8 per unit. Faults farther than 5 km away cannot drop the voltage level lower than 0.8 per unit. The results of the above can be expressed as a graph of the sag frequency versus the magnitude of the voltage sag.

It is important to note that the addition of a second feeder identical to F2 doubles the number of voltage sags, as seen by the industrial customer. The complete picture must also include the probability of voltage sags from the industrial plant distribution system and the transmission network (Fig. 21.6).

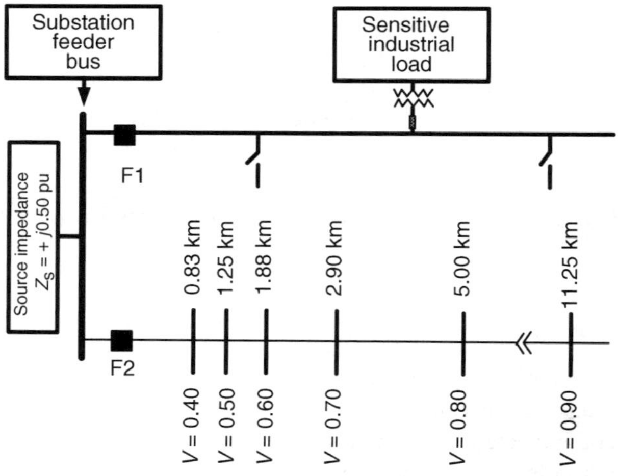


Figure 21.5. Radial distribution system single-line diagram.

TABLE 21.3. Results for Voltage Sags for the Radial Distribution System of Fig. 21.5

| Lowest Phase Sag Voltage Per Unit | Line Exposure (km) | Events/km/year | Number of Sags Less Than or Equal to Sag Voltage |
|---|---|---|---|
| 0.40 | 0.83 | 0.15 | 0.12 |
| 0.50 | 1.25 | 0.15 | 0.19 |
| 0.60 | 1.88 | 0.15 | 0.28 |
| 0.70 | 2.90 | 0.15 | 0.44 |
| 0.80 | 5.00 | 0.15 | 0.75 |
| 0.90 | 11.25 | 0.15 | 1.69 |

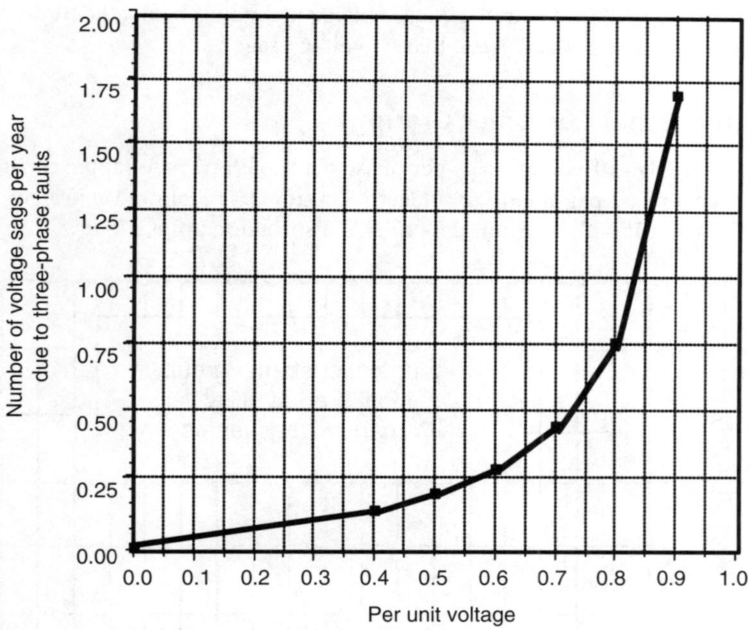


Figure 21.6. Number of voltage sags per year originating from feeder F2 affecting an industrial customer on feeder F1.

## 21.4 FREQUENCY OF VOLTAGE SAGS

In 1991, the Canadian Electricity Association (CEA) began a 3-year power quality survey. The main objective of the survey was to obtain an indication of the general levels of power quality that exist in Canada. The results would serve as a baseline against which future surveys could be compared to determine trends. A secondary objective was to increase utility expertise in making power quality measurements and interpreting the data.

Twenty-two utilities across Canada participated, with a total of 550 sites (industrial, commercial, and residential customer groups) monitored over a 3-year period. Each site was monitored for a nominal 25-day period, most at the customer's service entrance

TABLE 21.4. Monitor Thresholds for Voltage Sags

| Voltage Level (V) | Voltage Sag (80 ms–10 s) |
|---|---|
| 120 | 110 V rms |
| 120/208 | 110 V rms |
| 347/600 | 318 V rms |

panel (e.g., 120 and 347 V). Approximately 10% of the sites had primary metering to provide an indication of the power quality characteristics of the utility's distribution system serving their customers.

The power line threshold levels for voltage sags are shown in Table 21.4. The power line monitor (i.e., BMI) was set to capture voltage anomalies below the threshold level that lasted between 80 ms and 10 s), that is, voltage sags.

### 21.4.1 Industrial Customer Group

The average number of voltage sags per phase per month per site (approximately four voltage sags per phase per month) for industrial customers monitored at primary voltage levels is shown in Fig. 21.7. From the utility's distribution point of view, a significant

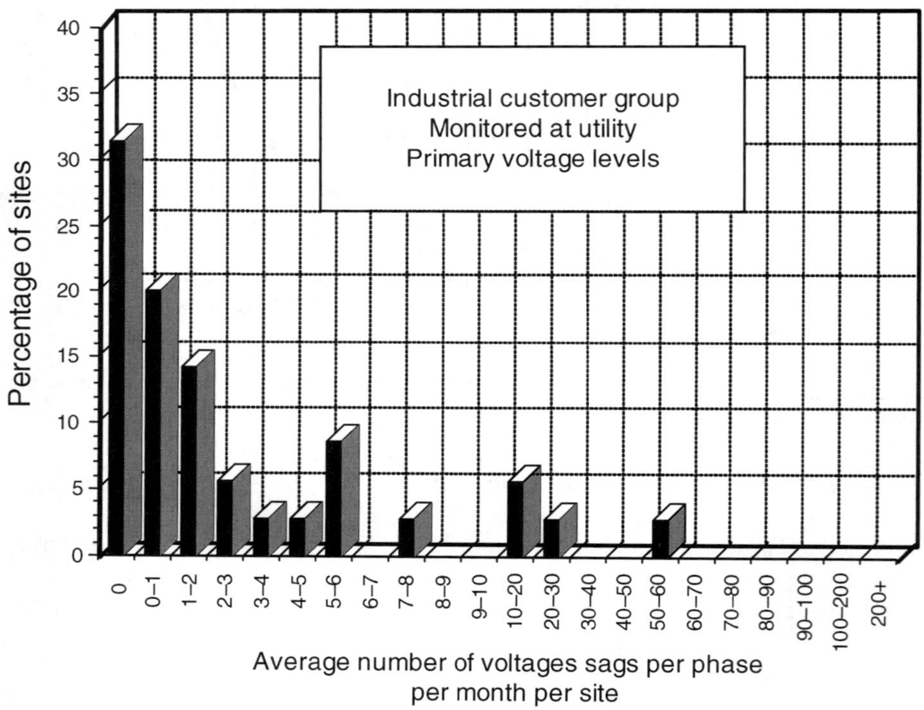


Figure 21.7. Average number of voltage sags per phase per month per site for industrial customer group monitored at utility primary voltage levels.

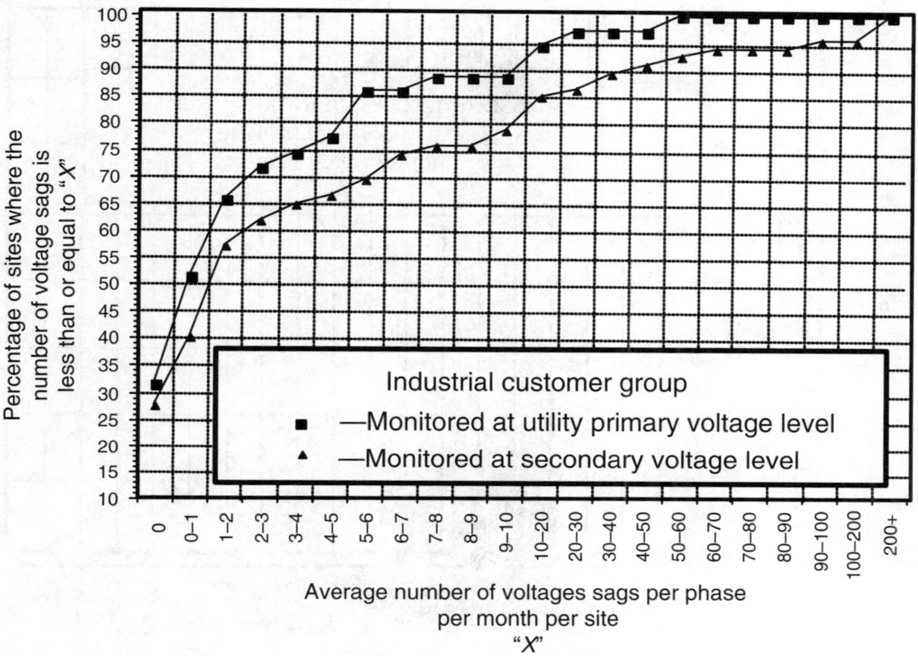


Figure 21.8. Cumulative distribution of voltage sags at industrial customer sites monitored at the primary and secondary voltages.

number of sites experienced no voltage sags on their primary during the monitoring period (i.e., approximately 31% of the sites).

The cumulative percentage of sites whose number of voltage sags is less than or equal to a specified value for primary and secondary monitored customers is shown in Fig. 21.8.

The average number of voltage sags monitored at industrial utilization voltage levels is significantly higher than those occurring at their primary voltage levels (e.g., 85% of sites will experience an average between 10 and 20 voltage sags at their utilization voltage level and an average of only 5–6 voltage sags on their primary implying that the origins of voltage sags are more likely to be on the industrial customer's utilization voltage levels. Based on the average values, clearly the number of voltage sags occurring at an industrial facility is significantly higher than those occurring on the utility primary (i.e., 38 compared to an average of four voltage sags per month per phase).

## 21.4.2 Commercial Customer Group

The frequency distribution of voltage sags monitored at the utility primary voltage levels is shown in Fig. 21.9. Note that 34% of the sites experienced no voltage sags during the 1-month monitoring period.

The average number of voltage sags per phase per month per site for commercial customers monitored at 120/208 V and at 347/600 V are shown in Figs 21.10 and 21.11,

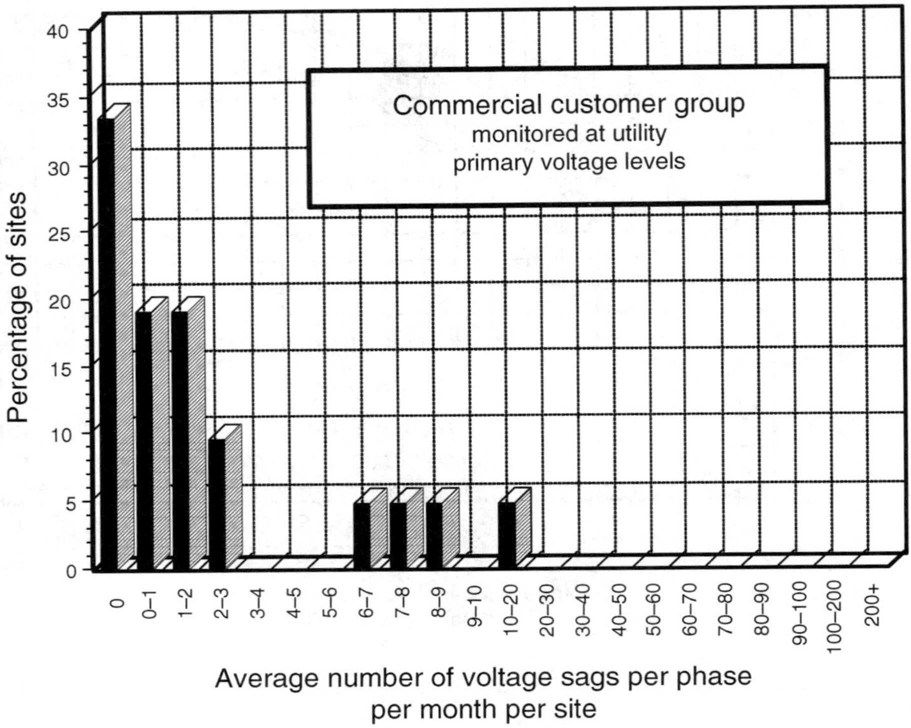


Figure 21.9. Average number of voltage sags per phase per month per site for commercial customers monitored at the utility primary voltages.

respectively. It is important to note that a significant number of commercial sites experienced no voltage sags during the monitoring period (i.e., approximately 23% for 120/208 V secondary monitored sites and 28% for 347/600 V sites). The commercial sites monitored at 120/208 V had more sites with a high frequency of voltage sags than those commercial sites monitored at 347/600 V. The cumulative distribution of voltage sags at commercial customer sites monitored at the primary and secondary voltages is shown in Fig. 21.12.

It is important to note that there are more voltage sags occurring at the secondary level than at the primary voltage level.

## 21.5 EXAMPLE VOLTAGE SAG PROBLEM: VOLTAGE SAG ANALYSIS OF UTILITY AND INDUSTRIAL DISTRIBUTION SYSTEMS

### 21.5.1 Utility Distribution Systems

A radial utility distribution system single-line diagram is shown in Fig. 21.13.

The average number of three-phase faults for this distribution feeder system is 0.20 faults/km/year and the feeder reactance is *j*0.40/unit/km.

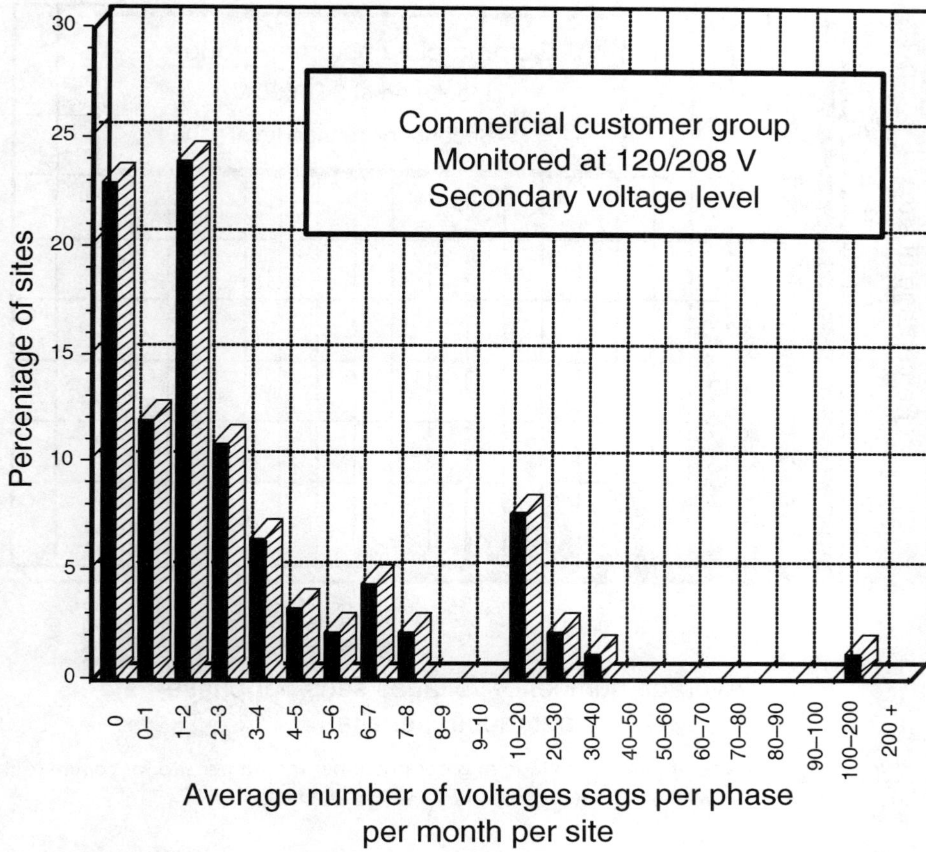


Figure 21.10. Average number of voltage sags per phase per month per site for commercial customers monitored at the customer's secondary voltage of 120/208 V.

Calculate the number of voltage sags per year caused by three-phase faults in feeders F1–F6 that are seen by the sensitive industrial load on feeder 3 that are less than or equal to a sag voltage on the substation feeder bus shown in the table below.

| Lowest Phase Sag Voltage Per Unit (Substation Bus) | Number of Voltage Sags Less Than or Equal to the Specified Sag Voltage |
|---|---|
| 0.70 | ? |
| 0.80 | ? |
| 0.90 | ? |

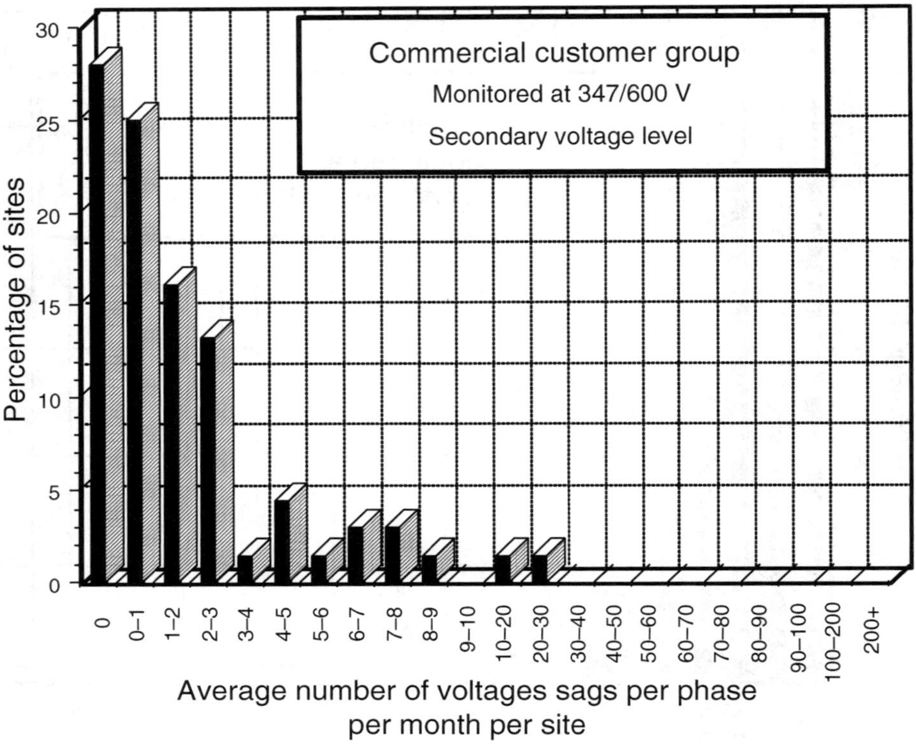


Figure 21.11. Average number of voltage sags per phase per month per site for commercial customers monitored at the customer's secondary voltage of 347/600 V.

*Solution*:

Part 1—Utility Distribution System

Consider the radial utility distribution system single-line diagram as shown in Fig. 21.13. The average number of three-phase faults for this distribution feeder system is 0.20 faults/km per year and the feeder reactance is $j0.40$/unit/km.

Calculate the number of voltage sags per year caused by three-phase faults on feeders F1–F6 that are seen by the sensitive industrial load on feeder 3 that are less than or equal to a sag voltage on the substation feeder bus shown in the table below. The calculated results are listed in the table below. Figure 21.14 shows voltage sag versus feeder physical length.

| Lowest Phase Sag Voltage Per Unit at the Substation Bus | Physical Length ($L$) of Line Exposure |
|---|---|
| 0.70 | 2.916667 |
| 0.80 | 5 |
| 0.90 | 8.75 |

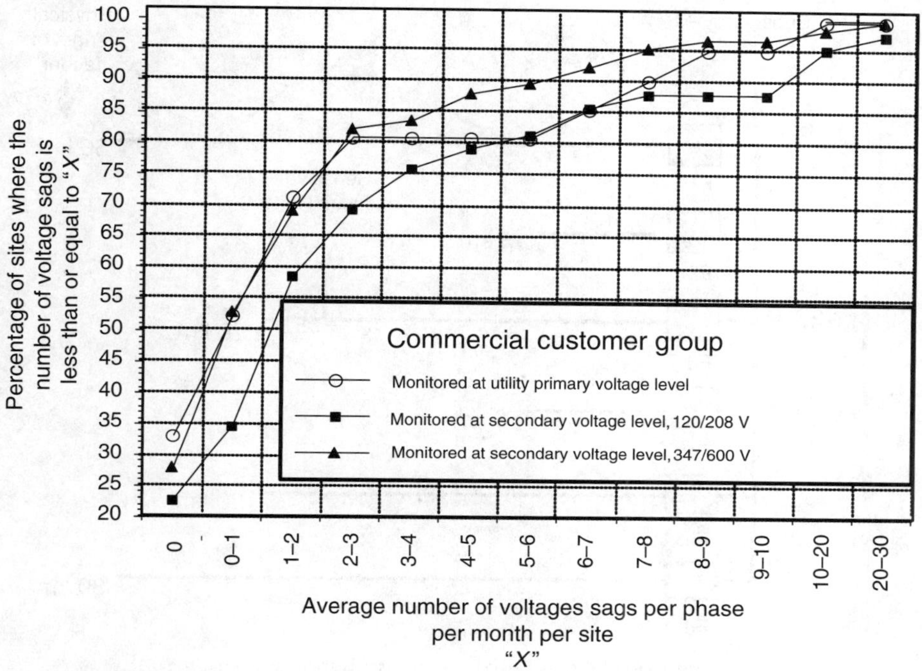


Figure 21.12. Cumulative distribution of voltage sags at commercial customer sites monitored at the primary and secondary voltages.

Given: $Z_s = 0.50$ $Z_f = 0.40$

$$L = V_{sag} \times (Z_s)/[Z_f \times (1 - V_{sag})]$$

*Notes*: Feeder faults close to the substation bus result in high fault currents and low sag voltages at the substation bus. As the feeder faults occur further from the substation bus, the fault current levels decrease and the voltage at the substation bus (sag voltage) increases. For example, faults occurring 2.916 km and less from the substation bus will result in a substation bus voltage less than or equal to 0.70 pu.

$$L = \frac{0.70 \times 0.50}{0.40 \times (1 - 0.70)} = 2.916 \text{ km for a sag voltage} = 0.70 \text{ pu}$$

$$L = \frac{0.80 \times 0.50}{0.40 \times (1 - 0.80)} = 5.000 \text{ km for a sag voltage} = 0.80 \text{ pu}$$

$$L = \frac{0.90 \times 0.50}{0.40 \times (1 - 0.90)} = 11.25 \text{ km for a sag voltage} = 0.90 \text{ pu}$$

By varying the voltage sag levels, a graph of the number of voltage sags less than a particular value can be plotted as shown in Fig. 21.14.

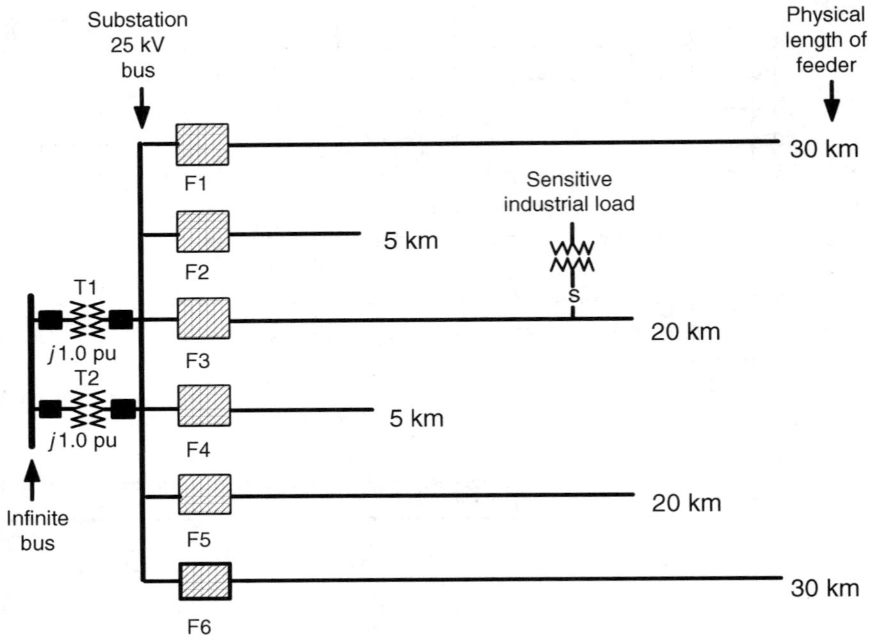


Figure 21.13. A radial utility distribution system.

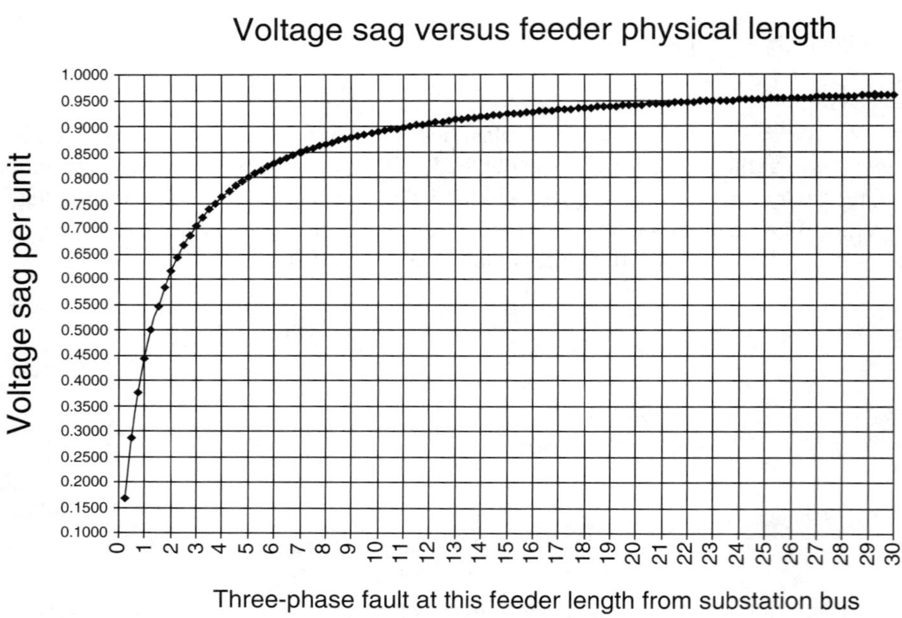


Figure 21.14. Voltage sag versus feeder physical length.

The number of voltage sags less than or equal to a specified value are calculated below:

Voltage sag at substation bus for $V_{sag} = 0.70$ pu

| Feeder Number | Physical Length ($L$) of Line Exposure | Number of Three-Phase Faults Per Year | Number of Voltage Sags Less Than or Equal to 0.7 pu |
|---|---|---|---|
| F1 | 2.916667 | 0.2 | 0.583333 |
| F2 | 2.916667 | 0.2 | 0.583333 |
| F4 | 2.916667 | 0.2 | 0.583333 |
| F5 | 2.916667 | 0.2 | 0.583333 |
| F6 | 2.916667 | 0.2 | 0.583333 |
| | | Total | 2.916667 |

Voltage sag at substation bus for $V_{sag} = 0.80$ pu

| Feeder Number | Physical Length ($L$) of Line Exposure | Number of Three-Phase Faults Per Year | Number of Voltage Sags Less Than or Equal to 0.8 pu |
|---|---|---|---|
| F1 | 5 | 0.2 | 1 |
| F2 | 5 | 0.2 | 1 |
| F4 | 5 | 0.2 | 1 |
| F5 | 5 | 0.2 | 1 |
| F6 | 5 | 0.2 | 1 |
| | | Total | 5 |

Voltage sag at substation bus for $V_{sag} = 0.90$ pu

| Feeder Number | Physical Length ($L$) of Line Exposure | Number of Three-Phase Faults Per Year | Number of Voltage Sags Less Than or Equal to 0.9 pu |
|---|---|---|---|
| F1 | 11.25 | 0.2 | 2.25 |
| F2 | 5 | 0.2 | 1 |
| F4 | 5 | 0.2 | 1 |
| F5 | 11.25 | 0.2 | 2.25 |
| F6 | 11.25 | 0.2 | 2.25 |
| | | Total | 8.75 |

| Lowest Phase Sag Voltage Per Unit (Substation Bus) | Number of Voltage Sags Less Than or Equal to the Specified Sag Voltage |
|---|---|
| 0.70 | 2.916667 |
| 0.80 | 5 |
| 0.90 | 8.75 |

### 21.5.2 Industrial Distribution System

An industrial power system is shown in Fig. 21.15.

The per unit impedances of all the electrical components are defined below.
Per Unit Impedances for all Electrical Components of Fig. 21.15

| Component | Per Unit Reactances |
|---|---|
| Equivalent 69 kV utility impedance | 0.2 |
| Generator 1 | 0.4 |
| Transformer 1 | 0.2 |
| Transformer 2 | 0.1 |
| Transformer 3 | 3.0 |
| Transformer 4 | 3.0 |
| Transformer 5 | 0.1 |
| Cable 1 | 0.1 |
| Cable 2 | 0.1 |
| Cable 3 | 0.2125 |
| Synchronous motor 3 | 3.0 |
| Induction motor 4 | 3.0 |

Calculate the voltage sag on bus 1 and bus 2 when a permanent three-phase fault occurs in the feeder as shown in Fig. 21.15.

*Solution*:

The one-line impedance diagram is shown in Fig. 21.16.

Definition of symbols: M, motor; T, transformer; C, Cable; and x, reactance

$$\text{x(utility + transformer 1)} = \text{xutility} + \text{xT1} = j0.20 + j0.20 = j0.40$$

$$\text{xgen 1} = j0.40$$

$$\begin{aligned} \text{xS1\&}S2 &= \text{x(utility+transformer 1)} \times \text{xgen1}/(\text{x(utility+transformer 1)} + \text{xgen1}) \\ &= j0.4 \times j0.4/(j0.4 + j0.4) \\ \text{xS1\&S2} &= 0.2000000 \text{ per unit} \end{aligned}$$

$$\begin{aligned} \text{xM3\&T3} &= \text{xM3} + \text{xT3} = j3.0 + j3.0 = j6.0 \\ \text{xM4\&T4} &= \text{xM4} + \text{xT4} = j3.0 + j3.0 = j6.0 \end{aligned}$$

$$\begin{aligned}
\text{xS3\&S4} &= \text{xM3\&T3} \times \text{xM4\&T4}/(\text{xM3\&T3} + \text{xM4\&T4}) \\
&= j0.6 \times j0.6/(j0.6 + j0.6) = j3.0
\end{aligned}$$

$$\begin{aligned}
\text{xS1\&S2\&S3\&S4} &= \text{xS1\&S2} \times \text{xS3\&S4}/(\text{xS1\&S2} + \text{xS3\&S4}) \\
&= j0.20 \times j3.0/(j0.20 + j3.0) \\
&= j0.1875
\end{aligned}$$

$$\begin{aligned}
\text{cablebranch1} &= \text{cable 1} + \text{xT5} = j0.10 + j0.10 = j0.20 \\
\text{cablebranch2} &= \text{cable 2} + \text{xT2} = j0.10 + j0.10 = j0.20
\end{aligned}$$

$$\begin{aligned}
\text{xcableq} &= \text{cablebranch1} \times \text{cablebranch 2}/(\text{cablebranch1} + \text{cablebranch 2}) \\
&= j0.20 \times j0.20/(j0.20 + j0.20) \\
&= j0.10
\end{aligned}$$

$$\begin{aligned}
\text{xtotal} &= \text{xS1\&S2\&S3\&S4} + \text{xcableq} + \text{cable 3} = j0.1875 + j0.10 + j0.2125 \\
&= j0.50
\end{aligned}$$

$$\begin{aligned}
\text{If3ph} &= 1.0/\text{xtotal} = 1.0/j0.50 = -j2.0 \\
\text{Vbus2} &= \text{If3ph} \times \text{cable3} \times -j2.0 \times j0.2125 = 0.4250\,\text{pu} \\
\text{Vbus1} &= \text{If3ph} \times (\text{cable3} + \text{xcableq}) = -j2.0 \times (j0.2125 + j0.10) = 0.6250\text{pu}
\end{aligned}$$

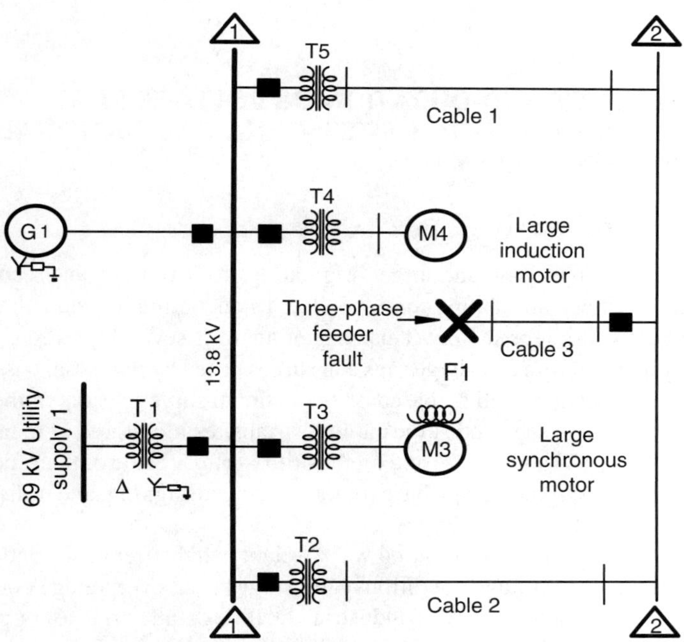


Figure 21.15. An industrial power distribution system.

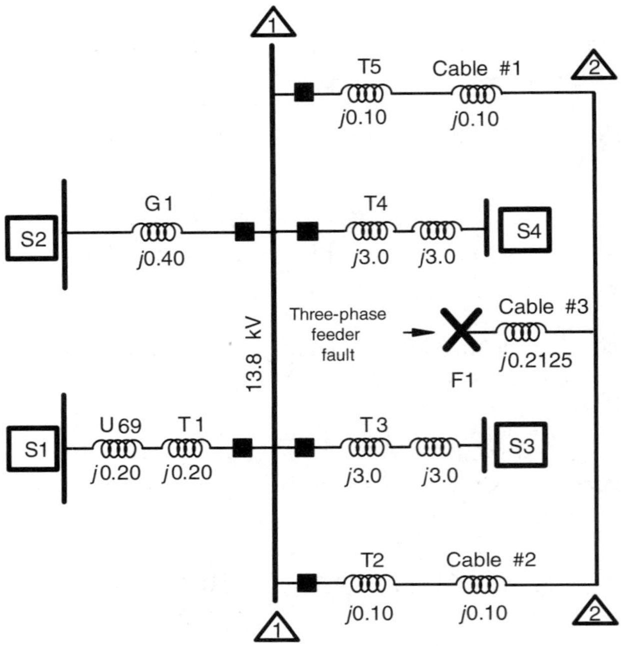


Figure 21.16. An industrial power distribution system single-line impedance diagram.

## 21.6 FREQUENCY AND DURATION OF VOLTAGE SAGS AND SURGES AT INDUSTRIAL SITES: CANADIAN NATIONAL POWER QUALITY SURVEY

### 21.6.1 Background

The occurrence of voltage sags and surges originating on the primary and secondary sides of industrial facilities can disrupt continuous and noncontinuous industrial computer processes, a costly issue for society. This chapter answers several questions concerning the frequency and duration of voltage sags and surges posed by industrial customers. The answers to these questions will be based on the national survey results of the frequency and duration of voltage sags and surges at industrial sites monitored at their utilization voltage levels (e.g., 120 and 347 V) and on the utility primary side of their facilities. The survey results provide a knowledge base for monitoring, designing, and utilizing voltage sag and surge mitigating technologies.

Before 1990, very little information was available on the frequency and duration of voltage sags (i.e., undervoltage conditions) and surges (i.e., overvoltage conditions) on the primary and secondary sides of industrial facilities. Industrial users posed many questions; for example, did the voltage anomalies that disrupted their processes originate on the utility's system or within the industrial facilities electrical system?

In 1991, the CEA took a proactive approach to power quality problems and initiated a 3-year Canadian National Power Quality Survey involving 22 utilities. This chapter will present and reference some of the detailed survey results of the frequency and duration of voltage sags and surges at industrial sites monitored at their utilization voltage levels (e.g., 120 and 347 V) and on the utility primary side of their facilities. The chapter answers several key questions posed by industrial users related to the power quality characteristics of their sites:

1. Do these power supply anomalies occur more frequently during certain times of the day or are they random events?
2. Is the frequency of voltage sags and surges clustered or random or dependent upon the day of the week?
3. Is the frequency of power line disturbances on the utility system significantly lower than those monitored at industrial facilities?
4. Is each site unique or are there significant similarities between industrial sites?
5. Are the coordinates of the magnitude and duration of voltage sags and surges clustered or random and is there a difference between primary and secondary occurrences?
6. Can the frequency and duration of voltage sags and surges at industrial sites be statistically modeled by a normal distribution?

The database of voltage sags and surges monitored at industrial sites was divided into two categories:

1. Primary (utility side)
2. Secondary (industrial side)

The division provides a means of revealing the unique power quality characteristics of utility primary delivery systems and industrial secondary electrical systems.

The results of the national survey will provide a guide to the number of the various types of utility and industrial site disturbances that can be expected to be captured at an industrial site for a monitoring period of 1 month. Answers to the above questions are based on 33 primary monitored and 66 secondary monitored industrial sites.

## 21.6.2 Voltage Sags and Surges (Time of Day)

The question posed by many utility industrial customers is: Do the primary (i.e., utility side) and secondary (i.e., customer side) voltage sags and surges follow a daily pattern? The voltage sags and surges monitored on the primary side of industrial sites as a function of the time of day are shown in Figs 21.17 and 21.18, respectively.

It is obvious from Figs 21.17 and 21.18 that the occurrence of voltage sags and surges is not a uniform distribution (i.e., a random event). It appears that the voltage sags tend to follow the daily loading patterns of the utility (i.e., the utilities response to changing demands appears to cause more voltage sags at certain times of the day). The voltage sags

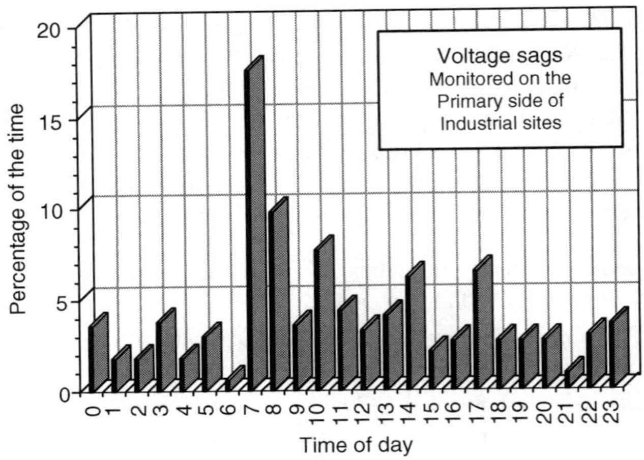


Figure 21.17. Voltage sags as a function of the time of the day monitored on the primary side of industrial facilities.

and surges monitored on the secondary side of the industrial sites as a function of the time of day are shown in Figs 21.19 and 21.20, respectively.

The occurrence of secondary voltage sags tended to occur more frequently in the early morning (i.e., 4–9 a.m.) during plant startup and shutdown periods.

Voltage surges exhibited a distinctive pattern and occurred more frequently during the daytime (i.e., 600–1500).

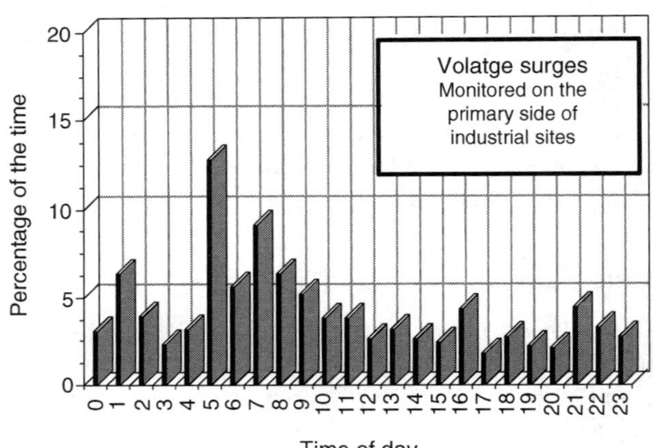


Figure 21.18. Voltage surges as a function of the time of the day monitored on the primary side of industrial facilities.

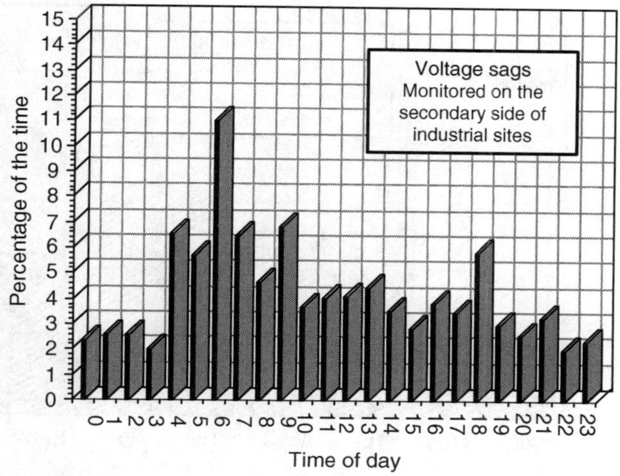


Figure 21.19. Voltage sags as a function of the time of the day monitored on the secondary side of industrial facilities.

## 21.6.3 Voltage Sags and Surges (Day of Week)

Another question posed by utility industrial customers is: Do the primary (i.e., utility side) and secondary (i.e., customer side) voltage sags and surges follow a weekly pattern?

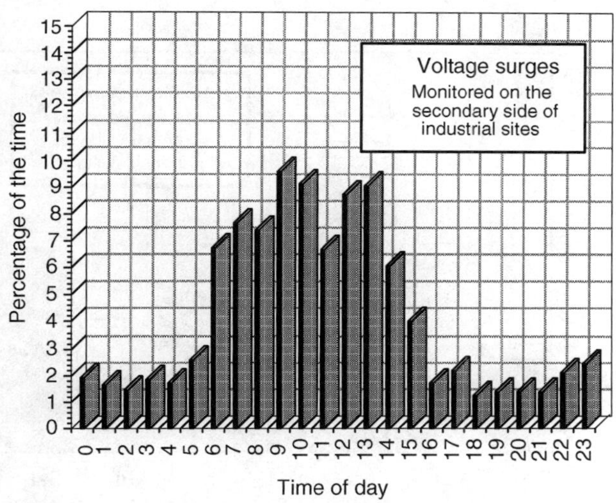


Figure 21.20. Voltage surges as a function of the time of the day monitored on the secondary side of industrial facilities.

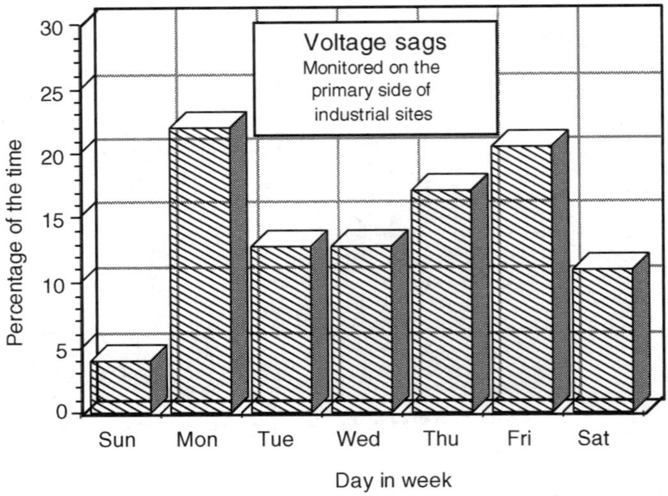


Figure 21.21. Voltage sags as a function of the day of the week monitored on the primary side of industrial facilities.

The voltage sags and surges monitored on the primary side of industrial sites as a function of the day of the week are shown in Figs 21.21 and 21.22, respectively.

The voltage sags and surges monitored on the secondary side of industrial sites as a function of the day of the week are shown in Figs 21.23 and 21.24, respectively.

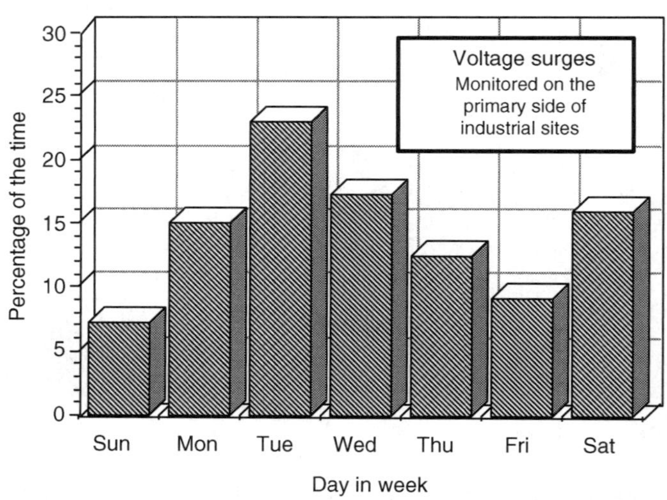


Figure 21.22. Voltage surges as a function of the day of the week monitored on the primary side of industrial facilities.

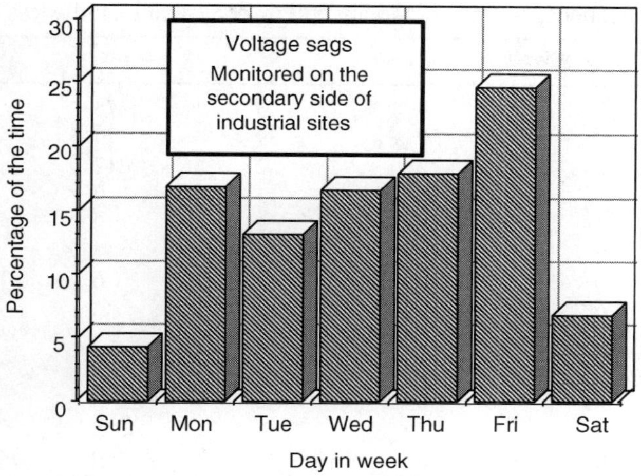


Figure 21.23. Voltage sags as a function of the day of the week monitored on the secondary side of industrial facilities.

The primary and secondary monitored voltage sags tended to occur for a larger percentage of the time during the normal working days (i.e., Monday to Friday) for many of the industrial plants.

The primary monitored voltage surges tended to occur more frequently during the week and on Saturdays, while the secondary monitored voltage surges tended to be more uniformly distributed.

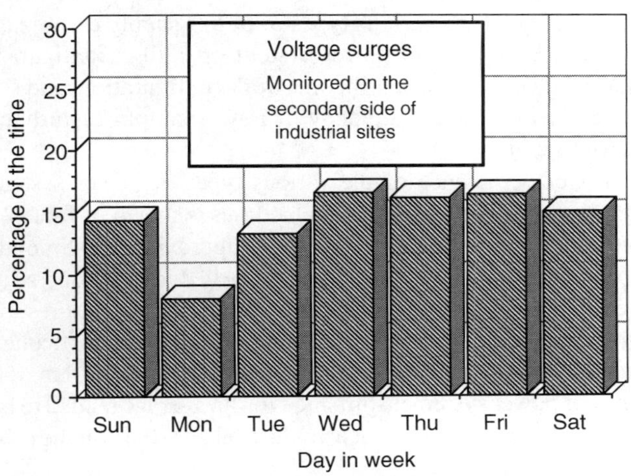


Figure 21.24. Voltage surges as a function of the day of the week monitored on the secondary side of industrial facilities.

TABLE 21.5. Frequency of Primary Monitored Power System Disturbances

| Site No. | Outage | Surge | Sag | Swell | Wave Shape |
|---|---|---|---|---|---|
| 1 | 0 | 0 | 3 | 78 | 0 |
| 2 | 0 | 60 | 39 | 175 | 1 |
| 3 | 6 | 22 | 2 | 212 | 0 |
| 4 | 1 | 83 | 5 | 0 | 1 |
| 5 | 2 | 18 | 6 | 0 | 8 |
| 6 | 0 | 98 | 0 | 72 | 0 |
| 7 | 0 | 0 | 8 | 0 | 5 |
| 8 | 0 | 57 | 0 | 30 | 2 |
| 9 | 0 | 10 | 116 | 2 | 52 |
| 10 | 0 | 11 | 4 | 0 | 3 |
| 11 | 0 | 0 | 0 | 3 | 0 |
| 12 | 0 | 8 | 1 | 45 | 1 |
| 13 | 6 | 395 | 17 | 0 | 0 |
| 14 | 0 | 0 | 0 | 0 | 0 |
| 15 | 0 | 17 | 2 | 0 | 0 |
| AV | 1 | 51.9 | 13.5 | 41.1 | 4.9 |

AV, avrage value per site per month.

### 21.6.4 Frequency of Disturbances Monitored on Primary and Secondary Sides of Industrial Sites

The frequency of occurrence of the various types of power system disturbances monitored on the primary side of industrial facilities for all three phases is shown in Table 21.5 for a sample of 15 sites, each with a monitoring period of approximately 1 month.

An examination of Table 21.5 clearly reveals the uniqueness of each industrial site in terms of the frequency of the various types of power supply disturbances monitored on the primary side. Variations in power system operating configurations, physical location of an industrial site within a utility network configuration, and so on are factors that can significantly affect the frequency of power supply disturbances seen by a particular industrial facility.

The frequency of occurrence of the various types of power system disturbances monitored on the secondary side of industrial facilities is shown in Table 21.6 for another sample of 15 sites, each with a monitoring period of approximately 1 month. The cause of these power supply disturbances in many cases is partly due to the operational patterns of various loads (e.g., motor starting, switching loads in and out in a response to process demands, starting up and shutting down of various processes) at a particular industrial site.

An examination of Tables 21.5 and 21.6 reveals that at many, but not all, industrial sites the frequency of power system disturbances on an average tended to be higher on the secondary side (i.e., utilization voltage at an industrial site) than on the primary side (i.e., utility supply).

It is very important to note that the power quality characteristics of each site are unique. At some of the sites where the frequency of voltage sags and surges was

TABLE 21.6. Frequency of Secondary Monitored Power System Disturbances

| Site No. | Outage | Surge | Sag | Swell | Wave Shape |
|---|---|---|---|---|---|
| 1 | 6 | 6 | 4 | 0 | 7 |
| 2 | 3 | 2560 | 211 | 16 | 2 |
| 3 | 0 | 29 | 9 | 0 | 0 |
| 4 | 0 | 0 | 1050 | 0 | 0 |
| 5 | 0 | 20 | 1 | 0 | 0 |
| 6 | 0 | 47 | 2 | 556 | 0 |
| 7 | 0 | 2 | 0 | 1066 | 4 |
| 8 | 6 | 13 | 25 | 0 | 1 |
| 9 | 1 | 181 | 403 | 0 | 0 |
| 10 | 0 | 4 | 2 | 451 | 2 |
| 11 | 0 | 1 | 13 | 0 | 0 |
| 12 | 0 | 12 | 239 | 365 | 4 |
| 13 | 2 | 20 | 37 | 423 | 77 |
| 14 | 0 | 0 | 81 | 0 | 2 |
| 15 | 3 | 257 | 63 | 24 | 8 |
| AV | 1.4 | 210.1 | 142.7 | 193.4 | 7.1 |

AV, average value per site per month.

extremely high, the primary causes of these anomalies were linked to poor wiring, inadequate grounding, poor design, and operating practices.

In the electric environment of industrial sites, there are many polluting and nonpolluting loads operating in distinctive cycles to meet the demands of the plant processes. Usually, transitions in their operating cycles generate local disturbances and can be specifically identified by a detailed on-site power quality audit. The audit provides a means of correlating the various on-site disturbances with particular electronic and electrical equipment transitional operating patterns.

## 21.7 SCATTER PLOTS OF VOLTAGE SAG LEVELS AS A FUNCTION OF DURATION

Scatter plots of the magnitude of voltage sags versus the duration of their existence are shown in Figs 21.25 and 21.26, respectively.

Comparing Figs 21.25 and 21.26, the voltage sags occurring on the secondary side of industrial sites were more frequent and "deeper" than on the primary side (i.e., voltage sags occurring below 60% of their nominal voltage were the lower limit on the sample of primary monitored voltage sags shown in Fig. 21.22).

## 21.8 SCATTER PLOTS OF VOLTAGE SURGE LEVELS AS A FUNCTION OF DURATION

Scatter plots of the magnitude of voltage surges versus the duration of their existence are shown in Figs 21.27 and 21.28, respectively.

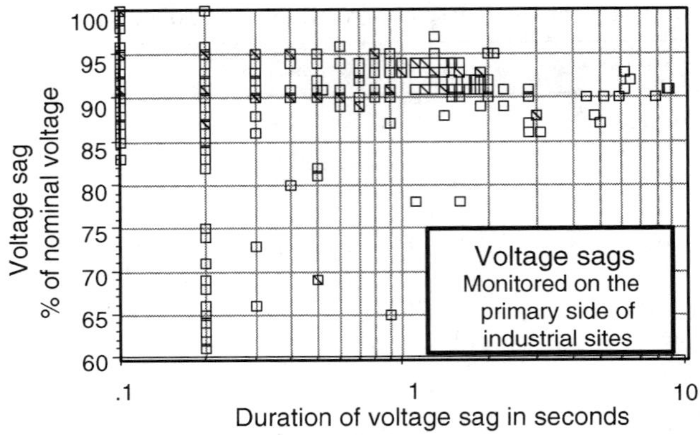


Figure 21.25. Magnitude of the primary monitored voltage sags as a function of their duration.

Note that the upper threshold limit of the power line monitors was set at 200% of the nominal voltage for surges. Voltage surges occurring between 0 and 200% (i.e., a very high-frequency event) were not captured due to the limited storage capability of the power line monitor's computer disks.

## 21.9 PRIMARY AND SECONDARY VOLTAGE SAGES STATISTICAL CHARACTERISTICS

The statistical distributions of the magnitude and duration of voltage sags monitored on the primary and secondary sides of industrial sites are shown in Figs 21.29 and 21.30, respectively.

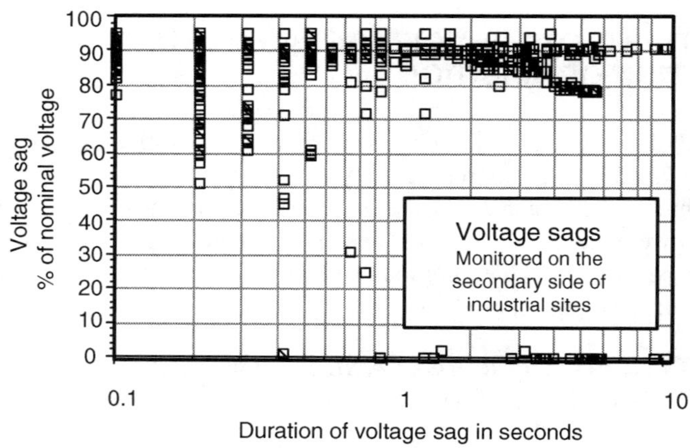


Figure 21.26. Magnitude of the secondary monitored voltage sags as a function of their duration.

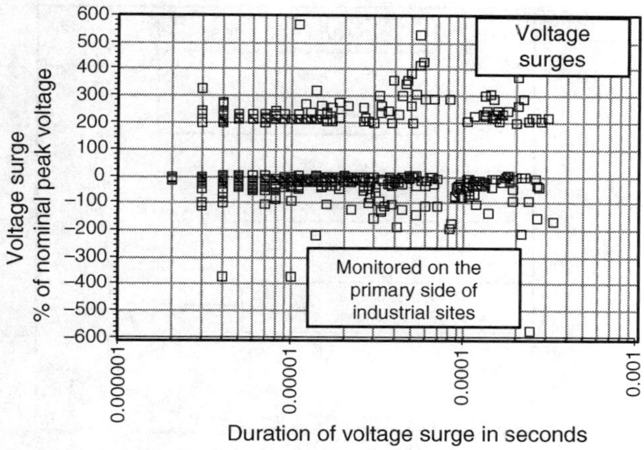


Figure 21.27. Magnitude of the primary monitored voltage surges as a function of their duration.

Note the similarities in the shape of the statistical distribution of primary and secondary voltage sags shown in Figs 21.29 and 21.30.

## 21.10 PRIMARY AND SECONDARY VOLTAGE SURGES STATISTICAL CHARACTERISTICS

The statistical distributions of the magnitude of positive and negative voltage surges monitored on the primary and secondary sides of industrial sites are shown in Figs 21.31 and 21.32, respectively.

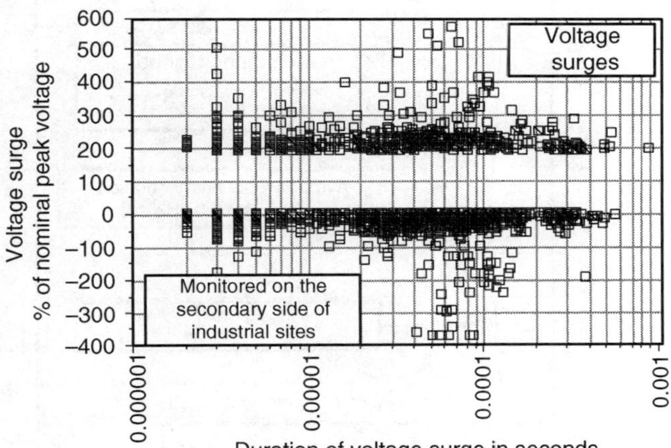


Figure 21.28. Magnitude of the secondary monitored voltage surges as a function of their duration.

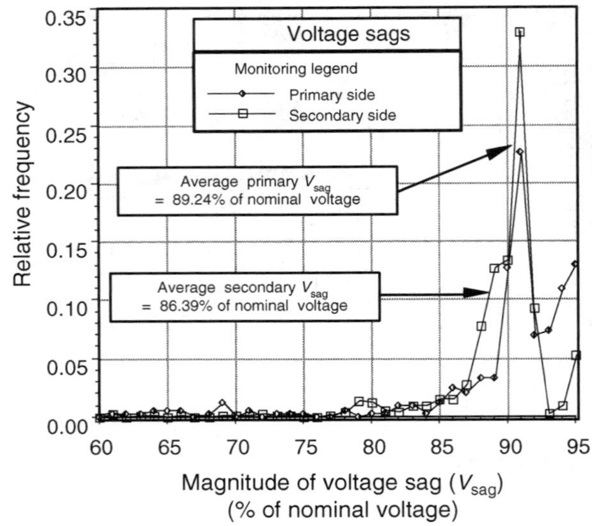


Figure 21.29. Distribution of the magnitude of the primary and secondary monitored voltage sags.

Note the similarities in the shape of the statistical distribution of primary and secondary surges shown in Figs 21.31 and 21.32. The statistical distributions of the duration of positive and negative voltage surges monitored on the primary and secondary sides of industrial sites are shown in Figs 21.33 and 21.34, respectively.

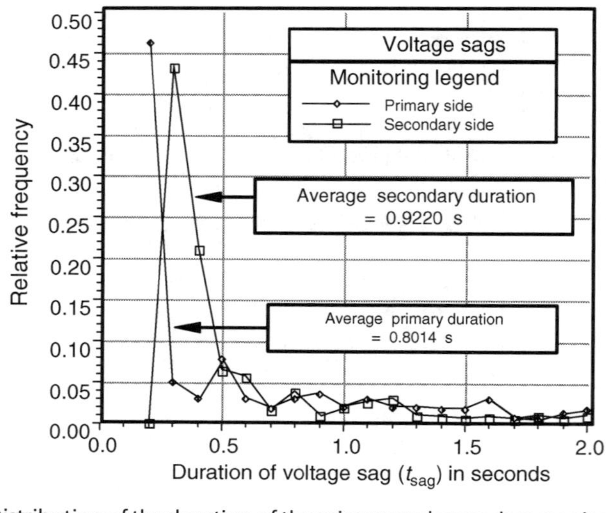


Figure 21.30. Distribution of the duration of the primary and secondary monitored voltage sags.

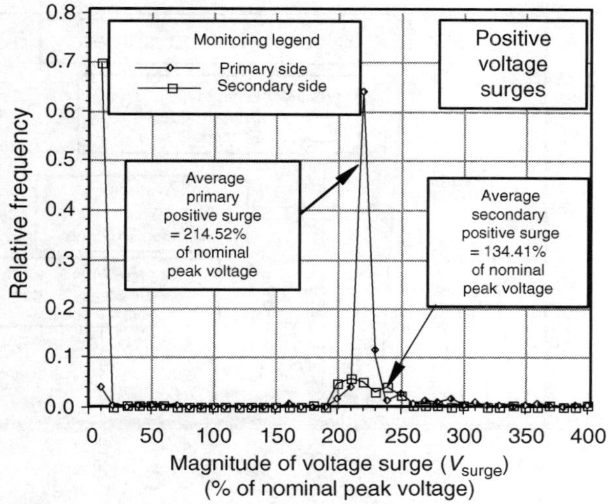


Figure 21.31. Distribution of the magnitude of the primary and secondary monitored positive voltage surges.

A summary of the statistical characteristics of the primary and secondary voltage sags is shown in Table 21.7, the positive voltage surges in Table 21.8, and the negative voltage surges in Table 21.9, respectively.

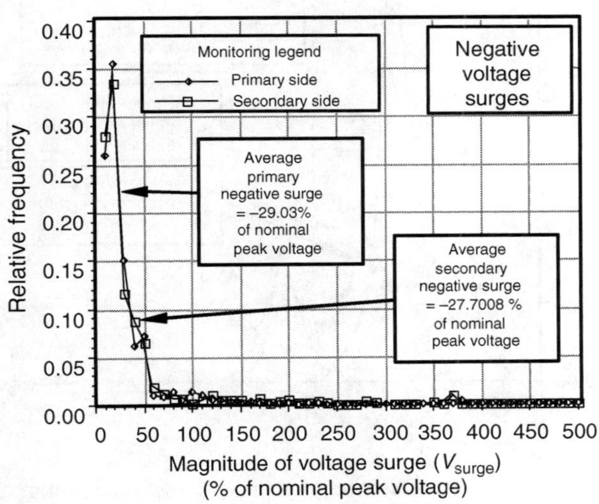


Figure 21.32. Distribution of the magnitude of the primary and secondary monitored negative voltage surges.

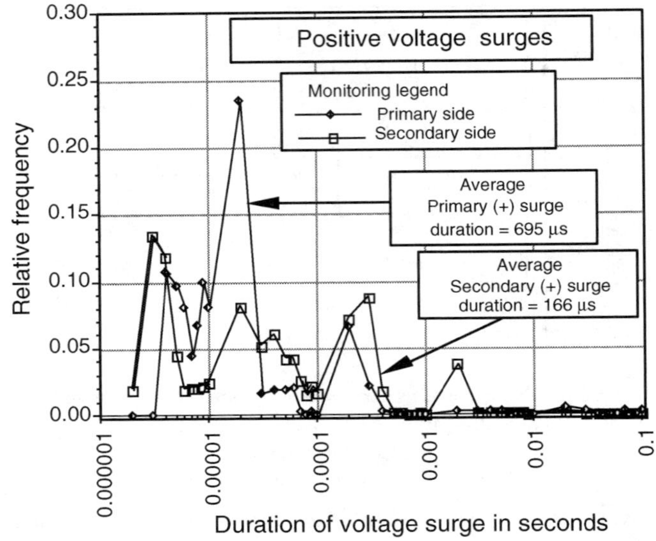


Figure 21.33. Distribution of the duration of the primary and secondary monitored positive voltage surges.

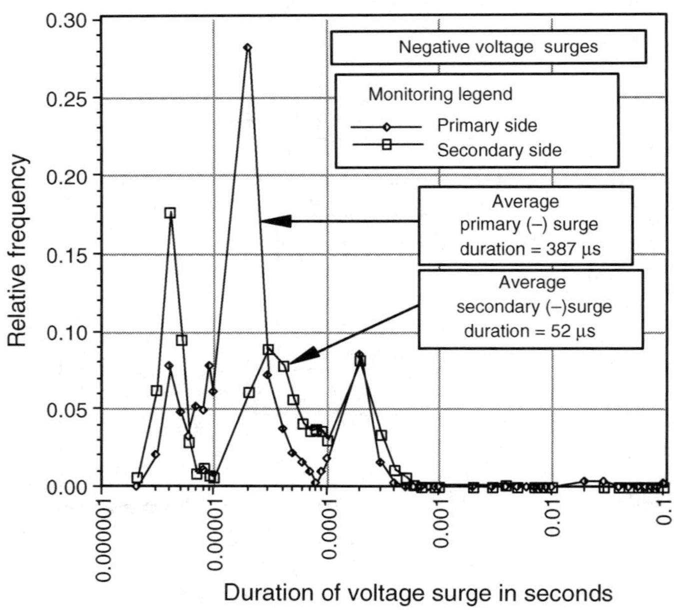


Figure 21.34. Distribution of the duration of the primary and secondary monitored negative voltage surges.

TABLE 21.7. Voltage Sags Statistical Characteristics (Primary and Secondary)

| Statistic | Primary | Secondary |
|---|---|---|
| Average magnitude | 89.3274%[a] | 86.3909%[a] |
| Standard deviation | 11.3427%[a] | 15.9453%[a] |
| Minimum magnitude | 0.0 | 0.0 |
| Maximum magnitude | 95% | 95% |
| Average duration | 0.8014 s | 0.9221 s |
| Standard deviation | 1.2245 s | 1.3659 s |
| Minimum duration | 0.1000 s | 0.1000 s |
| Maximum duration | 8.8 s | 9.7 s |

[a]The manitude of voltage sags is expressed as a percentage of the nominal voltage.

TABLE 21.8. Positive Voltage Surges Statistical Characteristics (Primary and Secondary)

| Statistic | Primary | Secondary |
|---|---|---|
| Average magnitude | 214.5182% | 134.4129% |
| Standard deviation | 62.7181% | 116.0626% |
| Minimum magnitude | 0.0 | 2.0% |
| Maximum magnitude | 571.0% | 571.0% |
| Average duration | 695 μs | 166 μs |
| Standard deviation | 6440 μs | 729 μs |
| Minimum duration | 3 μs | 2 μs |
| Maximum duration | 0.10 s | 17.2 ms |

*Note*: Te magnitude of voltage sags is expressed as a percentage of the nominal voltage.

TABLE 21.9. Negative Voltage Surges Statistical Characteristics (Primary and Secondary)

| Statistic | Primary | Secondary |
|---|---|---|
| Average magnitude | – 29.0396% | – 27.7000% |
| Standard deviation | 44.9848% | 46.6490% |
| Minimum magnitude | 1.0% | 1.0% |
| Maximum magnitude | 574.0% | 371.0% |
| Average duration | 387 μs | 52 μs |
| Standard deviation | 4793 μs | 126 μs |
| Minimum duration | 2 μs | 2 μs |
| Maximum duration | 0.10 s | 3.42 ms |

*Note*: The magntude of voltage surges is expressed as a percentage of the nominal peak voltage.

## 21.11 CONCLUSIONS

This chapter dealt with voltage sag and surge characteristics monitored on the primary and secondary sides of customer transformers. Several example calculations for voltage surge and sag for industrial and commercial customer installations have been presented. The chapter also revealed that primary (i.e., utility) generated voltage sags and surges were not uniformly distributed (i.e., random events) and tended to follow the daily loading patterns of the utility. Secondary industrial voltage surges exhibited distinctive patterns and occurred more frequently during the daytime (e.g., from 6 a.m. to 3 p.m.). Secondary voltage sags also tended to occur more frequently during the early morning. Primary and secondary voltage sags and surges tended to occur more frequently during the weekdays.

This chapter has attempted to answer several questions posed by a utility's industrial customers. The answers were based on the statistical characteristics of the Canadian National Power Quality Survey. It was found that the frequency of monitored voltage sags and surges on an average was higher on the secondary side of the industrial sites than the primary (i.e., utility side). The secondary side voltage sags were deeper than those of the primary side and on average lasted longer. The power quality characteristics of each site were unique.

The per unit magnitude of voltage surges on the primary side was on average higher than that monitored on the secondary side. The average duration of primary voltage surges was significantly longer than that monitored on the secondary.

## REFERENCES

1. CEA Report: 220 D711A, CEA National Power Quality Survey, prepared by J.S. Chan, Brent Hughes, Powertech Labs Inc., Canadian Electricity Association, Montreal, Quebec, Canada.
2. M. B. Hughes, J. S. Chan, and D.O. Koval, Distribution customer power quality experience, *IEEE Transactions on Industry Applications*, 1993, 1204–1211.
3. B. Roger, Power Quality Measurement Protocol: CEA Guide to Performing Power Quality Surveys, CEA 220 D711, Canadian Electricity Association, Montreal, Quebec, Canada.
4. IEEE P-1100, Powering and Grounding Sensitive Electronic Equipment (IEEE Emerald Book).
5. G. W. Allen and D. Segall, Monitoring of computer installations for power line disturbances, IEEE Winter Power Meeting Conference Paper WINPWR, C74, 199-6, 1974.
6. M. Goldstein and P. D. Speranza, The quality of U.S. commercial AC power, in INTELEC (IEEE International Telecommunications Energy Conference), 1992, pp. 28–33.
7. Voltage Sag Working Group, L. Conrad, Chairman, Proposed Chapter 9 for Predicting Voltage Sags (Dips) in Revision to IEEE Standard 493, *IEEE Transactions on Industry Applications*, **30** (3), 1994, 805–821.
8. IEEE Standard 493-1990, IEEE Recommended Practice for the Design of Reliable Industrial and Commercial Power Systems (ANSI).
9. IEEE Standard 446-1987, IEEE Recommended Practice for Emergency and Standby Power Systems for Industrial and Commercial Applications, 1987, 74–77.

10. Canadian Electricity Association, Power Quality Labels for Sensitive Electronic Equipment to Withstand Voltage Surges, Sags and Swells from the Utility System or Customer Sites, Canadian Electrical Association Report No. 9109 U 856, 1994.

11. Undervoltage Ride-Through Performance of Off-The-Shelf Personal Computers, EPRI Power Electronics Application Center, PQTN Brief No. 7, December 1992.

12. D. O. Koval, J. J. Leonard, and Z. J. Licsko, Power quality of small rural industries, *IEEE Transactions on Industry Applications*, **29** (5), 1993, 696–700.

13. D. O. Koval, Power system disturbance patterns, *IEEE Transactions on Industry Applications*, **26**(3), 1990, 556–562.

14. D. O. Koval, How long should power system disturbance site monitoring be in order to be significant, *IEEE Transactions on Industry Applications*, **26** (4), 1990, 556–562, 705–710.

# SELECTED PROBLEMS AND ANSWERS

## PROBLEM SET FOR CHAPTERS 2 AND 3

1. The following table shows the time in years between maintenance actions to 50 manual switches for a given utility.

| Years Between Maintenance Actions | Number |
|---|---|
| 1 | 3 |
| 2 | 3 |
| 3 | 5 |
| 4 | 8 |
| 5 | 8 |
| 6 | 11 |
| 8 | 4 |
| 9 | 3 |
| 11 | 1 |
| 15 | 2 |
| 18 | 1 |
| 20 | 1 |

   (a) What is the average number of years between maintenance actions of manual switches?

   (b) What is the standard deviation?

2. It is known that 10% of the insulators are defective. What is the probability of finding three or more defective insulators in a string of five?

3. The failure of power transformers is assumed to follow a Poisson probability distribution. Suppose on average, a transformer fails once every 10 years. What is the probability that it will not fail in the next 12 months? That it will fail once in the next 24 months?

*Power Distribution System Reliability.* By Ali A. Chowdhury and Don O. Koval

4. In a normal distribution, what percentage is within 1.5 standard deviations from the mean?
5. One thousand new OCRs are put in service. They have a constant failure rate of 0.05 per year.
   (a) How many units of the original 1000 will still be in service after 5 years?
   (b) How many of the original will fail in Year 5?
6. The results of ground testing in a station area with 450 ground rods are as follows:

| | | | | | | | | | |
|---|---|---|---|---|---|---|---|---|---|
| 17 | 25 | 32 | 10 | 6 | 5 | 8 | 9 | 12 | 17 |
| 46 | 64 | 83 | 70 | 10 | 15 | 2 | 8 | 29 | 11 |

   What is the 95% confidence interval?
7. One hundred poles out of 200 were tested and 4 were found to be rotten. We want to state that we are 95% confident that the number of rotten poles does not exceed a certain percentage. What is that percentage?
8. A utility opinion poll was taken to find out the percentage of population that supports a rate hike based on increased levels of service reliability. If the result is to be within three percentage points at 95% confidence level, what sample size should be used? (Assume that the support is around 30%.)
9. A utility random sample survey of 3000 apartments showed that 64 were vacant. The survey was conducted to estimate the amount of energy used by apartments. Estimate the vacancy rate for the entire community using 93% confidence level.
10. The average weight of a batch of 2000 screws is determined by sampling. If the standard deviation of the weights is 0.1 g, what sample size has to be used so that we are 99% sure that the sampling result is within 0.05 g of the true average.
11. A utility's policy is that if the average ground rod resistance in a station area is over 25 Ω, it will have to be retested the following year. In a station area with 1000 rods, 40 were tested. To be on the safe side, the area manager decided that, if the average of these 40 rods was over 20 Ω, the station area would be marked for retest the next year. Past year's records indicated that the standard deviation of rod resistances was 12 Ω. Using the area manager's rule, what was the type I error when the actual average resistance of the area was 20 Ω? What was the type II error when the actual average resistance was 23 Ω? What would the respective errors be if the retest criterion was set at 25 Ω (instead of 20 Ω)?
12. Develop the mathematical expression for the reliability of the following three system configurations assuming that *each component in the system is identical* and can exist in either an operational state or a failed state. The reliability of each component is given by the following expression:

$$R(t) = e^{-\lambda t}$$

For each system configuration calculate the system reliability given the following individual component parameters:

$$\lambda = 0.00439 \text{ failures/h} \qquad t = 24 \text{ h}$$

(a) *System configuration 1*: simple parallel redundant system.
*System success criterion*: one or more components operating required for system success.

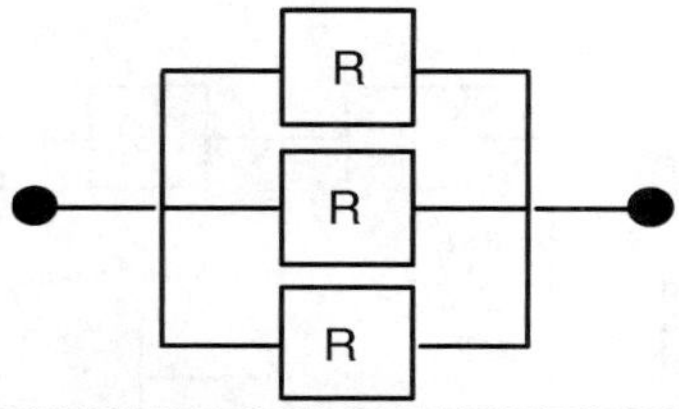


Simple parallel redundant system

(b) *System configuration 2*: bimodal parallel-to-series redundant system.
*Parallel subsystem success criterion*: one or more components operating required for subsystem success.
*System success criterion*: both subsystems operating successfully.

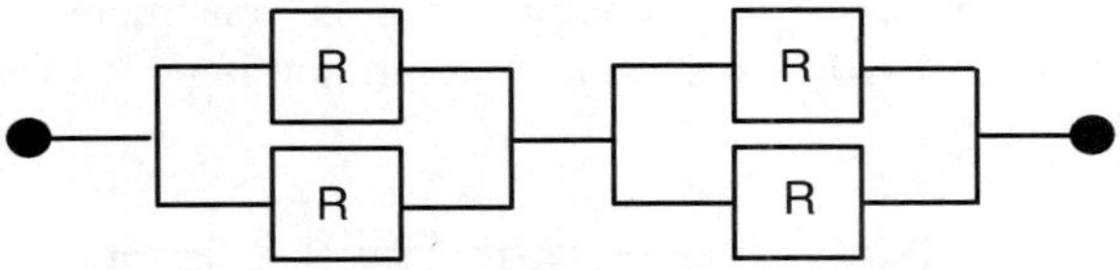


Bimodal parallel-to-series redundant system

(c) *System configuration 3*: bimodal series-to-parallel redundant system.
*Series subsystem success criterion*: both components operating for subsystem success.
*System success criterion*: one or more subsystems operating successfully.

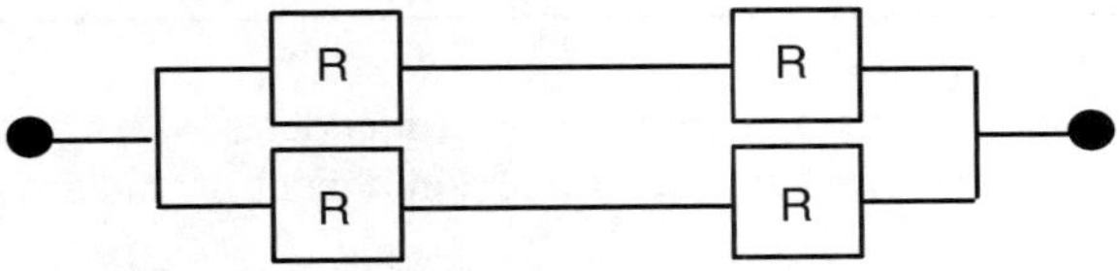


Bimodal series-to-parallel redundant system

13. Four independent and identical 1000 hp induction motors form a parallel system configuration in an industrial process plant. This motor system configuration is used in a cooling system for a large digital computer center. *Two motors functioning successfully are required for system success*. Each motor has a constant failure rate of 0.005 failures/h.

    Find the reliability of the motor system configuration for a 20 h workday.

14. A computer network shown below is made up of 10 identical components. For the group of components 8, 9, and 10, it is necessary that 2 out of the 3 work for subsystem success.

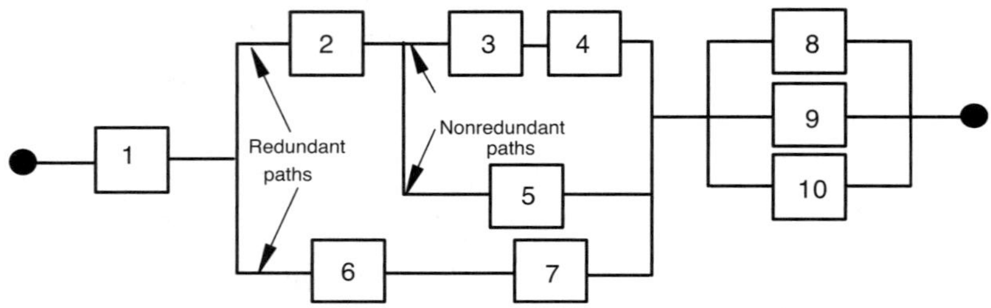


    (a) Develop a general expression for the reliability of the computer network assuming that each component has a reliability value equal to $R$.

    (b) Evaluate the reliability of the computer network if the reliability of each component is equal to 0.80.

15. A new power system configuration is constructed from 10 components, numbered 1–10. The system configuration is shown below.

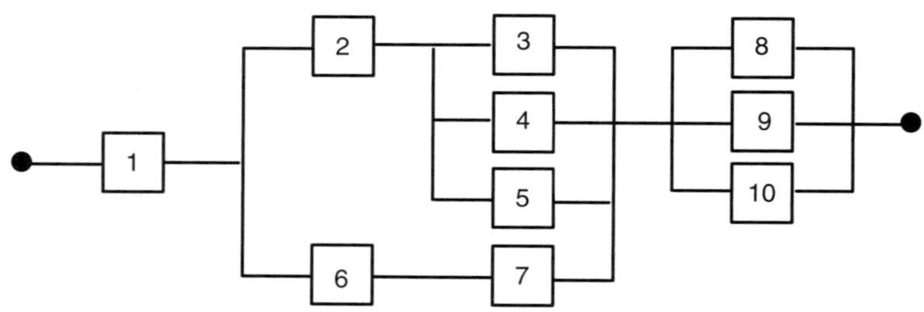


Subsystem (3–4–5)

Components 3, 4, and 5 are *not* identical

*Success criteria*: at least one component of this group must be available

Subsystem (8–9–10)

Components 8, 9, and 10 are identical

*Success criteria*: *two* out of three components must be available

(a) Write a general expression for the reliability of the above system expressed in *terms of R values only* (i.e., $R1 = R2 = R3, \ldots, R10 = R$).

(b) Evaluate the reliability of the above system if the reliability of each component is equal to 0.80.

16. A computer memory system consists of four identical memory units connected in parallel. The memory system's "system success criteria" require that at least *three* memory units *must* function.

    What is the probability of system success if the reliability of each memory unit is 0.90?

    System success criteria: 3 out of 4 memory units must operate.
17. A communication system has 10 identical components connected in series. If the overall system reliability must be at least 0.99 for the system to be marketable, then *what is the minimum reliability required for each component*?
18. A series system has identical components with a known reliability of 0.998. What is the maximum number of components that can be allowed if the *minimum* system reliability is to be 0.90?
19. A fully redundant parallel system has 10 identical components. If the overall system reliability must be at least 0.99, *how poor can these components be*?

## Answers to Problem Set for Chapters 2 and 3

1. 
   (a) 6

   (b) 3.94
2. Probability of three or more defective insulators = 0.00861
3. 
   (a) 0.9048

   (b) 0.1637
4. 86.64% is within 1.5 standard deviation from the mean.
5. 
   (a) 779 should survive

   (b) 40
6. ±9.856
7. 7.14%
8. 896.37
9. 1.65–2.61%
10. 25.935

11.
   (a) Probability of getting a sample of under 20 Ω = shaded area = 0.08%.
   (b) Probability of getting a sample of under 20 Ω = shaded area = 94.3%.
   (c) Probability of making type I error = 30%.
12.
   (a) $Rs = R^3 - 3R^2 + 3R = 0.999000014$.
   (b) $Rs = 4R^2 - 3R^3 + R^4 = 0.980103680$.
   (c) $Rs = R^2 + R^2 - R^4 = 0.96390133$.
13. $Rs = 0.995798891$
14.
   (a) $Rs = 3R^5 - 2R^6 + 3R^7 - 2R^8 - 3R^9 + 2R^{10}$
   (b) $Rs = 0.5644484$
15.
   (a) $Rs = 12R^5 - 17R^6 + 13R^8 - 9R^9 + 2R^{10}$
   (b) $Rs = 0.663538893$
16. 0.947700
17. $R = e^{-0.001005} \geq 0.998995 \approx 0.99$
18. $n = 56.62756001 \approx 57$
19. 0.369042656

## PROBLEM SET FOR CHAPTER 4

1. Twenty-six motors were tested for 200 h. Three motors failed during the test. The failures occurred after the following test times:

| | |
|---|---|
| Motor 1 | 50 h |
| Motor 2 | 61 h |
| Motor 3 | 146 h |

   What is the estimated failure rate?

2. Two hundred capacitors were installed and at the end of each year, the number of surviving units was tallied.

| End of | No. Remaining |
|---|---|
| Year 1 | 192 |
| Year 2 | 184 |
| Year 3 | 177 |
| Year 4 | 170 |
| Year 5 | 163 |

Based on these figures

(a) What is the reliability of the capacitors for 5 years?

(b) What is the annual reliability of Year 4?

(c) Assuming the reliability function is exponential, that is, $R = e^{-\lambda t}$, what is the failure rate for this formula?

3. One thousand lightning arresters are installed. Assuming they have a failure rate of 0.01 per year, how many units (of the original batch) are expected to fail in the 10th year of service?
4. There are 10 generators in a generating station. The unit's area is assumed to have a forced outage rate of 0.01 per year. What is the mean time between failures in that station?
5. What is the reliability of the following system?

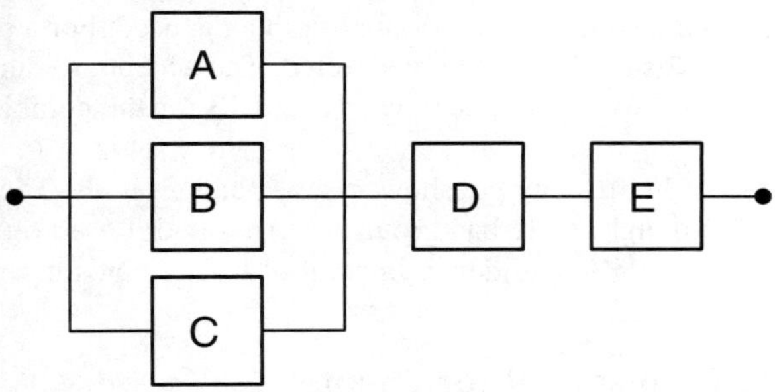


$R_A = R_B = R_C = 0.80$
$R_D = 0.95$
$R_E = 0.85$

6. Two diodes are connected in parallel as shown below:

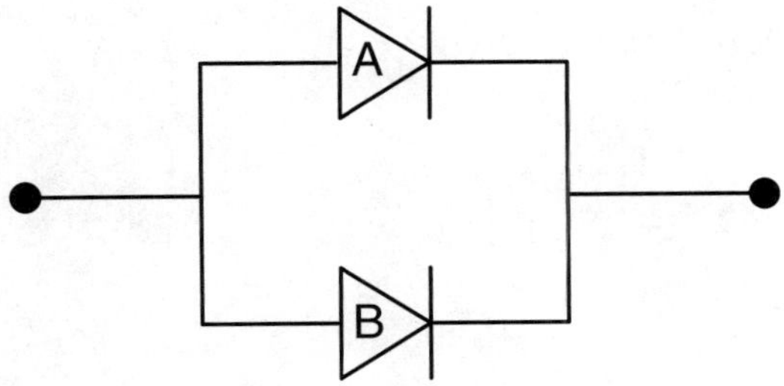


$p(\text{o.c.}) = 0.1$
$p(\text{s.c.}) = 0.2$

A diode may fail in one of the two ways: by short-circuiting or by open circuiting. What is the probability that two-diode arrangement will work as a diode?

7. We have 3 spare transformers supporting 50 single-bank stations. If the failure rate of a transformer in service is 0.05 per year, what is the probability of having no spares available in a year?

8. We have fifty 110–69 kV transformers and have experienced six failures in the past 10 years.
   (a) What is the estimated failure rate?
   (b) What is the 95% upper confidence limit of the failure rate?
   (c) A larger utility with 500 transformers experiences 60 failures in 10 years (for the same failure rate). What is the 95% upper confidence limit of their failure rate?
9. If there are three good insulators in a string of four, for 72 kV line, the probability of flashover is quite small (0.01%). On a 72 kV line, salvaged insulators with 10% defective are used. What is the probability that a string will have fewer than three good insulators? The engineering manager decides to add one unit to each string. How does that help?
10. In ground testing, a double sampling plan is used. For a station area with 400 grounds, 2 samples of 32 are selected. The acceptance and rejection numbers are 5 and 9 for the first sample; 12 and 13 for the combined sample (of 64).

    In a test, the first sample showed seven bad grounds. The second sample was tested and eight more bad grounds were found. Based on the sampling result, what is the 95% confidence interval of bad grounds in that station area?

## Answers to Problem Set for Chapter 4

1. 0.00062 failures/h
2. 
   (a) 0.815
   (b) 0.9605
   (c) 0.041
3. 9
4. 10 year
5. 0.801
6. 0.63
7. 0.2424
8. 
   (a) 0.012
   (b) 0.0236
   (c) 0.1395
9. 
   (a) 0.0523
   (b) 0.00856
10. $13.87\% \leq \bar{P} \leq 32.93\%$ at 95% confidence.

## PROBLEM SET FOR CHAPTER 5

1. A man sets up a fund for his new-born son's college education. He figures the son will go to college at 18 for a cost of $40,000. How much should be put in the fund if the interest rate is 10%.
2. A man retires with $250,000 at 65. He wants to convert it to an annuity that will look after him until he is 100. If interest rate is 8%, what is his annual retirement income?
3. A person wants to buy a $300,000 home in 10 years. He wants to save by depositing an equal amount of money each year into a special account. At 13.5% interest rate, what should his annual deposit be?
4. The major expenses of a 5-year project are as follows:

   Year 0—$120 million cofferdam and site construction
   Year 3—$40 million powerhouse construction
   Year 5—$100 million equipment
   Plus $10 million annually for operating expenses.

   A contractor offers to provide a turnkey project for $300 million, half to be paid at the beginning and half to be paid at the end of the 5-year project. If the interest rate is 7%, is that offer worth accepting?
5. A small computer manufacturing company is operated by four employees. It is known that one particular employee misses an average of 10 out of every 100 days.

   - Each of the other three employees is absent on an average of 5 out of every 100 days. Absences are random and independent.
   - The expenses of the company are $5000/day when operating and $4000/day when shut down. The income at *full production* is $8000/day.
   - The company can still operate if three employees are present; however, the income of the company drops to 60% of the income at full production.
   - If *more* than one employee is absent, *production stops*.

   *What is the expected daily profit* (i.e., *income–expenses*) *for this company*? *Hint*: Try the binomial distribution.

### Answers to Problem Set for Chapter 5

1. $7194.35
2. $21,450.82
3. $15,896.09
4. Total present value of project is $264.95 million, total cost of proposal is $256.95 million, and the contractor's proposal is cheaper.
5. $2190.25

## PROBLEM SET FOR CHAPTER 6

1. A microprocessor system shown in the figure below is used to control and monitor the lighting system at a theater.

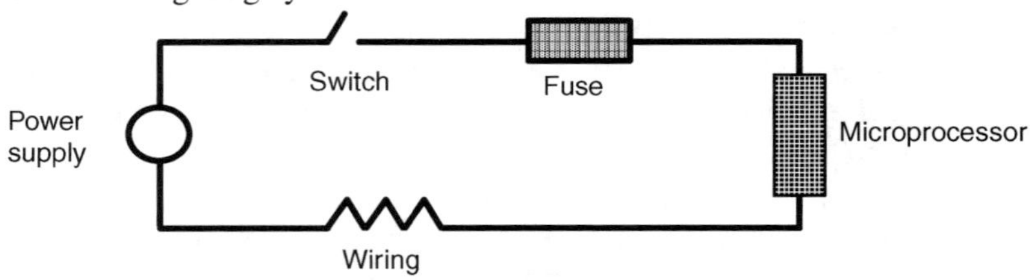


A reliability analysis of the above system revealed the following details:

(a) The major failure event that the microprocessor *will not operate* is dependent upon either the single failure event of known cause (i.e., the microprocessor fails) or the major failure event that there is no current being supplied to the microprocessor.

(b) The major failure event that there is no current being supplied to the microprocessor is dependent upon four mutually exclusive events as follows:

1. Major failure event—*switch open.*
2. Single failure event of known cause—*open-circuit wiring failure.*
3. Major failure event—*fuse fails to open.*
4. Single failure event of known cause—*power supply down.*

(c) The major failure event of the switch being open is dependent upon two mutually exclusive events as follows:

1. Single failure event of known cause—*switch failure open.*
2. Single failure event of unknown cause—*switch open.*

(d) The major event that the fuse fails open is dependent upon two mutually exclusive events as follows:

1. Single failure event of known cause—*fuse fails to open.*
2. Major failure event—*overload in circuit.*

(e) The major event of an overload in the circuit is dependent upon two mutually exclusive events as follows:

1. Single failure event of known cause—short circuit in wiring.
2. Single failure event of known cause—surge on power supply.

Construct a fault tree diagram for the microprocessor system by using the correct elementary fault tree symbols.
If all single failure events have a probability of failure equal to 0.01, calculate the probability that the microprocessor system will not operate.

2. A network configuration is shown below. All components in the network have the same mission reliability equal to "$p$."

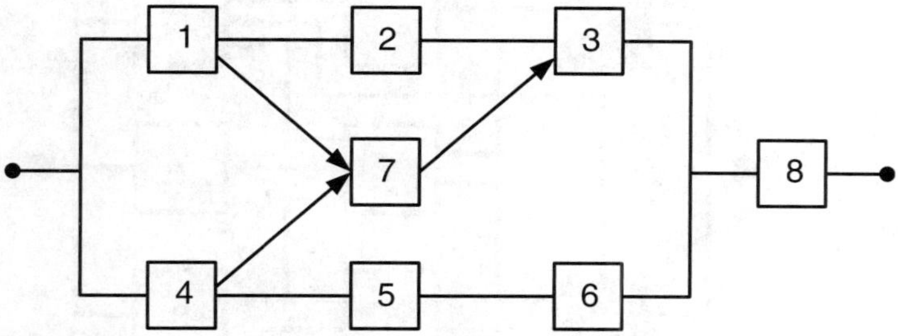


(a) Develop an expression for the reliability of the network configuration.

*Note*: The final reliability expression *must* be in the form of a polynomial in terms of "$p$" only (i.e., $R1 = R2 = R3 = R4 = R5 = R6 = R7 = R8 = p$).

(b) Calculate the system network reliability if $p = 0.90$.

3. A power system network configuration is shown below. All components in the network have the same mission reliability equal to "$p$."
   (a) Develop an expression for the reliability of the network configuration shown below. *Note*: The final reliability expression *must* be in the form of a polynomial in terms of "$p$" only.
   (b) Calculate the system network reliability if $p = 0.90$.

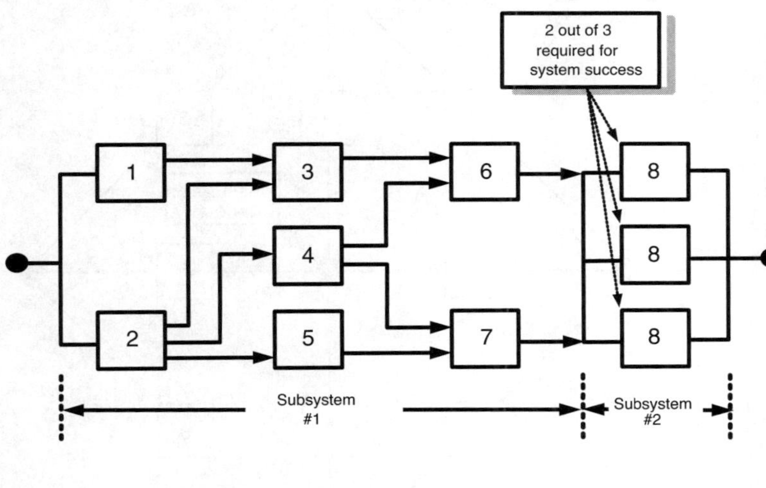


4. A network configuration is shown below. All components in the network have the same mission reliability equal to "$p$".
   (a) Develop an expression for the reliability of the network configuration shown below. *Note*: The final reliability expression *must* be in the form of a polynomial in terms of "$p$" only.
   (b) Calculate the system network reliability if $p = 0.90$.

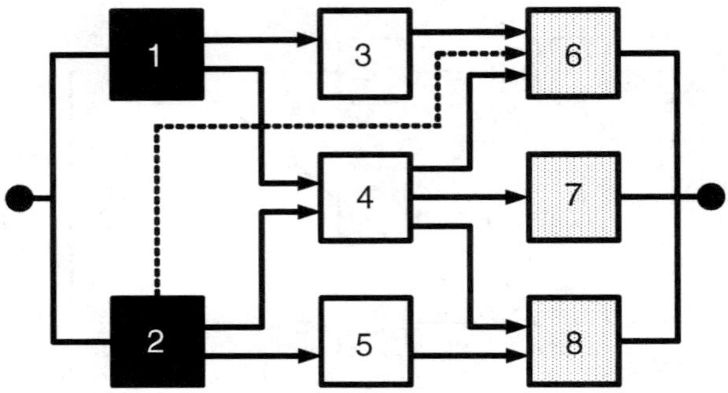


5. A network configuration is shown below. All components in the network have the same mission reliability equal to "$p$."
   (a) Develop an expression for the reliability of the network configuration shown below. *Note*: The final reliability expression *must* be in the form of a polynomial in terms of "$p$" only.
   (b) Calculate the network reliability if the reliability of each individual component is identical and equal to $p = 0.95$.

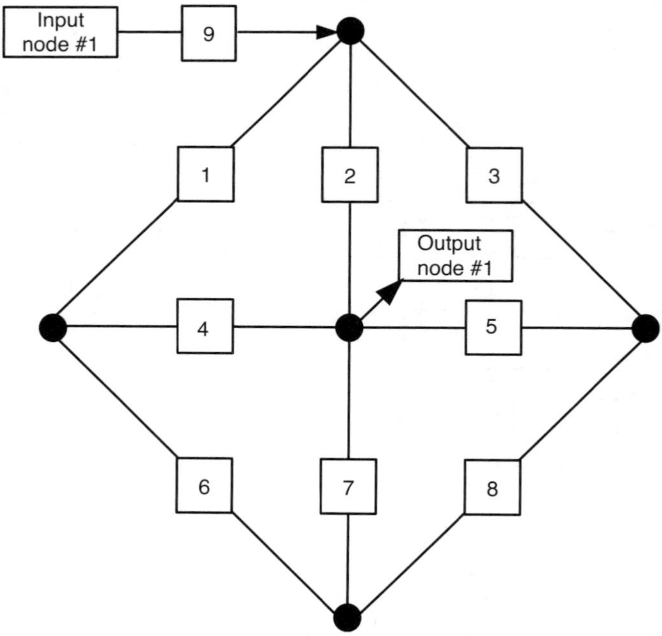


6. A single line diagram of an industrial substation configuration is shown in Fig. 1.

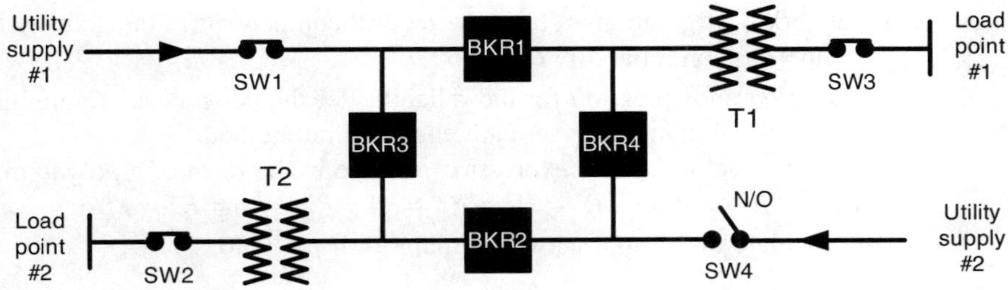


Figure 1. Network operating configuration 1.

The reliability data for the circuit breakers, transformers, manual switches, and utility supplies are shown in the table below. Note all other component failure rates are assumed to be zero (e.g., splices, bus connections, etc.). The breakers are assumed to interrupt all permanent faults 100% of the time.
Assume breakers 1–4 fail only in the open-circuit mode. Switch SW4 is locked out (i.e., cannot be closed during any component outage).

| Component | $\lambda$ (failures/year) | Average Repair Time ($r$) (h/failure) |
|---|---|---|
| Utility supply no. 1 | 0.5 | 0.5 |
| Utility supply no. 2 | 1.0 | 0.5 |
| BKR1 = BKR2 = BKR3 = BKR4 | 0.1 | 24.0 |
| SW1 = SW2 = SW3 = SW4 (manual switches) | 0.1 | 12.0 |
| T1 = T2 (transformers) | 0.05 | 24.0 |

(a) Calculate the frequency (i.e., failures per year) and the average duration of interruptions per outage (i.e., h/interruption) at load point 1.

(b) Calculate the reliability of load point 1 for network operating configuration 1.

(c) Calculate the reliability of load point 1 for network operating configuration 2. Assume the reliability of each electrical component is equal to 0.9999. Assume the reliability of utility supply no. 1 and 2 are also equal to 0.9999 (Fig. 2).

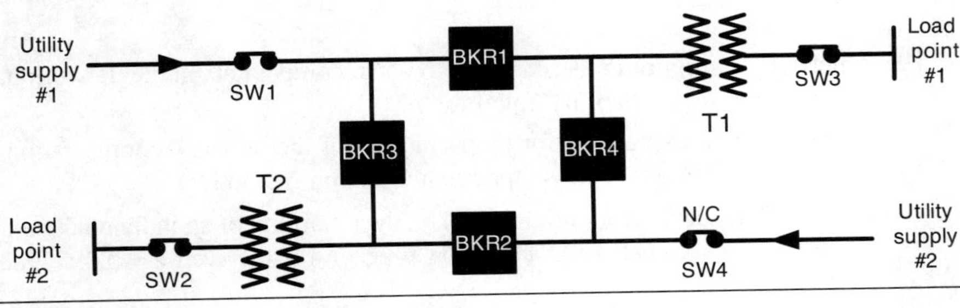


Figure 2. Network operating configuration 2.

7. A network configuration is shown below. All components in the network have the same mission reliability equal to "$p$."
   (a) Develop an expression for the reliability of the network configuration by using the minimum tie-set path enumeration method.
   *Note*: The final reliability expression *must* be in the form of a polynomial in terms of "$p$" only (i.e., $R1 = R2 = R3 = R4 = R5 = R6 = R7 = R8 = p$).
   (b) Calculate the system network reliability if $p = 0.90$.

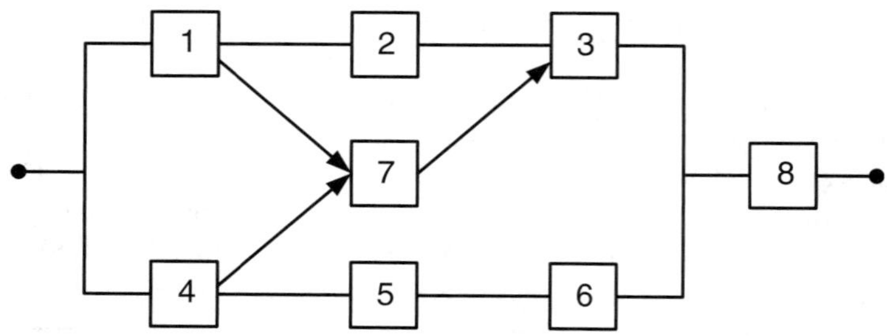


8. A network configuration is shown below. All components in the network have the same mission reliability equal to "$p$."
   (a) Develop an expression for the reliability of the network configuration by using the minimum cut-set methodology (third-order cut max).
   *Note*: The final reliability expression *must* be in the form of a polynomial in terms of "$p$" only (i.e., $R1 = R2 = R3 = R4 = R5 = R6 = R7 = R8 = p$).
   (b) Calculate the system network reliability if $p = 0.90$.

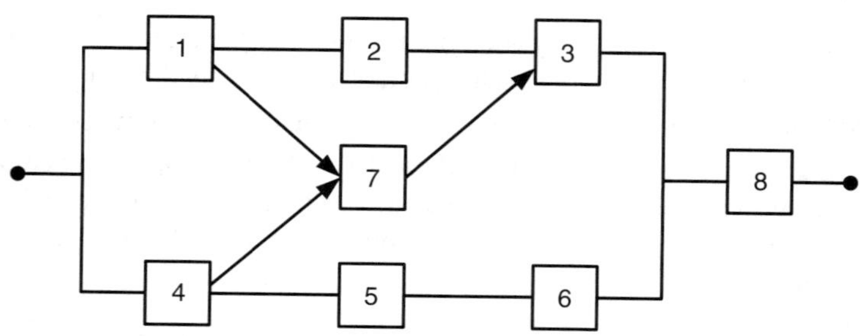


9. A network configuration is shown below. All components in the network have the same mission reliability equal to "$p$."
   (a) Develop an expression for the reliability of the network configuration by using the minimum tie-set path enumeration method.
   (b) Calculate the network reliability (i.e., the reliability of an individual component $p = \exp(-\lambda t_m)$ for a mission time of 168 h, given $\lambda = 5.493798319$ failures/year.

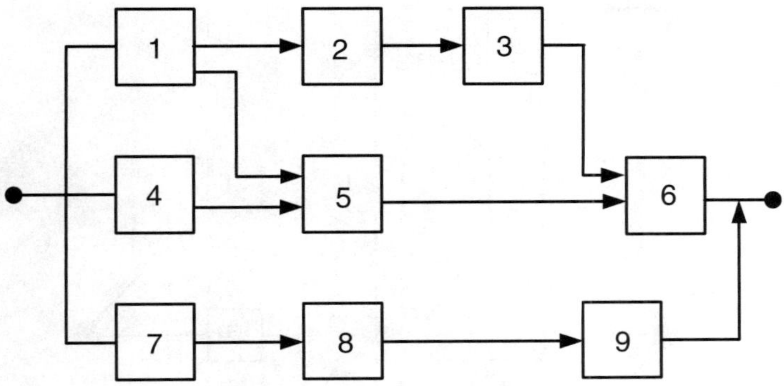


10. A network configuration is shown below. All components in the network have the same mission reliability equal to "$p$."
    (a) Develop an expression for the reliability of the network configuration using the minimum cut-set methodology.
    (b) Calculate the network reliability if the reliability of each individual component is identical and equal to $p = 0.90$.

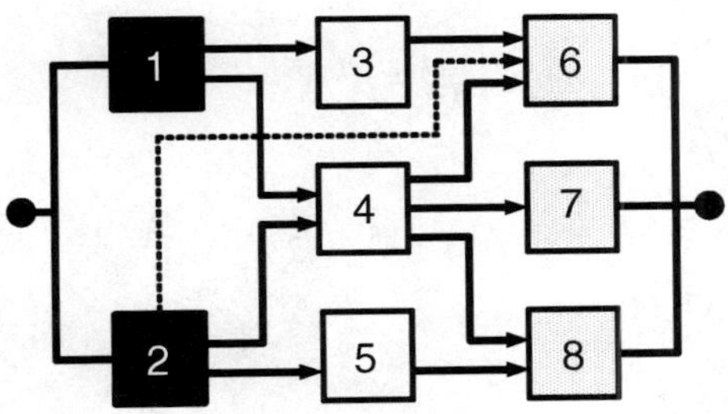


11. A network configuration is shown below. All components in the network have the same mission reliability equal to "$p$."
    (a) Develop an expression for the reliability of the network configuration shown below. *Note:* The final reliability expression *must* be in the form of a polynomial in terms of "$p$" only.
    (b) Calculate the network reliability if the reliability of each individual component is identical and equal to $p = 0.95$.

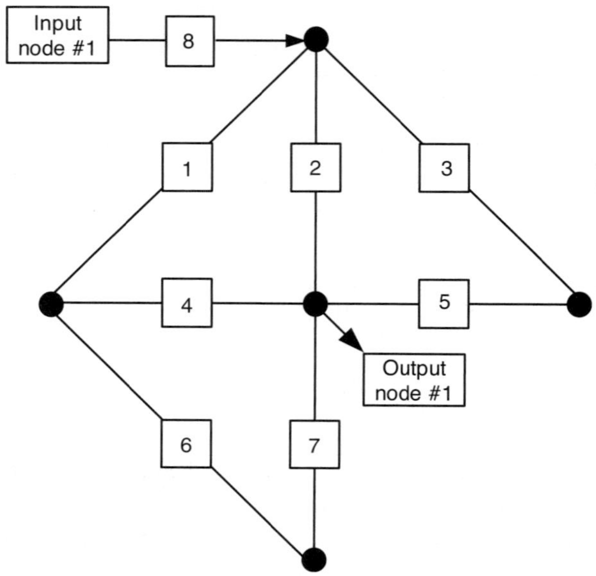


## Answers to Problem Set for Chapter 6

1. 0.08
2.
   (a) $Rs = 4p^4 - 2p^5 - p^6 - p^7 + p^8$
   (b) 0.864149310
3.
   (a) $Rs = 15p^5 - 22p^6 + 5p^7 + 5p^8 - 2p^9$
   (b) 0.93462757200000
4.
   (a) $Rs = p^2 + 7p^3 - 13p^4 + 7p^5 - p^6$
   (b) 0.985689
5.
   (a) $Rs = p^2 + 2p^3 - 3p^5 - 8p^6 + 19p^7 - 13p^8 + 3p^9$
   (b) 0.94986331069726
6.
   (a) 0.750164 interruptions/year, 5.134834 h/interruption
   (b) 0.9995604
   (c) 0.9997000499700

7.
(a) $4p^4 - 2p^5 - p^6 - p^7 + p^8$
(b) 0.86414931

8.
(a) $Rs = 36 - 164p + 291p^2 - 247p^3 + 101p^4 - 16p^5$
(b) 0.86526

9.
(a) $Rs = 3p^3 - p^5 - 2p^6 + p^8$
(b) $Rs = 0.96409521$

10.
(a) $Rs = 10 - 52p + 111p^2 - 113p^3 + 57p^4 - 13p^5 + p^6$
(b) $Rs = 0.985771$

11.
(a) $Rs = 10 - 52p + 111p^2 - 113p^3 + 57p^4 - 13p^5 + p^6$
(b) 0.949746989

## PROBLEM SET FOR CHAPTER 7

1. A "utility–diesel engine generator–fuel cell" system configuration is interconnected to deliver energy to a critical load point as shown in the figure below.

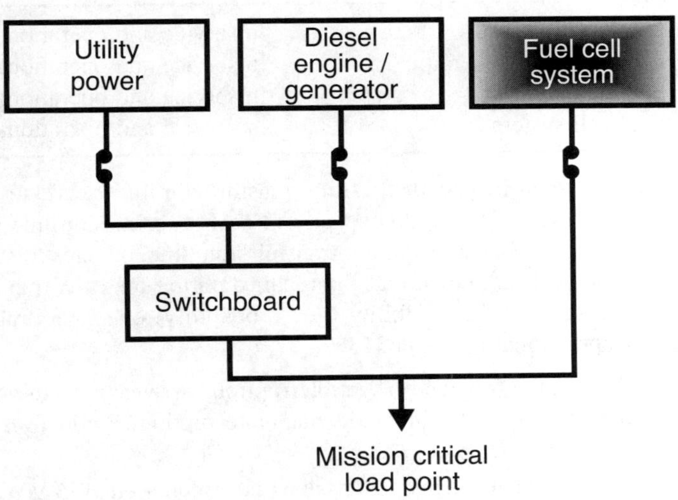

The reliability data for each subsystem is shown in the table below.

| Subsystem | MTTF (h) | MTTR (h) |
|---|---|---|
| Utility power | 4878.0 | 12.6 |
| Diesel engine/generator | 1150.0 | 37.5 |
| Switchboard | 25,033.0 | 9.9 |
| Fuel cell system | 8,760,000.0 | 11.0 |

*Note*: $\lambda = 1/\text{MTTF}$.

(a) What is the frequency (i.e., interruptions per year) and duration (i.e., hours per interruption) at the mission critical load point under the following system operation configuration:

Operating Configuration 1

| Subsystem | Operational Status |
|---|---|
| Utility power | In service and operational |
| Diesel engine/generator | Out of service |
| Switchboard | In service and operational |
| Fuel cell system | Out of service |

(b) What is the frequency (i.e., interruptions per year) and duration (i.e., hours per interruption) at the mission critical load point under the following system operation configuration:

Operating Configuration 2

| Subsystem | Operational Status |
|---|---|
| Utility power | In service and operational |
| Diesel engine/generator | In service and operational |
| Switchboard | In service and operational |
| Fuel cell system | In service and operational |

2. *Three identical* independent utility transmission lines serve an industrial plant load of 75 MW. The initial capacity of each transmission line is 25 MW. Any combination of one or more transmission line outages results in a plant interruption. It is known that the repair duration for *each* transmission line is 2.0 h/outage and the reliability (i.e., probability) of *all* the transmission lines being operational is 0.998631387.

   (a) What is the frequency (i.e., interruptions per year) and duration (i.e., hours per interruption) of industrial plant interruption? *Round your answers to two significant decimal places.*

   (b) The capacity of each transmission line is increased to 75 MW. Under the new transmission line expansion plan a plant interruption will occur only when *all* utility transmission lines are on outage.

What is the approximate frequency (i.e., interruptions per year) of industrial plant interruptions under the new expansion plan?

3. A commercial installation's low-voltage network configuration is shown in the figure below.

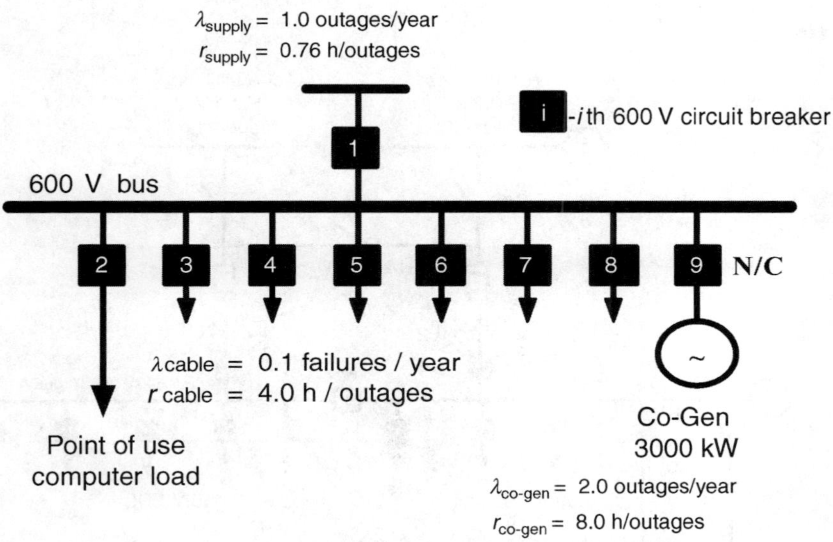


Equipment Failure Rates and Repair Duration

| Circuit Breaker Number | ($\lambda$) Total Failure Rate (failures/year) | ($\lambda$) Failed While Opening (failures/year) | ($r$) Repair Duration (h/outage) | Rsw Switching Duration (h) |
|---|---|---|---|---|
| 1 (ideal) | 0.0 | 0.0 | 0.0 | 0.0 |
| 2 | 0.005 | 0.0005 | 4.0 | 1.0 |
| 3 | 0.005 | 0.0005 | 4.0 | 1.0 |
| 4 | 0.005 | 0.0005 | 4.0 | 1.0 |
| 5 | 0.005 | 0.0005 | 4.0 | 1.0 |
| 6 | 0.005 | 0.0005 | 4.0 | 1.0 |
| 7 | 0.005 | 0.0005 | 4.0 | 1.0 |
| 8 | 0.005 | 0.0005 | 4.0 | 1.0 |
| 9 (ideal) | 0.0 | 0.0 | 0.0 | 0.0 |

*Note*: switchgear bus is assumed to be ideal.

$\lambda$, total failure rate of circuit breaker includes the failure rate $\lambda$ "failed while opening."

Calculate the following reliability indices for the computer load:

(a) $\lambda$—failures per year.

(b) $U$—annual duration of interruptions.

(c) $r$—average duration of interruptions.

4. A single-line diagram of the new University of Alberta Teaching and Learning Complex is shown in the figure below.

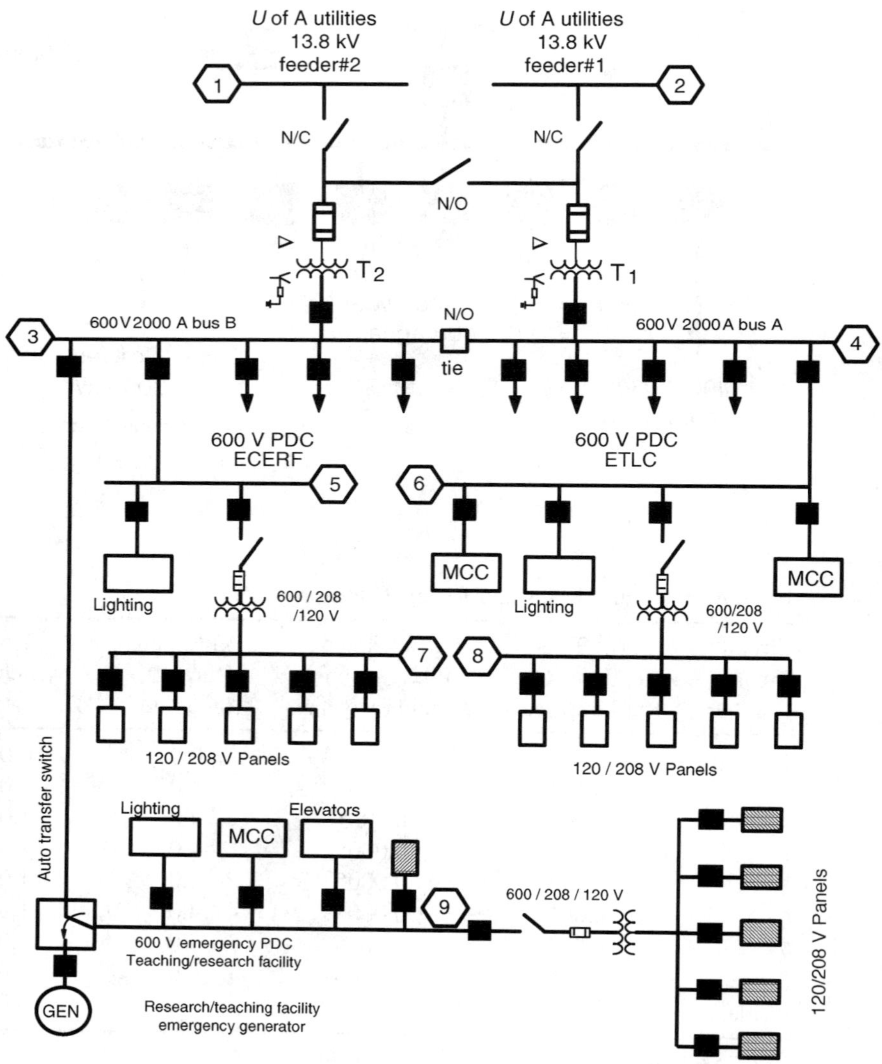


See Table 1. for equipment failure rates and average repair/restoration times for the electrical components in the circuit.

TABLE 1. Equipment Failure Rates and Average Repair/Restoration Times

| Component | $\lambda$ (failures/year) | ($r$) Average Repair Time (h/failure) |
|---|---|---|
| U of A utilities 13.8 kV feeder no. 1 | 0.2 | 1.0 |
| U of A utilities 13.8 kV feeder no. 2 | 0.2 | 1.0 |
| T1 = T2 (transformers) | 0.005 | 100.0 |
| All fuses | 0.005 | 1.0 |
| All closed circuit breakers | 0.005 | 25.0 |
| All opened circuit breakers | 0.0 | 0.0 |
| All switchgear bus (i.e., numbered buses) | 0.005 | 25.0 |
| Manual disconnect switches | 0.005 | 25.0 |
| All transformers (600/208/120 V) | 0.005 | 25.0 |
| Standby generator transfer switch | 0.005 | 25.0 |
| Standby generator | 0.0 | 168.0 |
| All other electrical equipment is assumed to have a zero failure rate | 0.0 | 0.0 |

*Note:*
1. "Automatic transfer switch"—switching time = 0.0 h.
2. All other manual switching or tiebreaker switching time is assumed to be 15 min.

(a) Calculate the frequency and average duration of interruptions to the "Elevators" connected to the 600 V emergency PDC—teaching/research facility.

(*Note:* Failure of feeder no. 2 supply results in the emergency and standby generator switching in.)

(b) Calculate the frequency and average duration of interruptions to the "120/208 V Panels" connected to bus 8.

(*Note:* Failure of feeder no. 1, N/O tiebreaker is manually closed.)

## Answers to Problem Set for Chapter 7

1.
   (a) 2.14575603991170 failures/year, 12.17547056266925 h/interruption.
   (b) $1.010713387317601 \times 10^{-6}$ failures/year, 5.18755121521138 h.

2.
   (a) 6 interruptions/year, 2.0 h/interruption.
   (b) $1.25101645 \times 10^{-6}$ failures/year.

3.
   (a) 0.110 interruptions/year.
   (b) 0.424388128 h/year.
   (c) 3.858073890 h/interruption.

4.
   (a) 0.2900 interruptions/year, 3.0196839008 h/interruption.
   (b) 0.325 interruptions/year, 7.49615384615384 h/interruption.

## PROBLEM SET FOR CHAPTER 8

1. A distribution radial primary feeder circuit services three industrial loads as shown in the figure below. The industrial feeder consists of 12 line sections. The source side of each line section has either a manual switch or a protective device installed. Each line section $(i)$ is characterized by its failure rate $\lambda(i)$.

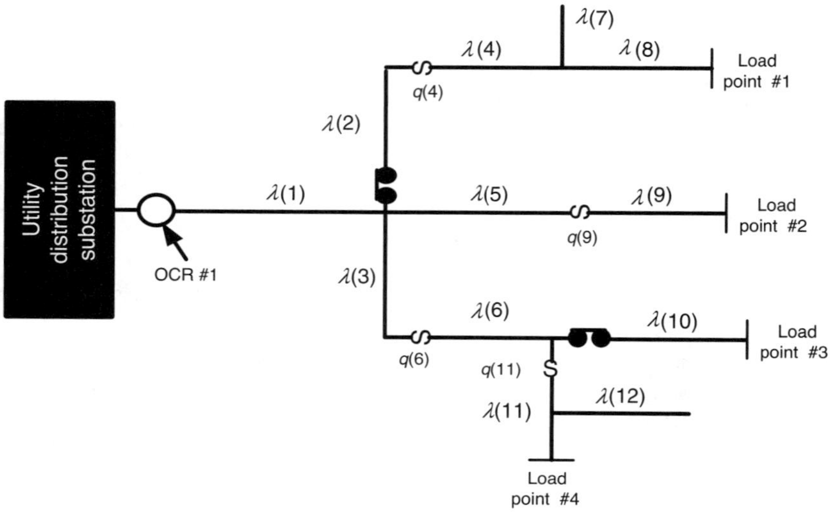


Given

$$\lambda(1) = \lambda(3) = \lambda(5) = \lambda(7) = \lambda(9) = \lambda(11) = 2.0 \text{ failures/year}$$
$$\lambda(2) = \lambda(4) = \lambda(6) = \lambda(8) = \lambda(10) = \lambda(12) = 1.0 \text{ failures/year}$$
$$q(4) = q(6) = q(9) = q(11) = 0.10$$

The duration of *repair* activities (i.e., $r(i)$) for each feeder section is equal to 4.0 h.

The average manual switching time for any isolating device is equal to 1.0 h. The failure rate of the utility distribution substation supply, disconnect switches, fuses, and OCRs are assumed to be equal to zero.

(a) Calculate the reliability indices for load points 1, 2, 3, and 4.

Place your results in the table below.

| Load Point | ($\lambda$) Failure rate (failures/year) | ($U$) Annual Duration of Interruptions (h) | ($r$) Average Interruption Duration (h/interruption) |
|---|---|---|---|
| 1 | | | |
| 2 | | | |
| 3 | | | |
| 4 | | | |

(b) If *all* the line sectionalizing devices (i.e., manual switches and nonideal fuses) are assumed to be ideal fuses (i.e., $q(i) = 0.0$), what is the frequency of interruptions for load points 1, 2, 3, and 4? Place your results in the table below.

| Load Point | ($\lambda$) Failure Rate (failures/year) |
|---|---|
| 1 | |
| 2 | |
| 3 | |
| 4 | |

2. A distribution looped radial system is shown in the figure below. The following reliability and load data are defined:

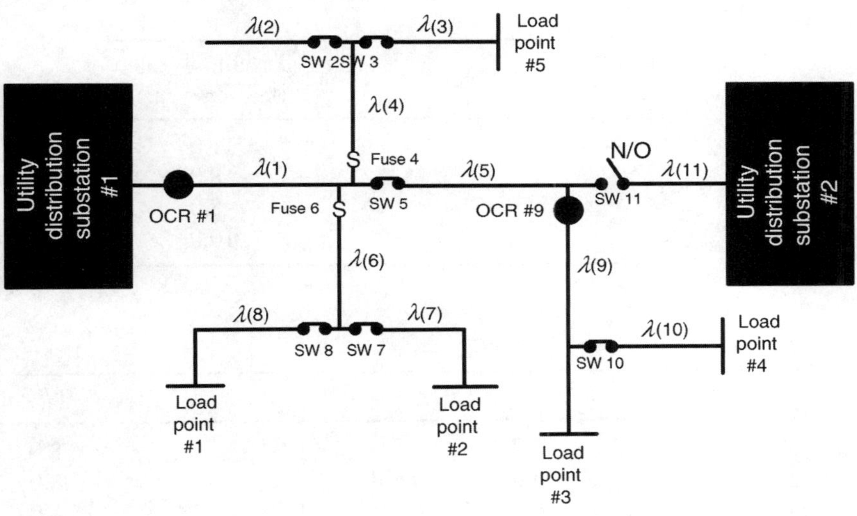


$$\lambda(\text{feeder section nunmber } 1) = \lambda(1) = \lambda(2) = \lambda(3) = \lambda(4) = \lambda(5) = \lambda(6)$$
$$= 1.0 \text{ failures/year}$$
$$\lambda(1) = \lambda(7) = \lambda(8) = \lambda(9) = \lambda(10) = \lambda(11)$$
$$= 1.0 \text{ failures/year}$$

Average time to repair each line section = 2.0 h for all sections.
Average switching and isolation time $r(\text{switching}) = 1.0$ h.
The OCRs and fuses are assumed to be ideal (i.e., $q$'s are equal to zero, failure rates are equal to zero). The failure rate of the manual disconnect switches is also zero.
Calculate the reliability indices ($\lambda$, $r$, $\lambda r$) for load points 1, 2, 3, 4, and 5.

## Answers to Problem Set for Chapter 8

1.

(a)

| Load Point | ($\lambda$) Failure Rate (failures/year) | ($U$) Annual Duration of Interruptions (h) | ($r$) Average Interruption Duration (h/interruption) |
|---|---|---|---|
| 1 | 11.43 | 44.43 | 3.88713910761155 |
| 2 | 9.63 | 33.63 | 3.49221183800623 |
| 3 | 9.90 | 33.90 | 3.42424242424243 |
| 4 | 12.60 | 42.60 | 3.38095238095238 |

(b)

| Load Point | ($\lambda$) Failure Rate (failures/year) |
|---|---|
| 1 | 11.0 |
| 2 | 8.0 |
| 3 | 8.0 |
| 4 | 10.0 |

2.

| Load Point | ($\lambda$) (failures/year) | $r$ (h) | $U$ (h/year) |
|---|---|---|---|
| 1 | 5.0 | 1.60 | 8.0 |
| 2 | 5.0 | 1.60 | 8.0 |
| 3 | 4.0 | 1.50 | 6.0 |
| 4 | 4.0 | 1.75 | 7.0 |
| 5 | 5.0 | 1.60 | 8.0 |

## PROBLEM SET FOR CHAPTER 21

1. A radial distribution system single-line diagram is shown in the figure below:

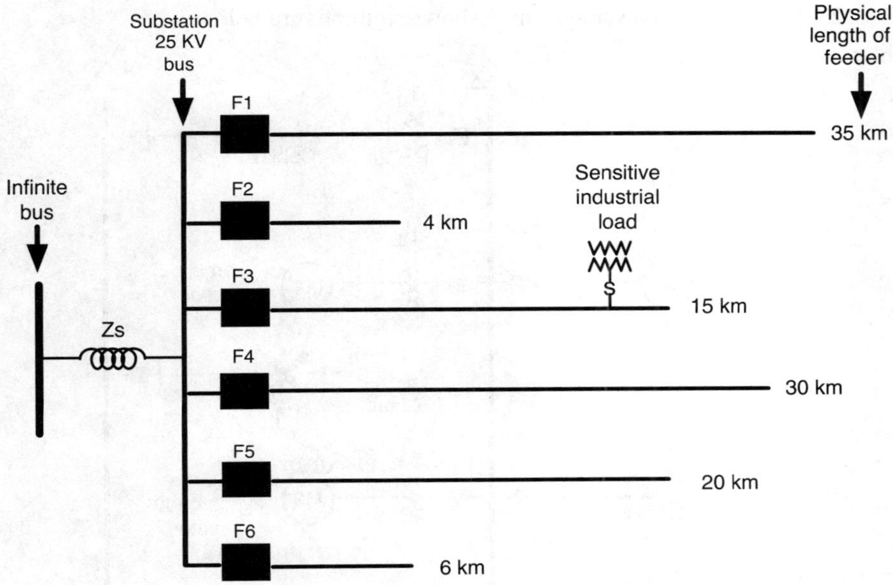


The feeder reactances/km are defined in the table below.

| Feeder number | F1 | F2 | F3 | F4 | F5 | F6 |
|---|---|---|---|---|---|---|
| Per unit feeder reactance/km | $j\,1.0$ | $j\,0.25$ | $j\,0.50$ | $j\,1.0$ | $j\,0.50$ | $j\,0.25$ |

The average number of three-phase faults for this distribution feeder system is 0.50 faults/km per year. The industrial load on feeder F3 is interrupted when voltage sags less than or equal to 0.8 per unit occur on the 25 kV bus.

*The number of voltage sags less than or equal to 0.8 per unit that were recorded at the substation 25 kV bus from three-phase faults on all six distribution feeder circuits except feeder circuit F3 was 11.0 V sags in 2001.*

*In 2002, the utility substation was significantly upgraded and the number of voltage sags recorded from three-phase faults on all six distribution feeder circuits except feeder circuit F3 was 6.0 in 2003.*

(a) Calculate the equivalent per unit utility system impedance Zs in 2001.

(b) Calculate the equivalent per unit utility system impedance Zs in 2003.

2. An industrial power system is shown in the figure below.

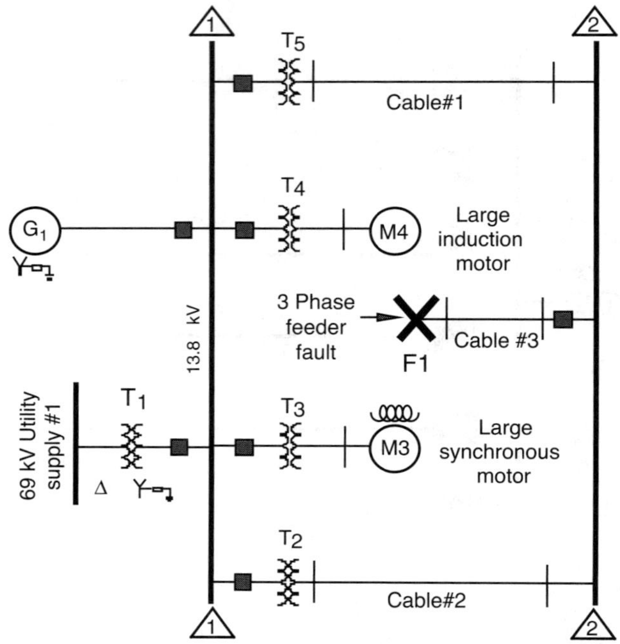


The per unit impedances of all the electrical components are defined in the table below:

| Component | Per Unit Reactances |
|---|---|
| Equivalent 69 kV utility impedance | 0.2 |
| Generator 1 | 0.4 |
| Transformer 1 | 0.2 |
| Transformer 2 | 0.1 |
| Transformer 3 | 3.0 |
| Transformer 4 | 3.0 |
| Transformer 5 | 0.1 |
| Cable 1 | 0.1 |
| Cable 2 | 0.1 |
| Cable 3 | 0.2125 |
| Synchronous motor 3 | 3.0 |
| Induction motor 4 | 3.0 |

Calculate the voltage sag on bus 1 and 2 when a permanent three-phase fault occurs on the feeder as shown in the figure above.

3. This problem will clearly demonstrate the importance of a neutral conductor. A three-phase unbalanced load is supplied by a balanced voltage supply as shown in the figure below. *Note*: the utility supply voltage is balanced, however, the phase voltages across the loads may or may not be balanced depending upon the impedance characteristics of the *neutral conductor*.

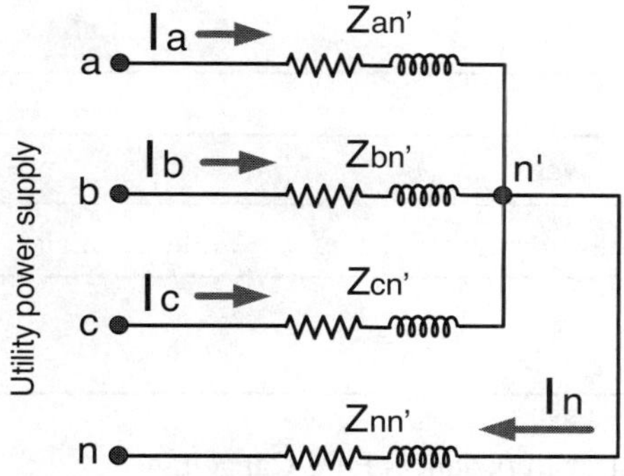


Utility power supply voltage characteristics

Van = 120.0/0°   Vbn = 120.0/– 120°   Vcn = 120.0/– 240°

Load impedance characteristics

Zan' = 10.0/0°   Zbn' = 10.0/45°   Zcn' = 10.0/45°   Zn'n' . . . variable

Calculate the load voltages (i.e., Van', Vbn', and Vcn'), the line and neutral currents (i.e., Ia, Ib, Ic, and In), and the voltage Vn'n for various values of the neutral impedance (i.e., Zn'n).
Complete the tables (two decimal places).

Part 1: Neutral Impedance–Pure Resistance

| Zn'n | Ia | | Ib | | Ic | | In | |
|---|---|---|---|---|---|---|---|---|
| | mag | angle | mag | angle | mag | angle | mag | angle |
| 0 | | | | | | | | |
| 10/0° | | | | | | | | |
| 100,000/0° | | | | | | | | |

| Zn'n | Van' | | Vbn' | | Vcn' | | Vn'n | |
|---|---|---|---|---|---|---|---|---|
| | mag | angle | mag | angle | mag | angle | mag | angle |
| 0 | | | | | | | | |
| 10/0° | | | | | | | | |
| 100,000/0° | | | | | | | | |

Part 2: Neutral Impedance–Pure Reactance

| Zn'n | Ia | | Ib | | Ic | | In | |
|---|---|---|---|---|---|---|---|---|
| | mag | angle | mag | angle | mag | angle | mag | angle |
| 10/90° | | | | | | | | |
| 100,000/90° | | | | | | | | |

| Zn'n | Van' | | Vbn' | | Vcn' | | Vn'n | |
|---|---|---|---|---|---|---|---|---|
| | mag | angle | mag | angle | mag | angle | mag | angle |
| 10/90° | | | | | | | | |
| 100,000/90° | | | | | | | | |

Part 3: Neutral Impedance–Pure Capacitance

| Zn'n | Ia | | Ib | | Ic | | In | |
|---|---|---|---|---|---|---|---|---|
| | mag | angle | mag | angle | mag | angle | mag | angle |
| 10/ – 90° | | | | | | | | |
| 100,000/ – 90° | | | | | | | | |

| Zn'n | Van' | | Vbn' | | Vcn' | | Vn'n | |
|---|---|---|---|---|---|---|---|---|
| | mag | angle | mag | angle | mag | angle | mag | angle |
| 10/ – 90° | | | | | | | | |
| 100,000/ – 90° | | | | | | | | |

## Answers to Problem Set for Chapter 21

1.
   (a) 0.75 per unit
   (b) 0.25 per unit
2.
   $V_{\text{bus 1}} = 0.6250\,\text{pu}$
   $V_{\text{bus 2}} = 0.4250\,\text{pu}$
3.

Part 1: Neutral Impedance–Pure Resistance

| Zn'n | Ia | | Ib | | Ic | | In | |
|---|---|---|---|---|---|---|---|---|
| | mag | angle | mag | angle | mag | angle | mag | angle |
| 0 | 12.0 | 0° | 12.000000 | – 165.00° | 12.0 | 75.00° | 9.184402 | 67.50° |
| 10/0° | 12.254657 | – 11.70° | 14.206766 | – 159.98° | 9.925775 | 82.19° | 2.485281 | 90.00° |
| 100,000/0° | 12.866549 | – 14.63° | 14.729956 | – 157.13° | 9.044558 | 82.86° | 0.000328 | 97.86° |

| Zn'n | Van' | | Vbn' | | Vcn' | | Vn'n | |
|---|---|---|---|---|---|---|---|---|
| | mag | angle | mag | angle | mag | angle | mag | angle |
| 0 | 120.0 | 0° | 120.0 | – 120.0° | 120.0 | 120.0° | 0.0 | 0° |
| 10/0° | 122.546572 | – 11.70° | 142.067669 | – 114.98° | 99.257755 | 127.19° | 24.852813 | 90.0° |
| 100,000/0° | 128.665492 | – 14.63° | 147.299566 | – 112.13° | 90.445585 | 127.86° | 32.824661 | 97.86° |

Part 2: Neutral Impedance–Pure Reactance

| Zn'n | Ia | | Ib | | Ic | | In | |
|---|---|---|---|---|---|---|---|---|
| | mag | angle | mag | angle | mag | angle | mag | angle |
| 10/90° | 13.264345 | – 10.79° | 13.803576 | – 156.10° | 9.339567 | 77.15° | 2.690049 | 22.50° |
| 100,000/90° | 12.866713 | – 14.64° | 14.729913 | – 157.14° | 9.044442 | 82.86° | 0.000328 | 7.86° |

| Zn'n | Van' | | Vbn' | | Vcn' | | Vn'n | |
|---|---|---|---|---|---|---|---|---|
| | mag | angle | mag | angle | mag | angle | mag | angle |
| 10/90° | 132.643453 | – 10.80° | 138.035764 | – 111.11° | 93.395671 | 122.15° | 26.900491 | 112.50° |
| 100,000/90° | 128.667136 | – 14.64° | 147.299127 | – 112.14° | 90.444419 | 127.86° | 32.825081 | 97.86° |

Part 3: Neutral Impedance–Pure Capacitance

| Zn'n | Ia | | Ib | | Ic | | In | |
|---|---|---|---|---|---|---|---|---|
| | mag | angle | mag | angle | mag | angle | mag | angle |
| 10/ – 90° | 11.754855 | – 18.12° | 15.620700 | – 160.92° | 9.591335 | 90.39° | 3.749516 | 167.23° |
| 100,000/ – 90° | 12.866572 | – 14.64° | 14.730117 | – 157.14° | 9.044440 | 82.86° | 0.000328 | – 172.14° |

| Zn'n | Van' | | Vbn' | | Vcn' | | Vn'n | |
|---|---|---|---|---|---|---|---|---|
| | mag | angle | mag | angle | mag | angle | mag | angle |
| 10/ – 90° | 117.548553 | – 18.12° | 156.207006° | – 115.92° | 95.913352 | 135.39° | 37.495165 | 77.24° |
| 100,000/ – 90° | 128.665719 | – 14.64° | 147.301166 | – 112.14° | 90.444405 | 127.86° | 32.826266 | 97.86° |

# INDEX